Genetic Expression in the Cell Cycle

Edited by

GEORGE M. PADILLA
Department of Physiology
Duke University Medical Center
Durham, North Carolina

KENNETH S. McCARTY, Sr.
Department of Biochemistry
Duke University Medical Center
Durham, North Carolina

1982

ACADEMIC PRESS

A Subsidiary of Harcourt Brace Jovanovich, Publishers
New York London
Paris San Diego San Francisco São Paulo Sydney Tokyo Toronto

ACADEMIC PRESS, INC.
111 Fifth Avenue, New York, New York 10003

United Kingdom Edition published by
ACADEMIC PRESS, INC. (LONDON) LTD.
24/28 Oval Road, London NW1 7DX

Library of Congress Cataloging in Publication Data
Main entry under title:

Genetic expression in the cell cycle.

(Cell biology)
Includes bibliographies and index.
1. Gene expression. 2. Cell cycle. I. Padilla, George M. II. McCarty, Kenneth Scott, Date. III. Series.
QH450.G464 574.87'322 82-3930
ISBN 0-12-543720-X AACR2

PRINTED IN THE UNITED STATES OF AMERICA

82 83 84 85 9 8 7 6 5 4 3 2 1

To Professor Thomas W. James, an inspiring teacher who by word and deed continues to guide us in the study of the cell cycle.

Contents

II. Genetic Expression and Posttranscriptional Modifications

III. Ionic and Membrane Modulations in the Cell Cycle

List of Contributors

Numbers in parentheses indicate the pages on which the authors' contributions begin.

Newell F. Bascomb[1] (199), Department of Biochemistry and Nutrition, Virginia Polytechnic Institute and State University, Blacksburg, Virginia 24061

Renato Baserga (231), Department of Pathology and Fels Research Institute, Temple University School of Medicine, Philadelphia, Pennsylvania 19140

D. P. Bedard (245), Faculty of Medicine, Dalhousie University, Halifax, Nova Scotia, Canada

E. M. Bradbury (31), Department of Biological Chemistry, School of Medicine, University of California, Davis, California 95616

I. L. Cameron (363), Department of Anatomy, The University of Texas Health Science Center at San Antonio, San Antonio, Texas 78284

Lee S. Chai (3), Departments of Genetics and Endocrinology, Division of Medicine, Roswell Park Memorial Institute, Buffalo, New York 14263

Paul A. Charp[2] (393), Department of Zoology, University of Tennessee, Knoxville, Tennessee 37916

Zbigniew Darzynkiewicz (103), Investigative Cytology Laboratory, Memorial Sloan–Kettering Cancer Center, New York, New York 10021

S. W. de Laat (337), Hubrecht Laboratory, International Embryological Institute, 3584 CT Utrecht, The Netherlands

[1] Present address: Department of Microbiology and Cell Science, University of Florida, Gainesville, Florida 32611.

[2] Present address: Division of Biology, Kansas State University, Manhattan, Kansas 66506.

Christopher N. Frantz (411), Harvard Medical School, Sidney Farber Cancer Institute, Boston, Massachusetts 02115

Jerrold Fried (289), Memorial Sloan–Kettering Cancer Center, New York, New York 10021

Luis Jimenez de Asua (315), Friedrich Miescher-Institut, CH-4002 Basel, Switzerland

G. C. Johnston (181, 245), Faculty of Medicine, Dalhousie University, Halifax, Nova Scotia, Canada

Drew N. Kelner (55), Department of Biochemistry, Duke University Medical Center, Durham, North Carolina 27710

Margarida O. Krause (151), Department of Biology, University of New Brunswick, Fredericton, New Brunswick, Canada

James J. Lynch[3] (199), Department of Biochemistry and Nutrition, Virginia Polytechnic Institute and State University, Blacksburg, Virginia 24061

Kenneth S. McCarty, Jr. (55), Departments of Pathology and Medicine, Duke University Medical Center, Durham, North Carolina 27710

Kenneth S. McCarty, Sr. (55), Department of Biochemistry, Duke University Medical Center, Durham, North Carolina 27710

H. R. Matthews (31), Department of Biological Chemistry, School of Medicine, University of California, Davis, California 95616

Harriet K. Meiss (129), Department of Cell Biology, New York University Medical Center, New York, New York 10016

William T. Molin[4] (199), Department of Biochemistry and Nutrition, Virginia Polytechnic Institute and State University, Blacksburg, Virginia 24061

Gertrude C. Moser (129), Institute of Toxicology, Federal Institute of Technology, University of Zurich, CH-8603 Schwerzenbach, Switzerland

James S. Murphy (289), The Rockefeller University, New York, New York 10021

John D. O'Connor (269), Department of Biology, University of California, Los Angeles, California 90024

Angela M. Otto (315), Friedrich Miescher-Institut, CH-4002 Basel, Switzerland

Lawrence M. Pfeffer (289), The Rockefeller University, New York, New York 10021

T. B. Pool (363), Department of Anatomy, The University of Texas Health Science Center at San Antonio, San Antonio, Texas 78284

[3] Present address: New England Biolabs, Beverly, Massachusetts 01915.

[4] Present address: Agronomy Department, University of Wisconsin, Madison, Wisconsin 53706.

Maurice J. Ringuette[5] (151), Department of Biology, University of New Brunswick, Fredericton, New Brunswick, Canada

Avery A. Sandberg (3), Departments of Genetics and Endocrinology, Division of Medicine, Roswell Park Memorial Institute, Buffalo, New York 14263

Robert R. Schmidt[6] (199), Department of Biochemistry and Nutrition, Virginia Polytechnic Institute and State University, Blacksburg, Virginia 24061

R. A. Singer (181, 245), Faculty of Medicine, Dalhousie University, Halifax, Nova Scotia, Canada

N. K. R. Smith (353), Department of Anatomy, The University of Texas Health Science Center at San Antonio, San Antonio, Texas 78284

R. L. Sparks[7] (363), Department of Anatomy, The University of Texas Health Science Center at San Antonio, San Antonio, Texas 78284

Bryn Stevens[8] (269), Department of Biology, University of California, Los Angeles, California 90024

Igor Tamm (289), The Rockefeller University, New York, New York 10021

Christopher F. Thurston[9] (199), Department of Biochemistry and Nutrition, Virginia Polytechnic Institute and State University, Blacksburg, Virginia 24061

Frank Traganos (103), Investigative Cytology Laboratory, Memorial Sloan–Kettering Cancer Center, New York, New York 10021

Katherine J. Turner[10] (199), Department of Biochemistry and Nutrition, Virginia Polytechnic Institute and State University, Blacksburg, Virginia 24061

P. T. van der Saag (337), Hubrecht Laboratory, International Embryological Institute, 3584 CT Utrecht, The Netherlands

Dieter E. Waechter[11] (231), Department of Pathology and Fels Research Institute, Temple University School of Medicine, Philadelphia, Pennsylvania 19140

[5] Present address: Department of Biochemistry, Queen's University, Kingston, K7L 3N6 Ontario, Canada.

[6] Present address: Department of Microbiology and Cell Science, University of Florida, Gainesville, Florida 32611.

[7] Present address: Division of Biophysics, School of Hygiene and Public Health, The Johns Hopkins University, Baltimore, Maryland 21205.

[8] Present address: Department of Cell Biology, Baylor College of Medicine, Houston, Texas 77030.

[9] Present address: Microbiology Department, Queen Elizabeth College, London W8 7AH, England.

[10] Present address: Rosenstiel Basic Medical Science Research Center, Brandeis University, Waltham, Massachusetts 02154.

[11] Present address: Friedrich Miescher-Institut, CH-4002 Basel, Switzerland.

Eugenia Wang (289), The Rockefeller University, New York, New York 10021

Gary L. Whitson (393), Department of Zoology, University of Tennessee, Knoxville, Tennessee 37996

Klaus Wilke (55), Department of Biochemistry, Duke University Medical Center, Durham, North Carolina 27710

Anthony T. Yeung[12] (199), Department of Biochemistry and Nutrition, Virginia Polytechnic Institute and State University, Blacksburg, Virginia 24061

[12] Present address: Department of Biochemistry, School of Hygiene and Public Health, The Johns Hopkins University, Baltimore, Maryland 21205.

Preface

An understanding of the molecular mechanisms that govern the expression of genetic information during the cell cycle requires full knowledge of how the genome is organized and the extent to which changes in its organization affect the ultimate synthesis and processing of RNA and other gene products. In this volume we have brought together investigators whose current research is directed toward several aspects of this central theme. The initial five chapters describe the intimate relationships between the supramolecular complexes that form the basic structure of chromatin. Emphasis is placed on the dynamics of cycle-dependent changes in the structural organization of some of these components.

The chromatosome, defined by neutron scatter, electron microscopy, and low resolution X-ray diffraction as a circular disk 11 nanometers in diameter and 5.5–6 nanometers in height, represents the primary subunit of chromatin. The first chapter introduces an extension of the details of our knowledge of this structure as a hexagonal bipartite disk stacked face-to-face and interconnected by axial histone H1, which has usually been considered as associated with the nucleosome linker region. The hexagonal bipartite disks appear to be aligned either in a continuous linear 100- to 140-Å-diameter nucleofilament or as 280-Å nucleofilaments achieved by a side-to-side association of the nucleosome hexagonal disks. This model proposes that the histone H1 is located at the axis in an optimum position to serve higher order packing and at the same time to provide postsynthetic modifications of histone H1 to accommodate the non-histone proteins, for example, the HMG proteins. The postsynthetic modifications of histone H1 and other histones are likely to play a major role in the transition from the extended state in interphase chromatin to the more contracted state in metaphase chromatin. These histone modifications are discussed in Chapter 2. This chapter reviews the details of histone acetylation and its effect on the structure and function of chromatin during the cell cycle. These studies exploit the naturally synchronous cell cycle of *Physarum polycephalum* in which there appear to be substantial changes in both quantity and quality of transcription during the cell cycle. A convincing argument is made that many of the diverse observations on chromatin structural behavior represent the consequence of the kinases, acetyltransferases, and deacetylases to coordinate postsynthetic modifications of phosphorylation and acetylation of histones. The third chapter extends this theme to the role of the HMG proteins in relation to eukaryotic gene activity during the cell cycle. The importance of the HMG proteins is evidenced by the fact that they are associated with specific transcriptionally active chromatin fractions. The characterization of the HMG proteins in terms of intracellular concentration, distribution between the nucleus and the cytoplasm, and tissue and species specificity is reviewed. A molecular mechanism for their role in RNA transcription is proposed.

The ability to measure biochemical features of individual cells (Chapter 4) provides an opportunity to close the gap in our knowledge between cellular metabolic events at the molecular level, to study the behavior of cell populations at precise phases of the cell cycle, and to examine some of the dynamic aspects of chromatin structure discussed in the first three chapters. This is accomplished by means of newer techniques of flow cytometry to study synchronized CHO cells and cycling lymphocytes. A two-parameter frequency histogram has the capacity to classify cells on the basis of RNA versus DNA in G_1, S, and G_2 + M phases. Several lines of evidence indicate that G_1 phase cells can be further subdivided into G_{1A} and G_{1B}, which appear to be functionally distinct. The metachromatic properties of acridine orange also provide an index of chromatin structure on the basis of DNA stability. The mechanism of dye interaction (quinacrine dihydrochloride) as reviewed in Chapter 5 also provides an opportunity to monitor some specific cytological aspects of chromatin. These techniques are particularly useful in the analyses of the cell cycle blocks induced in temperature-sensitive mutants.

The relationships between transcriptional and posttranscriptional events and cell cycle regulation are examined in the next four chapters, with special reference to specific RNAs and inducible enzymes as probes of genetic expression. Chapter 6 presents evidence to demonstrate that small nuclear RNAs (SnRNA) are actively involved in gene regulation in eukaryotic cells. The implication of these studies is that active SnRNAs interact with nuclear proteins, possibly HMGs, to stimulate transcription. A key element of the proposed mechanism is the base-pair formation between SnRNA and DNA at the promoter region to facilitate the entrance of RNA polymerase. Chapter 7 focuses on the relationship between cell cycle regulation in the yeast *S. cerevisiae* and transcription of ribosomal RNA genes. A detailed description of the use of G_1-arresting compounds together with an analysis of their effects on the production of precursors of ribosomal RNA is presented. The relationship between this aspect of RNA metabolism and cell cycle regulation is also discussed. In Chapter 8 a detailed experimental account is provided to show that the expression of the gene for the ammonium-inducible isozyme of glutamate dehydrogenase in *Chlorella* is regulated primarily at the post-transcriptional level. The central element of this model is that in the absence of the inducer, subunits of the enzyme form dimers, which are degraded by endogenous proteases to nonantigenic products. The extent to which this model serves to extend our understanding of cycle-dependent regulation of gene expression in this cell is discussed by the authors. Chapter 9 introduces the reader to the use of conditional lethal mutants (e.g., cycle-specific *ts* mutants) to study the regulation of the cell cycle of eukaryotic cells. It is shown that these mutants are useful to study progression through the G_1 phase, particularly with regard to the involvement of RNA polymerase II.

The impact of specific gene products and other agents on specific phases of the cell cycle is considered in detail in subsequent chapters. Chapter 10 presents the concepts and methodologies employed to isolate and study specific cell cycle mutants of *Saccharomyces cerevisiae*. Extensive evidence is presented to show that the cell cycle of this yeast is uniquely regulated at one point through the action of several gene products. The authors discuss qualitative and quantitative differences between resting and actively dividing cells in terms of the concept of the G_0 state and regulation of the yeast cell cycle. In Chapter 11 we are introduced to the use of cultured *Drosophila* cells, which are unique in that they are arrested in G_2 under the influence of ecdysteroids. This is a promising new experimental system utilizing an organism whose genetics and morphogenetic attributes are well documented. The antiproliferative effect of interferon on cultured human fibroblasts is evaluated in Chapter 12 in terms of the effects of this potent cellular inhibitor on the cell membrane, cytoskeletal components, and synthesis of macromolecules. The authors develop the notion that the response to interferon, while manifestly heterogeneous, operates through a common pathway resulting in impaired proliferative capacity for the treated cells. This section of the monograph closes with a detailed analysis in Chapter 13 of the complex pattern of interaction between insulin, hydrocortisone, prostaglandins, and two growth factors as determined by the kinetics of initiation of DNA synthesis in cultured mouse cells. This analysis serves to illustrate the complex program of genetic expression that governs this particular phase of the cell cycle.

One of the challenging questions in cell biology is: To what extent are the cell membrane and related subcellular elements involved in the control of proliferation, differentiation, and cell cycle kinetics? To be sure, a question of this magnitude deserves an extensive and thorough discussion. Chapters 14–17, which complete this monograph, highlight some of the most recent experimental approaches to this complex problem. The extent to which the dynamic properties of the cell membrane have an impact on the cell cycle of neuroblastoma cells is the subject of Chapter 14. Of particular relevance is the relationship between changes in cation transport and the ability of cells to progress toward cell division. The authors make use of synchronized cells and exogenous growth factors to show that electrical and ionic events at the cell membrane, such as the electroneutral Na^+–H^+ exchange, are prerequisites for cell proliferation. An extensive review of what is known of the role of ionic fluxes, as well as the activity of the Na^+,K^+-ATPase in the regulation of cell proliferation, differentiation, and transformation, is presented in Chapter 15. Having evaluated the extensive literature on this subject, the authors present their own studies that show that amiloride, a drug which blocks passive Na^+ influx, has an inhibitory effect on rapidly proliferating cells (normal or transformed), suggesting that Na^+ influx may have a regulatory function. Chapter 16 focuses on the role of calcium levels on cell division in synchronized *Tetrahymena*. It would appear that changes in Ca^{2+} influx may exert their influence not only through an interaction of Ca^{2+} with calmodulin but through the activation of microtubule disassembly and cortical changes associated with actin-like proteins. The correlation between stimulation of cell growth and stimulation of monovalent cation fluxes is examined in the last chapter, which summarizes studies in rat hepatocytes, human T lymphocytes, mouse neuroblastoma cells, and mouse 3T3 fibroblasts in particular. This chapter not only serves as a summary of the work discussed in this section of the monograph but provides us with a synthesis of the events in G_0 cells and points to the directions of future research in this area of cell biology.

The primary objective of this monograph is to formulate new concepts of the control of genetic expression in the cell cycle.

George M. Padilla

Kenneth S. McCarty, Sr.

I

Structure and Function of the Eukaryotic Genome

1

Organization of Nucleosomes in Chromatin and Chromosomes in Eukaryotic Cells

LEE S. CHAI AND AVERY A. SANDBERG

I. Introduction

Chromatin and chromosomes exhibit nucleofilaments measuring approximately 100 or 200 Å in diameter (*30,35,39,56,99,100,110*). The nucleofilaments of 100 Å in diameter are made of repeating units (*51,53, 81,85,128*). These units are known as nucleosomes and were initially characterized as spherical "beads-on-a-string," the string being DNA. They were later described as *spherical disks* (*29,33,34,57,89*) and are generally accepted as such at present. Further analysis of the nucleo-

GENETIC EXPRESSION IN THE CELL CYCLE

ISBN 0-12-543720-X

somes has uncovered subunits which consist of two heterotypic tetramer histones in an octamer complex $(H2A,H2B,H3,H4)_2$ (*53,116,117,122, 123*). Approximately 200 base pairs (bp) of DNA have been found to be associated with each nucleosome (*43,53,78,118,121,127*). However, the repeating length of DNA has been found to vary, depending on source and preparatory procedures (*3,4,17,42,43,78,79,105,108,118,121*). Histones exhibit specific interactions with each other in the formation of the nucleosomal core (*13,26,46,52,53,64–67,123*). In addition, histone H1 was also found along with the core histones in the repeating units but was not considered to be an integral part of the core histone complex (*2,11,41,53,76,125*).

Various proposals and models concerning the organization of nucleosomes have been advanced (*2,10,33,42,47,53,59,82,90,96,111,114,119, 122,129,130*). They were mostly depicted as spherical or round disks aligned either edge to edge or side by side (*33,111,114*). Core histones were conceived to be small spherical subunits within the larger disks (*10,130*). Noncore histone H1 was placed at the periphery of the repeating units, along with the internucleosomal DNA, and thought to compact the nucleosomes into a linear nucleofilament. In higher order packing, some investigators have proposed a "solenoid" model in which the string of repeating units was coiled to form a helix with a pitch of approximately 100 Å and an outer diameter of 300 Å (*33*). Each coil-bound segment of solenoid was considered to contain 6 to 8 nucleosomes. Another group of investigators has proposed the folding of nucleosomes at certain intervals into larger structures of approximately 200 Å in diameter (*45,96*). These structures were called "superbeads." Both solenoid and superbead structures have been observed from studies made with either purified and fragmented chromatin or chromatin subjected to certain concentrations of ions in buffer solutions. There is mounting evidence that the variable structural configuration of the nucleofilaments may be a function of the ionic concentration in solution. It is not clear, however, whether the solenoid or superbead configuration is present in native chromatin. Chromatin which has not been fragmented or subjected to unusual ionic conditions has shown quite a different structure, i.e., a thicker chromatin of 200–300 Å in diameter made of thinner 100 Å diameter units. This thicker unit was differentiated from solenoid and superbead structures by the side-by-side association of the two thinner 100 Å diameter units (*10,23,91,100*).

Our observation of chromatin in interphase nuclei and metaphase chromosomes in intact cells has revealed a number of new features. We find nucleosomes to be hexagonal bipartite disks, interconnected by a strand or strands and stacked face to face to form a linear array (*18,19*). In describing these new features, it is necessary to reevaluate earlier reports and modify, where needed, the structure and packing of nucleosomes in chromatin and chromosomes. We describe general features of nucleosomes; the conformation of DNA; possible structure of histone molecules and their interactions as well as DNA–histone interactions; the presence of axial structure; and alignment of nucleosomes in higher order packing. In conclusion, important aspects of the structural organization of nucleosomes are recapitulated, and the significance in transcription and replication processes is briefly discussed.

II. Hexagonal Bipartite Disk Structure of the Nucleosome

Electron microscopy of nucleosomes in chromatin had initially shown essentially spherical bodies of approximately 100 Å in diameter (*81, 85,128*). However, other investigators have indicated that the nucleosomes may not be strictly spherical but rather disk shaped (*57,89*). Furthermore, purified nucleosomes have shown bipartite characteristics (*29,34*). Our observations of intact interphase cell nuclei and intact metaphase cell chromosomes have revealed that nucleosomes are, indeed, disk shaped and possess bipartite characteristics (Fig. 1). However, as differentiated from a spherical or round disk, we found the nucleosomes to be slightly elongated hexagonal bipartite disks (Fig. 2). The overall dimensions of such a disk are 140 ± 25 Å in height, 100 ± 20 Å in width, and 55 ± 5 Å in thickness (Fig. 3). A hexagonal bipartite disk possesses four side facets measuring 70 ± 15 Å each and four upper and four lower facets measuring 35 ± 5 Å each. A unit, i.e., one disk of the double set, of a bipartite disk measures approximately 20 Å in thickness, with a center-to-center distance of approximately 35 Å. The overall dimensions are similar to those obtained by X-ray and neutron-scattering studies (*14,63,87,88,126*). The orientations of disk height, width, and thickness in this study are seen from different perspectives and differentiated from those in other reports. The bipartite disks are stacked face to face, connected by a strand or strands, and formed into a linear array of nucleofilaments.

III. The Conformation of DNA

A. Kinks

There is much evidence to support the concept that DNA is wound around a histone core (*6,61,77,90,112*). Spectral studies have indicated

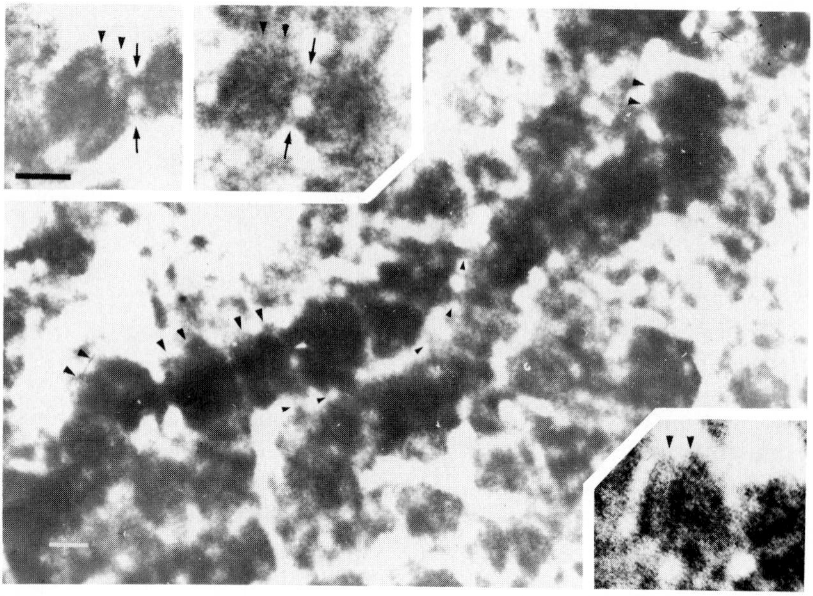

Fig. 1. Bipartite characteristics (double arrowheads in insets and other areas) of disks are evident. There are two connecting strands between bipartite disks (arrow in insets). One of the connecting strands is centrally located and the other is at the periphery of the disk. The bipartite disks are stacked face to face and aligned into a continuous linear array. The bipartite disks in one linear array are periodically cross-linked with the edges of bipartite disks of the adjoining array (small arrowheads). The bars indicate 100 Å. [Adapted from *Cancer Genet. Cytogenet.* (*19*).]

that the conformation of the DNA is of the B form (*24,32,37,115*). However, the configuration of the DNA supercoil around the histone core may produce bends or a continuous bending which would impart a small amount of twist (in a right-handed supercoil) or untwist (in a left-handed supercoil) in the DNA. Moreover, it is not clear whether the DNA in the nucleosome remains exactly in the normal B configuration. The asymmetric distribution of the frequency of DNase cuts from one end of a

---

Fig. 2. (a) A disk that has partially emerged from the plane evidently reveals polygonal characteristics (arrowheads). In this disk, the core and the strand that make up the polygonal perimeter are separated yet connected at a number of sites (white arrows). Another disk, which appears to be loose, shows an aggregate of four or five electron-dense structures (upper left with white bars). A rodlike structure extends through the aggregate. In addition, a curled structure encircles the aggregate. The inset shows a double coil which appears to be left-handed. (b) There are two hexagonal disks in the center area. One of these appears compact, whereas the other appears loose but reveals bridging structures between the core and the strand making up the hexagonal perimeter. (c) These disks also show elongated hexagons. The two opposing disks, which are in face view, show an association

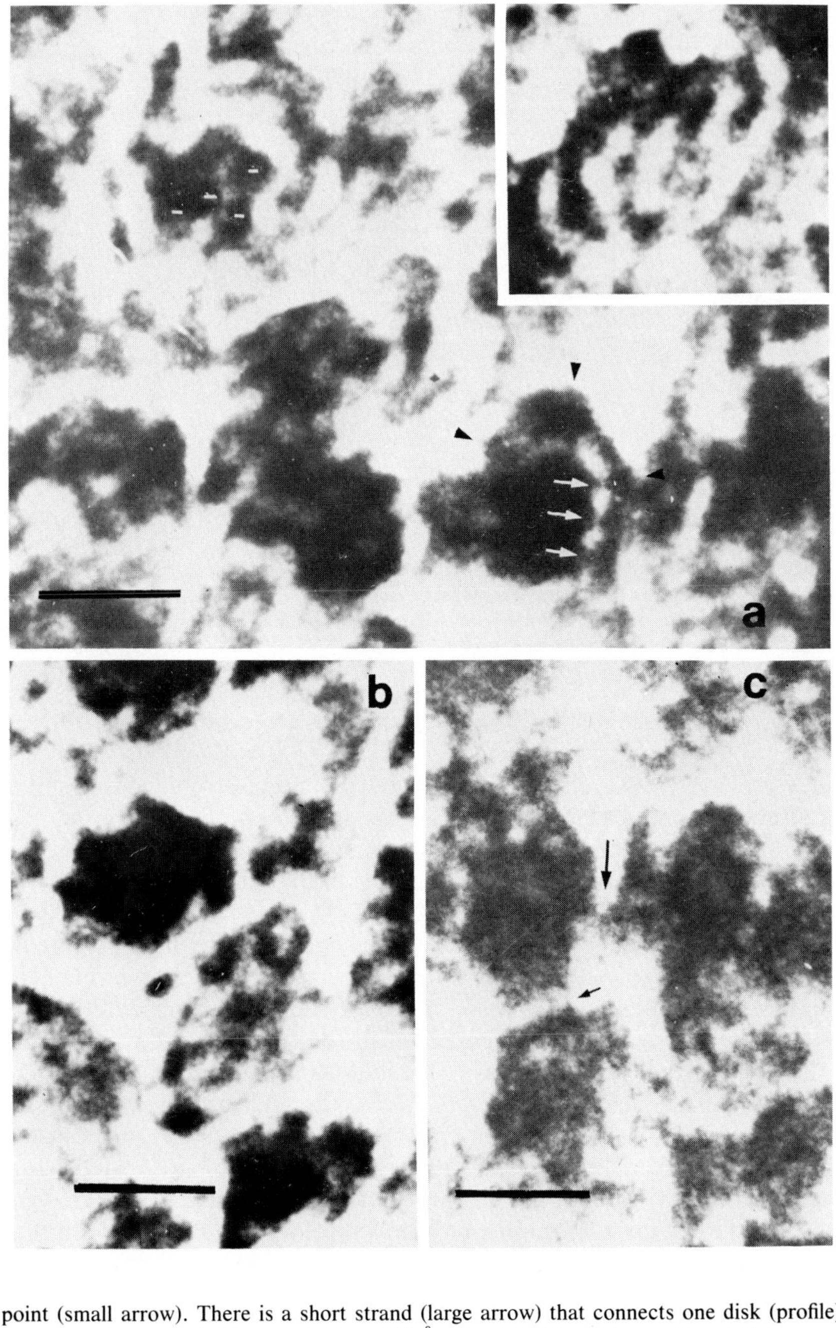

point (small arrow). There is a short strand (large arrow) that connects one disk (profile) to the next (face view). The bars indicate 100 Å. [Reprinted from *Cancer Genet. Cytogenet.* (*19*).]

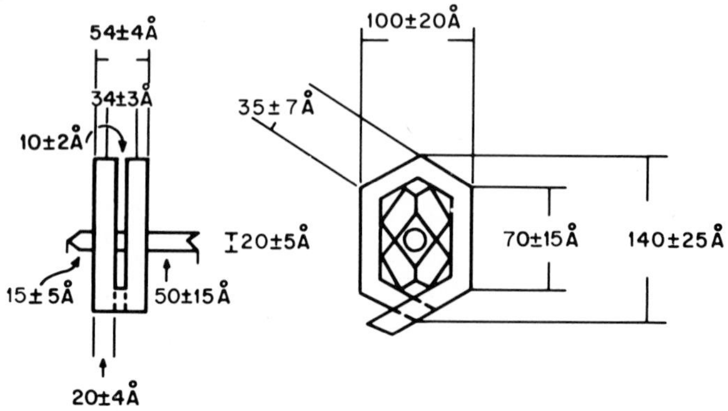

Fig. 3. Schematics of a nucleosome in a flat-face view and in profile. At right, the hexagonal perimeter represents the DNA. The four units within the hexagonal frame represent four histone complexes. (In a bipartite disk, there are octamer histone complexes.) The open circle at the center is the axis and represents histone H1. At left, a bipartite disk (in profile) possesses a narrow space of 10 Å or less within the bipartite. The center-to-center distance of the two units of a bipartite disk is about 34 Å. The axial arrow indicates one molecule of histone H1. The arrowhead depicts the amino-terminal region and the tail of the arrow the carboxyl-terminal region. [Reprinted from *Cancer Genet. Cytogenet.* (*19*).]

strand of DNA double helix to the other has indicated that the DNA is supercoiled in a left-handed manner (*62*). Thus, a small amount of untwist may have to be present in the left-handed supercoil. We have observed a strand which measured about 20 Å in diameter wound along the edge of a disk core and making up its hexagonal perimeter (Fig. 2). This strand was also seen in left-handed double coils in the other areas (Fig. 2). We interpret this structure to be DNA.

There has been considerable discussion about the manner in which the DNA may be wound around the histone core. DNA kinks at about 10 bp intervals have been proposed (*25,107*). Hence, we postulate that the DNA may be kinked at each corner of the hexagonal disk at a 60° angle. This was tested by using Corey–Pauling–Kolton (CPK) models. Initial results (R. Parthasarathy and L. S. Chai, unpublished data) have indicated that it is feasible, on the basis of these models, for the DNA to be wound with kinks at certain intervals around the core to form hexagonal disks. This analysis has also indicated that kinking would orient toward the direction of the major grooves with a small angle of unstacking. When the kinking was oriented toward the direction of the minor grooves, the DNA bases without support completely unstacked. The kink would occur by an alternation of pucker of the sugar coupled with some changes in phosphodiester angles. We have, therefore, adopted kinks toward the major grooves.

B. Intervals of Kink

A hexagonal bipartite disk possesses four sides 70 ± 15 Å each, equivalent to approximately 20 bp of DNA, and shorter upper and lower facets that are 35 ± 5 Å each, equivalent to approximately 10 bp. Each segment of DNA is placed in a corresponding position in the schematic of the bipartite disk drawn to hexagonal form (Fig. 4). Kinks occur at 10 and 20 bp intervals, with the total amount of DNA on the perimeter of a bipartite disk being equivalent to approximately 160 bp in two full turns or 140 bp in one and three-fourths turns. There are 16 sites (see Section IV,B) in an octamer that interact at every 10 bp of the DNA double helix or every 20 bp of a single strand (Figs. 4 and 5). Thus, 160 bp of DNA make up a full complement of nucleosomal DNA. This is in accordance with the 160 bp chromatosome concept described by others (*105*). Intervals of kinks postulated at multiples of 10 bp are in keeping with the 10 or 10.4 bp DNA fragments resulting fron nuclease treatment (*61,62,77,80,92,97,104*).

Digestion of chromatin with micrococcal nuclease produces mononucleosomes with variable lengths of DNA (*3,4,78,79,101,105,118,121, 125*). Furthermore, DNase I digestion of 140 bp mononucleosomes produces discrete patterns. Frequency of cuts is mostly 10, 20, 40, and 50 bp from the 5′-end and 130, 120, 100, and 90 bp from the 3′-end. These cuts are in symmetry (*62*). Nucleosomes with 160 bp also produced similar results (*105*). The presence of an extra 20 bp of DNA was seen as an extension of the core particle DNA of 140 bp by addition of a 10 bp segment to each end (*105*). The frequency of nuclease cuts, therefore, now becomes 10, 20, 30, 50, and 60 bp from the 5′-end and 150, 140, 130, 110, and 100 bp from the 3′-end; they are also in symmetry. When the kinetics of the DNase I cuts (*62,77,80,105*) and kinking of DNA in hexagonal bipartite patterns are superimposed, remarkable correlations are seen (see Fig. 4). Those bases which are not or infrequently cut are located at the top corner of the hexagonal bipartite disk, are stabilized by N-terminal regions, and correspond to bases 40 and 120 from the 5′-end. Bases 70 and 90 from the 5′-end are infrequently cut and are located at the inside of bases aligned with bases 10 and 150 on the outside. Base 80 is located in the middle of the nucleosomal DNA and is situated inside the bottom corner of the hexagonal bipartite disk. This region is infrequently cut by DNase I and is also stabilized by the N-terminal regions. Most of the DNase I cuts appear to be located on the broad facts of the hexagonal bipartite disk and are stabilized by C-terminal regions. The angulated top and bottom corners of the hexagonal disk appear to be protected either by DNA–histone binding patterns (between N-terminal regions) or by insusceptibility to certain enzyme action.

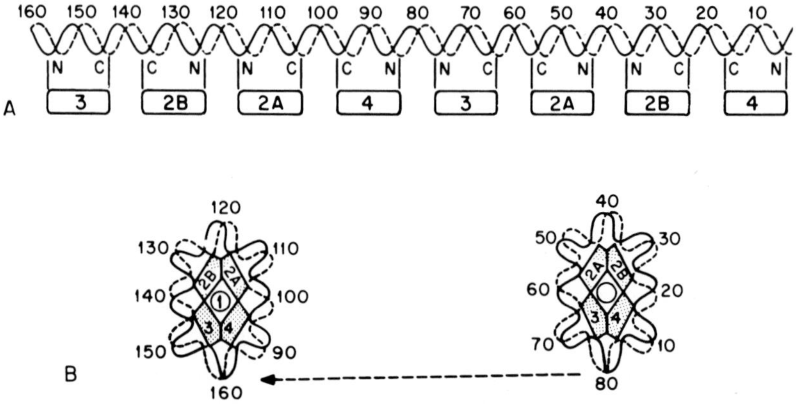

Fig. 4. (A) DNA–histone complexes are linearly extended. There are two points of interaction for each histone with the DNA double helix: one at the amino-terminal region and the other at the carboxyl-terminal region. Histones are aligned in such a way that the amino-terminals face each other (N–N) as do the carboxyl-terminals (C–C). Every 10 bp interval of the DNA double helix and every 20 base interval of the DNA single strand is stabilized by each interaction. Nuclease cuts may occur most frequently between facing carboxyl-terminals, e.g., bases 20, 60, 100, and 140 from the 5′-end, and may occur infrequently between facing amino-terminals, bases 40, 80, and 120 from the 5′-end. The rest of N–C may occur with intermittency. Nuclease cuts may produce free 10 or 10.4 bp segments which are not associated with histones. For example, free 10 bp segments may be found between bases 10 and 20, 50 and 60, 90 and 100, and 130 and 140 in the ------ strand and between bases 20 and 30, 60 and 70, 100 and 110, and 140 and 150 in the —— strand. The DNA double helix may be staggered at both ends of nucleosomal DNA. (B) conformation of DNA, location of histones and their interaction with each other, and sites of DNA–histone interaction are schematically illustrated. The position and sequence of octamer histones are postulated according to (*53*) and (*103*). DNA kinks at each hexagonal corner at a 60° angle with respect to the dyad axis of the double helix. The formation of kinks is related to the occurrence of the major groove; a portion of the minor groove is exposed to the outside. In histone–histone interactions, dimers (upper and lower dimers, respectively) appear to have a greater affinity for each other than tetramers. A split octamer shown in the figure is represented by displacement of the axis (histone H1) from the tetramer on the right. Histone H4 possibly stabilizes the incoming DNA, and histone H3 stabilizes the exiting DNA (*103*). The entering and exiting segments of DNA are not associated with histone H1. Histones H2A and H2B alternate within a bipartite disk (*53*). [Adapted and modified from *Cancer Genet. Cytogenet.* (*19*).]

IV. Histone–Histone and DNA–Histone Interactions

A. Histone–Histone Interactions

Our observations indicate that there are four subunits within a unit of bipartite disk (Fig. 2). This is consistent with the formation of a tetramer histone complex of H2A, H2B, H3, and H4. Each bipartite

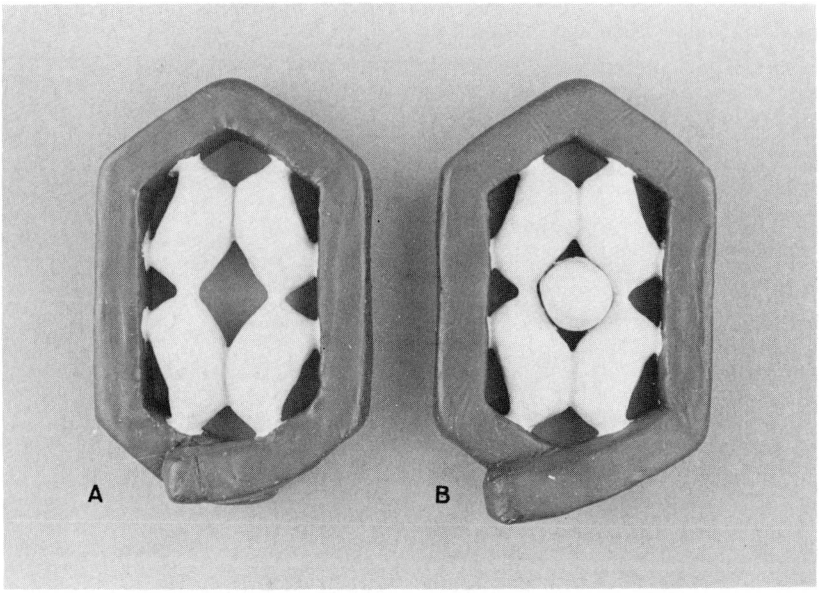

Fig. 5. Three-dimensional views of a split hexagonal bipartite disk and its internal organization are illustrated. A histone molecule may assume a cylinder-like structure. In the hierarchy of interaction between histones, dimers appear to have greater affinity for each other than tetramers. Note that histone molecules may not have to extend to make associations with DNA. However, at the site of DNA entry and exit, histones extend more than the rest of the interaction sites, thus accommodating the variable conformation of DNA. This variable extention is not shown in a split disk. Displacement of H1 has no disruptive effect on the core histone complexes (A). Histone H1 may establish ionic interaction with the histone core complexes at the axial position (B).

disk, therefore, would contain two of these histone complexes. It has been shown that there is a hierarchy of pairwise interactions between histones (*26,46,52,53,64–67,103,116,117,123*). For example, H2A–H2B and H3–H4 dimers and $(H3–H4)_2$ tetramers are preferentially cross-linked. In visual representations of these interactions, some investigators have proposed spherical globular structures (*10,130*). When the histone molecule is represented as a spherical globular structure, two such structures may establish contact across the width at the top and the bottom of the hexagonal disk. However, two respective pairs at the top and the bottom could not bridge across the length of an elongated hexagonal disk. Thus, it was necessary to picture the histone molecule as something other than a spherical globular structure. Modeling has revealed a cylindrical structure, which resembles an oil drum with an oblique angle cut at the top corner (Fig. 5). Circular dichroism, laser Raman (*84,115*), and infrared spectroscopy (*24*) studies have shown that the core histones possess a high content of α helices with little or no β sheets. The folded forms of histones possess no detectable β-sheet configuration (*46*). Since the histone complex in the core is folded and not aggregated, it is conceivable that the α helix may assume the cylindrical configuration as illustrated in our models (Fig. 5).

Nuclear magnetic resonance (nmr) studies have shown histones H2A and H2B to interact pairwise at the midregion of each respective molecule. The N-terminal regions and the short region near the C-terminus are not involved in complex formation (*7,13,74*). Cross-linking of histones by tetranitromethane produces an H2B–H4 complex (*64*), whereas uv cross-links H2A to H2B (*66*). H3 is linked to H4 by carbodiimide (*8*). These agents produce zero-length cross-links. H2A–H2B–H4 cross-linked trimers can be formed by sequential treatment of chromatin with uv and tetranitromethane (*65,67*). In this case, H2B is indicated to have three distinct regions: a certain portion of the C-terminal region, which interacts with H4; a midregion, which interacts with H2A; and the N-terminal region, which is free of interactions (*65,67*). Formaldehyde treatment of chromatin also cross-links H2B to H2A and H4 with relatively high specificity (*120*). It is interesting to note that the pair of H2A–H2B histone molecules interact at the midregions of the molecule. This interacting region is represented on the cylindrical structure by the oblique angle in our analysis. It is possible that H3–H4 may also interact at the midregion of the molecule in a similar manner as H2A–H2B. The two respective dimers, H2A–H2B and H3–H4, may then be opposed at or near the C-terminal regions and participate in the formation of a heterotypic tetramer. The model depicted in our studies (Figs. 4 and 5) is thus consistent with these cross-linking studies.

A histone molecule as represented in a cylindrical configuration may have multifacets and points of association with other molecules including the other histones and DNA. First, there is a facet at the midregion that participates in the formation of a dimer pair. Second, the two dimer pairs may then interact at the C-terminal regions or near them in the formation of a heterotypic tetramer. The third facet, which is not involved in the formation of a dimer pair or a heterotypic tetramer, may be aligned with hydrophobic residues on its side. The hydrophobic residues may align two heterotypic tetramers face to face and form an octamer. The fourth facet side (opposite the hydrophobic inside facet) may possess hydrophilic residues facing out on both sides of an octamer. The hydrophilic residues on its side wall may favor easy interaction with non-histone proteins and various other molecules. Acetylation, methylation, phosphorylation, and ADP-ribosylation may occur at this region (see Section

V). The fifth facet, which is on the opposite side from the dimer-forming facet, may be aligned with the N-terminus on one end and the C-terminal region on the other. The regions near the ends of both termini may complex with the phosphate backbone of the DNA double helix. Thus, histones may fasten the DNA to the core complex. The heterotypic tetramer and the octamer are complexed in such a way that the remaining portion inmediately below the dimer complex may form a hole that is at the center of the disk, where the axial histone may be inserted and transiently interact with all four histones (Figs. 3–8).

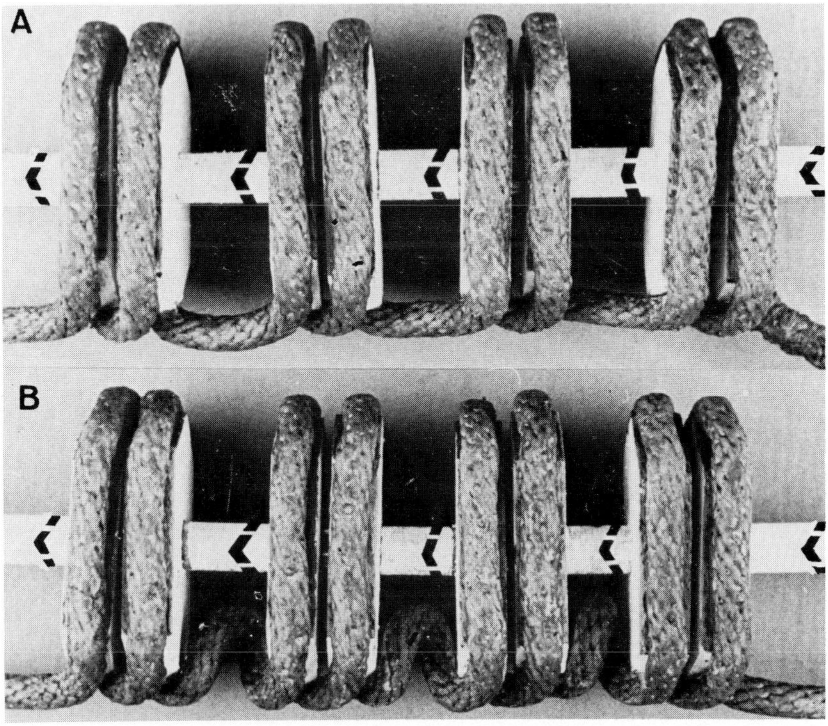

Fig. 6. Profile of bipartite disks and their interbipartite regions. (A) Where the interbipartite DNA strand is short, the DNA may be aligned parallel to the axis. (B) Where the interbipartite DNA strand is long and/or the axial structure has changed its conformation, the interbipartite DNA may recoil between two bipartite disks. It is possible that DNA may associate with the axial structure when the DNA is recoiled. Length and conformation of the interbipartite DNA strand may be correlated with variations of histone H1, and length of H1 may also be variable. There is directionality in the orientation of histone H1: The short arrowhead indicates the amino-terminal region and the long tail end indicates the carboxyl-terminal region. The connections between the head and the tail indicate some kind of interaction, possibly an ionic interaction.

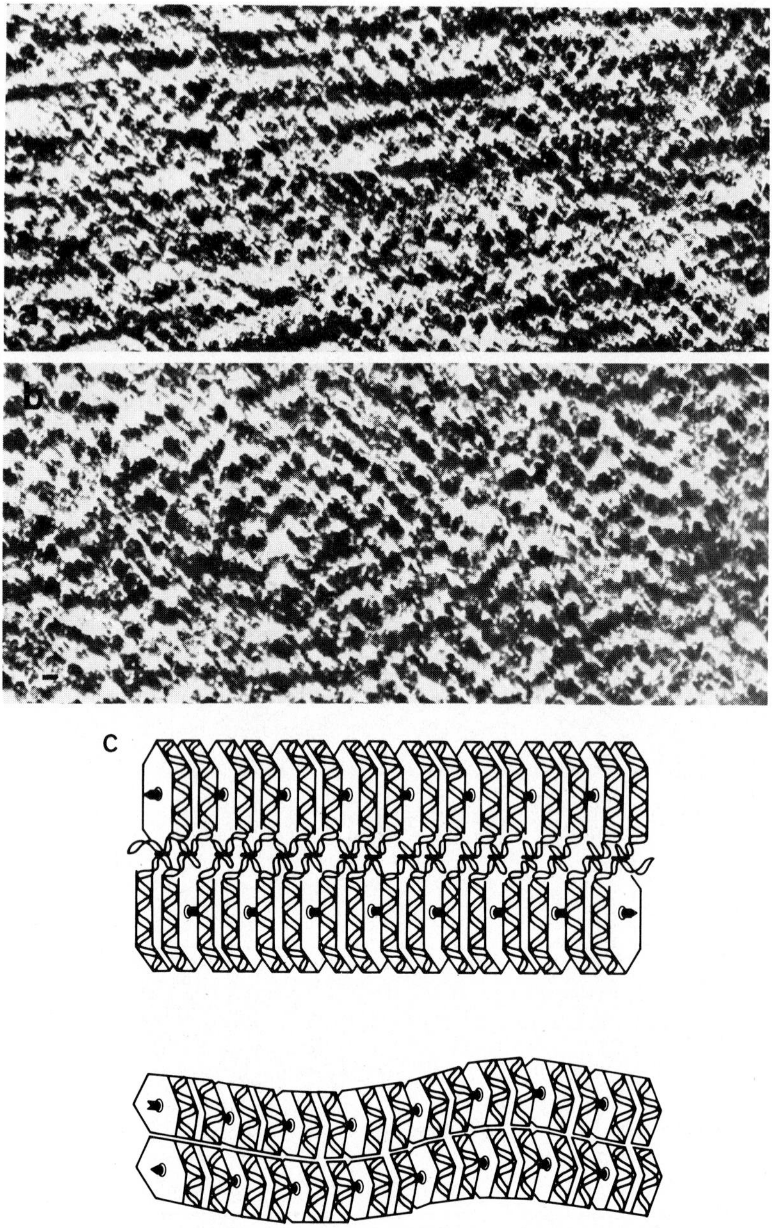

Fig. 7. Alignments of nucleosomes in linear (a) and wavy (b) orientations are clearly revealed when two identical original plates are superimposed on top of each other and aligned according to the run of grains. This process accentuates the subtlety of a single

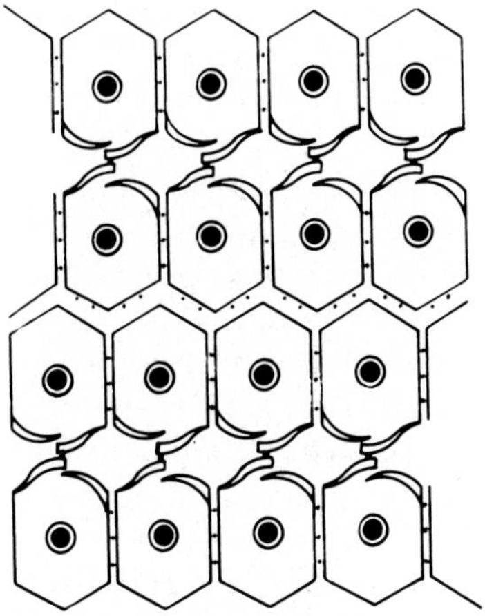

Fig. 8. A face-view alignment of bipartite disks in a paracrystalline lattice is illustrated. Entering and exiting DNA strands are identified by the horn-shaped forms. The possible ionic interaction is indicated by dots. Exposed hexagonal facets may also be neutralized by ionic interactions. It is possible that the lattice packing may not encompass the entire chromosome. It is estimated that 6–8 cm of DNA may be packed within a chromatid 5.0 × 10^4 Å long, if the DNA were organized according to our schematics. [Reprinted from *Cancer Genet. Cytogenet.* (*19*).]

B. DNA–Histone Interactions

Freeze-fracture studies of nucleosomes have shown that there are up to eight bridging structures between the core and the strand which make up the hexagonal perimeter of the bipartite disk (Fig. 2). The eight bridgings emanate from the tetramer complex, two from each histone molecule. Thus, there would be 16 sites for interaction in an octamer per unit bipartite disk (Figs. 3–5).

Thermal denaturation studies have revealed three components of the DNA melting curve: DNA free of histone (T_m 40°C), DNA bound to the less basic portion of the histone molecule (T_m 66°C), and DNA bound to the basic portion of the histone molecule (T_m 80°C) (*60*). Furthermore,

---

plate exposure. The two different orientations appear to occur perpendicular to each other and also depend on the plane of fractures. (c) A linear profile (top) and a wavy top view (bottom) are schematically illustrated. A 280 Å unit is formed by the doubling back of the 140 Å unit and by the bipartite disks interacting with each other at the areas where DNA enters and exits, thus exposing angulated hexagonal facets. The direction of the bipartite disk packing is indicated by the axial arrowhead. The short bars between adjoining DNA strands indicate possible ionic interactions. Bar represents 100 Å. [Figure 7c adapted and modified from *Cancer Genet. Cytogenet.* (*19*).]

if histone H2A is split, its basic N-terminal half and its less basic C-terminal half are bound separately to DNA. The basic N-terminal half of H2A stabilizes DNA at a melting temperature of about 70°C, whereas the C-terminal region of H2A does so at 57°C. This reveals two points of association for each histone molecule: one from the basic N-terminal region and the other from the less basic C-terminal region. This is a histone–DNA interaction in which N-terminal and C-terminal residues are associated with the different regions of DNA (*124*). Using cyanogen bromide cleavage of histone H4, it was found that the N-terminal region bound strongly to DNA, whereas the C-terminal fragment bound weakly to the DNA (*1*). Stabilization of 10 bp segments at both ends of the nucleosome by N-terminal regions of one molecule (H4) at the entering and another molecule (H3) at the exiting sites may have additional significance. The N-terminal region may freely extend like an arm without distorting the configuration of the rest of the histone core complex and possibly accommodate the variable configuration of the DNA molecule. It has been observed that the 10 bp segments at both ends of the chromatosome are in a precise manner related to the DNA of the core particle and do not interact with histone H1 (*105*). The eight association points with DNA in the tetramer unit of the bipartite disk, two from each histone molecule, are consistent with this observation.

C. Frequency of Histone Interaction with the DNA Double Helix

It has been reported that histone cross-links to DNA segments at about 10 bp intervals. One histone is cross-linked to each segment of the DNA, with the exception of two segments where two histones are cross-linked to the same 10 bp segment of DNA (*73*). In Section III we described DNA kinked at 10 and 20 bp intervals. When this B form of DNA is placed around a tetramer histone complex, DNA–histone interaction patterns emerge (Fig. 4). These interactions occur at 10 bp intervals of the DNA double helix or at 20 bp intervals of the DNA single strand. It was postulated earlier that the N-terminal region is associated with only one given strand of the DNA double helix, whereas the C-terminal region is associated with the opposing strand (*19*). In order to have such associations, histone molecules have to complex in the following manner: The N-terminal region of a preceding histone molecule may have to complex with the C-terminal region of the following histone molecule. Cross-linking studies have demonstrated the midregions of H2A and H2B to interact and form a dimer pair; the C-terminal region of H2B was shown to interact with the C-terminal region of H4, a nonpairing histone (*64,65,67*). DNA–histone interaction thus appears to be much more complicated than previously thought. Regardless of the polarity in orientation of the histone molecule, the frequency of interaction between DNA and histone remains the same (Fig. 4).

Schematically, histone molecules appear to shield about one-half or more of the major grooves and interact over minor grooves on alternating strands of the DNA double helix (Fig. 4). Other investigations have described histones located along the sugar phosphate backbone on the side of the major groove (*72,73,102*). Antibiotics have also been shown to interact preferentially at the minor grooves, which are free of histones (*72*). However, Raman scattering has indicated that histone bonding may occur in the minor groove (*38*). We concur, in part, that DNA kinks toward the major grooves are partly protected by histones and that the minor grooves are also partially covered by the histones. However, it should be pointed out that both major and minor grooves rotate 360° in a complete helical turn. The cylindrically shaped histone molecule may be located on one side (inside) of the helically rotating DNA and cover one-half or more of the major groove and/or minor grooves. Even though the length of this histone molecule may be shorter than the length of one helical turn of the DNA double helix, it is still capable of covering the entire length of one helical turn of DNA. This may be accomplished by the DNA kinking, which would shorten the distance between helical turns. The cylinder-like histone molecules could then bridge over the deeply recessed major grooves (not curving or following through the recessed grooves) and interact at 10 bp intervals on alternating strands (Fig. 4). Formation of kinks thus becomes a significant factor in both DNA–histone interactions and packing of DNA.

A single strand of DNA double helix is stabilized by histones at 20 bp intervals (Fig. 4). Thus, every other 10 bp segment of the single strand has no interaction with histones. When a nuclease "cuts" at about 10 bp intervals of the single strand, every other 10 bp segment, which is not complexed with histones, naturally becomes a free segment. As a result, both ends of the double helix of nucleosomal DNA become staggered (Fig. 4). Our analysis is thus consistent with the occurrence of free 10 bp segments present in nuclease cuts. However, other investigators have reported interaction of two different histones with the same 10 bp segment of DNA as well as more than two interaction sites with DNA from one histone molecule [see Fig. 5 of (*73*)]. We find a discrepancy between our interpretation and that of others (*73*).

V. Histone H1 and Alignment of Nucleosomes

A. Histone H1 in the Axial Position

Different methods of chromatin preparation have yielded configurations of nucleofilaments varying from "beads-on-a-string" to continuous nucleofilaments (*30,35,81,85,91,100*). Chromatin *in vitro* appears not to be preserved in its native conformation. It would, therefore, be desirable to observe native structures with minimum disruption. Intact cells processed by freeze-fracture have been shown to maintain their native chromatin structure with the least amount of disruption. Using this technique, the profile of bipartite disks and their interbipartite regions reveals a connecting strand or strands (Fig. 1). These were, however, not visualized where the bipartite disks were compact or not at right angles to the fracture. One of the connecting strands was located at the center of the disk and appeared as an axial structure (Fig. 1). We have postulated that histone H1 is a major component of this axial structure. The other strand, which is located in the periphery, is seen as DNA. This interpretation is derived from the following considerations. The tertiary structure of histone H1 consists of three regions: a relatively short hydrophilic segment located near the N-terminal region (residues 1–39), followed by a globular hydrophobic region (residues 40–115), and ending with a relatively long hydrophilic region (residues 116–215) composed of about one-half of the total H1 polypeptide (*12,21,22,31,41,44,46,93*). Histone H1 can thus be seen as a hydrophilic rod with a hydrophobic knob near one end. It has been suggested that the basic residues of the globular region of H1 may be on the surface; such a specific spatial configuration of positive charges may act as an anion receptor (*41*). It is therefore possible that the globular region of histone H1 and the core histones may have ionic interactions. Release of histone H1 in certain ionic solutions certainly indicates such a possibility.

Histone H1 located at the center of the disk as an axial structure is reinforced by the presence of a hole in the center of the bipartite disk (*19,57*). The integrity of the bipartite disk structure suggests that the axis is an independent entity easily removed without disrupting the core histone complex. For example, the continuous nucleofilament may be segmented at bisected interbipartite regions (Fig. 6; Section V,B). Once the continuous axial H1 link is segmented, H1 may be easily displaced from the core under certain ionic conditions. That histone H1 is inside of the nucleosome is also indicated by studies of H1 antibody binding patterns. H1 antibody is bound less to chromatin at high than at low ionic strength (*15*). There are also reports on cross-linking of histone H1 with histones

H2A (*9*), H2B, H3 (*98*), H2A, H4 (*48*), and all four histones (*11,98*). When H1 cross-links with all four histones, it has to be near the center area where the core histones face each other. An axial structure consisting of histone H1 spanning through the center of the disk core is certainly consistent with these observations.

B. Segmentation of Axis

Micrococcal nuclease digestion of chromatin results in a nucleosome with about 200 bp of DNA containing all five histones including H1. Further digestion of the nucleosomal DNA yielded a 170 bp DNA segment which had no histone H1 (*121*). In a similar study, histone H1 was present with a 160 bp DNA segment but absent from the nucleosome possessing 140 bp DNA (*42*). Nevertheless, both groups concluded that one H1 histone is present in one nucleosome (*42,121*). Structural analysis of the present study is in accord with the presence of one histone H1 to one octamer. Furthermore, chromatin fixed with glutaraldehyde (dialdehyde) is known to produce H1 polymers (*20,83*). Thus, formation of a continuous axial structure would require H1–H1 interactions and may form a long segmented polymer (Figs. 6 and 7). The rodlike structure of histone H1, which may be already complexed at the globular region with the histone core at the axial position, may also interact with similar nucleosomes. The short N-terminal region of H1 may have ionic interactions with the relatively long tail end of the C-terminal region and may in turn interact with the N-terminal region of the following H1 molecule. Formation of a continuous axial H1 molecule may be a consequence of sequential development and not random interactions. Interbipartite regions may be bisected by the short head of the N-terminal region (which belongs to the preceding bipartite disk) and the long tail end of the C-terminal region (which belongs to the following bipartite disk) (Fig. 6). It is, therefore, possible that the reconstitution of nucleosomes with H1 may require some modification of parts of the core histone complexes for insertion or accommodation of H1 at the axial position. Nonsequential or improper reconstitution may displace the H1 molecules.

Histone H1 as the axial structure may play a role in sister chromatid exchange. Such crossover or exchanges may develop at H1–H1 segmented interbipartite regions. Side-by-side association of two thinner nucleofilaments (100–140 Å in diameter) into a thicker nucleofilament (200–280 Å in diameter) may be a conducive structural alignment for such exchanges.

Whereas the C-terminal region is enriched with basic residues, there is a small cluster of basic residues found between the N-terminus and residue 40. It is therefore probable that the interbipartite DNA strand may interact with some portions on the axial H1 when the DNA strand recoils and is then located between the bipartite disks (Fig. 6).

It is interesting to note that the chromatin protein A24, a conjugate of histone H2A and ubiquitin (*36,69*) is indicated to be one of the structural polypeptides present in interphase chromatin but absent from mitotic chromosomes (*69*). A24 is also found to be proximal to histone H1 (*9*). It is further indicated that release of the ubiquitin coupled with H1 and H3 phosphorylation (*69*) may play a role in the condensation of chromosomes. Release of ubiquitin, which may be located at interbipartite regions, would certainly facilitate face-to-face stacking and compaction of disks into a linear nucleofilament. In addition, there are high mobility group (HMG) proteins which are found in large quantities and shown to associate with histone H1 (*70,71,106*). H1 and H3 phosphorylation (*40*), coupled with removal of ubiquitin, indicate that histone modification, including acetylation (*16,28,113*), methylation (*27,49, 68,86*), and other phosphorylations (*5,50,55,68,109*), may also occur at interbipartite regions. This may impart changes in size and charge to basic residues and to the side group and thus alter the structure of the histone molecules and play a necessary role in allowing variations in the histone interaction with DNA.

VI. Higher Order Packing

The variable diameter of the nucleofilament, ranging from 100 to 300 Å, is shown to be a function of ion concentration in buffer solutions and stems from utilization of various methods of isolation of chromatin (*33,34,81,85,91,100,114*). Fragmented chromatin prepared at low ionic strength and then adjusted to high ionic strength has shown a helical structure with striations (*33,114*). Based on this study and others, investigators have proposed a solenoid in which the string of nucleosomes is coiled to form a helix with a pitch of 110 Å and an outer diameter of about 300 Å (*33*). Another group of investigators has been able to induce a nucleofilament into a discrete structure under high ionic concentrations in the presence of histone H1. Their unit structures, called superbeads and measuring about 200 Å in diameter, are spaced at regular intervals (*45,96*). Since both studies obtained their materials from fragmented chromatin at variable ionic strengths, it is uncertain whether the same structures would occur or be present in the native state.

Unfragmented chromatin of interphase nuclei and metaphase chromosomes has also yielded a similar structure, with a diameter of 200 to

300 Å (*30,35,99,100*). However, whether nucleosomes are sequenced and aligned as a solenoid or as superbeads is not clear. Freeze-fracture studies of nuclei and metaphase chromosomes have revealed a nucleofilament with a 250 Å diameter consisting of two thinner nucleofilaments of 130 Å in diameter (*58*). Our freeze-fracture studies of nuclei of intact interphase cells and chromosomes of intact metaphase cells have shown that the bipartite disks in one linear array are periodically cross-linked with the edges of the bipartite disks of the adjoining linear array (Figs. 1 and 7) (*19*). We find that the thicker nucleofilament with a 280 Å diameter consists of side-by-side associations of two thinner nucleofilaments of approximately 140 Å in diameter. An axial structure goes through the center of the disks, maintaining the structure of disks which are stacked face to face in a compact linear array. Without this axial structure, the bipartite disks may become "beads-on-a-string" or may assume some other configuration.

Nucleosomes with about 140 bp of DNA in one and three-fourths turns, in the absence of a one-fourth turn of DNA, may become wedge shaped. Two incomplete turns would leave only one DNA double helix at the bottom portion of the bipartite disk. When such nucleosomes are stacked face to face in a continuous linear array, they may form either a single arc or concentric arcs (*29*). In cross section, multilayers of concentric arcs may appear as lattice packing. When bipartite disks pøssess two full turns of 160 bp of DNA, they may not be organized in arcs but in a continuous linear or wavy array.

There are two axes of orientation in these parallel arrays: a straight linear orientation (profile) and a wavy or zigzagging orientation (top view) (Fig. 7). The axes are perpendicular to each other. It is possible that the parallel alignment of two linear arrays may be formed by the doubling back of a single linear array on itself. This type of nucleofilament may be layered back and forth in chromosomes and/or in some portion of chromatin of interphase nuclei. A parallel doubled-back nucleofilament may be seen in unfragmented chromosomes or chromatin spreads (*54,75*).

Periodic interactions between parallel arrays appear to occur at inter- and intrabipartite regions where the DNA enters and exists (Figs. 7 and 8). The parallel array formed by doubling back of the thinner unit on itself may represent an inactive state of chromatin. Active and inactive states of chromatin may be expressed in zipping and unzipping of the parallel array prior to transcription and replication. The hexagonal bipartite disks may also interact with neighboring disks at the exposed hexagonal side and upper facets and form a paracrystalline lattice (Fig. 8). An analogous hexagonal lattice packing may be seen in herpes and Epstein–Barr virus crystals. Varying conditions may cause hexagonal bipartite disks to appear as round disks or as round disks in a hexagonal lattice. Furthermore, more than one parallel array may come together to form a bundle of parallel arrays. Cross sections of the multiple parallel arrays may appear in a lattice. In freeze-fracture, the hexagonal disks may be aligned in such a manner as to give the lattice packing. This is in accord with the crystal lattice observed in thin sections of purified nucleosomes (*34*). It is not clear whether the lattice packing encompasses the entire chromosome or not, especially with respect to non-histone chromosomal proteins that participate in chromosome organization. Nevertheless, the packing of the hexagonal bipartite disks stacked face to face in a linear array was seen from three different angles: a linear (profile), a wavy or zigzagged (top view) (Fig. 7), and a lattice (face view) orientation (Fig. 8).

VII. Interphase Chromatin and Metaphase Chromosomes

The structural organization of interphase nuclei was similar to that seen in metaphase chromosomes. Nucleosomes possessed hexagonal bipartite disks and were stacked face to face and formed into a linear array. Others have also observed similarities of certain features of nucleosomes in interphase chromatin and metaphase chromosomes (*94,95*). Differences between the chromatin of interphase cells and chromosomes were indicated to reside in the extent of compaction. The arm of a chromatid in cross section usually showed defined dimensions of about 1.0 to 1.5 $\times$ 10^4 Å in diameter, and the degree of compaction was very similar from one chromatid to another. In interphase nuclei, packing of nucleosomes in a lattice in some areas similar to that seen in metaphase chromosomes was observed. Other areas in the nuclei showed irregularly dispersed chromatin. Intact interphase cells treated with a nonionic detergent showed variable orientation of nucleosomes (unpublished data). One or more linear structures (organized in bundles) which varied in diameter from 100 to more than 320 Å were observed. The thicker nucleofilament was seen as a multiple of the thinner nucleofilament of 100 Å in diameter. Thus, the nucleofilament may be present as a multiple of the thinner unit or rearrangement of thicker units. Structures measuring 200 to 280 Å in diameter predominated (75–85%) in the total chromatin. Partial lattice packing in some areas of the nuclei was similar to that detected in freeze-fracture studies. Thus, packing of the nucleosomes in interphase nuclei was characterized by variables within the nucleus as well as by the stages of the cell cycle (see also Chapter 5).

A similar nucleosomal alignment was seen in studies of negatively stained unfragmented chromatin of interphase and metaphase chromosomes (*94,95*). Even though the nucleofilament presented in their result appear to us as a multiple of the thinner unit or rearrangement of thicker units, these investigators have interpreted otherwise (*94,95*).

VIII. Conclusion

Some significant aspects of the organization of nucleosomes in chromatin and chromosomes will now be recapitulated. Nucleosomes are hexagonal bipartite disks; DNA is wound at the periphery; nucleosomes are stacked face to face and interconnected by axial histone H1. The hexagonal bipartite disks are aligned into a continuous linear nucleofilament of 100 to 140 Å in diameter. This continuous nucleofilament doubles back on itself and forms side-by-side associations, which make up the thicker nucleofilament of 280 Å in diameter. The DNA is kinked toward the major grooves at 10 and 20 bp intervals and constitutes the hexagonal perimeter of the bipartite disks in which octamer histones are enclosed. The specific configuration of the histone molecule, a cylinder-like structure with the length of about one-half of a helical turn of DNA double helix cutting an oblique angle near the top, forms preferential interactions among the histones and becomes the core of the bipartite disks. Kinking may shorten the DNA double helix in such a way that the histone molecule may bridge deeply recessed major grooves. Each histone possesses two points of association with DNA, a total of 16 points in an octamer. Interaction occurs at intervals of about 10 bp on alternating strands of the DNA double helix or intervals of 20 bp on a single strand.

Hexagonal bipartite disk characteristics of the nucleosome may be an important factor for higher order packing in chromosomes. An analogous hexagonal lattice packing may be seen in viral crystals. Histones may modify their association with DNA to facilitate transcription and replication. During strand separation, a single strand of the double helix may be released while the other strand is being retained by histones. Histone H1 is located at the axis, which is an optimum position to serve higher order packing. The doubling back of a continuous thinner nucleofilament on itself may have several functions. The active and inactive state of chromatin may be controlled by the zipping and unzipping of the parallel array of the thicker nucleofilament. Histone H1 may also facilitate DNA crossover and sister chromatid exchange. The interbipartite regions may be one of the most active areas. During the course of conformational changes histone H1 may be modified to accommodate non-histone proteins or compact the interbipartite region either by the variable composition of the H1 molecule or by configuration changes.

In this analysis, attempts have been made to correlate what we have observed in intact cells with known evidence. Whether our postulates are valid or not, we hope that this new line of thinking will stimulate and promote the uncovering of the structural organization and function of chromatin and chromosomes.

Acknowledgment

We wish to thank Mrs. Anne Marie Block for her critical reading of the chapter.

References

1. Adler, A. J., Fulmer, A. W., and Fasman, G. D. (1975). Interaction of histone F2a1 fragment with deoxyribonucleic acid, circular dichroism and thermal denaturation studies. *Biochemistry* **14,** 1445–1454.
2. Allen, J., Hartman, P. G., Crane-Robinson, C., and Aviles, F. X. (1980). The structure of histone H1 and its location in chromatin. *Nature (London)* **288,** 18–25.
3. Axel, R. (1975). Cleavage of DNA in nuclei and chromatin with staphylococcal nuclease. *Biochemistry* **14,** 2921–2925.
4. Bakayev, V. V., Bakayeva, T. G., and Varshavsky, A. J. (1977). Nucleosomes and subnucleosomes: Heterogeneity and composition. *Cell* **11,** 619–629.
5. Balhon, R., Rieke, W. V., and Chalkley, R. (1971). Rapid electrophoretic analysis for histone phosphorylation. A reinvestigation of phosphorylation of lysine-rich histone during rat liver regeneration. *Biochemistry* **10,** 3952–3959.
6. Baudy, P., Bram, S., Vestel, D., and LePault, J. (1976). Chromatin subunit small angle neutron scattering: DNA rich coil surrounds a protein–DNA core. *Biochem. Biophys. Res. Commun.* **72,** 176–183.
7. Bohm, L., Hayashi, H., Cary, P. D., Moss, T., Crane-Robinson, C., and Bradbury, E. M. (1977). Sites of histone–histone interaction in the H3,H4 complex. *Eur. J. Biochem.* **77,** 487–493.
8. Bonner, W. M., and Pollard, H. B. (1975). The presence of F3–F2a1 dimers and F1 oligomers in chromatin. *Biochem. Biophys. Res. Commun.* **64,** 282–288.
9. Bonner, W. M., and Stadmen, J. D. (1979). Histone H1 is proximal to histone H2A and to A24. *Proc. Natl. Acad. Sci. U.S.A.* **76,** 2190–2194.
10. Bostock, C. J., and Sumner, A. T. (1978). "The Eukaryotic Chromosome," pp. 139–174. North-Holland Publ., Amsterdam.
11. Boulikas, T., Wiseman, J. M., and Garrard, W. T. (1980). Points of contact between histone H1 and the histone octamer. *Proc. Natl. Acad. Sci. U.S.A.* **77,** 127–131.
12. Bradbury, E. M., Cary, P. D., Chapman, G. E., Crane-Robinson, C., Danby, S. E., and Rattle, H. W. E. (1975). Studies on the role and mode of operation of the very lysine-rich histone H1 (F1). The conformation of histone H1. *Eur. J. Biochem.* **52,** 605–613.
13. Bradbury, E. M., Moss, T., Hayashi, H., Hjelm, R. P., Suau, P., Stephens, R. M.,

Baldwin, J. P., and Crane-Robinson, C. (1977). Nucleosomes, histone interactions and the role of histone H3 and H4. *Cold Spring Harbor Symp. Quant. Biol.* **42,** 277–286.
14. Bram, S., Butler-Brown, G., Bradbury, E. M., Baldwin, J. P., Reiss, C., and Ibel, K. (1974). Chromatin, neutron and x-ray diffraction studies and high resolution melting of DNA–histone complexes. *Biochimie* **56,** 987–994.
15. Bustin, M., Goldbalt, D., and Sperling, R. (1976). Chromatin structure visualized by immunoelectron microscopy. *Cell* **7,** 297–304.
16. Candido, E. P. M., and Dixon, G. H. (1971). Sites of *in vivo* acetylation in trout testis histone. IV. *J. Biol. Chem.* **246,** 3182–3188.
17. Carpenter, B. G., Baldwin, J. P., Bradbury, E. M., and Ibel, K. (1976). Organization of subunits in chromatin. *Nucleic Acids Res.* **3,** 1739–1746.
18. Chai, L. S., and Sandberg, A. A. (1979). Evidence of nucleosome *in situ* in chromatin and chromosomes of intact Chinese hamster cells. *J. Cell Biol.* **83,** 170a.
19. Chai, L. S., and Sandberg, A. A. (1980). Evidence of nucleosomes *in situ* and their organization in chromatin and chromosomes of Chinese hamster cells. *Cancer Genet. Cytogenet.* **2,** 361–380.
20. Chalkley, R., and Hunter, C. (1975). Histone–histone propinquity by aldehyde fixation of chromatin. *Proc. Natl. Acad. Sci. U.S.A.* **72,** 1304–1308.
21. Chapman, G. E., Hartman, P. G., and Bradbury, E. M. (1976). Isolation of globular and nonglobular regions of the histone H1 molecules. *Eur. J. Biochem.* **61,** 69–75.
22. Cole, R. D. (1978). Specific features of the structure of H1 histone. *In* "The Molecular Biology of the Mammalian Genetic Apparatus" (P. O. P. Ts'o, ed.), pp. 99–104. North-Holland Publ., Amsterdam.
23. Comings, D. E., and Okada, T. A. (1973). Some aspects of chromosome structure in eukaryotes. *Cold Spring Harbor Symp. Quant. Biol.* **38,** 145–153.
24. Cotter, K. I., and Lilley, D. M. J. (1977). The conformation of DNA and protein within chromatin subunits. *FEBS Lett.* **82,** 63–68.
25. Crick, F. H. C., and Klug, A. (1975). Kinky helix. *Nature* (*London*) **255,** 530–533.
26. D'Anna, J. A., and Isenberg, I. (1974). A histone cross-complexing pattern. *Biochemistry* **13,** 4987–4992.
27. DeLange, R. J., and Smith, E. L. (1975). Histone function and evolution as viewed by sequence studies. *Ciba Found. Symp.* [N.S.] **28,** 59–70.
28. Dixon, G. H., Candido, E. P. M., Honda, B. M., Lovie, A. J., MacLeod, A. R., and Sung, M. T. (1975). The biological roles of post-synthetic modifications of basic nuclear proteins. *Ciba Found. Symp.* [N.S.] **28,** 229–250.
29. Dubochet, J., and Noll, M. (1978). Nucleosome arcs and helices. *Science* **202,** 280–286.
30. DuPraw, E. J. (1968). "Cell and Molecular Biology," pp. 514–589. Academic Press, New York.
31. Elgin, S. C., and Weintraub, H. (1975). Chromosomal proteins and chromatin structure. *Annu. Rev. Biochem.* **44,** 725–774.
32. Feizon, J., and Kearns, D. R. (1979). ^{1}H NMR investigation of the conformational states of DNA in nucleosome core particles. *Nucleic Acids Res.* **6,** 2327–2337.
33. Finch, J. T., and Klug, A. (1976). Solenoidal model for superstructure in chromatin. *Proc. Natl. Acad. Sci. U.S.A.* **73,** 1897–1901.
34. Finch, J. T., Lutter, L. C., Rhodes, D., Brown, R. S., Rushton, B., Levitt, M., and Klug, A. (1977). Structure of nucleosomes core particles of chromatin. *Nature* (*London*) **269,** 29–36.
35. Gall, J. G. (1966). Chromsome fibers studied by a spreading technique. *Chromosoma* **20,** 221–233.
36. Goldknopf, I. L., Taylor, C. W., Baum, R. M., Yeoman, L. C., Olson, M. O. J.,

Prestaybo, A. W., and Busch, H. (1975). Isolation and characterization of protein A24, a "histone-like" non-histone chromosomal protein. *J. Biol. Chem.* **250,** 7182–7187.
37. Goodwin, D. C., and Brahms, J. (1978). Form of DNA and the nature of interactions with chromatin. *Nucleic Acids Res.* **5,** 835–850.
38. Goodwin, D. C., Vergne, J., Brahms, J., Defer, N., and Kruh, J. (1979). Nucleosome structure: Site of interaction of proteins in the DNA grooves as determined by raman scattering. *Biochemistry* **18,** 2057–2064.
39. Goyanes, V. J., Matsui, S. I., and Sandberg, A. A. (1980). The basis of chromatin fiber assembly within chromsomes studied by histone–DNA cross-linking followed by trypsin digestion. *Chromosoma* **78,** 123–135.
40. Gurley, L. R., Tobey, R. A., Walters, R. A., Hildebrand, C. E., Hohmann, P. G., D'Anna, J. A., Barham, S. S., and Deaven, L. L. (1978). Histone phosphorylation and chromatin structure in synchronized mammalian cells. *In* "Cell Cycle Regulation" (J. R. Jeter, G. M. Cameron, G. M. Padilla, and A. M. Zimmerman, eds.), pp. 37–60. Academic Press, New York.
41. Hartman, P. G., Chapman, G. E., Moss, T., and Bradbury, E. M. (1977). Studies on the role and mode of operation of the very lysine rich histone H1 in eukaryote chromatin. *Eur. J. Biochem.* **77,** 45–51.
42. Hayashi, K., Hofstaetter, T., and Takuwa, N. (1978). Asymmetry of chromatin subunits probed with histone H1 in an H1–DNA complex. *Biochemistry* **17,** 1880–1883.
43. Hewish, D. R., and Burgoryne, L. A. (1973). Chromatin structure: The digestion of chromatin at regularly spaced sites by a nuclear deoxyribonuclease. *Biochem. Biophys. Res. Commun.* **52,** 504–510.
44. Hohmann, P. (1978). The H1 class of histone and diversity in chromosomal structure. *Subcell. Biochem.* **5,** 87–127.
45. Hojier, J., Renz, M., and Nehles, R. (1977). The chromsome fiber: Evidence for a ordered superstructure of nucleosomes. *Chromosoma* 62, 301–317.
46. Isenberg, I. (1979). Histones. *Annu. Rev. Biochem.* **48,** 159–191.
47. Jackson, V., Hoffman, P., Hardison, R., Murphy, J., Eichner, M. E., and Chalkley, R. (1977). Some problems in dealing with chromatin structure. *In* "Molecular Biology of the Mammalian Genetic Apparatus" (P. O. P. Ts'o, ed.), pp. 281–300. North-Holland Publ., Amsterdam.
48. Kawashima, S., and Imakari, K. (1979). Studies on histone oligomers, I. Reconstitution and fractionation of homotypic oligomers. *J. Biochem.* (*Tokyo*) **85,** 197–202.
49. Kim, S., and Paik, W. K. (1965). Studies on the origin of ε-*N*-methyl-L-lysine in protein. *J. Biol. Chem.* **240,** 4629–4634.
50. Kleinsmith, L. J., Allfrey, V. G., and Mirsky, A. E. (1966). Phosphoprotein metabolism in isolated lymphocyte nuclei. *Proc. Natl. Acad. Sci. U.S.A.* **55,** 322–327.
51. Kornberg, R. D. (1974). Chromatin structure: A repeat unit of histones and DNA. *Science* **184,** 868–871.
52. Kornberg, R. D., and Thomas, J. O. (1974). Chromatin structure: Oligomer of the histones. *Science* **184,** 865–868.
53. Kornberg, R. D. (1977). Structure of chromatin. *Annu. Rev. Biochem.* **46,** 931–954.
54. Laemmli, U. K., Cheng, S. M., Adolph, K. W., Paulson, J. A., Brown, J. A., and Baumback, W. R. (1978). Metaphase chromosome structure: The role of non-histone proteins. *Cold Spring Harbor Symp. Quant. Biol.* **42,** 351–360.
55. Lake, R. S., Goidl, J. A., and Salzman, N. R. (1972). F1-histone modification at metaphase in Chinese hamster cells. *Exp. Cell Res.* **73,** 113–121.
56. Lampert, F. (1971). Coiled supercoiled DNA in critical point dried and thin sectioned human chromosome fiber. *Nature* (*London*), *New Biol.* **234,** 187–188.

57. Langmore, J. P., and Wooley, J. C. (1975). Chromatin architecture: Investigation of a subunits of chromatin by dark field electron microscopy. *Proc. Natl. Acad. Sci. U.S.A.* **72,** 2691–2695.
58. Lepault, J., Bram, S., Escaig, J., and Wray, W. (1980). Chromatin freeze-fracture electron microscopy: A comparative study of core particles, chromatin metaphase chromosome and nuclei. *Nucleic Acids Res.* **2,** 1275–1289.
59. Li, H. J. (1975). A model for chromatin structure. *Nucleic Acids Res.* **2,** 1275–1289.
60. Li, H. J., and Bonner, J. (1971). Interaction of histone half-molecules with deoxyribonucleic acid. *Biochemistry* **10,** 1461–1470.
61. Liu, L. F., and Wang, J. C. (1978). DNA–DNA gyrase complex: The wrapping of the DNA duplex outside the enzyme. *Cell* **15,** 979–984.
62. Lutter, L. C. (1978). Kinetic analysis of deoxyribonuclease I cleavages in the nucleosome core: Evidence for DNA superhelix. *J. Mol. Biol.* **124,** 391–420.
63. Luzzati, V., and Nicolaiff, A. (1963). The structure of nucleosomes and nucleoprotamines. *J. Mol. Biol.* **7,** 147–163.
64. Martinson, H. G., and McCarthy, B. J. (1975). Histone–histone association with chromatin cross-linking studies using tetranitromethane. *Biochemistry* **14,** 1073–1078.
65. Martinson, H. G., and McCarthy, B. J. (1976). Histone–histone interaction within chromatin. Preliminary characterization of presumptive H2B–H2A and H2B–H4 binding sites. *Biochemistry* **15,** 4126–4131.
66. Martinson, H. G., Shetlar, M. D., and McCarthy, B. J. (1976). Histone–histone interactions within chromatin. Cross-linking studies using ultraviolet light. *Biochemistry* **15,** 2002–2007.
67. Martinson, H. G., True, R., Lau, C. K., and Mehrabian, M. (1979). Histone–histone interaction within chromatin. Preliminary location of multiple contact sites between histone H2A, H2B, and H4. *Biochemistry* **18,** 1075–1082.
68. Marzluff, W. F., and McCarty, K. S. (1972). Structural studies of calf thymus F3 histone. II. Occurrence of phosphoserine and ε-*N*-acetylysine in thermolysine peptides. *Biochemistry* **11,** 2677–2681.
69. Matsui, S. I., Seon, B. K., and Sandberg, A. A. (1979). Disappearance of a structural chromatin protein A24 in mitosis: Implications for molecular basis of chromatin condensation. *Proc. Natl. Acad. Sci. U.S.A.* **76,** 6386–6390.
70. McCarty, K. S., and McCarty, K. S., Jr. (1978). Some aspects of chromatin structure and cell-cycle-related postsynthetic modification. *In* "Cell Cycle Regulation" (J. R. Jester, G. M. Cameron, G. M. Padilla, and A. M. Zimmerman, eds.), pp. 9–35. Academic Press, New York.
71. McCarty, K. S., and McCarty, K. S., Jr. (1974). Protein modification, metabolic control and their significance in transformation in eukaryotic cells. *JNCI, J. Natl. Cancer Inst.* **53,** 1509–1514.
72. Mirzabekov, A. D., Shick, V. V., Belyavsky, A. V., Karpov, V. L., and Bavykin, S. G. (1978). The structure of nucleosomes: The arrangement of histone in the DNA grooves and along the DNA chain. *Cold Spring Harbor Symp. Quant. Biol.* **42,** 149–155.
73. Mirzabekov, A. D., Shick, V. V., Belyavsky, A. V., and Bavykin, S. G. (1978). Primary organization of nucleosome core particles of chromatin: Sequence of histone arrangement along DNA. *Proc. Natl. Acad. Sci. U.S.A.* **75,** 4181–4188.
74. Moss, T., Cary, P. D., Abercrombie, B. D., Crane-Robinson, C., and Bradbury, E. M. (1976). A pH-dependent interaction between histones H2A and H2B involving secondary and tertiary folding. *Eur. J. Biochem.* **71,** 337–350.
75. Mullinger, A. M., and Johnson, R. T. (1979). The organization of supercoiled DNA from human chromosomes. *J. Cell Biol.* **38,** 369–389.

76. Nelson, P. P., Albright, S. C., Wiseman, J. M., and Garrad, W. T. (1979). Reassociation of histone H1w. *J. Biol. Chem.* **254,** 11751–11760.
77. Noll, M. (1974). Internal structure of the chromatin subunits. *Nucleic Acids Res.* **1,** 1573–1578.
78. Noll, M. (1976). Differences and similarities in chromatin structure of *Neurospora crassa* and higher eukaryotes. *Cell* **8,** 349–355.
79. Noll, M., and Kornberg, R. D. (1977). Action of micrococcal nuclease on chromatin and the location of histone H1. *J. Mol. Biol.* **109,** 393–404.
80. Noll, M. (1977). DNA folding in the nucleosome. *J. Mol. Biol.* **116,** 49–71.
81. Olins, A. L., and Olins, D. E. (1974). Spherical chromatin units (ν bodies). *Science* **183,** 330–332.
82. Olins, A. L., Breillatt, J. P., Carlson, R. D., Senior, M. B., Wright, E. B., and Olins, D. E. (1977). On nu models for chromatin structure. *In* "Molecular Biology of the Mammalian Genetic Apparatus" (P. O. P. Ts'o, ed.), pp. 211–234. North-Holland Publ., Amsterdam.
83. Olins, D. E., and Wright, E. B. (1973). Glutaraldehyde fixation of isolated eukaryotic nuclei. *J. Cell Biol.* **59,** 304–311.
84. Olins, D. E. (1979). Important hydrodynamic and spectroscopic techniques in the field of chromatin structure. *In* "Chromatin Structure and Function" (C. A. Nicolini, ed.), pp. 109–135. Plenum, New York.
85. Oudet, P., Gross-Bellard, M., and Chamborn, P. (1975). Electron microscopic and biochemical evidence that chromatin structure is a repeating unit. *Cell* **4,** 281–300.
86. Paik, W. K., and Kim, S. (1967). ε-*N*-dimethyllysine in histone. *Biochem. Biophys. Res. Commun.* **27,** 479–483.
87. Pardon, J. F., Richards, B. M., Skinner, L. G., and Ockey, C. H. (1973). X-ray diffraction from isolated metaphase chromosomes. *J. Mol. Biol.* **76,** 267–270.
88. Pardon, J. F., Worcester, D. L., Wooley, J. C., Tatchell, K., Van Holde, K. E., and Richards, B. M. (1975). Low-angle neutron scattering from chromatin subunit particles. *Nucleic Acids Res.* **2,** 2163–2176.
89. Pardon, J. F., Worcester, D. L., Wooley, J. C., Cotter, R. I., Lilley, D. M. J., and Richards, B. M. (1977). The structure of the chromatin core particle in solution. *Nucleic Acids Res.* **4,** 3199–3214.
90. Pardon, J. F., and Richards, B. M. (1980). Physical studies of chromatin. *In* "Cell Nucleus" (H. Busch, ed.), Vol. 7, pp. 371–411. Academic Press, New York.
91. Pooley, A. S., Pardon, J. F., and Richards, B. M. (1974). The relation between the unit thread of chromosomes and isolated nucleohistone. *J. Mol. Biol.* **85,** 533–549.
92. Prunell, A., and Kornberg, R. D. (1979). Periodicity of deoxyribonuclease I digestion of chromatin. *Science* **204,** 855–858.
93. Rall, S. C., and Cole, R. D. (1971). Amino-acid sequence and sequence variability of the amino-terminal regions of lysine-rich histones. *J. Biol. Chem.* **246,** 7175–7190.
94. Rattner, J. B., and Hamkalo, B. A. (1978). Higher order structure in metaphase chromosomes. II. The relationship between the 250 Å fiber, superbeads and bead-on-a-string. *Chromosoma* **69,** 373–379.
95. Rattner, J. B., and Hamkalo, B. A. (1979). Nucleosome packing in interphase chromatin. *J. Cell Biol.* **81,** 453–457.
96. Renz, M., Nehls, P., and Hozier, J. (1977). Histone H1 involvement in the structure of the chromosome fiber. *Proc. Natl. Acad. Sci. U.S.A.* **74,** 1879–1883.
97. Riley, D., and Weintraub, H. (1978). Nucleosomal DNA is digested to repeats of 10 bases by exonuclease. III. *Cell* **13,** 281–293.
98. Ring, D., and Cole, R. D. (1979). Chemical cross-linking of H1 histone to the nucleosomal histones. *J. Biol. Chem.* **25,** 11688–11695.

99. Ris, H., and Kubai, D. F. (1970). Chromatin structure. *Annu. Rev. Genet.* **4,** 263–294.
100. Ris, H. (1975). Chromosomal structure as seen by electron microscopy. *Ciba Found. Symp.* [N.S.] **28,** 7–23.
101. Shaw, B. R., Herman, T. M., Kovacic, R. T., Beaudreau, B. S., and Van Holde, K. E. (1976). Analysis of subunit organization in chicken erythrocyte chromatin. *Proc. Natl. Acad. Sci. U.S.A.* **73,** 505–509.
102. Simpson, R. T. (1970). Interaction of a reporter molecule with chromatin. Evidence suggesting that the proteins of chromatin do not occupy the minor groove of deoxyribonucleic acid. *Biochemistry* **9,** 4814–4819.
103. Simpson, R. T. (1976). Histone H3 and H4 interact with ends of nucleosomal DNA. *Proc. Natl. Acad. Sci. U.S.A.* **73,** 4400–4404.
104. Simpson, R. T., and Whitlock, J. P. (1976). Mapping DNase I susceptible sites in nucleosomes labeled at the 5′ ends. *Cell* **9,** 347–353.
105. Simpson, R. T. (1978). Structure of the chromatosome, a chromatin particle containing 160 base pairs of DNA and all the histones. *Biochemistry* **15,** 5524–5531.
106. Smerdon, M. J., and Isenberg, I. (1976). Interaction between the subfractions of calf thymus H1 and nonhistone chromosomal proteins HMG1 and HMG2. *Biochemistry* **15,** 4242–4247.
107. Sobell, H. M., Tsaie, C. C., Gilbert, S. G., Jain, S. C., and Sakore, T. D. (1976). Organization of DNA in chromatin. *Proc. Natl. Acad. Sci. U.S.A.* **73,** 3068–3072.
108. Sollner-Webb, B., and Felsenfeld, G. (1975). A comparison of the digestion nucleic and chromatin by staphylococcal nuclease. *Biochemistry* **14,** 2915–2920.
109. Stevely, W. S., and Stocken, L. A. (1966). Phosphorylation of rat-thymus histone. *Biochem. J.* **100,** 20c–21c.
110. Stubblefield, E. (1973). The structure of mammalian chromosome. *Int. Rev. Cytol.* **35,** 1–60.
111. Suau, P., Bradbury, E. M., and Baldwin, P. (1979). Higher-order structure of chromatin in solution. *Eur. J. Biochem.* **97,** 593–602.
112. Suau, P., Kneals, G. G., Braddock, G. W., Baldwin, J. P., and Bradbury, E. M. (1977). A low resolution model for the chromatin core particle by neutron scattering. *Nucleic Acids Res.* **4,** 3769–3786.
113. Sung, M. T., and Dixon, G. H. (1970). Modification of histones during spermiogenesis in trout: A molecular mechanism for altering histone binding to DNA. *Proc. Natl. Acad. Sci. U.S.A.* **67,** 1616–1623.
114. Thoma, F., Koller, T. H., and Klug, A. (1979). Involvement of histone H1 in the organization of the nucleosome and of the salt-dependent superstructure of chromatin. *J. Cell Biol.* **83,** 403–427.
115. Thomas, G. J., Prescott, B., and Olins, D. E. (1977). Secondary structure of histones and DNA in chromatin. *Science* **197,** 385–388.
116. Thomas, J. O., and Kornberg, R. D. (1975). An octamer of histones in chromatin and free in solution. *Proc. Natl. Acad. Sci. U.S.A.* **72,** 2626–2630.
117. Thomas, J. O., and Butler, P. J. G. (1977). Characterization of the octamer of histones free in solution. *J. Mol. Biol.* **116,** 769–781.
118. Todd, R. D., and Garrard, W. T. (1979). Overall pathway of mononucleosome production. *J. Biol. Chem.* **254,** 3074–3083.
119. Van Holde, K. E., Sahasrabudhe, C. G., and Shaw, B. R. (1974). A model for particulate structure in chromatin. *Nucleic Acids Res.* **1,** 1579–1586.
120. Van Lente, F., Jackson, J. E., and Weintraub, H. (1975). Identification of specific cross-linked histones after treatment of chromatin with formaldehyde. *Cell* **5,** 45–50.
121. Vershavsky, A. J., Bakayev, V. V., and Georgiev, G. P. (1976). Heterogeneity of chromatin subunits *in vitro* and location of histone H1. *Nucleic Acids Res.* **3,** 477–492.

122. Weintraub, H., Worcel, A., and Alberts, B. (1976). A model for chromatin based upon two symmetrically paired half-nucleosome. *Cell* **9,** 490–417.
123. Weintraub, H., Palter, K., and Van Lente, F. (1975). Histones H2a, H2b, H3, and H4 form a tetrameric complex in solution of high salt. *Cell* **6,** 85–110.
124. Weintraub, H., and Van Lente, F. (1974). Dissection of chromosome structure with tryspin and nuclease. *Proc. Natl. Acad. Sci. U.S.A.* **71,** 4249–4253.
125. Whitlack, J. P., and Simpson, R. T. (1976). Removal of histone H1 exposes a fifty base pair DNA segment between nucleosomes. *Biochemistry* **15,** 3307–3313.
126. Wilkins, M. H. F., Zubay, G., and Wilson, H. R. (1959). X-ray diffraction studies of the molecular structure of nucleohistone and chromosome. *J. Mol. Biol.* **1,** 179–185.
127. Williams, R. (1970). Properties of rapidly labeled deoxyribonucleic acid fragments isolated from the cytoplasm of primary cultures of embryonic mouse liver cells. *J. Mol. Biol.* **51,** 157–168.
128. Woodcock, C. L. F., Safer, J. P., and Stanchfield, J. E. (1976). Structural repeating units in chromatin I. Evidence for their general occurrence. *Exp. Cell Res.* **97,** 101–110.
129. Worcel, A. B., and Beyajati, C. (1977). Higher order coiling of DNA in chromatin. *Cell* **12,** 83–100.
130. Zinke, M. (1979). A new model of DNA and histone organization in chromatin: Explanation transitions in chromatin structure and their dissociation. *Stud. Biophys.* **75,** 107–130.

2

Cell Cycle Studies of Histone Acetylation and the Structure and Function of Chromatin

H. R. MATTHEWS AND E. M. BRADBURY

I. Introduction

Now that we have an understanding in outline of the structure of chromatin (Chapter 1), attention is directed to the roles of reversible postsynthetic modifications of histones in the structure and function of chromatin. There are two major reversible modifications: acetylation of lysines in the core histones H2A, H2B, H3, and H4 and phosphorylations of serines and threonines in histone H1 and to a lesser extent in histones H2A, H3, and H4. Acetylation converts a basic lysine residue to a neutral acetyllysine and reduces the net positive charge of the region of the polypeptide chain containing the modified residue. There are four sites of acetylation in the core histones H2B, H3, and H4 and one site, lysine

GENETIC EXPRESSION IN THE CELL CYCLE

ISBN 0-12-543720-X

5, in H2A. All of these sites are confined to the basic N-terminal regions up to residue 25 of all the core histones. In histone H4 the sites of acetylation are lysines 5, 8, 12, and 16—i.e., all of the lysines in the first 16 residues are acetylated in tetraacetylated H4 (Ac_4H4)—and this reduces the overall positive charge of this region from +5 to +1. The core histones have distinct domains: the basic N-terminal region, which is subjected to acetylation, and the central and C-terminal regions, which are the sites of interactions between the core histones and the DNA in the chromatin core particles. Reversible chemical modifications of histones are thought to be the mechanism for modulating histone interactions and controlling chromatin structure in response to cell functions and cell events, e.g., transcription, DNA replication, and chromosome condensation (Chapters 4 and 5). In this chapter the function of histone acetylation will be discussed in relation to chromatin structure.

II. Chromatin Structure

The basic repeating subunit of chromatin, the nucleosome, contains a variable length of DNA, about 195 base pairs (bp) in length, the histone octamer $(H2A,H2B,H3,H4)_2$, and one molecule of histone H1 [see (*27*) and Chapter 1]. Although the DNA content of the nucleosome is variable, *subnucleosome particles* have been characterized, which appear to have constant DNA lengths for most eukaryotes. These are the "*chromatosome,*" which contains 165–168 bp DNA, the histone octamer, and one H1 molecule (*53,64,65,74,77,79*), and the *chromatin core particle,* which contains 146 bp DNA and the histone octamer (*53,80*). Subsets of nucleosomes, chromatosomes, and core particles have also been identified, which contain variable amounts of the "high mobility group" (HMG) proteins, a subset of the non-histone proteins (see Chapter 3).

The structure of the core particle is understood in outline from neutron scattering (*7,9,67,69,75*) and electron microscopy combined with low resolution X-ray diffraction (*29,30*). It is a flat circular disk 11 nm in diameter and 5.5 to 6.0 nm in thickness, with 1.7 turns of DNA coiled with a pitch of 28 Å (obtained by X-ray diffraction) or 30 Å (obtained by neutron scattering) around a core of the histone octamer. A proposed model for the chromatosome (*2,22,78*) is based on: (1) the structure of the core particle outlined above; (2) the properties of the H1 histone (*8,18,40*); (3) electron microscopy of extended chromatin (*78*); and (4) nuclease digestion of chromatin reconstituted by adding back modified H1 and H1 peptides to H1-depleted chromatin (*2*). The model for the

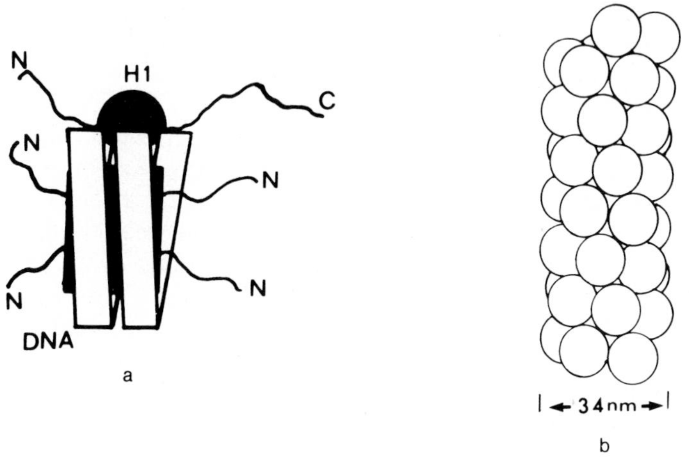

Fig. 1. (a) Model for the structure of the nucleosome. The path of the DNA linker between nucleosomes is not known. (b) Model for the supercoil or solenoid of nucleosomes. The pitch of the supercoil is 10–11 nm.

chromatosome is given in Fig. 1a and consists of two full turns of DNA coiled around the core of the histone octamer with the conserved globular region of H1 sealing off the two turns of DNA. The nucleosome consists of the chromatosome with linker DNA joining adjacent nucleosomes. The length of the linker DNA is variable depending on the organism, and its path outside of the chromatosome is not known. There are two proposals for the arrangement of the disk-shaped chromatosomes in extended chromatin at low ionic strength: (1) edge to edge and arranged roughly linearly in the 10 nm filament (*76*); and (2) edge to edge on a regular flat zigzag (*78*). It is probable that the above arrangements are different states of chromatin structure depending on solution conditions. With increase of ionic strength, the extended chromatin structure coils into the next higher order of chromatin structure, defined as a supercoil of nucleosomes. In its most compact form this supercoil (Fig. 1b) has an outside diameter of 34 nm (*76*) and a pitch of 10–11 nm, with 6 to 7 nucleosomes per turn of the coil (*15,28,76*). The interactions involved in generating this supercoil are not understood at present, though they probably involve the N-terminal region of the core histones as well as the histone H1, which is known to stabilize the supercoil. The effects of acetylation of lysines in these N-terminal regions on the stability of the nucleosome and the 34 nm supercoil are of considerable interest to the understanding of the functions of acetylation.

III. Cell Cycle Studies of Histone Acetylation Using *Physarum polycephalum* as a Model System

The studies described below exploit the naturally synchronous cell cycle of *Physarum polycephalum*. *Physarum* is a true slime mold unlike the cellular slime molds, such as *Dictyostelium discoideum*. In the plasmodial stage of the life cycle, *Physarum* has no cell walls between nuclei and so all the nuclei (about 10^8 per plasmodium) share a common cytoplasm. The result is a highly synchronous mitotic cycle with about 98% of nuclei passing through metaphase within a 5 min period in an 8 to 12 hr division cycle. Like very rapidly dividing mammalian cells, *Physarum* plasmodia enter S phase shortly after metaphase. S phase lasts about 2 hr and is followed by a G_2 phase of about 6 hr (*72*).

Physarum is of particular interest for studies of control of transcription, because there appear to be substantial changes in both quantity and quality of transcription during the cell cycle. [The ribosomal genes also provide an interesting system, but that is described elsewhere (*58*)]. In mitosis the rate of transcription is very low; in S phase the rate of transcription is high. In G_2 phase there is an early drop in transcription followed by a peak in mid-G_2 phase. S phase transcription is predominantly hnRNA, whereas G_2 phase transcription is predominantly rRNA [reviewed by Matthews (*56*), and Braun and Seebeck (*10*)].

Physarum has five major histone fractions, as in other eukaryotes. Mohberg and Rusch (*60*) analyzed total *Physarum* histone by polyacrylamide gel electrophoresis in acetic acid and urea. They identified four major bands and several minor bands. One of the major bands could be fractionated with ethanol–HCl into two components, so that a total of five major components were described. Three of these components were isolated by preparative gel electrophoresis and their amino acid compositions determined. Subsequently, similar gel electrophoresis profiles were obtained by Bradbury *et al.* (*6*) and Jockhusch and Walker (*45*), who published SDS–gel electrophoresis data from which they obtained values for the molecular weights of the major components. More recently, Chahal *et al.* (*17*) have carried out a comprehensive gel electrophoresis analysis including the use of gels containing Triton X-100. Figure 2 shows a two-dimensional gel electrophoresis pattern of *Physarum* total histone in which Triton X-100 was used in the first dimension and cetyltrimethylammonium bromide in the second dimension (*5,84*). Notice the major off-diagonal spot, which is typical of histone H2A (*39*), the separation of the spots assigned to H2B and H3, and the excellent separation of acetylated species of H4.

Total histone has been fractionated by preparative gel electrophoresis

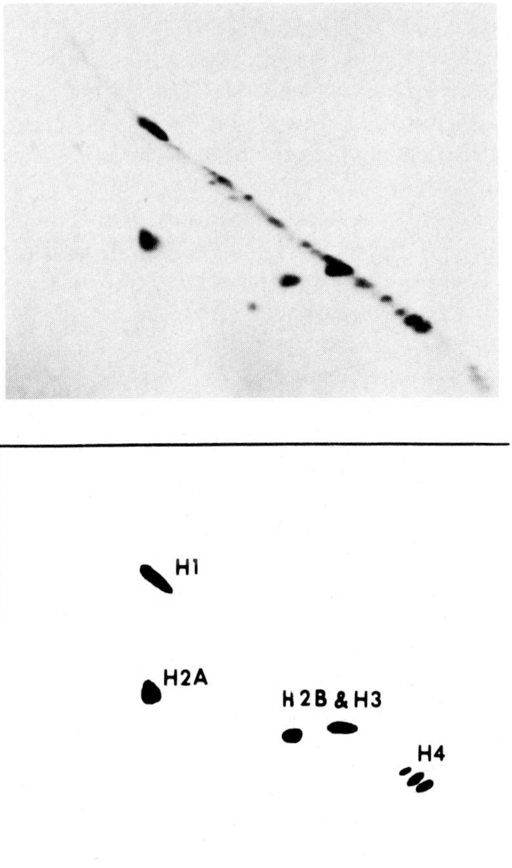

Fig. 2. Two-dimensional gel electrophoresis of $CaCl_2$-soluble nuclear proteins. The first dimension, horizontal, contained 8 *M* urea and 8 m*M* Triton X-100 in 1 *M* acetic acid, and the second dimension contained 6 *M* urea and 0.15% cetyltrimethylammonium bromide also in 1 *M* acetic acid (*5*). H1, H2A, and H4 are clearly identified, but the distinction between H3 and H2B is not yet clear. Several of the other spots correspond to spots observed by Bonner *et al.* (*5*) in electrophoresis of HeLa nuclear acid-extracted proteins, for example A24 [a ubiquitin–H2A complex described by Goldknopf and Busch (*32*)] and "z," which was noted but not identified by Bonner *et al.* (*5*).

or by chromatography on Bio-Gel P-10 or P-60 columns eluted with 10 to 20 m*M* HCl containing up to 0.1 *M* NaCl (*19*). H1 can be satisfactorily resolved by elution with 20 m*M* HCl from a fairly short column (60 cm) but resolution of H4 requires 75 m*M* NaCl–20 m*M* HCl and a longer column (100 cm) (*17*). Histone H2A can be resolved on a 1.5 m column, but H3 and H2B elute together under the conditions used so far.

Histone H4 can be isolated by chromatography of *Physarum* total histone on Bio-Gel P-10 in 20 m*M* HCl 75 m*M* NaCl (*17*). It comigrates with calf H4 in all the gel electrophoresis systems used to date and was identified by Mohberg and Rusch (*60*) as an H4 histone. Table I shows its amino acid composition (mean of three independent preparations). The similarity between this and the amino acid composition of calf thymus H4 (*46*) is striking.

In gel electrophoresis systems that are sensitive to the charge on a protein, H4 splits into five bands representing 0 to 4 acetyl groups on the N-terminal lysines. *Physarum* H4 shows this behavior. The proportion of H4 in the more highly acetylated species of H4 from cultured cells can be dramatically increased by the addition of 7 m*M* sodium butyrate to the cell culture medium (*70*). This concentration of butyrate interferes with the normal nuclear preparation in *Physarum,* so its effect on H4 acetylation has not been determined. However, even 2 m*M* bu-

TABLE I

Amino Acid Composition of *Physarum* and Calf Histone H4

	Amino acid levels in H4 (mole/100 mole)	
Amino acid	Calf[a]	*Physarum*[b]
Asp	5.2	6.7
Thr	6.3	7.9
Ser	2.2	3.8
Glu	6.9	7.8
Pro	1.5	2.5
Gly	14.9	13.7
Ala	7.7	3.2
Cys	0.0	0.3[c]
Val	8.2	5.0
Met	1.0	1.0
Ile	5.7	4.2
Leu	8.2	6.3
Tyr	3.8	3.6
Phe	2.1	2.3
His	2.2	2.7
Lys	11.4[d]	9.4
Arg	12.8[d]	13.7

[a] From Johns (*46*).

[b] S. Miller and H. R. Matthews, unpublished data quoted in Matthews and Bradbury (*58*).

[c] This should probably be 0.0.

[d] Sequence gives 10.8 for Lys and 13.7 for Arg.

tyrate has a substantial effect in increasing the proportion of H4 in the more highly acetylated forms (*17*). This further confirms the identification of H4 in *Physarum*.

IV. Acetate Content of H4 in the Cell Cycle

Histone acetylation has been correlated with important aspects of DNA processing: transcription (*3,4*), DNA replication (*44,70*), and spermogenesis (*26*). Its precise function is not yet understood, nor is the effect of acetylation on chromatin structure known. The finding that the addition of butyrate to cultured cell lines leads to a much enhanced level of histone acetylation (*38,70*) through the inactivation of histone deacetylase (*82*) has been used to investigate the effects of these levels of acetylation on chromatin structure and function (*54,82*). So far little effect has been detected either at the level of the nucleosome structure (*82*) or on the rates of transcription of minichromosomes assembled from SV40 DNA and acetylated histones from butyrate-treated cells (*54*). A possible explanation is that for the full effect of acetylation to be observed all the core histones have to be fully and not partially acetylated. A more probable explanation is that acetylation acts at the level of chromatin structure above the nucleosome (i.e., the 34 nm supercoil of nucleosomes) in order to make the extended chromatin structure available for processing. In keeping with this suggestion are the reports that core-particle-like structures can be reassembled with DNA and histones that lack N-terminal regions (*87*), and the basic N-terminal regions of histones can be released from core particles by salt (*16*) or trypsin digestion without unfolding the core particle. Furthermore, the highly acetylated regions of chromatin from butyrate-treated cells are more sensitive to DNase I digestion (*82*), suggesting that, similar to "active" chromatin (*85*), histone acetylation renders the DNA in these structures more accessible to nuclease attack. In this chapter we are concerned with precise studies of histone acetylation through the cell cycle and correlations of acetylation with structure and function.

Physarum H4 splits into five bands on long polyacrylamide gels with acetic acid–urea or Triton X-100 but not with SDS. Only very small amounts of ^{32}P are incorporated into *Physarum* H4 in plasmodia (*6*), certainly not enough to account for the observed heterogeneity. By analogy with other H4 histones, the heterogeneity was attributed to acetylation, and this was confirmed by two groups of experiments. First, *Physarum* plasmodia, prelabeled with [^{3}H]lysine, were grown in the presence of [^{14}C]acetate for 10 min during S phase. H4 was isolated and analyzed by gel electrophoresis. The fastest four bands contained $^{14}C:^{3}H$ ratios (acetate:lysine ratios) of 0, 1.0, 2.2, and 2.8 after a small correction for ^{14}C incorporated directly into amino acids. This corresponds to 0, 1, 2 and 3 acetates per molecule of H4. In a pulse-label experiment with [^{3}H]acetate in G_2 phase, high incorporation of acetate into the slow-moving bands was observed with no significant incorporation into the fastest moving band, which is consistent with the five bands being due to 0, 1, 2, 3, or 4 acetates per H4 molecule. Second, *Physarum* microplasmodia were grown in the presence of various amounts of sodium butyrate, which inhibits H4 deacetylase (*38,82*). We were unable to purify histones from plasmodia grown in 5 or 7 m*M*, but in 2 m*M* butyrate a substantial increase in the proportion of the slower moving bands of H4 was observed. These results strongly imply that the heterogeneity of *Physarum* H4 is due to acetylation, as in higher eukaryotes.

V. H4 Acetate Content Varies during the Cell Cycle

The proportion of *Physarum* H4 in each of the five bands was determined by polyacrylamide gel electrophoresis in acetic acid–urea (*66*). The H4 was the first purified by chromatography on Bio-Gel P-10. This purification was necessary in order to reduce the amount of protein loaded on the gels and to remove a minor component that interfered with the measurement of tetraacetylated H4, Ac_4H4. In scans of stained gels containing H4 from defined stages of the cell cycle, five bands were present, but the proportions at a specific stage were reproducible. The proportions were quantitated as follows. We assumed that each band could be represented by a Gaussian peak and hence that the total profile of five bands could be represented as the sum of five Gaussian peaks, with the area of each Gaussian peak being proportional to the mass of protein in its equivalent band. The initial parameters for the five Gaussian peaks were based on the heights, widths, and positions of the bands measured from the scan and then refined by trial and error until the computer plot of the five Gaussians could be superimposed exactly on the experimental scan. The areas of the five Gaussians were calculated. This procedure assumes that stain intensity is proportional to mass of protein. The accuracy of the assumption is supported by the excellent fit of experimental data to five Gaussian peaks, but it is difficult to test directly for the different molecular species. (The staining procedure is linear for total calf thymus H4, data not shown.) Any small variation from linearity will have small quantitative effect on the results, but the qualitative changes and interrelationships will remain as reported below.

The peak width used for all the acetylated bands was constant for each gel and from one gel to another, as expected for similar molecules run under similar conditions. The peak width for the nonacetylated H4 band was also constant from gel to gel, but it was narrower than the peaks for the acetylated H4 bands. The acetylated bands were equally spaced as expected, but the nonacetylated band was unexpectedly close to the monoacetylated band. These effects were not a consequence of nonlinearity of migration due to the high loading of protein, since the same effects were observed on underloaded gels. The effects were also seen in H4 isolated from proliferating HeLa cells (A. W. Thorne, unpublished data). Whatever the cause of the narrower peak for nonacetylated H4, these observations show that any measurement of amount of H4 in the different components based on peak heights will seriously overestimate the amount of nonacetylated H4. For this reason we have used only areas.

Gel scans of *Physarum* H4 from specific stages of the cell cycle have been analyzed in terms of five Gaussian peaks, and the area of each peak was calculated. The areas were thus fully corrected for the overlap between bands. The area corresponding to each species of H4 is plotted in Fig. 3A as a function of stage of the cell cycle.

It is clear that at all stages the most abundant species is monoacetylated H4, Ac_1H4. During G_2 phase the proportion of Ac_1H4 rises to about 62% and then falls to about 55%. The fall occurs during the middle part of G_2 phase, from 5 to about 9 hr after mitosis in the 11 hr cycle. During the period 6 to about 9 hr after mitosis, there are complementary increases in proportion of Ac_2H4 (16–23%), Ac_3H4 (2.5% to 4.3%), and Ac_4H4 (1% to 3%). The data imply an overall increase in acetylation during mid-G_2 phase. Late G_2 phase is characterized by a sharp drop in Ac_4H4 (3% to 1%) followed in early prophase by a drop in Ac_3H4 (4% to 2%). Ac_2H4 drops sharply through late G_2 phase and prophase (23% to 11%). In early prophase the overall proportion of highly acetylated H4 (2–4 acetates per molecule) is at its minimum value (13.5%). During late mitosis and early S phase the increase in proportion of highly acetylated H4 is balanced by drops in both non- and monoacetylated H4, so that in mid-S phase the proportion of highly acetylated H4 reaches its maximum (33.6%). In late S phase the proportion of highly acetylated H4 falls to a plateau level (about 26%) before the minimum (19.4%) in early G_2 phase. The most dramatic changes in an individual H4 species occur in Ac_4H4, and these are shown on an expanded scale in Fig. 3B. There is a clear minimum (0.4%) in early prophase, a sharp maximum (3.6%) in mid-S phase, and a second maximum (up to 3%) in mid-G_2 phase.

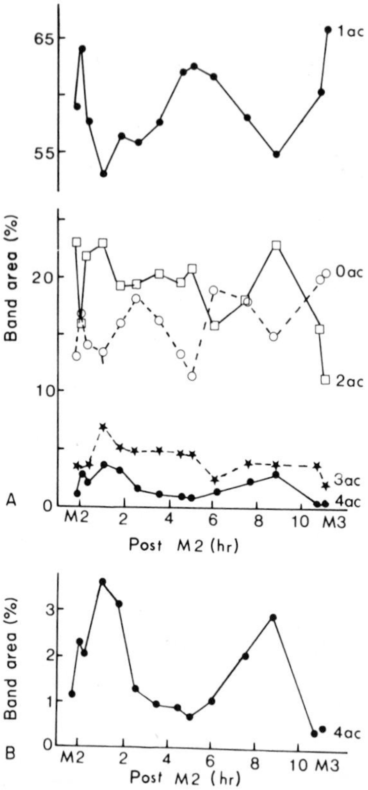

Fig. 3. Acetylated H4 through the cell cycle. (A) Percentage of total H4 in the non-acetylated form (0 ac); percentage of total H4 in the monoacetylated form, Ac_1H4 (1 ac); percentage of total H4 in the diacetylated form, Ac_2H4 (2 ac); percentage of total H4 in the triacetylated form, Ac_3H4 (3 ac); percentage of total H4 in the tetraacetylated form, Ac_4H4(4 ac). (B) Percentage of Ac_4H4 (4 ac) shown on a larger scale. M2 is the second mitosis after fusion.

VI. Acetate Turnover on H4 in the Cell Cycle

Long labeling times with [H^3]acetate lead to incorporation of label into stable positions in amino acids in H4, but short pulses (2–5 min) label only the acetate groups (*19*). *Physarum* macroplasmodia were pulse-labeled with sodium [^3H]acetate at different stages of the cell cycle by transferring the plasmodium, on its filter paper, to 1 ml of growth medium supplemented with 1 mCi sodium [^3H]acetate and then incubated for 5 min. After incubation, growth was stopped by plunging the plasmodium into liquid nitrogen. H4 was isolated, as previously described (*17*) using *Physarum* H4 as carrier, and analyzed by gel electrophoresis.

The bands were stained with Coomassie Blue and cut out by hand. On some gels the highly acetylated bands could not be clearly seen, so the appropriate regions were estimated by comparison with other gels. Figure 4 shows the radioactivity per band at a number of stages of the cell cycle (*84*).

In Fig. 4 there is some small incorporation into nonacetylated H4 in S phase. This may represent labeling of the N-terminal acetylserine during synthesis of H4. Only background labeling is observed at other times in the nonacetylated band. The overall incorporation is clearly much higher in S phase than at other times, particularly in the monoacetylated band. Notice also the relatively high incorporation in Ac_4H4 in S phase and late G_2 phase (M + 7 hr) correlating with high levels of Ac_4H4 at these times.

VII. Histone Deacetylase Activity in the Cell Cycle

A. Histone Deacetylase

Histones are domain proteins with structurally distinct regions [reviewed by Matthews and Bradbury (*58*)]. In the case of histone H4 the N-terminal region is not structured, whereas the remainder of the molecule will coil into a globular conformation. In the nucleosome core particle, the globular region of H4 is required for coiling the DNA (*20,87*). All the sites of postsynthetic modification of H4 are in the N-terminal region (*26*). These data suggest that studies of the isolated N-terminal region of H4 are likely to reflect the properties of this peptide in chromatin, particularly with respect to its interaction with histone acetyltransferase and deacetylase enzymes. This is supported by studies with synthetic peptides simulating only part of the N-terminal region of H4 and lacking its globular part. These were substrates for bovine histone deacetylase (*47*) and acetyltransferase (*48*) enzymes.

The N-terminal region of calf H4 can be isolated as peptide 1–23, which has the sequence

$$\text{ac-Ser·Gly·Arg·Gly·}\underset{5}{\text{Lys}}\text{·Gly·Gly·}\underset{8}{\text{Lys}}\text{·Gly·Leu·Gly·}\underset{12}{\text{Lys}}\text{·Gly·}$$
$$\text{Gly·Ala·}\underset{16}{\text{Lys}}\text{·Arg·His·Arg·Me-}\underset{20}{\text{Lys}}\text{·Val·Leu·Arg}$$

This peptide has lysines at 5, 8, 12, 16, and 20, but the lysine-20 has a stable modification, namely methylation, and so only lysines-5, -8, -12, and -16 are available for acetylation. These are identical to the four lysine sites that are acetylated *in vivo* in H4 (*25,26*). The sequence of

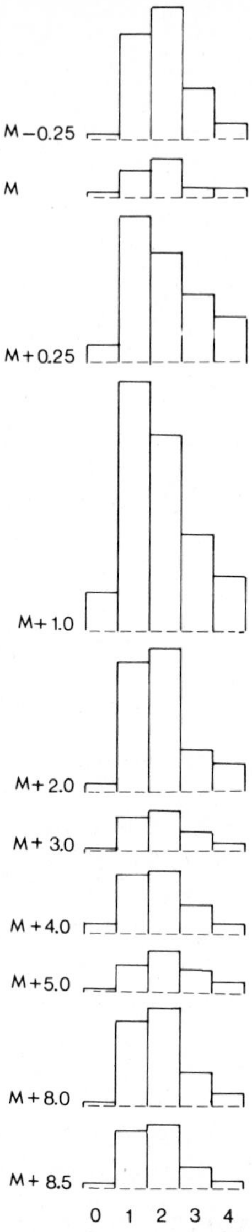

Fig. 4. *Physarum* plasmodia were labeled for 5 min with [^{3}H]acetate and harvested, and their histone H4 was isolated, purified, and analyzed by gel electrophoresis as described by Chahal *et al.* (*17*). The gels were stained, to locate the bands, and then sliced so that each band occupied one slice. For the highly acetylated species the position of the bands are approximate. The gel slices were bleached and digested with hydrogen peroxide and their ^{3}H-radioactivity determined. The figure shows the count rates observed in each slice for nine stages of the mitotic cycle. "M" refers to the second metaphase after fusion, and the numbers along the left hand side of the figure are the times, in hr, after M when the plasmodia were labeled. The third metaphase occurred 11 hr after M in these experiments. The numbers along the bottom (0–4) give the approximate positions of Ac_0H4, Ac_1H4, Ac_2H4, Ac_3H4 and Ac_4H4, respectively. All the data are plotted on the same scale.

H4 is highly conserved between different organisms [reviewed by Isenberg (*42*)], so peptide 1–23 prepared from calf thymus histone H4 can be used as a substrate in other systems. H4 from *Physarum* has not yet been sequenced, but proteolytic digestion patterns (*58*) and the occurrence of five H4 species with 0 to 4 acetylated lysines (*17,19*) show that *Physarum* H4 is very similar to calf H4. Deacetylases generally have low substrate specificity and all are active on acetylated H4 (*12,41,49*).

Previous work on histone deacetylases has used as substrates mixtures of histones, acetylated *in vitro* with purified acetyltransferases (*47,52*) or acetylated *in vivo,* sometimes in the presence of butyrate (*12,14, 21,63,68*). These substrates are only available in small amounts of low specific activity and they are likely to vary in degree of acetylation from batch to batch. We have overcome these problems by isolating peptide 1–23 and acetylating it *in vitro* with labeled acetic anhydride (*84*).

B. *Physarum* Histone Deacetylase

Histone deacetylase activity was measured in *Physarum* nuclei and cytoplasm. The activity was located specifically in the nuclei and was not released by washing them in homogenization medium. The activity could be partially extracted from nuclei by high-salt washing (2 *M* KCl) and almost completely by sonication in high salt, but none of these methods changed the total amount of deacetylase activity measurable. The enzyme has not been purified further.

The deacetylase activity in isolated *Physarum* microplasmodial nuclei showed a very broad pH optimum from 7 to 8 (Tris buffer) with 60% activity at pH 6.4 in sodium cacodylate or at pH 9.5 in glycine buffer. At identical pH values these three buffers gave similar levels of activity. Phosphate buffer was less effective and gave on average only 60% of the activity, relative to Tris.

The presence of –SH reagents such as 2-mercaptoethanol or dithiothreitol was essential to retain enzyme activity. The mercury compounds *p*-chloromercuribenzoate and *p*-chloromercuriphenylsulfonic acid strongly inhibited activity when they were present in the assay mixture at higher molar concentrations than 2-mercaptoethanol. Typically, only 60% activity remained when these compounds were present at a molar excess of 0.5 m*M*.

Divalent cations $MgCl_2$, $MnCl_2$, and $CaCl_2$ and the chelator EDTA only slightly affected the activity. At 5 m*M* all these compounds caused less than 20% inhibition. The salts NaCl, KCl, and sodium acetate are successively more effective inhibitors with 50% inhibition at 350, 170, and 80 m*M*, respectively (*84*).

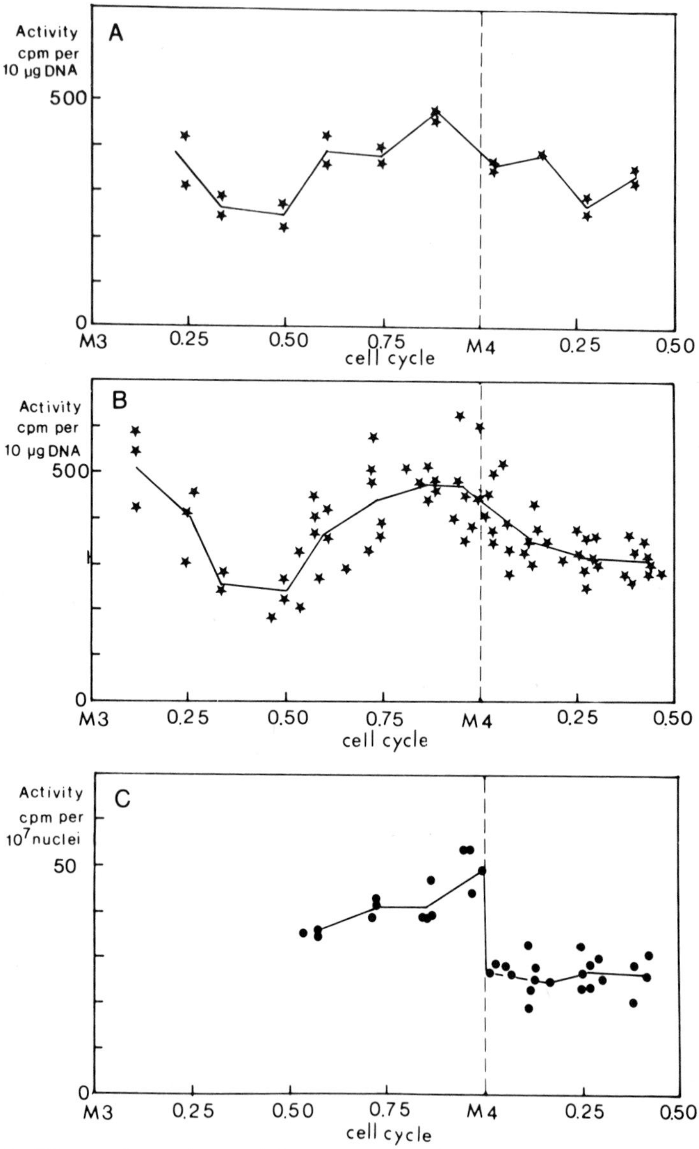

Fig. 5. Histone deacetylase activity was measured in triplicate in *Physarum* nuclei. The nuclei were isolated from macroplasmodia by Potter homogenization of each plasmodium in 50 ml homogenization buffer (30 m*M* NaCl, 1 m*M* KCl, 5 m*M* $MgCl_2$ (or 5 m*M* $CaCl_2$), 0.1% (w/v) Triton X-100, 10 m*M* Tris-HCl pH 7.5, 1 m*M* PMSF (phenylmethylsulfonyl fluoride), collection of the nuclei by centrifugation for 10 min at 1500, and two washes with homogenization buffer. Drained nuclear pellets containing 5 to 15 × 10^6 nuclei

Butyrate is a strong noncompetitive inhibitor for histone deacetylases, but the *Physarum* enzyme is less strongly inhibited than calf thymus deacetylase I. At 2, 10, and 50 m*M* the activity of the calf enzyme is reduced to 5, 2, and 1% (*21*), whereas the *Physarum* enzyme still retains 95, 75, and 30% activity at these concentrations (*84*). Butyrate also appears to inhibit the histone deacetylase activity *in vivo* in *Physarum* (*17*), although butyrate concentrations above 2 m*M* have serious effects on cellular and nuclear membranes in *Physarum*, leading rapidly to cell lysis and death.

C. Cell Cycle Dependence of Histone Deacetylase

Nuclei were isolated from single macroplasmodia, diameters 4 to 7 cm, harvested at specific times in the cell cycle, and the histone deacetylase activity was determined (*84*). Figure 5 shows the cell cycle dependence of deacetylase activity. Clearly, there are no major changes of histone deacetylase activity. The activity per nucleus appears to remain approximately constant during S phase, to double during G_2 phase, and then mitosis returns the activity to the S phase level. Similarly, the activity per unit mass of DNA falls during S phase, as DNA is synthesized, and rises again during G_2 phase.

VIII. Role of Histone Acetylation

Comparison with Acetylation of Histone H4 in Other Systems

Histone H4 is extensively modified in the cytoplasm before entry into the nucleus in duck erythroid cells (*71*) and hepatoma cell culture (*44*). The N-terminal acetylserine of H4 is phosphorylated and one lysine

---

were used in the deacetylase assay (*84*). The enzyme activity measured was correlated with the number of nuclei per assay, counted by hematocytometer (C), or with the amount of DNA present in the assay (A and B). The DNA content of the nuclei was determined according to Burton (*13*) in a 0.5 *N* PCA hydrolysate (70 min at 70°C), after preextraction on ice with 0.5 *N* PCA–50% ethanol and for 10 min at 70°C with ethanol–ether (3:1,v/v). Each data point shown is the average of the data obtained from a single macroplasmodium (cell cycle 7.3 to 8.9 hr) between the third and the fifth mitosis (M3 and M5) after fusion. (A) An example of one set of plasmodia with the specific activity of the deacetylase based on the DNA mass per assay (∗). (B) The combined results from six sets of experiments, as in (A). (C) The combined results from three sets of experiments with the specific activity of the deacetylase based on the number of nuclei per assay (•).

residue is acetylated in duck erythroid cells. The phosphate and acetate groups are removed soon after the histone enters the nucleus to leave unmodified H4. (The N-terminal acetylserine is not reversible.) Further modifications occur thereafter, within the nucleus, where up to four acetate groups can be added or removed and appear to turn over rapidly. Combinations of acetate and phosphate were observed, but since the amounts were very small it was not clear whether phosphorylation occurred in the nucleus as well as the cytoplasm (*71*). In hepatoma tissue culture (HTC) cells a similar pattern was observed, except that very little phosphorylation was observed and newly synthesized H4 was rapidly converted to a diacetyl form, presumably in the cytoplasm, and then returned to the unmodified form in the nucleus where further turnover occurred (*43,44*). In *Physarum* the lack of ^{32}P incorporation (*6*) argues against the presence of phosphorylated H4 in the nucleus, but H4 could still be phosphorylated in the cytoplasm if the modification were rapidly reversed on H4 entering the nucleus. However, Fig. 3 shows a peak of Ac_2H4 in S phase nuclei, which is consistent with the notion that newly synthesized H4 enters the nucleus in the diacetyl form and is then converted to the umodified form, which shows an increase in late S phase (M + 1 hr to M + 1/2 hr) following the increase in Ac_2H4 (M to M + 1 hr). This suggestion is analogous to the situation in HTC cells, but pulse-label experiments are required to confirm this interpretation in *Physarum*. The pulse-label data of Fig. 4 show a higher incorporation of the labeled acetate into all H4 species in S phase. However, it is the monoacetylated species that shows the greatest increase of incorporation. More detailed time-course studies at specific points in the cycle are required to clarify the role of mono- or diacetyl H4 in chromosome replication.

The overall level of acetylation is determined, most of the time, by the rates of acetylation and deacetylation occurring in the nucleus. The overall levels are substantially different in different systems. For example, dividing avian erythroblasts have most H4 modified and very little with more than one acetate (*71*); exponentially growing HeLa cells have about equal amounts of unmodified and Ac_1H4; and *Physarum* has, on average, about 58% of Ac_1H4 and approximately equal amounts of unmodified and Ac_2H4. In all these systems the amounts of Ac_3H4 and Ac_4H4 are small (about 4% in *Physarum*) and very small (about 2% in *Physarum*), respectively. Increase in acetylation is correlated, in various cell types, with transcription of greater regions of the genome (*3,4*). The above differences in acetate levels on H4 may be correlated with differences in the amount of the genome being expressed. However, *Physarum* nuclei have a substantial DNA content, 1.2 pg per plasmodial nucleus (*61*), and several different cell types so it is most unlikely that the amount of Ac_1H4 is a direct measure of the proportion of the genome that is active. That would require about 58% of the genome to be active during growth of the plasmodium. It is more likely that transcription is associated with highly acetylated H4, maybe even Ac_4H4, that is turning over rapidly. Then the increases in Ac_1H4 and Ac_2H4 between erythroblasts and *Physarum* reflect the overall increase in net acetylation that might accompany activation of a higher proportion of the genome.

The cell cycle dependence of H4 deacetylase (*84*) shows that the overall level of deacetylase activity does not change enough to account for the changes in acetylation levels (Figs. 3 and 5). The changes may be due to overall changes in acetyltransferase or to changes in the H4 substrate accessibility. The level of histone deacetylase activity is clearly crucial because 2 m*M* butyrate substantially affects H4 acetate content, although this concentration only inhibits *Physarum* deacetylase by about 10% (*17,84*). The data are consistent with the suggestion that local regions of chromatin have a structure or composition that encourages acetylation, such as local inhibition of deacetylase by the non-histone proteins HMG 14 and 17 (*68*).

The use of deoxyribonuclease I (DNase I) has also provided evidence linking H4 acetylation with transcription. First, DNase I digests chromatin in such a way that transcriptionally active chromatin is destroyed at early digestion times before the bulk of the chromatin (*31,85*) releases histone and non-histone proteins (*51,81*); at early digestion times the histone H4 released is more acetylated than the bulk H4, as if acetylated H4 were associated with transcriptionally active chromatin (*63*). DNase 1 digestion has also been used to link HMG non-histone proteins with transcription, particularly HMG 14 and 17 (equivalent to H6 in trout) (*51*). Second, as mentioned above cells grown in the presence of butyrate acquire a high level of acetylated histones (*38,70*). Chromatin from butyrate-treated cells shows a high degree of DNase I sensitivity, and acetylated histones are preferentially released (*82*).

Transcription in *Physarum* during the cell cycle is probably biphasic (*11,24,35,59*) with a very low level in mitosis, a high level in S phase, a low in early G_2 phase and the second high in mid to late G_2 phase. Several lines of evidence suggest that the S phase peak is predominantly mRNA synthesis and the G_2 phase peak is predominantly rRNA synthesis although the distinction, if correct, is not absolute [for reviews, see (*10,55,73*)]. The biphasic pattern correlates closely with the pattern of Ac_4H4 seen in Fig. 3b. This supports the correlation between acetylated histones and transcription discussed above and suggests the following, more stringent, test of the correlation. About 2% of *Physarum* DNA contains genes for rRNA and associated spacer sequences (rRNA) (*62,83*). If the G_2 phase transcription is predominantly rRNA and associated with Ac_4H4, then the G_2 phase increase in Ac_4H4 from 1 to 3% implies that most of the H4 associated with rDNA would become tetra-acetylated in G_2 phase. This prediction can be tested directly.

D'Anna *et al.* (*23*) found low levels of H4 acetate in mitosis in Chinese hamster ovary (CHO) cells and in the present experiments the lowest values for the proportion of highly acetylated H4 also occur in mitosis, particularly in prophase, where, at M3, the proportion of Ac_4H4 falls as low as 0.4%. The absolute values at M2 are rather different from those at M3 but the direction of change in proportion is the same. The absolute difference may reflect a change in growth pattern as the plasmodium gets larger and the composition of the medium changes. These factors add a slow monotonic change underlying the larger cell cycle-dependent changes. The amount of Ac_4H4 in prophase is inversely correlated with phosphorylation of histone H1. The phosphate content of H1 reaches its maximum about 20 min before metaphase in *Physarum* (*6*). Histone H1 also has a high phosphate content at mitosis in HeLa cells (*50*) and CHO cells (*37*). There is substantial circumstantial evidence linking H1 phosphorylation in late G_2 phase and prophase with chromosome condensation [for a review, see Matthews (*56*)]. Consequently, it now appears that chromosome condensation may involve the coordinated modification of H1 by phosphorylation and of H4 by deacetylation. It would follow from this that the extension of chromosomes after mitosis requires histone acetylation, and this may explain the very rapid acetylation of histone H4 immediately following metaphase. Modification of H3 by phosphorylation has been reported at mitosis in CHO cells (*36*); deacetylation of H3 may also occur.

So far, little or no effect of acetylation has been observed in the structure of the nucleosome or core particle. A reasonable assumption is that interactions of the basic N-terminal regions of core histones are involved in the generation of the 34 nm diameter supercoil of nucleosomes and thus acetylation of histones acts at this level of chromatin structure. A scheme we have proposed (Fig. 6) is based on the assumption that the basic structural form of inactive chromatin is the 34 nm supercoil. It follows that all higher order chromatin structures are also genetically inactive. The transition from inactive to active chromatin involves the destabilization of this supercoil and its transition to a string of nucleosomes. This process requires the acetylation of the N-terminal regions of the core histones and the possible displacement of histone H1. HMG proteins, characterized by Johns' group, (*33,34*) have been increasingly implicated in the structure of active chromatin (*51,81,86*), and the basic

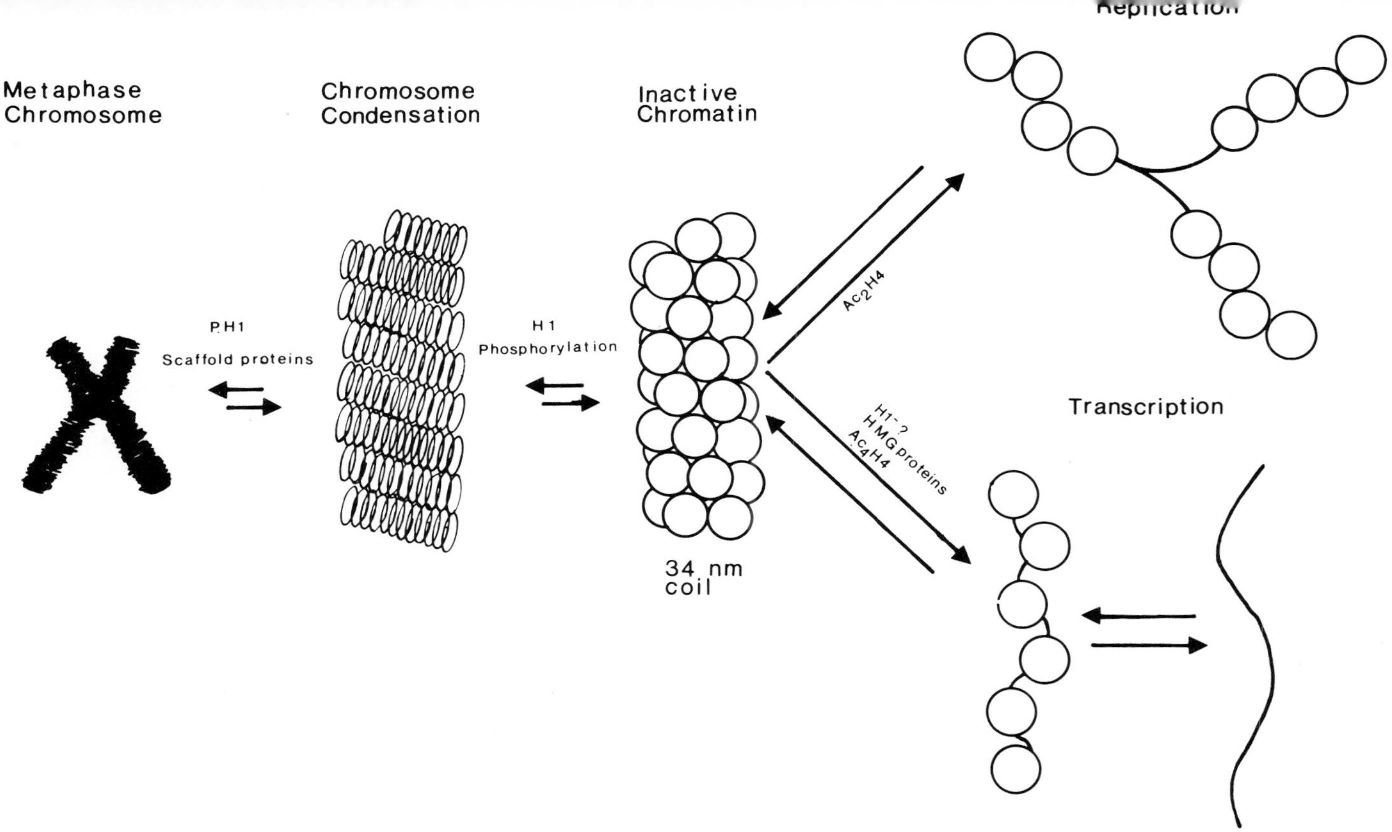

Fig. 6. Model representing the major structural transitions in chromatin. The different structural states are shown in diagrammatic form only.

region of these proteins may bind to the DNA binding sites exposed by the displacement of H1 and the N-terminal regions of the core histones. It is not clear at this stage whether the transition to active chromatin requires the full unfolding of DNA to the linear form or whether RNA polymerase can circumvent or utilize nucleosomes in reading DNA. The rigid sequence conservation of histones H3 and H4 may be required for functional aspects of chromosomes in addition to their more obvious structural roles. Chromosome condensation is controlled by H1 phosphorylation (*6,57*) and there must also be an essential involvement of the "scaffold" proteins in the structure of the metaphase chromosome (*1*). Overall, the scheme provides explanations for diverse observations on chromatin structural behavior, coordinating both phosphorylation and acetylation and their effects on chromosome structure.

Acknowledgment

This work is supported by a grant from the National Institutes of Health (GM 26901).

References

1. Adolph, K. W., Cheng, S. M., and Laemmli, U. K. (1977). Role of nonhistone proteins in metaphase chromosome structure. *Cell* **12,** 805–816.
2. Allan, J., Hartman, P. G., Crane-Robinson, C., and Aviles, F. X. (1980). The structure of histone H1 and its location in chromatin. *Nature* (*London*) **288,** 675–679.
3. Allfrey, V. G. (1977). Post-synthetic modifications of histone structure. *In* "Chromatin and Chromosome Structure" (H. J. Li and R. A. Eckhardt, eds.), p. 167. Academic Press, New York.
4. Allfrey, V. G., Faulkener, R. M., and Mirsky, A. E. (1964). Acetylation and methylation of histones and their possible role in the regulation of RNA synthesis. *Proc. Natl. Acad. Sci. U.S.A.* **51,** 786–794.
5. Bonner, W. M., West, M. H. P., and Stedman, J. D. (1980). Two-dimensional gel analysis of histones in acid extracts of nuclei, cells and tissues. *Eur. J. Biochem.* **109,** 17–23.
6. Bradbury, E. M., Inglis, R. J., Matthews, H. R., and Sarner, H. (1973). Phosphorylation of very-lysine-rich histone in *Physarum polycephalum:* Correlation with chromosome condensation. *Eur. J. Biochem.* **33,** 131–139.
7. Bradbury, E. M., Baldwin, J. P., Carpenter, B. G., Hjelm, R. P., Hancock, R., and Ibel, K. (1975). Neutron scattering studies of chromatin. *Brookhaven Symp. Biol.* **27,** 97–117.
8. Bradbury, E. M., Chapman, G. E., Danby, S. E., Hartman, P. G., and Riches, P. L. (1975). Studies on the role and mode of operation of the very-lysine-rich histone H1 (F1) in eukaryote chromatin. The properties of the N-terminal and C-terminal halves of histone H1. *Eur. J. Biochem.* **57,** 521–528.

9. Braddock, G. W., Baldwin, J. P., and Bradbury, E. M. (1981). Neutron scattering studies of the structure of chromatin core particles in solution. *Biopolymers* **20,** 327–343.
10. Braun, R., and Seebeck, T. (1982). RNA metabolism. *In* "Cell Biology of *Physarum* and *Didymium*" (H. H. Aldrich, ed.), pp. 393–436. Academic Press, New York.
11. Braun, R., Mittermayer, C., and Rusch, H. P. (1966). Ribonucleic acid synthesis *in vivo* in the synchronously dividing *Physarum polycephalum* studied by cell fractionation. *Biochim. Biophys. Acta* **114,** 527–535.
12. Burger, S. (1976). Isolierung und Characterisierung von Histon-Deacetylasen aus Kalbsthymus 9. Ph.D. Thesis, Univ. of Marburg, Germany.
13. Burton, K. (1956). A study of the conditions and mechanism of the diphenylamine reaction for the colorimetric estimation of deoxyribonucleic acid. *Biochem. J.* **62,** 315–323.
14. Candido, E. P. M., Reeves, R., and Davie, R. (1978). Sodium butyrate inhibits histone deacetylation in cultured cells. *Cell* **14,** 105–114.
15. Carpenter, B. G., Baldwin, J. P., Bradbury, E. M., and Ibel, K. (1976). Organization of subunits in chromatin. *Nucleic Acids Res.* **3,** 1739–1746.
16. Cary, P. D., Moss, T., and Bradbury, E. M. (1978). High resolution proton magnetic resonance studies of chromatin core particles. *Eur. J. Biochem.* **89,** 475–482.
17. Chahal, S. S., Matthews, H. R., and Bradbury, E. M. (1980). Acetylation of histone H4 and its role in chromatin structure and functions. *Nature* (*London*) **287,** 76–79.
18. Chapman, G. E., Hartman, P. G., and Bradbury, E. M. (1976). Studies on the role and mode of operation of the very-lysine-rich histone H1 in eukaryote chromatin. The isolation of the globular and non-globular regions of the histone H1 molecule. *Eur. J. Biochem.* **61,** 69–75.
19. Corbett, S., Miller, S., Robinson, V. J., Matthews, H. R., and Bradbury, E. M. (1977). *Physarum polycephalum* histones. *Biochem. Soc. Trans.* **5,** 943–946.
20. Couppez, M., Sautiere, P., Brahmachari, S. K., Brahms, J., Liquier, J., and Taillandier, E. (1980). Site and role of the N-terminal fragment of the nucleosomal core histones in their binding to DNA as determined by vibrational spectroscopy. *Biochemistry* **19,** 3358–3363.
21. Cousens, D. G., Gallwitz, D., and Alberts, B. M. (1979). Different accessibilities in chromatin to histone acetylase. *J. Biol. Chem.* **254,** 1716–1723.
22. Crane-Robinson, C., Bohm, L., Puigdomenech, P., Cary, P. D., Hartman, P. G., and Bradbury, E. M. (1980). Structural domains in histones. *In* FEBS "DNA-Recombination Interactions and Repair" (S. Zadrazil and J. Sponar, eds.), pp. 293–300. Pergamon, Oxford.
23. D'Anna, J. A., Tobey, R. A., Barham, S. S., and Gurley, L. R. (1977). A reduction in the degree of H4 acetylation during mitosis in Chinese hamster cells. *Biochem. Biophys. Res. Commun.* **77,** 187–202.
24. Davies, K. E., and Walker, I. O. (1977). *In vitro* transcription of RNA in nuclei, nucleoli and chromatin from *Physarum polycephalum*. *J. Cell Sci.* **26,** 267–279.
25. DeLange, R. J., Famborough, D. M., Smith, E. L., and Bonner, J. (1969). Calf and pea histone. II. The complete amino acid sequence of calf thymus histone IV. Presence of ε*N*-acetyllysine acetylation. *J. Biol. Chem.* **244,** 319–334.
26. Dixon, G. H., Candido, E. P. M., Honda, B. M., Louie, A. J., McLeod, A. R., and Sung, M. T. (1975). The structure and function of chromatin. *CIBA Found. Symp.* [N.S.] **28,** 220–240.
27. Felsenfeld, G. (1978). Chromatin. *Nature* (*London*) **271,** 115–122.

28. Finch, J. T., and Klug, A. (1976). Solenoidal model for superstructure in chromatin. *Proc. Natl. Acad. Sci. U.S.A.* **73,** 1897–1901.
29. Finch, J. T., and Klug, A. (1978). X-ray and electron microscope analyses of crystals of nucleosome cores. *Cold Spring Harbor Symp. Quant. Biol.* **42,** 1–11.
30. Finch, J. T., Lutter, L. C., Rhodes, D., Brown, R. S., Rushton, B., Levitt, M., and Klug, A. (1977). Structure of nucleosome core particles of chromatin. *Nature (London)* **269,** 29–36.
31. Garel, A., and Axel, R. (1976). Selective digestion of transcriptionally active ovalbumin genes from oviduct nuclei. *Proc. Natl. Acad. Sci. U.S.A.* **73,** 3966–3970.
32. Goldknopf, I. L., and Busch, H. (1977). Isopeptide linkage between nonhistone and histone 2A polypeptides of chromosomal conjugate-protein A24. *Proc. Natl. Acad. Sci. U.S.A.* **74,** 864–868.
33. Goodwin, G. H., and Johns, E. W. (1973). Isolation and characterization of two calf thymus non histone proteins with high contents of acidic and basic amino acids. *Eur. J. Biochem.* **40,** 215–219.
34. Goodwin, G. H., Walker, J. M., and Johns, E. W. (1978). The high mobility group (HMG) chromosomal non-histone proteins. *In* "The Cell Nucleus" (H. Busch, ed.), Vol. 4, Part A. Academic Press, New York.
35. Grant, W. D. (1972). The effect of alpha-amanitin and (NH4)2 SO4 on RNA synthesis in nuclei and nucleoli isolated from *Physarum polycephalum* at different times during the cell cycle. *Eur. J. Biochem.* **2,** 94–98.
36. Gurley, L. R., Walters, R. A., and Tobey, R. A. (1975). Sequential phosphorylation of histone sub-fractions in the Chinese hamster cell cycle. *J. Biol. Chem.* **250,** 3936–3944.
37. Gurley, L. R., D'Anna, J. A., Barham, S. S., Deaven, L. L., and Tobey, R. A. (1978). Histone phosphorylation and chromatin structure during mitosis in Chinese hamster cells. *Eur. J. Biochem.* **84,** 1–16.
38. Hagopian, H. R., Riggs, M. G., Swartz, L. A., and Ingram, V. M. (1977). Effect of *n*-butyrate on DNA synthesis in chick fibroblasts and HeLa cells. *Cell* **12,** 855–860.
39. Halleck, M. S., and Gurley, L. R. (1980). Histone H2A subfractions and their phosphorylation in cultured *Peromyscus* cells. *Exp. Cell Res.* **125,** 377–388.
40. Hartman, P. G., Chapman, G. E., Moss, T., and Bradbury, E. M. (1977). Studies on the role and mode of operation of the very-lysine-rich histone H1 in eukaryote chromatin. The three structural regions of the histone H1 molecule. *Eur. J. Biochem.* **77,** 45–51.
41. Inoue, A., and Fujimoto, D. (1972). Substrate specificity of histone deacetylase from calf thymus. *J. Biochem. (Tokyo)* **72,** 427–431.
42. Isenberg, I. (1979). Histones. *Annu. Rev. Biochem.* **48,** 159–191.
43. Jackson, V., Shires, A., Chalkley, R., and Granner, D. K. (1975). Studies on highly metabolically active acetylation and phosphorylation of histones. *J. Biol. Chem.* **250,** 4856–4863.
44. Jackson, V., Shires, A., Tanphaichitr, N., and Chalkley, R. (1976). Modifications to histones immediately after synthesis. *J. Mol. Biol.* **104,** 471–483.
45. Jockusch, B. M., and Walker, I. O. (1974). The preparation and preliminary characterisation of chromatin in from the slime mold *Physarum polycephalum. Eur. J. Biochem.* **48,** 417–425.
46. Johns, E. W. (1976). Fractionation and isolation of histones. *In* "Subnuclear Components" (G. D. Bernie, ed.), pp. 187–208. Butterworth, London.
47. Kervabon, A., Parello, J., and Mery, J. (1979a). Chemical studies on histone acetylation using a synthetic peptide fragment of histone H4. *FEBS Lett.* **98,** 152–156.

48. Kervabon, A., Mery, J., and Parello, J. (1979b). Enzymatic deacetylation of a synthetic peptide fragment of histone H4. *FEBS Lett.* **106,** 93–96.
49. Kikuchi, H., and Fujimoto, D. (1973). Multiplicity of histone deacetylase from calf thymus. *FEBS Lett.* **29,** 280–282.
50. Lake, R. S., and Salzman, N. P. (1972). Occurrence and properties of a chromatin-associated F-1 histone phosphokinase in mitotic Chinese hamster cells. *Biochemistry* **11,** 4817–4826.
51. Levy-Wilson, B., Wong, N. C. W., and Dixon, G. H. (1977). Selective association of the trout-specific H6 protein with chromatin regions susceptible to DNase II: Possible location of HMG-T in the spacer region between core nucleosomes. *Proc. Natl. Acad. Sci. U.S.A.* **74,** 2810–2814.
52. Libby, P. R., and Bertram, J. S. (1980). Biphasic effect of polyamines on chromatin-bound histone deacetylase. *Arch. Biochem. Biophys.* **201,** 359–361.
53. Lutter, L. C. (1979). Precise location of DNase I cutting sites in the nucleosome core determined by high resolution gel electrophoresis. *Nucleic Acids Res.* **6,** 41–56.
54. Mathis, D. J., Oudet, P., Wasylyk, B., and Chambon, P. (1978). Effect of acetylation on structure and *in vitro* transcription of chromatin. *Nucleic Acids Res.* **5,** 3523–3547.
55. Matthews, H. R. (1980). Modification of histone H1 by reversible phosphorylation and its relation to chromosome condensation and mitosis. *In* "Protein Phosphorylation in Regulation" (P. Cohen, ed.), pp. 235–254. Elsevier/North-Holland, Amsterdam.
56. Matthews, H. R. (1981). Chromatin proteins and progress through the cell cycle. *In* "The Cell Cycle" (P. John, ed.), pp. 223–246. Cambridge Univ. Press, London and New York.
57. Matthews, H. R., and Bradbury, E. M. (1978). The role of histone H1 phosphorylation in the cell cycle: Turbidity studies of H1–DNA interaction. *Exp. Cell Res.* **111,** 343–351.
58. Matthews, H. R., and Bradbury, E. M. (1982). Chromsome organization and chromosomal proteins in *Physarum polycephalum*. *In* "The Cell Biology of *Physarum* and *Didymium*" (H. Aldrich, ed.), pp. 317–369. Academic Press, New York.
59. Mittermeyer, C., Braun, R., and Rusch, H. P. (1964). RNA synthesis in the mitotic cycle of *Physarum polycephalum*. *Biochim. Biophys. Acta* **91,** 399–405.
60. Mohberg, J., and Rusch, H. P. (1969). Isolation of the nuclear histones from the myxomycete *Physarum polycephalum*. *Arch. Biochem. Biophys.* **134,** 577–589.
61. Mohberg, J., Babcock, K. L., Haugli, F. B., and Rusch, H. P. (1973). Nuclear DNA content and chromosome numbers in the myxomyute *Physarum polycephalum*. *Dev. Biol.* **34,** 228–245.
62. Molgaard, H. V., Matthews, H. R., and Bradbury, E. M. (1976). Organisation of genes for ribosomal RNA in *Physarum polycephalum*. *Eur. J. Biochem.* **68,** 541–549.
63. Nelson, D. A., Perry, M., Sealy, L., and Chalkley, R. (1978). DNase I preferentially digests chromatin containing hyperacetylated histones. *Biochem. Biophys. Res. Commun.* **82,** 1346–1353.
64. Nelson, P. P., Albright, S. C., Wiseman, J. M., and Garrard, W. T. (1979). Reassociation of histone H1 with nucleosomes. *J. Biol. Chem.* **254,** 11751–11760.
65. Noll, M., and Kornberg, R. (1977). Action of micrococcal nuclease on chromatin and the location of histone H1. *J. Mol. Biol.* **109,** 393–404.
66. Panyim, S., and Chalkley, R. (1969). High resolution acrylamide gel electrophoresis of histones. *Arch. Biochem. Biophys.* **130,** 337–346.
67. Pardon, J. F., Worcester, D. L., Wooley, J. C., Tatchell, K., Van Holde, K. E., and Richards, B. M. (1975). Low angle neutron scattering from chromatin subunit particles. *Nucleic Acids Res.* **2,** 2163–2176.

68. Reeves, R., and Candido, E. P. M. (1980). Partial inhibition of histone deacetylase in active chromatin by HMG14 and HMG17. *Nucleic Acids Res.* **8,** 1947–1963.
69. Richards, B. M., Pardon, J. F., Lilley, D., Cotter, R., and Wooley, J. C. (1977). The sub-structure of nucleosomes. *Cell Biol. Int. Rep.* **1,** 107–116.
70. Riggs, M. G., Whittaker, R. G., Neumann, J. R., and Ingram, V. M. (1977). *n*-Butyrate causes histone modification in HeLa and Friend erythroleukaemia cells. *Nature (London)* **268,** 462–464.
71. Ruiz-Carrillo, A., Waugh, L. J., and Allfrey, V. G. (1975). Processing of newly synthesised histone molecules: Nascent histone H4 chains are reversibly phosphorylated and acetylated. *Science* **190,** 117–128.
72. Rusch, H. P. (1970). Some biochemical events in the life cycle of *Physarum polycephalum. Adv. Cell Biol.* **1,** 297–327.
73. Sauer, H. W. (1978). Regulation of gene expression in the cell cycle of physarum. *In* "Cell Cycle Regulation" (J. R. Jeter, J. L. Cameron, G. M. Padilla, and A. M. Zimmerman, eds.), pp. 149–156. Academic Press, New York.
74. Simpson, R. T. (1978). Structure of the chromatosome, a chromatin core particle containing 160 base pairs of DNA and all the histones. *Biochemistry* **17,** 5524–5531.
75. Suau, P., Kneale, G. G., Braddock, G. W., Baldwin, J. P., and Bradbury, E. M. (1977). A low resolution model for the chromatin core particle by neutron scattering. *Nucleic Acids Res.* **4,** 3769–3786.
76. Suau, P., Bradbury, E. M., and Baldwin, J. P. (1979). Higher order structures of chromatin in solution. *Eur. J. Biochem.* **97,** 593–602.
77. Tatchell, K., and Van Holde, K. E. (1978). Compact oligomers and nucleosome phasing. *Proc. Natl. Acad. Sci. U.S.A.* **75,** 3583–3587.
78. Thoma, F., Koller, T., and Klug, A. (1979). Involvement of histone H1 in the organisation of the nucleosome and of the salt-dependent superstructure of chromatin. *J. Cell Biol.* **83,** 403–427.
79. Todd, R. D., and Garrard, W. T. (1979). Overall pathway of mononucleosome production. *J. Biol. Chem.* **254,** 3074–3083.
80. Van Holde, K. E., and Isenberg, I. (1975). Nucleosome chromatin structure. *Acc. Chem. Res.* **8,** 327.
81. Vidali, G., Boffa, L. C., and Allfrey, V. G. (1977). Selective release of chromosomal proteins during limited DNase 1 digestion of avian erythrocyte chromatin. *Cell* **12,** 409–415.
82. Vidali, G., Boffa, L. C., Bradbury, E. M., and Allfrey, V. G. (1978). Butyrate suppression of histone deacetylation leads to an accumulation of multiacetylated forms of histone H3 and H4 and increased Dnase 1 sensitivity of the associated DNA sequences. *Proc. Natl. Acad. Sci. U.S.A.* **75,** 2239–2243.
83. Vogt, V., and Braun, R. (1976). Repeated structure of chromatin in metaphase nuclei of *Physarum. FEBS Lett.* **64,** 190–192.
84. Waterborg, J. H., Chahal, S. S., Muller, R. D., and Matthews, H. R. (1981). Histone acetylation in the *Physarum* cell cycle. *Proc. Eur. Physarum Conf. 1981* p. 160.
85. Weintraub, H., and Groudine, M. (1976). Chromosomal subunits in active genes have an altered conformation. *Science* **193,** 848–856.
86. Weisbrod, S., and Weintraub, H. (1979). Isolation of a sub class of nuclear proteins responsible for conferring a DNase I-sensitive structure on globin chromatin. *Proc. Natl. Acad. Sci. U.S.A.* **76,** 630–635.
87. Whitlock, J. P., and Stein, A. (1978). Folding of DNA by histones which lack their NH-terminal regions. *J. Biol. Chem.* **253,** 3857–3861.

3

Role of HMG–Nucleosome Complexes in Eukaryotic Gene Activity

KENNETH S. McCARTY, SR., DREW N. KELNER,
KLAUS WILKE, AND KENNETH S. McCARTY, JR.

I. Introduction

An understanding of the molecular mechanisms involved in the regulation of gene expression forms the basis of one of the fundamental problems of modern molecular biology. The acquisition of this knowledge is essential before we can begin to comprehend the processes which dictate the differing rates, frequencies, and specificities of RNA synthesis characteristic of eukaryotic cells.

The genetic material of eukaryotic cells is packaged in such a way as to achieve a 10^5-fold DNA compaction ratio. This compaction is accomplished at several levels of complexity requiring both protein–protein and protein–DNA interactions. The fundamental unit of complexity re-

GENETIC EXPRESSION IN THE CELL CYCLE

ISBN 0-12-543720-X

quires the precise interaction of two copies of four different histones with two complete DNA loops folded on the surface. This unit retains a six- to sevenfold DNA compaction ratio and is often referred to as the nucleosome, ν body or chromatosome. The additional DNA compaction requires histone H1 and perhaps other factors, such as postsynthetic modifications of chromosomal proteins. In any event, it should be emphasized that at all levels of complexity, the nucleosomal DNA complex is far more than an inert storage reservoir of genetic information which undergoes dynamic processes to permit access to enzymes required for RNA transcription, DNA replication, recombination, and repair. We still face a number of major impediments to our understanding of how the necessary enzymes gain access to the nucleosomal DNA complex.

Research on the control of transcription has focused on regulatory processes occurring at the level of the nucleosome core particle. This approach has been stimulated by two observations. The first suggests that postsynthetic modifications of nucleosomal histones are associated with mRNA synthesis [for review, (*118*)]. The second observation suggests that a group of non-histone proteins, first described in 1964 by Johns as high mobility group (HMG) proteins [for review, (*42*)], are associated with nucleosome core particles obtained from active regions of chromatin.

This chapter discusses the control of transcription at the level of the nucleosome core particle, with emphasis on the HMG proteins, properties of HMG nucleoprotein complexes, acetyltransferase function, and the molecular consequences of acetylation as a postsynthetic modification of HMG nucleosome complexes. This represents an extension of the material presented by Chai and Sandberg (*24*) on recent concepts of nucleosome structure and that of Matthews and Bradbury (see Chapter 2), Chahal *et al.* (*23*), and Levy-Wilson *et al.* (*93*) on acetylation of histones with emphasis on histone H4. Our primary objective is to focus attention on the rapidly accumulating evidence that the HMG proteins are essential, critical components in the organization and expression of the eukaryotic genome.

II. Nucleosome Core Particles

A. Structure of Nucleosome Cores

There is ample evidence that the nucleosome is a repeating structural nucleoprotein subunit that may be isolated from either nuclei or chromatin by the action of endonucleases. This structure is fundamental in the packaging of DNA in the eukaryotic chromosome (see Chapter 2) (*18,24,96,117–119,122,128*).

For the sake of clarity, the *nucleosome core particle* (*128*) should be clearly distinguished from the *chromatosome,* a term proposed by Simpson (*154*). Thus, the nucleosome core particle is limited to 146 DNA base pairs (bp) permitting 1.75 superhelical DNA turns coiled around two heterotypic histone tetramers forming a spherical disk composed of an octamer histone complex $(H2A,H2B,H3,H4)_2$. The chromatosome is associated with an additional histone H1, partially protecting the ends of two complete superhelical turns of 166 DNA bp, which includes 20 DNA bp extending into the linker region. This accounts for the observation that when histone H1 remains associated with nucleosomes, a portion of the terminal nucleosomal DNA is protected from nuclease attack (*4,142*).

For many studies the *nucleosome core* particle has been more useful than the chromatosome because it is (1) chemically and physically better defined, (2) stoichiometrically associated with HMG 14 and HMG 17, (3) easily reconstructed from its components, and (4) useful for studies requiring the absence of histone H1.

The nucleosome core particle model for the histone complex is best described as an inner domain of globular components and an outer domain of mobile amino terminal tails which overlap the DNA (*96*). This is in agreement with the initial neutron-scattering studies, which showed within the accuracy of the data that the protein and the DNA were concentric (*129*).

B. Enzymatic Studies of Nucleosome Cores

As reviewed in Chapter 1, nuclease probes have been useful to obtain information on the internal structure of the nucleosome core and some details of the DNA conformation. The enzymes most frequently selected for these studies are DNase I, DNase II, and staphlococcal nuclease, all of which cleave sites on opposite strands of the nucleosomal DNA. The shortest stagger is produced by DNase I and staphlococcal nuclease (two bp), whereas DNase II leaves a four-bp stagger. These enzymes all cleave the nucleosomal DNA symmetrically with preferential cuts at 10-bp intervals, suggesting that the enzyme may recognize the same site at 10-bp intervals, cleaving, however, at one or two bp distant from the initial DNA cleavage site (*29,158*).

Enzymatic digestion of nucleosome core particles with DNase I represents an exceedingly useful technique to define protein–DNA interactions. The precise work of Lutter (*97*) and others (*158*) are in general agreement with the observation that a rapid cleavage of DNA is observed at 10, 20, 40, 50, and 130 nucleotides from the 5′-terminus. The degree of symmetry that is observed is emphasized by the demonstration of specific cleavage sites of the DNA at 130, 120, 100, and 90 bp from the 3′-end. In the chromatosome, consisting of 160 bp, the frequency now becomes 10, 20, 30, 50, and 60 bp from the 5′-end, and 150, 140, 130, 110, and 100 bp from the 3′-end are also in symmetry. It should be noted that those bases that are not cut (or are only infrequently cut) are located at the top corner of the hexagonal bipartite disk proposed by Chai and Sandberg (*24*), in agreement with early works of Klug *et al.* (*46*). These sites are stabilized by N-terminal histone regions (*121,122*) corresponding to DNA bases 40 and 120 from the 5′-end. Protein masking of nuclease sensitivity is useful to assign the position of HMG proteins in relation to DNA, as exemplified by the experiments of Mardian *et al.* (*100*), Billet (*12*), Kuehl (*78*), Levy *et al.* (*88,90,91*), Weisbrod and Weintraub (*197*), Vidali *et al.* (*144*), Abercrombie *et al.* (*1*), and Rabbani *et al.* (*133*).

C. Association of HMG Proteins with Nucleosome Cores

Recent observations implicating the HMG proteins in the structure and function of nucleosome cores are (1) their high concentration (10^5 to 10^6 molecules per cell) (*49*); (2) their location in both the nucleus and cytoplasm (*20*); (3) their presence in a variety of eukaryotic cells (duck, chicken, trout testis, insects, and *Tetrahymena*) (*140,160,162*); (4) their existence as integral components of chromatin (*3,7,89,105,106*); (5) their nonrandom stoichiometric distribution on the nucleosome core particle (*43,100,146*); (6) their ability to alter the T_m of DNA (*122*); (7) their association with specific transcriptionally active chromatin fractions (*196–198*); (8) their capacity to modulate deacetylase activity (*136*); and (9) their potential to be postsynthetically modified (*59,73,74,86,145,163,164*).

The chromatosome has provided valuable evidence that HMG 1 interacts specifically with histone H1 (*22,30*). As an extension of these studies, Yu and Spring (*204*) have used immobilized histone H1 to fractionate HMG 1 and HMG 2. The HMG 1 and/or 2 proteins are associated with the chromatosome and can be crosslinked to histone H1 (*16,22*). The interaction of histone H1 and HMG 1 has many important implications in the nucleosome superstructure.

HMG 14 and 17 can be stoichiometrically bound to nucleosomes (*7–9,41,100,106,146*). The interactions of the amino termini of the HMG proteins with DNA resemble histone–DNA interactions (*79,122*). DNase I digestion patterns have been exceedingly useful in the elegant experiments from the laboratory of Olins to demonstrate the specific binding of HMG 14 and 17 to the DNA termini of nucleosomes reconstituted *in vitro* (*100*).

D. Properties of HMG–Nucleosome Complexes

1. T_m Measurements

Of the physical properties which help to define the nucleosome core particle, thermal denaturation is most useful for estimating the relative amounts of regions of differing thermal stability caused by localized interactions between the histones and the DNA. Heating causes two prominent transitions in core particles. The first (generally centered at 62°C) is thought to reflect a denaturation at the ends of the 146 bp DNA, and the second represents a final and irreversible thermal collapse at 75°C which appears to be highly cooperative. This technique has been useful to provide an estimate of the regions of unbound DNA within the modified and control nucleosome core particles (*13*).

HMG 1 and HMG 2 proteins lower the melting temperature of DNA (*67*), while HMG 14 and/or 17 stabilizes DNA against denaturation. This stabilization is also observed when HMG 14 and/or 17 is bound to nucleosomes (*146*).

2. Circular Dichroism Measurements

A nucleosome circular dichroism spectrum may be arbitrarily divided into two regions. The spectral range above 240 nm is almost exclusively determined by the DNA conformation, while the ellipticity at 223 nm is conditioned by the α-helical content of the histone. Changes in the ellipticity below 260 nm should reveal some useful information (*13,153*). HMG 17 has been studied using circular dichroism, nmr, and small-angle scattering (*1*).

These studies show little or no secondary or tertiary structure in free solution. This is in contrast to HMG 1 and 2, which exhibit highly ordered structures (*21*). Amino acid sequences are given in Fig. 1. The nmr data suggest that the principal DNA-binding segment of HMG 17 is that between residues 15 and 40 (Fig. 2).

3. Electrophoresis

Both Mardian *et al.* (*100*) and Sandeen *et al.* (*146*) have demonstrated that those chicken erythrocyte nucleosome core particles which are complexed with 2 moles of HMG 14 and/or 17 undergo a dramatic change in their electrophoretic mobility (slower). We have also confirmed these observations, as illustrated in Fig. 3 (*74,124*).

5 10 15 20
HMG 1) Gly-Lys-Gly-Asp-Pro-Lys-Lys-Pro-Arg-Gly-Lys-Met-Ser-Ser-Tyr-Ala-Phe-Phe-Val-Gln-
HMG 2) Gly-Lys-Gly-Asp-Pro-Asn-Lys-Pro-Arg-Gly-Lys-Met-Ser-Ser-Tyr-Ala-Phe-Phe-Val-Gln-
25 30 35 40
HMG 1) Thr-Ser-Arg-Glu-Glu-His-Lys-Lys-Lys-His-Pro-Asp-Ala-Ser-Val-Asn-Phe-Ser-Glu-(phe,
HMG 2) Thr-Ser-Arg-Glu-Glu-His-Lys-Lys-Lys-His-Pro-Asp-Ala-Ser-Val-Asx-Phe-Ser-Glu-(phe,
45 50 55 60
HMG 1) ser,lys,lys,cys,ser,glu,ser,gly,ala,tyr,lys,lys,glu,glu)Arg-Trp-Lys-Thr-Met-Ser-
HMG 2) ser,lys,lys,cys, - ,glu,val,gly,ala,tyr,lys, - , - , -)Arg-Trp-Lys-Thr-Met-Ser-
65 70 75 80
HMG 1) Ala-Lys-Glu-Lys-Gly-Lys-Phe-Glu-Asp-Met-Ala-Lys-Ala-Asp-Lys-Ala-Arg-Try-Glu-Arg-
HMG 2) Ala-Lys-Glu-Lys-Ser-Lys-Phe-Glu-Asp-Met-Ala-Lys-Ser-Asp-Lys-Ala-Arg-Try-Asp-Arg-
85 90 95 100
HMG 1) Glu-Met-Lys-Thr-Tyr-Ile-Pro-Pro-Lys-Gly-Glu-Thr-Lys-Lys-Lys-Phe-Lys-Asp-Pro-Asn-
HMG 2) Glu-Met-Lys-Asn-Tyr-Val-Pro-Pro-Lys-Gly-Asp-Lys-Lys-Gly-Lys-Lys-Lys-Asp-Pro-Asn
105 110 115 120
HMG 1) Ala-Pro-Lys-Arg-Arg-Pro-Pro-Ser-Ala-Phe-Leu-Phe-Ala-Ser-Glu-Tyr-Arg-Pro-Lys- Ile -
HMG 2) Ala-Pro-Lys-Arg-Arg-Pro-Pro-Ser-Ala-Phe-Leu-Phe-Ser-Ala-Glu-His-Arg-Pro-Lys-Ile-
125 130 135 140
HMG 1) Lys-Gly-Glu-His-Pro-Gly-Leu-Ser-Ile-Gly-Asp-Val-Ala-Lys-Lys-Leu-Gly-Glu-Met-Trp-
HMG 2) Lys-Ala-Glu-His-Pro-Gly-Leu-Ser-Ile-Gly-Asp-Thr-Ala-Lys-Lys-Leu-Gly-Glu-Met-Trp-
145 150 155 160
HMG 1) Asn-Asn-Thr-Ala-Ala-Asp-Asp-Lys-Gln-Pro-Tyr-Glu-Lys-Lys-Ala-Ala-Lys-Leu-Lys-Glu-
HMG 2) Ser-Gly-G-Asp-Lys-Gln-Pro-Tyr-Glu-Gln-Lys-Ala-Ser-Lys-Leu-Lys-Glu-
165 170 175 180
HMG 1) Lys-Tyr-Glu-Lys- ? -Ala-Ala-Tyr-Arg-Ala-Lys-Gly-Lys-Pro-Asp-Ala-Ala-Lys-Lys-Gly-
HMG 2) Lys-Tyr-Gly-Lys- ? -Ala-Ala-Tyr-Arg-Ala-Lys-Gly-Lys-Ser-Glu-Ala-Gly-Lys-Lys-Gly-
185 190 195 200
HMG 1) Val-Val-Lys-Ala-Glu-Lys-Ser-Lys-Lys-Lys-Lys-Glu-Glu-Glu-Glu-Asp-Glu-Glu-Asp-Asp-
HMG 2) Arg-Pro-Thr-Gly-Ser-Lys-Lys-Lys-Asn-Glu-Pro-Glu-Asp-Glu-Gly-Asp-Asp-Glu-
HMG 1) Glu-(glu22,lys,asp,ile,ala2,tyr,thr,pro,ala,leu,phe,arg,ser2,glu3,gly3,lys5)-
HMG 2) Glu-(glu21,asp5,pro,lys,ala2,tyr,thr,pro,leu,arg,ser2,glu3,gly3,lys5)-
HMG 1) Phe-Ala-Lys
HMG 2) Phe-Ala-Lys

Fig. 1. Amino acid sequence of non-histone proteins HMG 1 and 2 isolated from calf thymus. [From Walker *et al.* (*177*).]

4. *Acetylation of Nucleosome Core Proteins and Chromatin Activity*

Postsynthetic modifications of histones have been implicated for many years as modulators of gene activation (*6,35,101–104,110–115,120,149,201*). Among the known postsynthetic modifications in the eukaryotic nucleus (methylation, acetylation, phosphorylation, and ADP-ribosylation), nucleosomal acetylation has received considerable attention in recent years, largely because it can easily by envisioned how this process might afford greater accessibility of the transcriptional enzymes to the nucleosomal DNA by weakening the electrostatic interactions between the ε-amino groups of lysines on the core histones and the phosphates on the DNA.

5 10 15 20

Calf thymus HMG 14) Pro-Lys-Arg-Lys-Val-Ser-Ser-Ala-Glu-Gly-Ala-Ala-Lys-Glu-Glu-Pro-Lys-Arg-Arg-Ser-
Chick HMG 14) Pro-Lys-Arg-Lys-Ala-Pro - - - - - Ala-Glu-Gly-Glu-Ala-Lys-Glu-Glu-Pro-Lys-Arg-Arg-Ser-
Calf thymus HMG 17) Pro-Lys-Arg-Lys-Ala-Glu-Gly-Asp-Ala-Lys-Gly-Asp-Lys-Ala-Lys-Val-Lys-Asp-Glu-Pro-
Chick HMG 17) Pro-Lys-Arg-Lys-Ala-Glu-Gly-Asp-Thr-Lys-Gly-Asp-Lys-Ala-Lys-Val-Lys-Asp-Glu-Pro-
Fish H6) Pro-Lys-Arg-Lys-Ser-Ala-Thr-Lys-Gly-Asp-Glu-Pro-Ala-Arg-Arg-Ser-Ala-Arg-Leu-Ser-

25 30 35 40

Calf thymus HMG 14) Ala-Arg-Leu-Ser-Ala-Lys-Pro-Ala-Pro-Ala-Lys-Val-Glu-Thr-Lys-Pro-Lys-Lys-Ala-Ala-
Chick 14) Ala-Arg-Leu-Ser-Ala-Lys- - - -Ala-Pro ....
Calf thymus HMG 17) Gln-Arg-Arg-Ser-Ala-Arg-Leu-Ser-Ala-Lys-Pro-Ala-Pro-Pro-Lys-Pro-Glu-Pro-Lys-Pro-
Chick HMG 17) Gln-Arg-Arg-Ser-Ala-Arg-Leu-Ser-Ala-Lys-Pro-Ala-Pro-Pro-Lys-Pro-Glu-Pro-
Fish H6) Ala-Arg-Pro-Val-Pro-Lys-Pro-Ala-Ala-Lys-Pro-Lys-Lys-Ala-Ala-Ala-Prp-Lys-Lys-Ala-

45 50 55 60

Calf thymus HMG 14) Calf thymus HMG 14) Gly-Lys-Asp-Lys-Ser-Ser-Asp-Lys-Lys-Val-Gln-Thr-Lys-Gly-Lys-Arg-Gly-Ala-Lys-Gly-
Chick HMG 14) ....
Calf thymus HMG 17) Ala-Lys-Lys-Ala-Pro-Ala-Lys-Lys-Gly-Glu-Lys-Val-Pro-Lys-Gly-Lys-Lys-Gly-Lys-Ala-Asp-
Chick HMG 17) Fish H6) --
Val-Lys-Gly-Lys-Lys-Ala-Ala-Glu-Asn-Gly-Asp-Ala-Lys-Ala-Glu-Ala-Lys-Val-Gln-Ala-

65 70 75 80

Calf thymus HMG 14) Lys-Gln-Ala-Glu-Val-Ala-Asn-Gln-Glu-Thr-Lys-Glu-Asp-Leu-Pro-Ala-Glu-Asn-Gly-Glu-
Chick HMG 14) ....
Calf thymus HMG 17) Ala-Gly-Lys-Asx-Gly-Asx-Asx-Pro-Ala-Glx-Asx-Gly-Asx-Ala-Lys-Thr-Asx-Glx-Ala-Glx-
Chick HMG 17) Thr-Lys-Asn-Glu-Glu-Ser-Pro-Ala-Ser-Asp-Glu-Ala-Glu-Glu-Lys-Glu-Ala-Lys-Ser-Asp-
Fish H6) Ala-Gly-Asp-Gly-Ala-Gly-Asn-Ala-Lys- ....

85 90 95 100

Calf thymus HMG 14) Thr-Lys-Asn-Glu-Glu-Ser-Pro-Ala-Ser-Asp-Glu-Ala-Glu-Glu-Lys-Glu-Ala-Lys-Ser-Asp-
Chick HMG 14) --
Calf thymus HMG 17) Lys-Ala-Glu-Gly-Ala-Gly-Asp-Ala-Lys
Chick HMG 17) ....
Fish H6) ....

Fig. 2. Amino acid sequences of non-histone proteins HMG 14 and 17 isolated from calf thymus and chick and of the amino terminal of H6 isolated from fish. [From Walker *et al.* (*182,184,186*) and Walker and Johns, (*183*).]

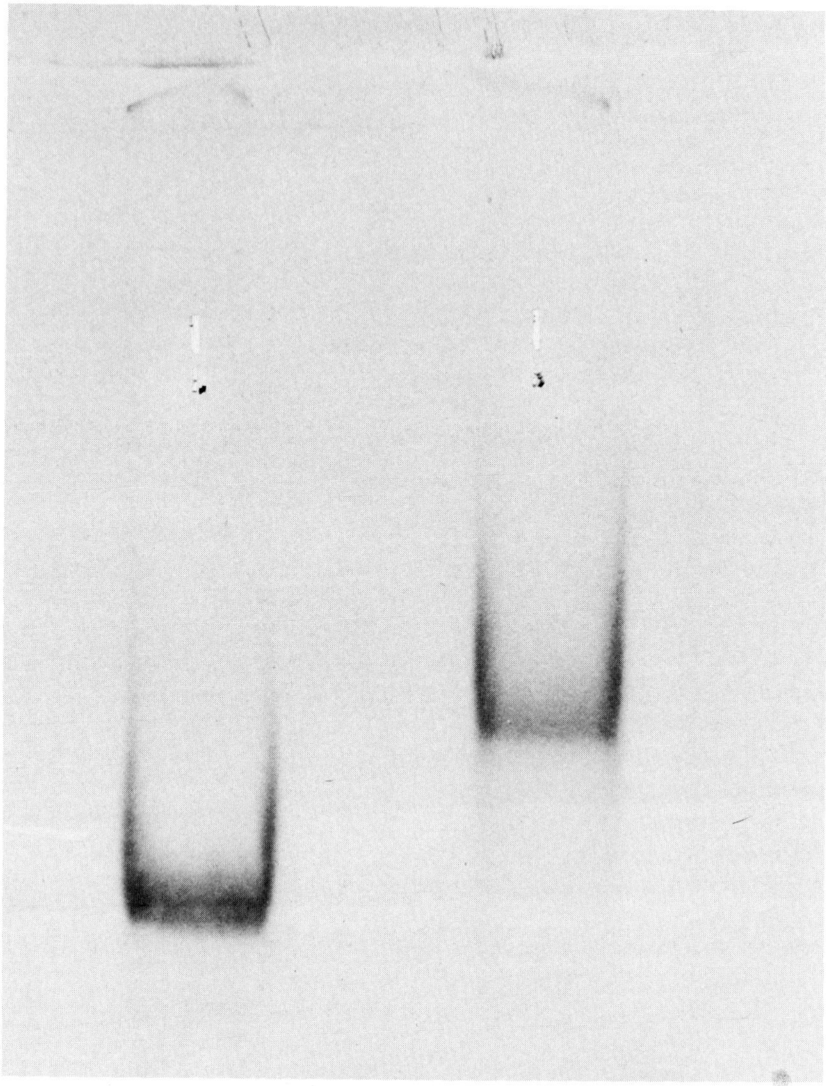

Fig. 3. Enzymatic acetylation of nucleosome core particles. Purified chicken erythrocyte core nucleosomes (500 μg lane 1) and 500 μg of core nucleosomes titrated with purified chicken erythrocyte HMG 14 (2 HMG 14/core, lane 2) were incubated with 500 μl of the hydroxylapatite pool (250 μg protein) and 250 nCi of [1-^{14}C]acetyl-COA at 22°C for 1 hr. The reaction mixtures (1 ml) were concentrated to 100 μl at 40°C on a microprodicon model 120 (Bio-Molecular Dynamics), applied to a core particle gel (5.8% polyacrylamide-0.2% bisacrylamide in 90 m*M* Tris, pH 8.3, 90 m*M* borate, 2.5 m*M* EDTA) and run for 3 hr at 100 V. The stained gel was impregnated with enhance (NEN), dried on a Bio-Rad gel slab dryer (model 224), and fluorographed (below) on Kodak X-Omat R film for 36 hr at −70°C.

The observation of Riggs *et al.* (*138*) that 5 m*M* sodium *n*-butyrate inhibits histone deacetylase without inhibiting histone acetyltransferase has provided a technique for hyperacetylation of histones in cell cultures. Several groups have shown that this hyperacetylation results in an increased sensitivity to both DNase I (*150*) and micrococcal nuclease (*153–155*). The selective release of acetate from HeLa nuclei by DNase I digestion (*174*) gives further evidence that acetylation is important in gene regulation. Unfortunately, it has not yet been possible to detect differences in template activity between normal and hyperacetylated chromatin *in vitro* (*96,108*). There is little question, however, that *in vivo,* the acetylation of histone H4 is preferentially associated with template active chromatin (*27*).

The evidence that transcription reads through nucleosomes *in vitro* is derived from experiments using cDNA prepared from polysomal mRNA. In these studies it was possible to hybridize the cDNA to DNA from purified nucleosome particles (*81*). The observation that pancreatic DNase I preferentially digests fetal globin genes (*187,198,206*), but not ovalbumin sequences from chick embryo red blood cells (*195*), and that the ovalbumin sequences are rendered acid soluble in the chick oviduct nuclei (*32*), suggests a preferential DNase I sensitivity of active DNA sequences. DNase II and micrococcal nuclease have also been shown to be associated with the enrichment of actively transcribed sequences by a number of authors (*32,34,42,84,92,174,195*). Although the details of DNase sensitivity are not available at this time, it is apparent that active genes must reflect a change in conformation or structure of the nucleosome. In fact, the capacity to enrich actively transcribed DNA sequences provides a unique opportunity to examine the nature of chromosomal proteins fractionated from active chromatin (*9,27,32,34,72, 80,87,89,153,196,195*).

III. Characterization of the High Mobility Group Proteins

A. Intracellular Concentration

It is now generally accepted that the non-histone proteins increase in quantity with evidence of increasing RNA transcription. For example, the overall composition of chromatin from relatively inactive cells such as thymocytes is 45% DNA, 45% histones, and 10% non-histone proteins (very little evidence of RNA). In liver, a relatively active tissue, the non-histone proteins are equal to or more abundant than the histones (*18*).

The group of non-histone proteins extracted from chromatin using 0.35

M NaCl have been termed high mobility group proteins (HMG) by Johns *et al.* (*70*). Since there are up to a hundred different proteins in this salt extraction, Johns resorted to 2% trichloroacetic acid precipitation to fractionate the extract. This partial fractionation renders insoluble a large fraction of the proteins leaving the four major HMG proteins in addition to at least 20 additional proteins. The HMG 1 and HMG 2 are then fractionated from HMG 14 and HMG 17 by precitation with 10% TCA. The classification of the HMG proteins is unfortunately entirely empirical, meeting the amino acid composition criteria for non-histone proteins (high aspartic I and glutamic acids and acid/basic amino acids = 1.2–0.8, see Table I). Those non-histone proteins, which precipitate with 2% TCA, migrate on polyacrylamide gel electrophoresis with lower mobility than HMG proteins and therefore have been designated as low mobility group (LMG) proteins. [For review of these proteins, see (*36,37,104,125*).]

HMG proteins (HMG 1 and 2) appear to be present in all the tissues so far examined (*131,132*) and HMG 14 and/or 17 in several tissues including rabbit, calf, and chicken. All of these proteins have been fractionated to homogeneity. An estimate of the quantity of HMG proteins associated with chromatin is difficult to obtain in view of the fact that these proteins may become soluble in the process of their extraction (*26*). This may account for the early estimate of Johns *et al.* (*71*) of 10^5 compared to later reports of 10^6 copies of HMG proteins per cell (*42*). Garrard *et al.* (*33*) were among the first to use quantitative disc electrophoresis to determine the stoichiometry of total non-histone proteins from rat liver chromatin. They concluded that there are about 4.8×10^7 non-histone polypeptide molecules present per diploid genome and that these proteins ranged in frequency from 8.4×10^3 (the limit of sensitivity)

TABLE I

Molecular Weights of HMG Proteins from Calf Thymus

	MW derived from			B/A[b]	
HMG	SDS–polyacrylamide gel electrophoresis	Sedimentation equilibrium	Primary structure[a]	Calf	All
1	26,000 27,000	26,500	29,900	1.1	1.2
2	24,000	26,000 28,000	29,000	1.1	1.1
14	18,500		10,390	0.9	0.9
17	17,000		9,247	0.8	0.8

[a] Determined from incomplete sequence.
[b] Ratio of basic amino acid residues divided by acidic amino acid residues.

to 3.4 × 10^6 copies per diploid nucleus. The total number of HMG 1,2,14, and 17 proteins as estimated by Goodwin *et al.* (*42*) represents only 3% by weight of the histone or DNA content. The total of HMG 1 and 2 has been estimated to constitute 1–2% by weight of the histone or DNA content of chromatin based on the assumption that the histone concentration is about 2.7 × 10^8 molecules per diploid genome (*33*). The HMG proteins in hen oviduct may be as high as 4.8% (*169*).

The concentration of HMG proteins might be considered as high when compared to a gene regulatory protein (e.g., specific gene repressor), where only a very limited number of molecules per structural gene are required. If we assume, however, that there are 3.4 × 10^7 nucleosomes per diploid genome (2.7 × 10^8 histones/8 histones/nucleosome) and that four HMG proteins may be associated per nucleosome, then only 2.5% of the total nucleosome population could be associated with HMG proteins [(3.4 × 10^6 HMG 4) – (3.4 × 10^7 nucleosomes)]. If we assume, in addition, that postsynthetic modifications of HMG 14 and 17 represent a prerequisite for gene modulation, then the fraction of potentially active nucleosomes is even smaller.

B. Distribution between the Nucleus and the Cytoplasm

In contrast to the histones, the HMG proteins do not appear to be limited to the nucleus. Bustin's laboratory (*20,167*) made microscopic observations of cells stained with fluorescent-labeled, affinity-purified antibodies against HMG 1. They presented evidence that HMG 1 or other proteins that are immunologically cross-reactive with HMG 1 occur in both the cytoplasm and in the nucleus of several types of cultured mammalian cells (Chinese hamster, rat liver, and bovine trachea). These observations by themselves, however, cannot be accepted as definitive proof that HMG proteins are present in the cytoplasm, since HMG precursors or cross-reactive protein sequences also have the potential to interact with these antibodies. These authors have demonstrated that rat liver cytosol contains a protein which demonstrates the same electrophoretic mobility as HMG 1 in two gel systems. The results of Bustin and Neihart (*20*) have been corroborated by observations of Seyedin and Kistler (*152*), and Smith *et al.* (*156*), who found that, although the bulk of HMG 1 and 2 appear to be associated with chromatin from tissues of three rat organs, a measurable fraction of each protein also appeared in the cytoplasmic fraction.

Additional support for the concept that authentic HMG 1 is at least partially responsible for the cytoplasmic fluorescence observed with antibodies against HMG 1 comes from the experiments of Isackson *et al.*

(*60*). Using sequential chromatography on columns containing immobilized double- and single-stranded DNA, these authors have been able to purify an HMG protein from the cytosol of cultured rat hepatoma cells. In these studies, analyses of DNA-binding properties, electrophoretic mobilities, amino acid compositions, and immunochemical reactivities revealed that the protein was identical to HMG 1 isolated from purified chromatin of the same cell line. Experiments similar to those of Bustin and Neihart (*20*) have also been performed by Alfageme *et al.* *lop3*) in order to determine the distribution of D1 protein of salivary gland polytene chromosomes of the fruit fly *Drosophila melanogaster*. The D1 chromosomal protein resembles HMG proteins of higher vertebrates. Alfageme *et al.* studied the immunological specificity of anti-D1 serum by comparing its reactivity with *D. melanogaster* and *D. virilis* chromosome spreads to those with whole salivary glands. These authors concluded that D1 HMG-like protein is also abundant in both the cytoplasm and nucleus.

Bustin and Neihart (*20*) have suggested from their experiments that the HMG proteins, rather than functioning in the nucleus alone, may represent important structural elements in the entire cell. Since there are indications for cell cycle-dependent association of HMG proteins with chromatin and the involvement of HMG 1 and 2 proteins in DNA replication as helix-destabilizing proteins (*11,60,63,65,66*), it has been proposed that HMG 1 proteins shuttle between the nucleus and cytoplasm in response to a demand for DNA-unwinding proteins. Support for this concept has been provided by Rechensteiner and Kuehl (*135*), who introduced ^{125}I-labeled HMG 1 into the cytoplasm of HeLa cells and bovine fibroblasts by erythrocyte-mediated microinjection. Autoradiography of thin sections demonstrated that HMG 1 is rapidly associated with the nuclei and remains complexed with metaphase chromosomes. When uninjected HeLa cells or bovine fibroblasts were fused with ^{125}I-HMG 1-injected HeLa cells, or ^{125}I-HMG 1-injected fibroblasts, respectively, the labeled molecules equilibriated between the nuclei within 12 hr, indicating a dynamic equilibrium within living cells.

It is unlikely that the mechanism by which HMG 1 moves from a presumed cytoplasmic site of synthesis to the nucleus is due to selective transfer across the nuclear membrane. The translocation is more likely to involve a selective binding of HMG 1 protein to intranuclear structures (*135*). Kuehl *et al.* (*80*) incubated ^{125}I-labeled HMG-T (the trout testis analog of HMG 1) with trout testis nuclei under conditions approximating those *in vivo* and found that 70–90% of the radioactivity was bound to the nuclei and was not released at low ionic strength. These observations are consistent with the hypothesis that the equilibrium between cytoplasmic and chromatin-bound HMG 1 observed *in vivo* is a passive process that is dependent on the diffusion coefficient of HMG 1 between cytoplasm and nucleus. The translocation is modulated by its binding to chromatin.

C. Tissue and Species Specificity

Of all the non-histone proteins, the high mobility group proteins demanded the most attention because they were frequently observed as contaminants in salt-extracted histone fractionation procedures (*68*). Thus, Johns was one of the first to recognize the significance of these proteins in the early 1970s, and his laboratory has devoted much of their efforts on the fractionation and structure of this class of nonhistone proteins (*11,38–45,48,53,54,69–71,105,147,175–186*). A number of other laboratories have contributed to the fractionation of HMG proteins (*62,68,97,148*).

HMG proteins have been isolated from a variety of eukaryotic tissues and are widely distributed throughout eukaryotes. They have now been shown to be universal in all four eukaryotic "kingdoms": animals, plants, fungi, and protista, according to the five-kingdom classification of Whittaker (*199*). As reviewed here, a number of chromosomal non-histone proteins appear to be related to the HMG proteins. Calf thymus will be used as an HMG prototype.

1. Mammalian HMG Proteins

HMG proteins were first isolated from calf thymus by Goodwin and Johns (*382*) and Goodwin *et al.* (*38,44*) as a group of proteins which could be extracted from chromatin using 0.35 *M* sodium chloride. The HMG proteins isolated from calf thymus, HMG 1,2,14, and 17, have been well characterized and intensively studied [for review, see (*46*)]. The amino acid sequences of the four proteins have been determined (*177,182,184*), revealing a remarkable degree of similarity between the primary structural features of HMG 1 and 2, as well as considerable amino acid and sequence homology between the N-terminal regions of HMG 14 and 17 (Tables II–IV; Figs. 1 and 2).

The immunological relatedness between HMG 1 and other HMG proteins was determined by Bustin *et al.* (*20*), who elicited antibodies against HMG 1 from calf thymus in rabbits and measured the immunological distance via quantitative microcomplement fixation. From the immunological studies, they calculated a sequence homology between HMG 1 and 2 of 94%, between HMG 1 and 14 of 80%, and between HMG 1 and 17 of 84%. Using this method, but with a different set of antibodies

TABLE II

High Mobility Group Proteins: Type 1

Amino acid residue	Calf						Sheep thymus	Chicken[b]		Rat		Mouse myelo.[b]
	Thymus	Thymus	Kidney	Spleen	Liver	Fetal		Ery.	Ovi.	Liver	Hep.[b]	
Asx	11.0	12.9	10.9	10.5	11.6	11.0	11.7	11.5	10.6	10.6	11.0	10.1
Thr	3.0	2.1	2.9	2.7	2.6	2.5	2.6	2.7	4.5	3.7	3.4	3.6
Ser	5.3	6.0	5.3	5.1	5.3	5.5	5.9	5.8	7.4	8.1	6.1	7.3
Glx	17.8	18.2	16.6	17.1	16.5	18.5	17.8	18.1	15.8	17.2	19.0	20.2
Pro	5.2	5.7	6.5	6.0	6.3	6.2	4.1	5.1	8.4	4.8	5.7	6.3
Gly	5.2	5.5	6.2	6.3	5.7	5.6	4.7	5.9	7.0	9.5	9.3	7.0
Ala	8.2	8.3	8.9	8.2	8.9	9.0	8.6	8.7	8.3	8.2	9.5	9.9
Val	2.3	3.4	2.1	2.4	2.3	2.6	2.3	3.5	4.3	2.5	2.4	3.6
Cys	3.0	1.3	1.3	1.8	1.4	0.8	0.3	0.9	Tr[c]	N.D.[c]	N.D.	N.D.
Met	1.1	2.1	0.7	0.8	1.7	2.0	1.8	0.5	1.1	1.4	N.D.	N.D.
Ile	1.9	1.4	2.0	1.9	2.0	1.7	1.9	1.8	1.8	2.8	2.0	1.9
Leu	2.2	2.5	2.4	2.4	2.6	2.0	2.2	2.7	3.9	4.5	2.9	3.6
Tyr	2.8	2.5	3.1	3.1	3.2	3.2	2.9	2.2	2.8	2.8	1.4	1.6
Phe	5.6	4.3	4.3	5.4	4.7	3.7	3.9	4.3	2.9	2.9	3.6	2.0
Lys	16.8	18.6	19.9	18.8	18.7	20.4	23.9	18.2	14.3	13.3	17.0	18.0
His	1.7	1.5	1.5	1.5	1.7	1.4	1.1	1.7	1.7	1.3	2.5	0.7
Arg	5.5	3.6	3.9	4.6	4.3	3.9	4.0	4.0	4.8	5.3	3.8	4.0
Ref.[a]	*(132)*	*(164)*	*(132)*	*(132)*	*(132)*	*(89)*	*(157)*	*(131)*	*(169)*	*(28)*	*(11)*	*(62)*

[a] Reference number in parentheses.
[b] Ery, erythrocyte; ovi., oviduct; hep., hepatocytes; myelo., myeloma.
[c] Tr, trace; N.D., not detected.

and antigens, Romani *et al.* (*140*) were able to calculate a homology between HMG 1 and 2 of 92%. This homology has, of course, been confirmed by sequence homology (Figs. 1 and 2).

Two HMG proteins have been isolated which appear to be breakdown products of other nuclear proteins. HMG 3 has been observed in different calf tissues (*132*) and hen oviduct (*171*) and appears to be the N-terminal two-thirds of HMG 1 (*45*). This is confirmed by the observation that HMG 1 and 3 are immunologically indistinguishable (*19*). It has been suggested that HMG 8 is a degradation product of histone H1, comprising the N-terminal half of that protein (*45*).

Four additional HMG proteins have been extracted from calf thymus. They have been designated HMG 18, 19A, 19B, and 20 (*38,182*). HMG 18 is very basic, resembling both histones H1 and H5 in its lysine and alanine content. The amino acid compositions of proteins HMG 19A and 19B are similar. The high contents of basic and acidic amino acids of HMG 19A and 19B resemble HMG 1, 2, 14, and 17. HMG 19B also shares sequence similarities with HMG 17 (*38*).

HMG 20 is identical to ubiquitin, as determined by amino acid analysis, N-terminal sequence, electrophoretic mobility, and molecular weight. The molecular weight was determined as 7800, compared to 8400 for ubiquitin (*182*). The function of ubiquitin is not clear (*56,182*), but it was found that a non-histone chromosomal protein, A24, from calf thymus chromatin (*37*) consists of histone H2A covalently attached to ubiquitin via an isopeptide linkage (*36*). The amino acid sequence of the first 37 residues from the amino terminal end of A24 and ubiquitin are identical (*56,126*).

The four major HMG proteins have also been isolated from fetal calf thymus and have been shown to be very similar in amino acid composition to those of calf thymus (*92*) (Tables II and III). HMG 1 and 2 have been isolated from calf liver, spleen, and kidney (*132*). Their electrophoretic mobilities, amino acid analyses, and microheterogeneities upon isoelectric focusing are very similar to each other and to HMG 1 and 2 from calf thymus. The only detectable difference is that HMG 2 proteins from liver and spleen have slightly different mobilities in the CM-Sephadex ion exchange chromatography.

HMG 14 has been isolated from calf kidney (*42*), liver, and spleen (*132*). The calf kidney protein is very similar to HMG 14 from pig thymus, as judged by its amino acid composition (*42*). HMG 17 was isolated from both calf thymus and kidney, with amino acid composition similar to that of calf thymus HMG 14 (Table IV). The HMG proteins are similar, if not identical, in calf thymus, kidney, spleen, and fetal tissues, resembling the histones in their lack of tissue specificity (*132*).

TABLE III

High Mobility Group Proteins: Type 2

Amino acid residue	Calf						Chicken							
	Thymus	Thymus	Kidney	Spleen	Liver	Fetal	Ery.[b] HMG 2b	Ery. HMG 2a	Ery. HMG 2	Ovi.[b]	Thymus HMG 2a	Thymus HMG 2b	Rat	Mouse myelo.[b]
Asx	9.6	11.9	10.1	9.7	10.6	11.2	9.8	12.1	12.9	9.7	11.4	8.3	8.4	10.1
Thr	2.7	2.4	3.3	2.5	2.3	2.9	2.0	2.3	2.5	3.7	3.3	2.3	3.4	3.6
Ser	7.5	7.3	7.6	7.2	6.1	8.7	6.6	5.3	5.7	7.2	6.4	7.2	6.6	7.3
Glx	17.1	17.8	16.6	16.9	16.1	19.8	18.8	16.4	15.3	15.1	14.7	18.5	16.0	20.2
Pro	7.7	7.5	6.9	7.5	5.7	8.0	7.8	7.2	6.1	8.8	8.0	8.4	5.6	6.3
Gly	6.3	6.0	7.7	6.9	6.5	6.7	6.7	6.7	7.2	6.3	7.8	7.6	14.0	7.0
Ala	7.3	7.9	8.3	7.0	11.6	7.4	9.8	9.2	10.0	7.6	11.4	10.2	7.9	9.9
Val	2.1	3.2	2.6	2.0	3.3	2.2	2.5	3.4	3.5	6.7	3.2	2.1	4.4	3.6
Cys	1.6	1.2	N.D.[c]	1.6	N.D.	N.D.	0.7	0.8	0.4	Tr[c]	0.7	0.7	N.D.	N.D.
Met	1.2	1.6	1.2	1.2	0.2	N.D.	1.5	1.7	0.4	1.5	1.3	1.5	0.5	N.D.
Ile	1.4	1.2	1.7	1.6	1.8	1.0	1.4	1.5	1.7	1.5	1.3	1.4	3.2	1.9
Leu	2.1	2.5	3.0	2.4	2.4	2.0	2.1	2.0	2.3	2.8	2.1	2.1	4.2	3.6
Tyr	2.6	2.5	2.4	2.9	3.6	2.3	2.4	2.7	2.5	1.4	1.9	2.6	1.9	1.6
Phe	4.2	3.5	4.4	4.1	4.0	3.5	3.5	4.9	4.9	2.4	3.4	3.0	2.8	2.0
Lys	19.1	18.0	16.8	18.4	18.7	19.5	18.5	19.6	17.3	18.2	18.6	18.0	16.0	18.0
His	1.7	1.4	1.8	2.0	1.8	1.9	1.5	0.7	0.9	2.4	1.0	1.7	1.5	0.7
Arg	5.1	4.1	5.2	4.8	5.5	4.7	4.5	3.6	4.2	4.6	3.5	4.4	3.8	4.0
Ref.[a]	*(135)*	*(164)*	*(132)*	*(132)*	*(132)*	*(89)*	*(106)*	*(106)*	*(135)*	*(175)*	*(106)*	*(106)*	*(13)*	*(62)*

[a] Reference number in parentheses.
[b] Ery, erythrocyte; ovi., oviduct; hep., hepatocytes; myelo., myeloma.
[c] Tr, trace; N.D., not detected.

HMG proteins have been isolated from a variety of other mammalian species (Tables II–IV). All four major HMG proteins have been purified from mouse brain chromatin and show qualitative similarities in their electrophoretic mobilities with those isolated from calf thymus. HMG 1 and HMG 2 have also been isolated from mouse myeloma cells (*65*). Romani *et al.* (*140*) purified HMG 1 from mouse liver to homogeneity and elicited antibodies in rabbits against this protein. Antibodies were also elicited against HMG 1 from calf thymus. The interaction between the antibodies and the immunogens was measured by passive hemagglutination and by quantitative microcomplement fixation. The latter method revealed an immunological distance between the two HMG proteins corresponding to about 3% sequence difference.

Eight different HMG proteins have been detected in mouse L cells. Two of these, HMG-E and HMG-G, were purified (*9*) and probably correspond to HMG 14 and 17, respectively, in calf thymus.

Elgin and Bonner (*28*) isolated a protein (E) from rat liver which is similar to calf thymus HMG 1 and 2 in having about the same size (32 K dalton), the same N-terminal glycine, and a high content of basic and acidic residues, although it has a much lower lysine content and a more acidic p*K* (5.6) than HMG 1 and 2 from calf thymus (*47*). Bidney and Reeck (*11*) purified two proteins, NH-1 and NH-2, from cultured rat hepatoma cells (Tables II and III). They resemble HMG 1 and 2 in their extractability from chromatin at moderate NaCl concentrations, solubility in 2% trichloroacetic acid, and insolubility in 10% trichloracetic acid. Furthermore, their molecular weights (approximately 28 K) are close to those of HMG 1 and 2, and their amino acid compositions resemble those of HMG 1 and 2 in their high content of acidic and basic amino acids. Isackson *et al.* (*60*) isolated a protein from the cytoplasm of cultured rat hepatoma cells. Analysis of DNA-binding properties, electrophoretic mobility, amino acid composition (Table II), and immunochemical reactivities revealed that this protein is the same as HMG 1 isolated from chromatin by Bidney and Reeck (*11*).

All four major HMG proteins have been isolated from rabbit thymus (*40*). A protein termed P1, with similar amino acid composition to calf thymus HMG 1 and 2, has been isolated from sheep thymus (*157*).

2. *Avian HMG Proteins*

HMG 1 and 2 from duck red blood cells were first isolated by Sterner *et al.* (*162*) and shown to have the same electrophoretic mobilities as the corresponding proteins from calf thymus, Romani *et al.* (*140*) elicited antibodies against pure HMG 1 from calf thymus, mouse liver, and duck erythrocytes and showed by means of quantitative microcomplement

TABLE IV

High Mobility Group Proteins: Types 14 and 17

Amino acid residue	HMG type 14							HMG type 17								
	Calf			Pig thymus	Chick		Mouse HMG-G	Calf					Pig thymus	Chick		Mouse HMG-G
	Thymus	Kidney	Fetal		Ery.[b]	Ery.		Thymus	Thymus	Liver	Kidney	Fetal		Ery.[b]	Ery.	
Asp	8.1	6.8	8.8	5.6	11.2	9.3	5.3	12.0	12.4	11.6	9.6	10.1	11.7	10.5	9.1	9.6
Thr	4.2	5.2	3.5	2.8	4.2	4.6	5.2	1.2	1.7	2.3	3.0	1.1	1.4	3.1	3.0	2.9
Ser	7.8	7.8	7.8	7.3	5.1	5.2	8.2	2.3	3.0	5.5	4.5	2.1	2.3	4.1	4.3	5.4
Glu	17.1	15.4	16.5	16.3	13.2	15.6	16.0	10.5	10.8	12.2	12.8	10.0	9.4	12.5	11.7	10.8
Pro	8.5	9.3	8.5	8.3	10.2	10.5	9.1	12.9	10.0	15.7	12.2	11.8	12.1	10.1	12.1	11.2
Gly	6.5	7.6	6.5	7.2	6.6	5.6	11.8	11.2	11.7	11.8	10.7	11.2	10.8	9.4	10.0	10.6
Ala	14.5	3.9	14.5	18.2	18.6	18.0	9.9	18.4	18.0	11.6	16.3	18.4	18.3	16.5	17.2	14.2
Val	4.2	14.2	4.2	3.7	1.0	0.3	4.2	2.0	1.9	3.9	3.0	2.0	2.3	2.6	2.2	4.7
Cys	0.7	N.D.[c]	N.D.	N.D.	0.7	N.D.	N.D.	N.D.	N.D.	N.D.	N.D.	N.D.	N.D.	1.0	N.D.	N.D.
Met	N.D.	0.3	N.D.	0.4	0.3	N.D.	N.D.	N.D.	N.D.	0.2	0.1	N.D.	N.D.	N.D.	N.D.	N.D.
Ile	0.5	0.6	0.4	N.D.	N.D.	N.D.	Tr.[c]	N.D.	0.2	0.3	0.2	N.D.	N.D.	N.D.	N.D.	Tr.
Leu	2.0	2.4	1.9	1.9	1.0	1.1	2.8	1.0	1.8	1.6	1.5	1.0	1.1	1.2	1.2	2.1
Tyr	0.4	0.5	0.3	0.3	N.D.	0.2	Tr.	N.D.	N.D.	0.1	0.1	N.D.	0.2	N.D.	0.2	N.D.
Phe	0.6	0.7	10.5	0.1	N.D.	0.1	Tr.	N.D.	N.D.	0.2	0.1	N.D.	0.1	N.D.	0.1	N.D.
Lys	19.0	18.5	19.2	20.6	17.8	24.0	19.2	24.3	22.6	17.6	20.9	24.3	25.3	22.6	23.6	18.5
His	0.3	0.4	0.4	0.7	1.0	1.1	Tr.	N.D.	N.D.	0.3	0.2	N.D.	0.1	0.4	0.2	N.D.
Arg	5.6	5.0	4.6	6.5	4.3	4.1	8.0	4.1	4.1	5.0	4.8	4.1	4.9	5.1	4.6	9.7
Ref.[a]	*(47)*	*(45)*	*(89)*	*(45)*	*(131)*	*(183)*	*(9)*	*(47)*	*(163)*	*(132)*	*(132)*	*(92)*	*(45)*	*(126)*	*(183)*	*(9)*

[a] Reference number in parentheses.
[b] Ery., erythrocytes.
[c] Tr., trace; N.D., not detected.

fixation assays that these proteins have an immunological distance corresponding to 3% sequence difference. These results were confirmed by Romani *et al.* (*141*), who made use of a solid-phase radioimmunoassay for the serological analysis of chromosomal components and found that HMG 1 from duck erythrocytes and calf thymus are very similar.

In addition to HMG 1 and 2, Sterner *et al.* (*162*) also detected a third component migrating significantly ahead of the HMG 2 position in SDS polyacrylamide gel electrophoresis but comigrating with HMG 2 in the acetic acid/urea system. They designated it HMG-E and identified it as an HMG protein by amino acid analysis. Tryptic peptide maps suggested significant differences from HMG 1 of duck erythrocytes and from HMG 2 of calf thymus. Using immunomicrocomplement fixation assays, Romani *et al.* (*140*) determined that HMG-E and calf thymus HMG 1 are immunologically distinct, corresponding to a sequence difference of 13% of 19 position differences in the amino acid sequence. Rabbani *et al.* (*131*) suggested that HMG-E may correspond to HMG 2 from calf thymus. The R_fs of these proteins are the same in the acetic acid/urea gel (*162*). Serological analyses support the notion that there are significant differences between the two proteins (*140*). Romani *et al.* (*141*) used solid-phase radioimmunoassays in order to determine the serological similarity between the two proteins. They concluded that HMG-E is similar to although distinct from calf thymus HMG 2. HMG-E has not been found by Sterner *et al.* (*162*) in other avian and mammalian sources, suggesting that it is unique to avian erythrocytes. In partial contradiction, Gordon *et al.* (*49*) later showed that a corresponding protein is present in four chicken tissues, but they did not detect it in other duck tissues.

Six different HMG proteins have been found in chicken tissues and erythrocytes (Tables II–IV). Rabbani *et al.* (*131*) and Sterner *et al.* (*162*) isolated HMG 1 and 2 from chicken thymus and chicken erythrocyte nuclei and demonstrated electrophoretic identity with the corresponding proteins from calf thymus. Their amino acid compositions are very similar to those of HMG 1 and 2 from calf thymus. Furthermore, they have the same N-terminal amino acid, glycine, and a similar isoelectric-focusing pattern (*131*). Mathew *et al.* (*105*) showed that protein HMG 2 from chicken thymus and erythrocytes can be resolved by chromatography into two fractions, HMG 2a and 2b. The HMG 2a fraction predominates in erythrocytes while the reverse is true for the thymus. Both proteins show structural similarity to HMG 2 from calf thymus, as judged from their amino acid analyses (Table III), peptide maps and isoelectric-focusing patterns. Furthermore, HMG 2a is analogous to the protein HMG-E isolated by Sterner *et al.* (*131*) from duck erythrocytes. Gordon *et al.* (*49*) showed that a protein corresponding to HMG-E from duck erythrocytes was found in at least four other chicken tissues (liver, brain, thymus, and erythrocytes). Therefore, HMG 2a is not erythrocyte specific, as is HMG-E in the duck (*162*).

HMG 14 and 17 from avian tissues were first isolated by Rabbani *et al.* (*131*). These proteins have similar electrophoretic mobilities to the corresponding proteins in pig thymus. The amino acid composition resembles the thymus HMG 14 and 17 in having a high amount of acidic and basic residues, little or no aromatic residues, quite high values of proline, alanine, and glycine, and the same N-terminal residue, proline (*131*) (Table IV; Fig. 2). The amino acid composition was also determined by Walker and Johns (*183*). They calculated that avian HMG 14 shows at least 11 amino acid changes in comparison to HMG 14 from calf thymus, corresponding to a difference in the amino acid composition in the two species of 11%. HMG 14 from chicken erythrocytes is lacking the hydrophobic amino acids valine and leucine present in calf thymus HMG 14 and differs in four amino acids in the N-terminal sequence. Walker and Johns (*183*) made a similar comparison between HMG 17 from chicken erythrocytes and calf thymus and concluded that these proteins differ in at least five amino acids, corresponding to 5% sequence variation between the two species. Their results were confirmed by the work of Walker *et al.* (*186*), who determined the primary structure of HMG 17 from chicken erythrocytes. They observed five differences in the sequences between HMG 17 from chicken erythrocytes and calf thymus (Fig. 2). These differences are essentially conservative and do not appear to effect the overall architecture of the molecule (Fig. 2).

A protein designated 38K, with a molecular weight (38 K) exceeding that previously reported for HMG proteins, has been identified by Gordon *et al.* (*49*). It is extracted from avian nuclei with 0.35 *M* NaCl, is soluble at low acid concentrations, and partitions with the HMG proteins in acetone–HCl.

All four major HMG proteins have been isolated from hen oviduct by Teng *et al.* (*169,170*). The molecular weights of HMG 1 and 2 have been determined as 28,000 and 27,000, respectively. Their amino acid compositions are very similar to those of the corresponding proteins in calf thymus (Tables II–IV). HMG 3, which probably is a degradation product of HMG 1 (*170*), was also observed in hen oviduct.

Another HMG protein with a high molecular weight of 95 K, designated 95K protein, has been isolated by Teng *et al.* (*169,170*). This protein constitutes a large proportion of the population of HMG proteins in hen oviduct and appears to be tissue specific in that it is not found in liver, brain, or thymus from chick or calf. This is supported by the observation of Teng and Teng (*171*) that antibodies specific for the 95K protein did not react with other HMG proteins of chick oviduct, brain, or liver. The amino acid composition of the 95K protein, especially the high content of acidic and basic amino acids, identifies it as an HMG protein, and its isoelectric point (7.2–7.6) is comparable to that of other HMG proteins. The molecular weight and lack of specificity appears at this time to violate the criteria for HMG proteins.

3. *Pisces HMG Proteins*

A protein similar to HMG 1 and 2 (protein R) has been isolated from rainbow trout testis by Huntley and Dixon (*57*) and was later renamed HMG-T by Watson and co-workers (*191*). A protein (protein T) with the same electrophoretic mobility and similar amino acid analysis has also been isolated from trout liver (*134*). HMG-T has a molecular weight of approximately 28.7 K as determined by SDS gel electrophoresis. It was extracted with 0.35 *M* NaCl (*57*) and migrates similarly to HMG 1 and 2 from calf thymus on polyacrylamide gels (*191*). The amino acid analysis (Table V) reveals a high content of both acidic and basic residues, but also some large differences from the HMG proteins of calf thymus (*57,191*). Because of its high glycine content, Goodwin *et al.* (*47*) suggested that HMG-T may have very different secondary and tertiary structure from the calf thymus HMG proteins. HMG-T has two cysteine residues which are capable of forming intramolecular disulfide bonds. Its amino acid sequence for the first 29 residues shows that it is distinct from all trout testis histones yet sequenced. Yet, the N-terminal sequence bears considerable similarity to the HMG 1 and 2 proteins of calf thymus, identical sequences occurring through-out the N-terminal region. In contrast to the HMG 1 and 2 proteins, HMG-T appears to be a single polypeptide, showing no microheterogeneity upon isoelectric focusing due to sequence heterogeneity as with HMG 1 and 2 of calf thymus (*39,179*).

Another HMG protein has been isolated from rainbow trout testis by Wigle and Dixon (*200*) which is also present in other trout tissue such as liver, spleen, kidney, and erythrocyte. Because of its basic net charge and its high electrophoretic mobility, it was first classified as a histone (histone T) (Table V) and was renamed H6 in order to conform with the histone nomenclature adapted at the Ciba Symposium in 1974 (*17,134*). Protein H6 can be extracted from chromatin with 0.35 *M* NaCl, appearing together with the two other non-histone proteins, protein R (*57*), which was renamed HMG-T (*190,191*), and protein S, which was identified as ubiquitin (*190*). It can also be extracted with 5% trichloroacetic acid along with histone H1 (*192*). Protein H6 comprises about 10% of the histones in trout testis chromatin. Peptide maps of tryptic peptides

TABLE V

Amino Acid Analysis of HMG Proteins in Diptera and Pisces

	Fruit fly			Trout testis		Trout liver			
Amino acid residue	*Drosophila melanogaster* D1	*Ceratitis capitata* CMC 1	Flounder H6	H6	HMG-T	HMG-C	HMG-D	HMG-E	HMG-T
Asx	16.0	17.2	5.1	6.7	11.3	6.0	8.0	6.1	10.1
Thr	3.1	4.8	6.6	1.6	3.0	2.6	4.1	1.6	4.6
Ser	10.6	7.7	4.5	5.6	4.5	4.4	4.3	5.1	8.8
Glx	10.7	10.2	9.7	6.1	9.1	23.8	21.9	6.6	7.3
Pro	8.0	7.2	9.5	12.3	7.6	10.9	8.4	10.0	8.2
Gly	13.2	11.4	5.1	7.4	17.4	2.8	3.3	6.2	10.3
Ala	9.6	8.3	23.3	25.4	8.3	16.0	16.6	26.5	11.3
Val	4.8	2.3	4.8	3.4	3.8	4.2	2.5	4.2	3.9
Cys	0.4	N.D.[b]	N.D.	N.D.	0.8	N.D.	N.D.	N.D.	N.D.
Met	0.2	0.4	N.D.	N.D.	1.9	0.1	0.3	N.D.	1.6
Ile	1.7	1.8	1.7	N.D.	1.5	0.6	N.D.	N.D.	1.3
Leu	1.5	1.5	1.4	1.2	2.6	0.8	1.0	1.2	2.6
Tyr	0.7	0.8	N.D.	N.D.	2.3	0.2	N.D.	N.D.	1.8
Phe	<0.1	Tr.[b]	N.D.	N.D.	3.4	0.1	0.1	N.D.	2.9
Lys	11.5	17.6	22.1	23.1	15.5	19.6	23.7	23.4	17.1
His	1.3	0.6	N.D.	N.D.	0.4	0.2	0.9	N.D.	1.2
Arg	7.4	8.2	6.3	7.2	5.3	4.6	4.2	8.6	6.4
Ref.[a]	*(139)*	*(30)*	*(75)*	*(200)*	*(191)*	*(134)*	*(134)*	*(134)*	*(134)*

[a] Reference number in parentheses.
[b] Tr., trace; N.D., not detected.

showed 13 unique peptides and thus ruled out that it could have originated by degradation of one of the nucleosome core histones (*200*). The amino acid analysis revealed that the protein is devoid of seven amino acids, namely the aromatic and sulfur-containing amino acids and isoleucine. The amino acid composition of H6 was described as having features in common with histone H1, histone IIb_2 (*200*), and HMG 14 and 17 from calf thymus (*38*). The final proof that H6 is related to HMG 17 from calf thymus came from its sequence analysis (Fig. 2). Not only does its N-terminal region (*57*) demonstrate regions of homology with that of HMG 17 (*47,179,184*), but the total sequence analysis of the protein shows that 75% of the positions of the 69-residue long H6 are identical compared to the 89-residue long HMG 17 and that a further 16% are related by a single base change in their codons. The total sequence homology between the two proteins is close to 90%. The molecular weight of H6, derived from the amino acid sequence, is 7.2 K, making it the smallest chromosomal protein so far described with the exception of the sperm protamines (*192*). The molecular weight was previously determined by SDS–polyacrylamide gel electrophoresis as 14.5 K.

Rabbani *et al.* (*134*) have isolated two additional proteins (proteins C and D) from trout liver and have detected corresponding proteins with the same electrophoretic mobilities in trout testis. Proteins C and D have similar electrophoretic mobilities to calf thymus HMG 14 and 17 on SDS–polyacrylamide gels and run closely together on acetic acid polyacrylamide gels in the position of calf thymus HMG 14. The amino acid sequence for the first four N-terminal residues of protein D is identical to the common N-terminal sequence of HMG 14, HMG 17, and H6, and the sequence in protein C has three out of four amino acids in common with that sequence. It has been suggested that H6 is the analogous protein to HMG 14 and 17 of mammals, in the same way as HMG-T resembles HMG 1 and 2). In contrast, Rabbani *et al.* (*134*) suggested from their observations that proteins C and D correspond to mammalian HMG 14 and 17 and that H6 is a truly trout-specific protein (not found in calf tissues). It is more likely, however, that H6 corresponds to HMG 17 and that proteins C and D are both variants corresponding to HMG 14. Proteins corresponding to HMG-T and H6 from trout testis have also been isolated from winter flounder testis by Kennedy and Davies (*75*). These proteins have similar R_fs on acid/urea polyacrylamide gels (*127*), yet the mobilities of HMG-T from flounder and trout differ on Triton gels (*205*). Flounder H6 has very similar amino acid composition to trout H6. In both molecules, lysine and alanine each comprise almost one-quarter of the amino acid residues, with no histidine, sulfur-containing, or aromatic residues (Table V). The main difference from its trout counterpart is that it contains leucine and a higher amount of threonine and glutamic acid residues (*75*).

4. *Diptera HMG Proteins*

A non-histone chromosomal protein, D1, has been isolated from embryos of the fruit fly *Drosophila melanogaster* by Alfageme and coworkers (*3*) (Table V). It is similar to HMG proteins from vertebrates by virtue of its extractability from chromatin at low salt, solubility in dilute perchloric and trichloracetic acid (*139*), quantity in nuclei, high contents of acidic and basic amino acids, and low contents of hydrophobic amino acids. Protein D1 differs from mammalian HMG proteins, however, in having aspartic rather than glutamic acid as its most abundant amino acid, a lower electrophoretic mobility, and higher molecular weight, which was estimated by SDS–gel electrophoresis at about 50 K.

Franco *et al.* (*30*) have isolated a protein from the fruit fly *Ceratitis capitata* with similar electrophoretic mobility to HMG 1 and 2 from calf thymus. Its amino acid composition bears some resemblance to calf thymus HMG 1 and 2 and trout HMG-T.

5. *Plant HMG Proteins*

Four proteins have been extracted from wheat germ chromatin with 0.35 *M* NaCl by Spiker *et al.* (*160*). These proteins, labeled HMG-a, -b, -c and -d, are soluble in dilute trichloroacetic acid and have electrophoretic mobilities on SDS–polyacrylamide gel electrophoresis similar to those of HMG proteins from mammalian tissues. The amino acid composition of HMG-b (Table V) is characteristic of HMG proteins in having a high content of acidic and basic residues.

6. *Fungal HMG Proteins*

Spiker *et al.* (*160*) have isolated a protein from chromatin of the yeast *Saccharomyces cerevisiae* which has similar electrophoretic mobility to the mammalian HMG proteins (Table VI). This protein, termed HMG-a by Weber and Isenberg (*194*), is not extracted by salt, but rather by 0.25 *N* HCl. It was first observed as an extra band in preparations of core histones (*98,99*). Its amino acid analysis reveals the high content of acidic and basic residues characteristic of HMG proteins (Table VI). Sommer (*159*) has isolated a protein (band I protein) from *S. cerevisiae* by extraction with 0.2 *N* HCl resembling HMG 2 from calf thymus in amino acid composition and molecular weight. It has a low lysine content, similar to histone H1, and is not extracted by 5% perchloric acid (Table VI). It is present in the cell in higher concentration than the mammalian HMG proteins and constitutes a major component of the basic chromatin proteins, together with the core histones. Sommer (*159*) suggested that

TABLE VI

Amino Acid Analysis of HMG Proteins in Plants and Fungi

Amino acid residue	Wheat HMG-B	*Saccharomyces cerevisiae* Band I	HMG-a	Fraction S1	Fraction S2	Fraction S3	Fraction S4
Asx	11.7	10	8.5	7.8	9.9	6.6	9.5
Thr	1.4	2.9	3.4	5.9	5.5	7.3	5.6
Ser	8.3	7.1	7.5	10.4	9.6	9.9	8.0
Glx	12.7	15.5	15.6	9.7	12.5	10.2	11.7
Pro	4.7	5.9	5.9	N.D.	N.D.	N.D.	N.D.
Gly	13.7	7.7	3.6	7.7	13.5	5.6	11.4
Ala	15.9	7.4	8.8	9.9	9.1	10.8	9.5
Val	3.6	3.2	2.3	5.5	8.3	5.5	4.8
Cys	N.D.[b]	Tr.[b]	N.D.	0.7	N.D.	N.D.	N.D.
Met	N.D.	<1	N.D.	N.D.	N.D.	N.D.	N.D.
Ile	1.5	5.0	6.5	3.8	0.6	5.7	3.0
Leu	2.2	7.3	7.5	6.5	8.5	5.0	6.7
Tyr	0.8	4.2	4.4	2.0	0.7	4.9	0.7
Phe	2.7	3.0	2.8	3.8	1.7	4.7	5.7
Lys	17.0	14.7	15.9	19.2	8.8	12.2	14.1
His	1.3	1.3	1.3	4.9	7.8	7.2	4.6
Arg	2.5	6.7	5.5	3.5	3.9	4.7	5.1
Ref.[a]	(*160*)	(*159*)	(*160*)	(*194*)	(*194*)	(*194*)	(*200*)

[a] Reference numbers in parentheses.
[b] Tr., trace; N.D., not detected.

this protein might represent an early form of histone H1 and of HMG proteins. Weber and Isenberg (*194*) noticed that HMG-a and band I protein have very similar amino acid compositions and similar mobilities in SDS and acetic acid/urea gels. They suggested that the proteins are identical.

Four other proteins have been isolated by Weber and Isenberg (*194*) from *S. cerevisiae* (Table VI). Only three of them are extractable from chromatin by 0.35 *M* NaCl. The fourth has been extracted by 0.25 *N* HCl, indicating that extractability at low salt is not a necessary criterion for the identification of HMG proteins. Proteins S1, S3, and S4 have high and nearly equal amounts of acidic and basic residues characteristic of HMG proteins. S2 differs in having fewer basic residues. They also differ in their mobilities on SDS and acetic acid/urea gels, although S3 does not differ much from HMG17 in this respect. Three proteins (bands 1,3, and 4) have been extracted with $CaCl_2$ from nuclei of the myxomycete, *Physarum polycephalum* (*123*). Their amino acid compositions resemble HMG proteins in their high content of acidic and basic amino acids (*159*).

Charlesworth and Parish (*25*) isolated a protein (fraction 1) from the cellular slime mold *Dictyostelium discoideum* by $CaCl_2$ extraction which is similar to band I protein from *S. cerevisiae*. It resembles mammalian histone H1 in being soluble in perchloric acid and comigrates with H1 during acetic acid/urea gel electrophoresis, but not in the SDS–gel system. It is similar to HMG proteins in its high content of acidic and basic residues and its relatively low lysine/arginine ratio. Its molecular weight (about 20 K) is similar to that of HMG proteins. It differs from HMG proteins in its concentration in the cell, being the most prominent band among the basic nucleoproteins. The tryptic peptide map of the protein revealed striking correspondence to that of H1. In comparison to H1 from calf thymus, it contains methionine and more histidine, less alanine and proline, and a high content of acidic residues (Table VII).

7. *Protista HMG Proteins*

Hamana and Iwai (*51*) have extracted two histone-like proteins (LG-1 and LG-2) with 0.25 *N* HCl from the chromatin of each of two strains, V1 and GL (Table VII), of the ciliated protozoan *Tetrahymena pyriformis*. LG-1 has previously been detected by Hamana and Iwai (*52*) in histone preparations from the GL strain. Both components belong to the class of HMG proteins, based on several criteria, demonstrating that HMG proteins also exist in the fourth eukaryotic kingdom, Protista (*51,199*). These proteins are extractable from chromatin with 0.35 *M* NaCl or 0.5 *N* $HClO_4$ and have low molecular weights. The molecular weights of LG-1s of the V1 and GL strains have been estimated at 13 and 12.5 K, respectively. The molecular weight of LG-2 of the GL strain is about 15 K. These proteins are present in chromatin in relatively high concentration and appear to be approximately equimolar to histone H1. They have high mobilities on polyacrylamide gel electrophoresis, LG-1 resembling calf thymus HMG 17 and LG-2 resembling HMG 1 in electrophoretic mobility. They are rich in acidic and basic residues and have a very high lysine content. In respect to their amino acid compositions, LG-1 is rather similar to HMG 1 and 2, while LG-2 resembles HMG 14 and 17 (Table VII).

8. *Prokaryotic-like HMG Proteins*

It is still uncertain that HMG proteins exist in prokaryotes (Table VII). Spiker *et al.* (*160*) noted that several DNA-binding proteins isolated from prokaryotic sources resemble HMG proteins in their high content of acidic and basic amino acids. These authors suggested that HMG proteins may exist in all biological species. Among these proteins are a nucleo-

TABLE VII

Amino Acid Analysis of HMG Proteins in Procaryotes and Protista

Amino acid residue	*Dictyostelium discoideum* fraction 1 protein	*Physarum polycephalum*			*Tetrahymena pyriformis*			*Thermoplasma acidophilum*	*E. coli* HU protein	Blue-green alga *Aphano capsa* HU protein
		Band 1	Band 3	Band 4	V1 strain LG-1	GL strain LG-1	LG-2			
Asx	9.8	5.4	5.9	4.7	7.5	7.1	1.9	6.0	8.1	8.0
Thr	8.9	5.4	2.0	5.1	3.5	3.7	3.8	5.7	6.0	6.0
Ser	7.8	6.2	2.6	4.1	5.9	5.2	6.9	7.7	4.4	8.1
Glx	10.6	9.1	8.9	9.4	15.1	14.5	15.3	13.8	9.6	8.1
Pro	6.8	10.0	3.3	3.8	6.0	6.8	5.7	2.6	3.0	4.5
Gly	7.4	6.7	9.5	8.0	5.7	4.7	7.8	6.6	7.4	7.5
Ala	10.5	17.4	7.2	10.6	7.7	7.5	11.9	8.0	16.3	11.0
Val	3.7	3.3	4.9	7.4	3.2	4.8	2.4	9.4	6.0	10.0
Cys	N.D.[b]	N.D.	N.D.	N.D.	N.D.	N.D.	N.D.	<0.2	N.D.	N.D.
Met	1.0	N.D.	N.D.	N.D.	1.4	1.7	0.5	0.3	1.5	2.5
Ile	3.0	2.5	4.9	3.8	2.8	3.1	3.0	7.7	6.0	3.3
Leu	3.0	3.3	7.2	5.3	3.9	3.9	5.2	3.7	6.6	3.4
Tyr	1.4	N.D.	N.D.	1.8	3.3	4.0	N.D.	1.0	N.D.	N.D.
Phe	0.6	1.2	0.9	1.8	2.8	2.9	1.8	5.0	3.0	4.0
Lys	19.2	18.2	21.9	17.9	26.5	26.9	23.5	15.8	14.0	14.0
His	2.7	2.5	4.9	3.8	2.9	2.2	1.0	<0.2	1.5	1.5
Arg	4.3	1.7	9.5	5.0	1.7	1.3	3.3	6.7	5.1	5.4
Ref.[a]	*(25)*	*(123)*	*(123)*	*(123)*	*(51)*	*(51)*	*(51)*	*(151)*	*(142)*	*(54)*

[a] Reference numbers in parentheses.
[b] N.D., not detected.

protein isolated from *Thermoplasma acidophilum* (*151*) and DNA-binding proteins (HU) from *Escherichia coli* (*142*) and cyanobacteria (blue-green algae) (*54,143*). These proteins have previously been described as "histone-like" proteins, but may be designated as HMG proteins based on their amino acid compositions. Spiker *et al.* (*160*) noted that this classification only implies a semantic preference as long as the functions of the HMG proteins are now known.

IV. Fractionation and Characterization of Acetyltransferases

An understanding of the biological role of acetylation of nucleosomal proteins in eukaryotic cells will require detailed biochemical analysis of the enzymes. This information should include an analysis of the number of distinct acetyltransferase activities in different tissues, the physicochemical properties of the enzymes, and data pertaining to the mechanism of their regulation *in vivo*.

The highly selective nature of nucleosome acetylation *in vivo* produces substantial nucleosome heterogeneity (*27,153,174*). These studies suggest that some mechanism must exist which allows the acetyltransferases to distinguish certain classes of nucleosomes as proper substrates within a specific tissue. It is clear that an investigation of the acetyltransferase activities in eukaryotic cells is essential to provide an understanding of the function of these highly integrated DNA–protein interactions.

Some recent experimental observations on mammalian histone acetyltransferases are reviewed in Table VIII. The major strategy that has evolved to fractionate the acetyltransferases has utilized the extraction of crude nuclear preparations or whole tissues in buffers containing high salt concentrations (0.4–2 *M*), followed by chromatography of the crude extracts using ion-exchange, hydroxylapatite, gel filtration, and DNA cellulose chromatography.

The assay for acetyltransferase activity has been facilitated by a filter binding technique (*55*). The requirement for salt concentrations in excess of 0.5 *M* KCl in the extraction buffers strongly implies that the nuclear histone acetyltransferase activity is bound to chromatin. This hypothesis is substantiated by the work of Belikoff *et al.* (*10*), who found an increase in the amount of extractable acetyltransferase activity in calf thymus nuclei as the salt concentration in the extraction buffer was raised from 0.5 to 2 *M*. The fact that Libby (*94,95*) has been able to extract most of the activity from calf and rat liver in 0.4 *M* KCl prompted Belikoff *et al.* (*10*) to suggest that this represents the consequence of endogenous liver nuclease action, rendering the enzyme activity more easily accessible to extraction. Our laboratory has confirmed the observation that most of the acetyltransferase activity is extracted from hog liver nuclei at 0.5 *M* KCl (*73*).

More definitive evidence supporting the concept that the nuclear acetyltransferase activity is closely associated with chromatin in a salt-labile manner has recently been provided by Bohm *et al.* (*15*), who demonstrated that a histone-specific acetyltransferase is closely associated with bovine lymphocyte nucleosomes. The acetyltransferase activity sedimented with the nucleosomal particles in the absence of salt in sucrose gradients, whereas the activity sedimented much slower than the nucleosomes in the presence of 0.5 *M* NaCl. Dialysis of salt-treated nucleosome preparations against low-salt buffers led to a reassociation of the acetyltransferase activity with the nucleosomes, demonstrating the reversible nature of the salt lability of the acetyltransferase activity.

The number of distinct acetyltransferase activities not only appears to vary from organism to organism, but also within different tissue types of the same organism. Thus, Libby (*95*) demonstrated two distinct acetyltransferase activities in calf liver nuclei resolved on the basis of their molecular weight, *in vitro* substrate specificity, and heat inactivation profiles. In contrast to the observations of Belikoff *et al.* (*10*), who have been unable to detect more than one acetyltransferase activity in calf thymus, Sures and Gallwitz (*166*) have demonstrated three distinct activities. As shown in Table VIII, acetyltransferase B is most likely the cytoplasmic enzyme responsible for acetylating nascent amino terminal serine residues of histone H4. The data on acetyltransferase C, which appears to have the same substrate specificity as acetyltransferase A and a molecular weight in excess of 200,000, is suggestive of an aggregated form of acetyltransferase A (*15*).

The work of Böhm *et al.* (*15*) on the bovine lymphocyte acetyltransferase activity demonstrated the presence of two distinct nucleases. The acetyltransferase A, which acetylates all five histones *in vitro,* is similar in substrate specificity to the nuclear enzyme previously described by Sures and Gallwitz (*166*) in calf thymus. A new acetyltransferase activity, termed DB by these authors, was distinguished by its ability to acetylate the nucleosomal core histones (free or associated in nucleosomes) but not histone H1.

In view of the difficulty in obtaining sufficient quantities of purified acetyltransferases, detailed biochemical analysis will have to be deferred. The estimated molecular weights range from 70 K for the calf thymus nuclear enzyme to 175 K for the acetyltransferase B activity of calf liver nuclei. The enzyme activities are highly sensitive to sulfhydryl oxidation, as evidenced by the strong inhibition exhibited by *p*-chloromercuribenzoate and/or *N*-ethylmaleimide, as well as a diminished enzyme activity

TABLE VIII

Fractionation of Mammalian Histone Acetyltransferases

System	Extraction method	HAT activity	Purif. factor (X)	MW (K)	Substrate: Specificity	Substrate: Inhibition	References
Calf liver nuclei	0.4 *M* KCl extract of crude nuclei	A	6,000	150	H4>H2A> H3>H2B>H1	EDTA(weak) *p*-CMB(strong)	(*94*)
		B	7,000	175	H4–H3>H2A> H2B>H1	EDTA(weak) *p*-CMB(strong)	(*94*)
Rat liver nuclei	0.4 *M* KCl extract of crude nuclei	A B	500	N.D.[a]	H3>H2A–H2B> H1,H4	A:CT DNA(strong) B:CT DNA(weak)	(*95*)
Calf thymus	Sonication of whole tissue homogenate in 1 *M* ammonium sulfate	A B C	200 200 200	120 98 >200	All 5 histones H4,H2A All 5 histones	Strong inhib. by oxidizing agents	(*166*)
Calf thymus nuclei	PEG precipitation of enzyme activity from 2 *M* NaCl extract of crude nuclei	A	30,000	70	Histones: H4>H3>>H2B >H2A Nucleosomes: H4>H2A–H2B> >H3	Inhib. by salt, NEM, urea	(*10*) (*31*)
			—				

Hog liver nuclei	0.5 *M* KCl extract of crude nuclei	A	2,000	N.D.	Histones: H3>H4>H2B> H2A Nucleosomes: H4–H2B>H2A> >H3	Salt(weak) Divalent cation(strong) CT DNA(strong)	(*73,74*)
Bovine lymphocyte nuclei	0.5 *M* KCl extract	A		N.D.	All 5 histones		(*15*)
		B		N.D.	H4		
		DB		N.D.	Histones: H3>H4>H2B> H2A Nucleosomes: H2B,H3,H4> H2A	Salt 50% at 100 m*M* NaCl 90% at 250 m*M* NaCl *N*-ethylmaleimide 95% at 10 μ*M*	

[a] N.D., not determined.

when the isolation procedure is performed in the absence of DTT. We have demonstrated a 50% inhibition of the hog liver nuclear acetyltransferase activity by 100 m*M* KCl, millimolar concentrations of divalent cations, and 40 μg/ml of calf thymus DNA (*73,74*).

The substrate specificities of the isolated histone acetyltransferase activities are of particular interest in the characterization of the enzyme activities, since this information is fundamental to our understanding of *in vivo* function. The cytoplasmic enzyme, which is coextracted with the nuclear form in whole tissue preparations, may be recognized by its preference for histones H4 and H2A. The nuclear enzymes will generally acetylate all five histones *in vitro,* although acetylation levels of histone H1 are generally quite low compared to the core histones.

Although much of the substrate specificity data to date has been obtained using free histones in the incubation mixture, it is likely that more meaningful information will be gleaned using nucleosome preparations. Belikoff's calf thymus nuclear enzyme, which at this time appears to be the most highly purified preparation available, shows a clear difference in substrate specificity depending on whether free histone or purified mononucleosomes are used as substrates (Table VIII). Work in our laboratory using the hog liver nuclear enzyme has also demonstrated a change in substrate specificity, depending on whether free chicken erythrocyte histones or highly purified chicken erythrocyte nucleosome monomers are used as substrates for the *in vitro* acetylation reactions (Table VIII and Fig. 4) (*74*).

Fig. 4. Substrate specificity of the acetyltransferase. Lanes 1 and 2: Purified chicken erythrocyte HMG 14 (100 μg, lane 1) and HMG 17 (lane 2) were incubated with 200 μl of the hydroxylapatite pool (100 μg protein) and 100 nCi of [1-^{14}C]acetyl-COA at 22°C for 2 hr in a total volume of 0.400 ml. The reaction mixture was precipitated with ice-cold 100% (w/v) TCA to 25% and spun for 15 min in a Beckman Microfuge B. The pellet was washed with acidified acetone followed by acetone, dried under nitrogen, and solubilized in electrophoresis buffer. Lane 3: Unreacted chicken erythrocyte core histones (20 μg). Lanes 4–7: Chicken erythrocyte core histones (20 μg) were incubated with 40 μl of the hydroxylapatite pool (20 μg protein) and 20 nCi of [1-^{14}C]acetyl-COA at 22°C in a total volume of 80 μl. At each time point, the reaction was terminated by the addition of 50% acetic acid to 5%. The solution was evaporated to dryness on a Speed-Vac concentrator (Savant) and the residue was solubilized in electrophoresis buffer. Lanes 8–11: Purified chicken erythrocyte core nucleosomes (40 μg) were incubated in the same reaction mixture used for the core histones. At each time point, the reaction mixture was added to four volumes of absolute ethanol at −20°C. After storage at −20°C for 12 hr, the samples were spun for 15 min in a Beckman Microfuge B. The pellets were dried under nitrogen and solubilized in electrophoresis buffer. Triton/acid/urea polyacrylamide gel electrophoresis was performed on a 12% gel for 10.5 hr at 5 mA according to Zweidler (*209*). The sample buffer contained 1% salmon sperm protamine sulfate (Sigma) to release the histones from the nucleosome cores. The gel was stained with Coomassie Blue (panel A) and treated for fluorography as in Fig. 3. The fluorograph (panel B) was exposed for 14 days.

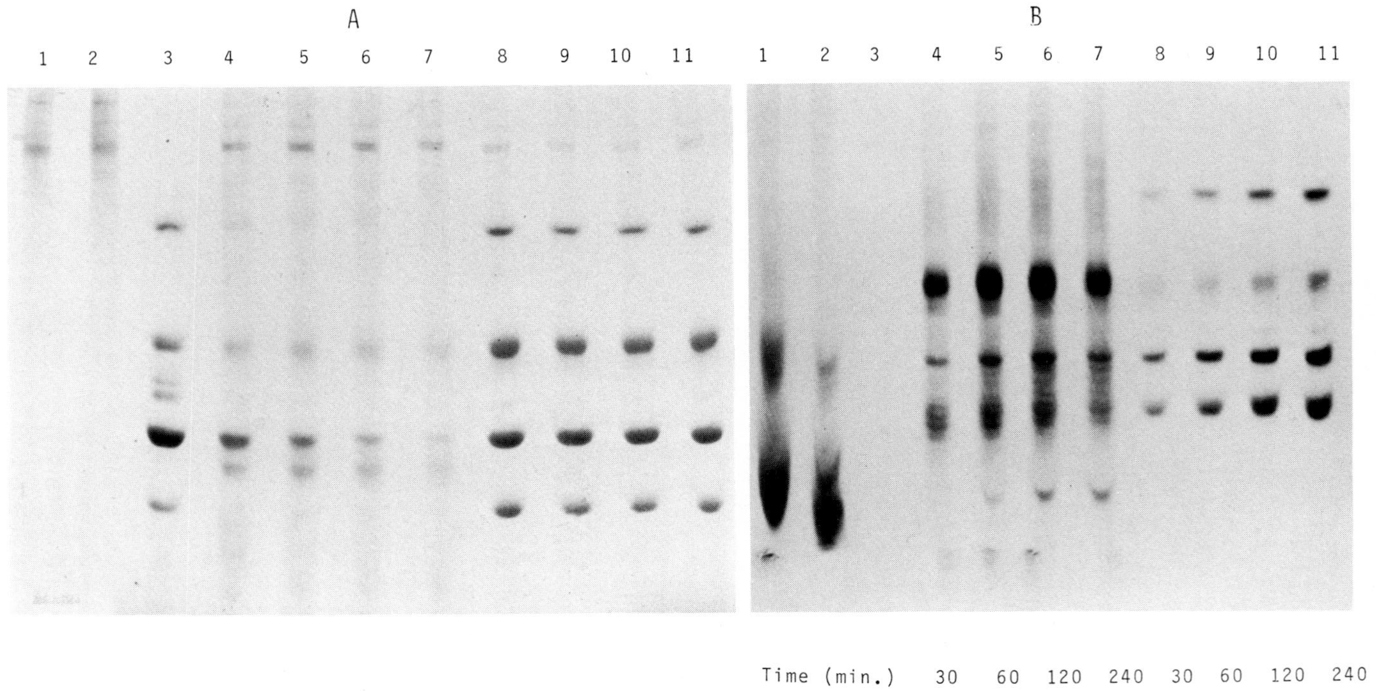
A
B
1 2 3 4 5 6 7 8 9 10 11
1 2 3 4 5 6 7 8 9 10 11
Time (min.) 30 60 120 240 30 60 120 240

As shown in Fig. 4, histone H3 is acetylated quite readily by the hog liver enzyme when the histone is free in solution, whereas H3 acetylation is very poor when the H3 molecules are complexed in nucleosome cores. Garcea *et al.* (*31*) found a similar result using their highly purified calf thymus enzyme. These authors suggest that some nuclear factor(s) may be lost during the preparation of the enzyme. It is proposed that these factors may be responsible for regulating the *in vivo* acetylation of histone H3. Confirmation of this hypothesis awaits further experiments.

Figure 4 also illustrates the *in vitro* acetylation of free HMG 14 and HMG 17 by the hog liver acetyltransferase. To the best of our knowledge, this represents the first demonstration of HMG 14 and 17 acetylation *in vitro* by histone acetyltransferase (*74*). Our preliminary experiments have been unable to demonstrate HMG 14 and/or 17 acetylation, however, when these proteins are stoichiometrically bound to chicken erythrocyte nucleosome core particles according to Mardian *et al.* (*100*). The presence of HMG 14 on the core particle does not qualitatively alter the acetylation pattern of the core histones (*74*) under the conditions of our standard nucleosome incubation mixture. It should be emphasized that our preparations show no evidence of deacetylase activity.

Many innovative experiments are clearly indicated in order to understand the complex biochemical details of nucleosomal protein acetylation. Future investigations should include, in addition to continued characterization of the acetyltransferases themselves, fractionation of the deacetylase enzymes and an analysis of the regulatory role of nucleosomal deacetylation in transcriptional control. The selective inhibition of the deacetylase activity by millimolar concentrations of sodium butyrate affords an opportunity for extensive *in vitro* studies of the delicate balance between enzymatic acetylation and deacetylation. Only in this manner will we be able to approach an understanding of the regulation and biological significance of nucleosomal postsynthetic modifications in the eukaryotic nucleus.

V. Proposed Mechanisms of HMG-Induced RNA Transcription

Electrophoresis of chromatosomes and nucleosome core particles reveals an impressive degree of heterogeneity, in which it is possible to predict a microheterogeneity with 648 possible unique octamers, if ubiquitin (*176*), HMG 14 and 17, and histone postsynthetic modifications are included. An illustration of this nucleosome microheterogeneity is presented in Fig. 5A, which shows diagramatically an example of six

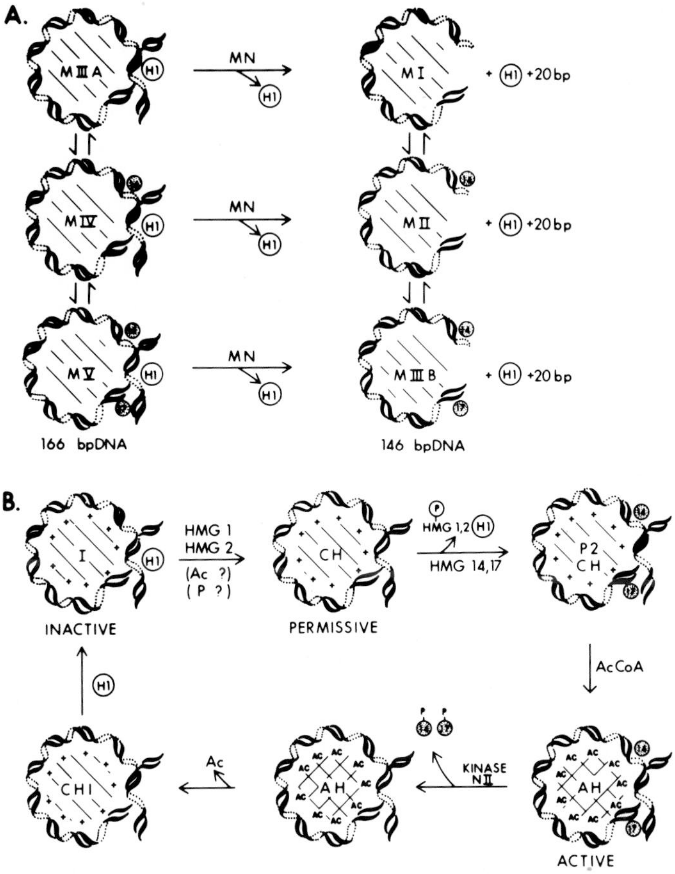

Fig. 5. Proposed mechanisms of HMG-induced RNA transcription. The chromatosome (166 bp DNA—MIIIA, MIV, and MV) and the nucleosome core particles (146 bp DNA—MI, MII, and MIIIB) are represented schematically. These subunits have been demonstrated electrophoretically (*2*) (see text). (B) A schematic illustration of our model of the potential control of chromatosome core histone acetylation. I represents the inactive deacetylated state, CH and CH-P2 the permissive state, AH the active acetylated state, and CHI a return to the inactive deacetylated state (see text for model).

major nucleosome and chromatosome configurations (*2*). The chromatosome particles MIIIA, MIV, and MV are composed of 166 bp of DNA and various combinations of HMG 14 and 17 and histone H1. The digestion of these chromatosomes with monococcal nuclease (MN) releases histone H1, leaving a nucleosome core particle of 146 DNA bp. Thus, MI is derived from MIIIA, MII is derived from MIV, and MIIIB is derived from MV. These six nucleosome configurations can be resolved by polyacrylamide gel electrophoresis (*2*). These particles can be reconstituted *in vitro*. In fact, several methods have been used to reconstitute nucleosome core particles in high yield (80–90%) that are fully native by a number of criteria (*18a,161,168,202*). Nucleosones have also been reconstructed using poly(dA-dT) (*137,155*), SV40 DNA (*189*), and cross-linked histone octamers (*188*).

Thus it is now feasible to reconstruct nucleosomes with defined protein and DNA composition. This then provides an opportunity to reconstruct nucleosomes with hyperacetylated HMG proteins. In future studies, it is likely that nucleosome assembly factors (*82,83*) and two-dimensional hybridization mapping of nucleosomes (*85*) will more clearly define the role nucleosomal microheterogeneity and the role of HMG proteins in the control of RNA transcription. Until these experiments are complete we can only speculate.

This discussion was intended, however, to emphasize many of the compositional and structural similarities of the HMG proteins found in all four kingdoms. All four of the main types of the HMG proteins (HMG 1,2,14, and 17) are characterized by a high concentration of basic and acidic amino acids, whose ratio is close to 1.0. The subclassification of the HMG proteins divides the four main classes into HMG 1 and 2 and HMG 14 and 17, on the basis of not only their molecular weights (Table I), but also their composition (Tables II–IV), and amino acid sequences (Figs. 1 and 2). Progress in the characterization of HMG proteins has been very rapid since it was last reviewed in 1978 (*47*), thus providing the impetus to devote a major section of this chapter to the most recent observations.

The HMG proteins do not displace the core histones (H2A,H2B,H3, and H4) or the binding of the fifth histone H1. HMG 1 and 2 have many amino acid sequences in common, have been isolated from a number of sources, and are believed to be present in all tissues. HMG 1 and 2 have in common the capacity to reduce the linking number of circular DNA and exhibit selective affinity for single-stranded DNA (*61,66,203*). It is suggested that these proteins function as a DNA unwindase. The HMG 1 and 2 proteins are highly structured and globular. The most significant observation about HMG 1 and 2 is the fact that they have a high affinity for histone H1 and can be cross-linked to this protein when associated with the nucleosome (*173*). This is an important observation in view of the fact that histone H1 can influence the transcription of chromatin (*172*) and cell cycle control (*109,130*).

It would appear that the function of HMG 14 and 17 differs from that of the HMG 1 and 2 proteins. These proteins have a random structural configuration and resemble each other in a number of physiochemical characteristics (*64*) and have closely related amino acid sequences (Fig. 2). Most important is the fact that in addition to being subject to postsynthetic modifications such as methylation (*14*), phosphorylation (*86*), and acetylation (*165*), they also function as inhibitors of deacetylase activity (*136*).

To propose a mechanism for the control of RNA transcription in view of the paucity of information is perhaps premature, but hypotheses are sometimes useful in the design of future experiments.

If one assumes that histone acetylation is associated with active nucleosomes, then the control of this postsynthetic modification would be primary in the modulation of this process. As shown in Fig. 5B, inactive nucleosomes (I) are characterized by the presence of deacetylated histones and histone H1. The loss of histone H1 would provide a permissive state in which the core histones (CH) are still deacetylated. The addition of HMG 14 and/or 17 could play a key role in the process of RNA transcription for the following reasons: (1) These non-histone proteins are associated with active regions of chromatin; (2) HMG 14 and/or 17 can act to inhibit deacetylase activity. Thus, in the presence of HMG 14 and/or 17, there would be unopposed acetyltransferase activity, and as a result the core histones would become hyperacetylated (AH) and achieve an active state of RNA transcription. Upon loss of HMG 14 and/or 17, perhaps by phosphorylation, the acetylated lysine residues would assume the deacetylated, inactive state (CH1).

VI. Summary

On the basis of a number of *in vivo* observations that have been presented here, a strong case can made for the hypothesis that those nucleosomes that are either actively engaged in transcription or are in a competent state to be transcribed are characterized by (1) the presence of HMG proteins, (2) acetylated core histones, and also (3) a lack of histone H1. It is also reasonable to postulate that those histones associated with active genes are postsynthetically modified *in situ* as a prerequisite for achieving the open configuration presumed to be required for the initiation of gene activation or representing the consequence of the process of gene transcription.

In either case, we propose that the process of postsynthetic modification (acetylation) is modulated by specific HMG nucleosomal complexes associated with active genes. This hypothesis is consistent with the literature review presented here, which suggests that the role of the HMG proteins is to provide a specific signal to inactivate the deacetylase.

References

1. Abercrombie, B. D., Kneale, C., Crane-Robins, C., Bradbury, E. M., Goodwin, G. H., Walker, J. H., and Johns, E. W. (1978). Studies on the conformational properties of the HMG chromosomal protein HMG 17 and interaction with DNA. *Eur. J. Biochem.* **84,** 173–177.
2. Albright, S. C., Wiseman, J. M., Lange, R. A., and Garrard, W. T. (1980). Subunit structures of different electrophoretic forms of nucleosomes. *J. Biol. Chem.* **255,** 3673–3684.
3. Alfageme, C. R., Rudkin, G. T., and Cohen, L. H. (1976). Locations of chromosomal proteins in polytene chromosomes. *Proc. Natl. Acad. Sci. U.S.A.* **73,** 2038–2042.
4. Allan, J., Hartman, P. G., Crane-Robins, C., and Aviles, F. (1980). The structure of histone H1 and its location in chromatin. *Nature* (*London*) **288,** 675.
5. Allan, J., Staynov, D. Z., and Gould, H. (1980). Reversible dissociation of linker histone from chromatin with preservation of internucleosomal repeat. *Proc. Natl. Acad. Sci. U.S.A.* **77,** 885–889.
6. Allfrey, V. G. (1977). Post-synthetic modifications of histone structure: A mechanism for the control of chromosome structure by the modulation of histone–DNA interactions. *In* "Chromatin and Chromosome Structure" (H. J. Li and R. A. Eckhardt, eds.), pp. 167–192. Academic Press, New York.
7. Bakayev, V. V., Bakayeva, T. G., Schmatchenko, V. V., and Georgiev, G. P. (1978). Non-histone proteins in Mononucleosomes and subnucleosomes. *Eur. J. Biochem.* **91,** 291–301.
8. Bakayev, V. V., Schmatchenko, V. V., and Georgiev, G. P. (1979). Subnucleosomes, HMG-type proteins and active chromatin. *Dokl. Akad. Nauk SSSR* **245,** 734–736.
9. Bakayev, V. V., Schmatchenko, V. V., and Georgiev, G. P. (1979). Subnucleosome particles containing high mobility group proteins HMG-E and HMG-G originate from transcriptionally active chromatin. *Nucleic Acids Res.* **24,** 1525–1540.
10. Belikoff, E., Wong, L. J., and Alberts, B. M. (1980). Extensive purification of histone acetylase A., the major histone *N*-acetyl transferase activity detected in mammalian cell nuclei. *J. Biol. Chem.* **255,** 11440–11453.
11. Bidney, D. L., and Reeck, G. R. (1978). Purification from cultured hepatoma cells of two non-histone chromatin proteins with preferential affinity for single-stranded DNA: Apparent analogy with calf thymus HMG proteins. *Biochem. Biophys. Res. Commun.* **85,** 1211–1218.
12. Billett, M. A. (1979). The release of high mobility group non-histone proteins fron nuclei during digestion with deoxyribonuclease II. *Biochem. Soc. Trans.* **2,** 381–382.
13. Bode, J., Henco, K., and Wingender, E. (1980). Modulation of the nucleosome structure by histone acetylation. *Eur. J. Biochem.* **110,** 143–152.
14. Boffa, L. C., Sterner, R., Vidali, G., and Allfrey, V. G. (1979). Post-synthetic modifications of nuclear proteins. High mobility group proteins are methylated. *Biochem. Biophys. Res. Commun.* **89,** 1322–1327.
15. Böhm, T., Schaeger, E., and Knippers, R. (1980). Acetylation of nucleosome histones *in vitro*. *Eur. J. Biochem.* **112,** 353–362.
16. Boulikas, T. B., Wiseman, J. M., and Garrad, W. T. (1980). Points of contact between histone H1 and the histone octamer. *Proc. Natl. Acad. Sci. U.S.A.* **77,** 127–133.
17. Bradbury, E. M. (1975). Foreword: Histone nomenclature. 1. *Ciba Found. Symp.* [N.S.] **28,** 1–4.
18. Bradbury, E. M., and Javaherian, K. (1977). The Organization and Expression of the Eukaryotic Genome." Academic Press, New York.

18a. Bryan, P. N., Wright, E. B., Hsie, M. H., Olins, A. L., and Olins, D. E. (1978). Physical properties of inner histone DNA complexes. *Nucleic Acids Res.* **5,** 3603–3617.
19. Bustin, M., Hopkins, R. B., and Isenberg, I. A. (1978). Immunological relatedness of high mobility group chromosomal proteins from calf thymus. *J. Biol. Chem.* **253,** 1694–1699.
20. Bustin, M., and Neihart, N. K. (1979). Antibodies against chromosomal HMG proteins stain the cytoplasm of mammalian cells. *Cell* **16,** 181–189.
21. Cary, P. D., Crane-Robins, C., Bradbury, E. M., Javaherian, K., Goodwin, G. H., and Johns, E. W. (1976). Conformation studies of two NH chromosomal proteins and their interactions with DNA. *Eur. J. Biochem.* **62,** 583–590.
22. Cary, P. D., Shooter, K. V., Johns, E. W., Olayemi, J. Y., Hartman, P. G., and Bradbury, E. M. (1979). Does high-mobility-group, non-histone protein HMG 1 interact specifically with histone H1 subfractions? *Biochem. J.* **183,** 657–662.
23. Chahal, S. S., Matthews, H. R., and Bradbury, E. M. (1980). Acetylation of histone H4 and its role in chromatin structure and function. *Nature* (*London*) **287,** 76–79.
24. Chai, L. S., and Sandberg, A. A. (1981). Organization of nucleosomes. *In* "Chromatin and Chromosomes in Eukaryotic Cells" (K. S. McCarty and G. Padilla, eds.). Academic Press, New York (in press).
25. Charlesworth, M. C., and Parish, R. W. (1975). Further studies on basic nucleoproteins from the cellular slime mold *Dictostelium discoideum. Eur. J. Biochem.* **75,** 241–250.
26. Comings, D. E., and Harris, D. C. (1976). Nuclear proteins. II. Similarity of nonhistone proteins in nuclear sap and chromatin, and essential absence of contractile proteins from mouse liver nuclei. *J Cell Biol.* **70,** 440–452.
27. Davie, J. R., and Candido, E. P. M. (1978). Acetylated histone H4 is preferentially associated with template active chromatin. *Proc. Natl. Acad. Sci. U.S.A.* **75,** 3574–3577.
28. Elgin, S. C. R., and Bonner, J. (1972). Partial fractionation and chemical characterization of the major nonhistone chromosomal proteins. *Biochemistry* **11,** 772–781.
29. Felsenfeld, G. (1978). Chromatin. *Nature* (*London*) **271,** 115–122.
30. Franco, L., Montero, F., and Rodriguez-Molina, J. J. (1977). Purification of the histone H1 from the fruit fly *Ceratitis capitata.* Isolation of a high mobility group (HMG) nonhistone protein and aggregation of H1 through a disulphide bridge. *FEBS Lett.* **78,** 317–320.
31. Garcea, R. L., and Alberts, B. M. (1980). Comparative studies of histone acetylation in nucleosomes, nuclei and intact cells. *J. Biol. Chem.* **255,** 11457–11463.
32. Garel, A., and Axel, R. (1976). Selective digestion of transcriptionally active ovalbumin genes from oviduct nuclei. *Proc. Natl. Acad. Sci. U.S.A.* **75,** 3966–3970.
33. Garrard, W. T., Pearson, W. R., Wake, S. K., and Bonner, J. (1974). Stoichiometry of chromatin proteins. *Biochem. Biophys. Res. Commun.* **58,** 50–57.
34. Gazit, B., Panet, A., and Cedar, H. (1980). Reconstitution of a deoxyribonuclease I-sensitive structure on active genes. *Proc. Natl. Acad. Sci. U.S.A.* **77,** 1787–1790.
35. Gershey, E. L., Vidali, G., and Allfrey, V. G. (1968). Chemical studies of histone acetylation: The occurence of epsilon-*N*-acetyl lysine in histones. *J. Biol. Chem.* **243,** 5018–5022.
36. Goldknopf, I. L., and Busch, H. (1977). Isopeptide linkage between nonhistone and histone 2A polypeptides of chromosomal conjugate-protein A24. *Proc. Natl. Acad. Sci. U.S.A.* **74,** 864–868.
37. Goldknopf, I. L., Taylor, C. W., Baum, R. M., Yedman, L. C., and Olson, M. O. J. (1975). of Protein A24, a "histone-like" non-histone chromosomal protein. *J. Biol. Chem.* **250,** 7182–7187.
38. Goodwin, G. H., Brown, E., Walker, J. M., and Johns, E. W. (1980). The isolation

of three new high mobility group nuclear proteins. *Biochim. Biophys. Acta* **623,** 329–338.

38a. Goodwin, G. H., and Johns, E. W. (1969).

39. Goodwin, G. H., and Johns, E. W. (1973). Isolation and characterization of two calf thymus non-histone proteins with high content of acidic and basic amino acids. *Eur. J. Biochem.* **40,** 215–219.

40. Goodwin, G. H., and Johns, E. W. (1977). The isolation and purification of the high mobility group (HMG) non-histone chromosomal proteins. *Methods Cell Biol.* **16,** 257–267.

41. Goodwin, G. H., and Johns, E. W. (1978). Are the high mobility group non-histone chromosomal proteins associated with active chromatin? *Biochim. Biophys. Acta* **519,** 279–284.

42. Goodwin, G. H., Johns, E. W., and Walker, J. M. (1977). Further characterization of HMG non-histone proteins. In "The Organization and Expression of the Eurkaryotic Genome" (E. M. Bradbury and K. Javaherian, eds.), pp. 43–50. Academic Press, New York.

43. Goodwin, G. H., Mathew, C. G., Wright, C. A., Venkov, C. D., and Johns, E. W. (1979). Analysis of the high mobility group proteins associated with salt-soluble nucleosomes. *Nucleic Acids Res.* **7,** 1815–1835.

44. Goodwin, G. H., Nicolas, R. H., and Johns, E. W. (1975). An improved large scale fractionation of high mobility group non-histone chromatin proteins. *Biochim. Biophys. Acta* **405,** 280–291.

45. Goodwin, G. H., Rabbani, A., Nicolas, P. H., and Johns, E. W. (1977). The isolation of the high mobility group non-histone chromosomal protein HMG 14. *FEBS Lett.* **80,** 413–416.

46. Goodwin, G. H., Walker, J. M., and Johns, E. W. (1978). Studies on the degradation of high mobility group non-histone chromosomal proteins. *Biochim. Biophys. Acta* **519,** 233–242.

47. Goodwin, G. H., and Walker, J. M., and Johns, E. W. (1978). The high mobility group (HMG) nonhistone chromosomal proteins. *In* "The Cell Nucleus" (H. Busch, ed.), Vol. 6, Part C, pp. 182–219. Academic Press, New York.

48. Goodwin, G. H., Woodhead, L., and Johns, E. W. (1977). The presence of high mobility group non-histone chromatin proteins in isolated nucleosomes. *FEBS Lett.* **73,** 85–88.

49. Gordon, J. S., Rosenfeld, B. I., Kaufman, R., and Williams, D. L. (1980). Evidence for a quantitative tissue-specific distribution of high mobility group chromosomal proteins. *Biochemistry* **19,** 4395–4402.

50. Griffith, J. D., and Christiansen, G. (1978). The multifunctional role of histone H1, probed with the SV40 minichromosome. *Cold Spring Harbor Symp. Quant. Biol.* **42,** 215–226.

51. Hamana, K., and Iwai, K. (1979). High mobility group non-histone chromosomal proteins also exist in tetrahymena. *J. Biochem.* (*Tokyo*) **86,** 789–794.

52. Hamana, K., and Iwai, K. (1974). Gel chromatography and gel electrophoresis of histones in denaturing solvents. *J. Biol. Chem.* **76,** 503–512.

53. Hancock, R. (1969). Conservation of histones in chromatin during growth and mitosis *in vitro*. *J. Mol. Biol.* **40,** 457–466.

54. Haselkorn, R., and Rouviere-Yaniv, J. (1976). Cyanobacterial DNA-binding protein related to *Escheriochia coli* HY. *Proc. Natl. Acad. Sci. U.S.A.* **73,** 1917–1920.

55. Horiuchi, K., and Fujimoto, D. (1975). Use of phospho-cellulose paper disks for the assay of histone acetyltransferase. *Anal. Biochem.* **69,** 491–496.

56. Hunt, L. T., and Dayhoff, M. O. (1977). Amino-terminal sequence identity of ubiquitin and the nonhistone component of nuclear protein A24. *Biochem. Biophys. Res. Commun.* **74,** 650–655.
57. Huntley, G. H., and Dixon, G. H. (1972). The primary structure of the NH_2-terminal region of histone T. *J. Biol. Chem.* **247,** 4916–4919.
58. Hutcheon, T., Dixon, G. H., and Levy-Wilson, B. (1980). Transcriptionally active mononucleosomes from trout testis are heterogeneous in composition. *J. Biol. Chem.* **255,** 681–685.
59. Inoue, A., Tei, Y., Hasuma, T., Yukioka, M., and Morisawa, S. (1980). Phosphorylation of HMG 17 by protein kinase NII from rat liver cell nuclei. *FEBS Lett.* **117,** 68–72.
60. Isackson, P. J., Bidney, D. L., Reeck, G. R., Neihart, N. K., and Bustin, M. (1980). High mobility group chromosomal proteins isolated from nuclei and cytosol of cultured hepatoma cells are similar. *Biochemistry* **19,** 4466–4471.
61. Isackson, P. J., Fishback, J. L., Bidney, D. L., and Reeck, G. R. (1979). Preferential affinity of high molecular weight high mobility group non-histone chromatin proteins for single-stranded DNA. *J. Biol. Chem.* **254,** 5569–5572.
62. Jackson, J. B., Pollock, J. M., Jr., and Rill, R. L. (1979). Chromatin fractionation procedure that yields nucleosomes containing near-stoichiometric amounts of high mobility group non-histone chromosomal proteins. *Biochemistry* **18,** 3739–3748.
63. Javaherian, K. (1977). Conformations on HMG non-histone proteins and their interactions with DNA and histones. *In* "The Organization and Expression of the Eukaryotic Genome" (E. M. Bradbury and K. Javaherian, eds.), pp. 51–65. Academic Press, New York.
64. Javaherian K., and Amini, S. (1978). Conformation study of calf thymus HMG 14 non-histone protein. *Biochem. Biophys. Res. Commun.* **85,** 1385–1391.
65. Javaherian, K., and Amini, S. (1977). Physiochemical studies of non-histone protein HMG 17 with DNA. *Biochim. Biophys. Acta* **478,** 295–304.
66. Javaherian, K., Liu, J. F., and Wang, J. C. (1978). Non-histone proteins HMG 1 and HMG 2 change the DNA helical structure. *Science* **199,** 1345–1346.
67. Javaherian, K., Sadeghi, M., and Liu, L. F. (1979). Non-histone proteins HMG 1 and HMG 2 unwind DNA double helix. *Nucleic Acids Res.* **6,** 3569–3580.
68. Jenson, J. C., Chin-Lin, P., Gerber-Jenson, B., and Litman, G. W. (1980). Structurally unique basic protein coextracted with histones from calf thymus chromatin. *Proc. Natl. Acad. Sci. U.S.A.* **77,** 1389–1393.
69. Johns, E. W., and Forrester, S. (1969). Studies on nuclear proteins. The binding of extra acidic proteins to deoxyribonucleoprotein during the preparation of nuclear proteins. *Eur. J. Biochem.* **8,** 547–551.
70. Johns, E. W., Goodwin, G. H., Hastings, J. R., and Walker, J. M. (1977). The histones and some histone-like chromosomal proteins. *In* "The Organization and Expression of the Eukaryotic Genome" (E. M. Bradbury and K. Javaherian, eds.), pp. 3–19. Academic Press, New York.
71. Johns, E. W., Goodwin, G. H., Walker, J. M., and Sanders, C. (1975). The Structure and function of chromatin. *Ciba Found. Symp.* [N.S.] **28,** 95–108.
72. Karabanov, A. A., Afanas'ev, B. V., and Chestkov, V. V. (1979). Chromatin HMG-protein, located in regions sensitive to the action of micrococcal nuclease. *Dokl. Akad. Nauk SSSR* **246,** 1239–1243.
73. Kelner, D. N., Paton, A. E., Olins, D. E., and McCarty, K. S., Sr. (1981). Characterization of a nuclear histone acetyltransferase activity from hog liver. *Fed. Proc., Fed. Am. Soc. Exp. Biol.*

74. Kelner, D. N., Paton, A. E., Olins, D. E., McCarty, K. S., Jr., and McCarty, K. S., Sr. (1982). In preparation.
75. Kennedy, B. P., and Davies, P. L. (1980). Acid-soluble nuclear proteins of the testis during spermatogenesis in the winter flounder. Loss of the high mobility group proteins. *J. Biol. Chem.* **255,** 2533–2539.
76. Klug, A., Rhodes, D., Smith, J., Finch, J. T., and Thomas, J. O. (1980). A low resolution structure for the histone core of the nucleosome. *Nature* (*London*) **287,** 509.
77. Kootstra, A., Shah, Y. B., and Slaga, T. J. (1980). Binding of B[A]P biol-epoxide (Anti) to nucleosomes containing high mobility group proteins. *FEBS Lett.* **116,** 62–66.
78. Kuehl, L., Lyness, T., Dixon, G. H., and Levy-Wilson, B. (1980). Distribution of high mobility group proteins among domains of trout testis chromatin differing in their susceptibility to micrococcal nuclease. *J. Biol. Chem.* **255,** 1090–1095.
79. Kuehl, L., Lyness, T., Watson, D. C., and Dixon, G. H. (1979). Binding of HMG-T to trout testis chromatin. *Biochem. Biophys. Res. Commun.* **90,** 391–397.
80. Kuehl, L. (1979). Synthesis of high mobility group proteins in regenerating rat liver. *J. Biol. Chem.* **254,** 7276–7281.
81. Lacy, E., and Axel, R. (1975). Analysis of DNA of isolated chromatin subunits. *Proc. Natl. Acad. Sci. U.S.A.* **72,** 3978–3982.
82. Laskey, R. A., and Earnshaw, W. C. (1980). Nucleosome assembly. *Nature* (*London*) **286,** 763–767.
83. Laskey, R. A., Honda, D. M., Mills, A. D., and Finch, J. T. (1978). Nucleosomes are assembled by an acidic protein which binds histones and transfers them to DNA. *Nature* (*London*) **275,** 416–420.
84. Leffak, I. M., Grainger, R., and Weintraub, H. (1977). Conservative assembly and segregation of nucleosomal histones. *Cell* **12,** 837–845.
85. Levinger, L., Balsoum, J., and Varshavsky, A. (1981). Two-dimensional hybridization mapping of nucleosomes—Comparison of DNA and protein patterns. *J. Mol. Biol.* **146,** 287–304.
86. Levy, W. B. (1981). Enhanced Phosphorylation of high-mobility group proteins in nuclease sensitive mononucleosomes from bytyrate-treated Hela cells. *Proc. Natl. Acad. Sci. U.S.A.* **78,** 2189–2193.
87. Levy, W. B., Connor, W., and Dixon, G. H. (1979). A subset of trout testis nucleosomes enriched in transcribed DNA sequences contains high mobility group proteins as major structural components. *J. Biol. Chem.* **254,** 609–620.
88. Levy, W. B., and Dixon, G. H. (1978). Partial purification of transcriptionally active nucleosomes from trout testis cells. *Nucleic Acids Res.* **5,** 4155–4163.
89. Levy, W. B., and Dixon, G. H. (1978). A Study of the localization of high mobility group proteins in chromatin. *Can. J. Biochem.* **56,** 480–491.
90. Levy, W. B., Wong, N. C., and Dixon, G. H. (1977). Selective association of the trout-specific H6 protein with chromatin regions susceptible to DNase I and DNase II: Possible location of HMG-T in the spacer region between core nucleosomes. *Proc. Natl. Acad. Sci. U.S.A.* **74,** 2810–2814.
91. Levy, W. B., Wong, N. C., Watson, D. C., Peters, E. H., and Dixon, G. H. (1977). Strcuture and function of the low-salt extractable chromosomal proteins. Preferential association of trout testis proteins H6 and HMG-T with chromatin regions selectively sensitive to nucleases. *Cold Spring Harbor Symp. Quant. Biol.* **42,** 793–801.
92. Levy-Wilson, B., and Dixon, G. H. (1979). Limited action of micrococcal nuclease on trout testis nuclei generates two mononucleosome subsets enriched in transcribed DNA sequences. *Proc. Natl. Acad. Sci. U.S.A.* **76,** 1682–1686.

93. Levy-Wilson, B., Watson, D. C., and Dixon, G. H. (1979). Multiacetylated forms of H4 are found in a putative transcriptionally competent chromatin fraction from trout testis. *Nucleic Acids Res.* **6,** 259–274.
94. Libby, P. R. (1978). Calf liver nuclear *N*-acetyltransferases. *J. Biol. Chem.* **253,** 233–237.
95. Libby, P. R. (1980). Rat liver nuclear *N*-acetyltransferases: Separation of two enzymes with both histone and spermidine acetyltransferase activity. *Arch. Biochem. Biophys.* **203,** 384–389.
96. Lilley, D. M. J., and Pardon, J. F. (1979). Structure and function of chromatin. *Annu. Rev. Genet.* **13,** 197–233.
97. Lutter, L. C. (1978). Kinetic Analysis of deoxyribonuclease I cleavages in the nucleosome core: Evidence for a DNA superhelix. *J. Mol. Biol.* **124,** 391–420.
98. Mardian, J. K. W., and Isenberg, I. (1978). Yeast inner histones and the evolutionary conservation of histone–histone interactions. *Biochemistry* **17,** 3825–3833.
99. Mardian, J. K. W., and Isenberg, I. (1978). Preparative gel electrophoresis: Detection, excision, and elution of protein bands from unstained gels. *Anal. Biochem.* **91,** 1–12.
100. Mardian, J. K. W., Paton, A. E., Bunick, G. J., and Olins, D. E. (1980). Nucleosome cores have two specific binding sites for non-histone chromosomal proteins HMG 14 and HMG 17. *Science* **209,** 1534.
101. Marzluff, W. F., and McCarty, K. S. (1970). Two classes of histone acetylation in developing mouse mammary gland. *J. Biol. Chem.* **245,** 5635–5642.
102. Marzluff, W. F., and McCarty, K. S., Sr. (1972). Structural studies of calf thymus F3 histone. II. Occurence of phosphoserine and ε-*N*-acetyllysine in thermolysine peptides. *Biochemistry* **11,** 2677–2681.
103. Marzluff, W. F., Sanders, L. A., Miller, D. M., and McCarty, K. S., Sr. (1972). Two chemically and metabolically distinct forms of calf thymus histone F3. *J. Biol. Chem.* **247,** 2026–2033.
104. Marzluff, W. F., Miller, D. M., and McCarty, K. S. (1972). Occurrence of ε-*N*-acetyllysine in calf thymus histone F2b. *Arch. Biochem. Biophys.* **152,** 472–474.
105. Mathew, C. G., Goodwin, G. H., Gooderham, K., Walker, J. M., and Johns, E. W. (1979). A comparison of the high mobility group non-histone chromatin protein HMG 2 in chicken thymus and erythrocytes. *Biochem. Biophys. Res. Commun.* **87,** 1243–1251.
106. Mathew, C. G., Goodwin, G. H., and Johns, E. W. (1979). Studies on the association of the high mobility group non-histone chromatin proteins with isolated nucleosomes. *Nucleic Acids Res.* **6,** 167–179.
107. Mathis, D. J., Oudet, P., and Chambon, P. (1980). Structure of transcribing chromatin. *Prog. Nucleic Acid Res. Mol. Biol.* **24,** 1–55.
108. Mathis, D. J., Oudet, P., Wasylyk, B., and Chambon, P. (1978). Effect of histone acetylation on structure and *in vitro* transcription of chromatin. *Nucleic Acids Res.* **5,** 3523–2547.
109. Matsumoto, Y., Yasuda, H., Mita, S., Marunouchi, T., and Yamada, M. (1980). Evidence for the involvement of H1 histone phosphorylation in chromosome condensation. *Nature* (*London*) **284,** 181–183.
110. McCarty, K. S., Sr. (1972). The specificity of hormone-induced acetylation in mammary gland cultures. *In Vitro* **7,** 244.
111. McCarty, K. S., Sr., Jones, R. F., and McCarty, K. S., Jr. (1973). Hormone induction of phosphorylation of chromosomal proteins. *Am. Soc. Cell Biol., 1973.*
112. McCarty, K. S., Jr., and McCarty, K. S., Sr. (1975). Evidence of hormone induction

of milk proteins in normal and malignant mammary gland cultures. *In* "Electron Microscopic Concepts of Secretion Ultrastructure of Endocrine and Reproductive Organs" (M. Hess, ed.), p. 129. Wiley, New York.
113. McCarty, K. S., Jr., and McCarty, K. S., Sr. (1977). Steroid hormone receptors in the regulation of differentiation. *Am. J. Pathol.* **86,** 740.
114. McCarty, K. S., Sr., and McCarty, K. S., Jr. (1974). Protein modification, metabolic control and their significance in transformation in eukaryotic cells. *JNCI, J. Natl. Cancer Inst.* **53,** 1509.
115. McCarty, K. S., Sr., and McCarty, K. S., Jr. (1975). Early mammary gland responses to hormones. *J. Dairy Sci.* **58,** 1–22.
116. McCarty, K. S., Sr., and McCarty, K. S., Jr. (1976). Hormonal induction of post-synthetic modifications of chromosomal proteins in mamnary neoplasia. *In* "Control Mechanisms in Cancer" (W. E. Criss, T. Ono, and J. R. Sabine, eds.), p. 37. Raven, New York.
117. McCarty, K. S., Sr., and McCarty, K. S., Jr. (1978). Some aspects of chromatin structure and cell cycle related post-synthetic modifications. *In* "Cell Cycle Regulation" (J. R. Jeter, I. L. Cameron, G. M. Padilla, and A. M. Zimmerman, eds.), pp. 9–35. Academic Press, New York.
118. McCarty, K. S., Sr., and McCarty, K. S., Jr. (1981). Structure and function. *Adv. Cell Biol.* **12,** 1029–1052.
119. McGhee, J. D., and Felsenfeld, G. (1980). Nucleosome structure. *Annu. Rev. Biochem.* **49,** 1115–1156.
120. Miller, D. M., Williams, R., and McCarty, K. S. (1973). Localization and *in vitro* specificity of histone acetylation. *Biochim. Biophys. Acta* **316** , 437.
121. Mirzabekov, A. D., Shick, V. V., Belyavsky, A. V., and Bavykin, S. G. (1978). Primary organization of nucleosome core particles of chromatin sequence of histone arrangement along DNA. *Proc. Natl. Acad. Sci. U.S.A.* **75,** 4181–4188.
122. Mirzabekov, A. D. (1980). Nucleosomes structure and its dynamic transitions. *Q. Rev. Biophys.* **13,** 255–295.
123. Mohlberg, J., and Rush, H. P. (1969). Isolation of the nuclear histones from the myxomycete, *Physarum polycephalum. Arch. Biochem. Biophys.* **134,** 577–589.
124. Olins, D. E. (1981). In preparation.
125. Olson, M. O. J., Goldknopf, I. L., Gutetzow, K. A., James, G. T., Hawkins, T. C., Mays-Rothberg, C. J., and Busch, H. (1976). The NH_2 and COOH-terminal amino acid sequence of nuclear protein A24. *J. Biol. Chem.* **251,** 5901–5903.
126. Otto, B., Bohm, J., and Knippers, R. (1980). A histone-specific acetyltransferase is associated with simian-virus 40 chromatin. *Eur. J. Biochem.* **112,** 363–366.
127. Panyim, S., and Chalkley, R. (1969). High resolution acrylamide gel electrophoresis of histones. *Arch. Biochem. Biophys.* **130,** 337–346.
128. Pardon, J. F. (1979). Physical studies of chromatin. *In* "The Cell Nucleus" (H. Busch, ed.), Part D, pp. 371–411. Academic Press, New York.
129. Pardon, J. F., Cotter, R. I., Lilley, D. M. J., Worcester, D. L., Campbell, A. M., Wooley, J. I., and Richards, B. M. (1978). Scattering studies of chromatin subunits. *Cold Spring Harbor Symp. Quant. Biol.* **42,** 11–22.
130. Pehrson, J.,and Cole, R. D. (1980). Histone H1 accumulates in growth-inhibited cultured cells. *Nature (London)* **285,** 43–44.
131. Rabbani, A., Goodwin, G. H., and Johns, E. W. (1978). High mobility group non-histone chromosomal proteins from chicken erythrocytes. *Biochem. Biophys. Res. Commun.* **81,** 351–358.
132. Rabbani, A., Goodwin, G. H., and Johns, E. W. (1978). Studies on the tissue spec-

ificity of the high-mobility group non-histone chromosomal proteins from calf. *Biochem. J.* **173,** 497–505.
133. Rabbani, A., Goodwin, G. H., and Johns, E. W. (1980). Structural studies on two HMG proteins from calf thymus HMG-14 and HMG-20 (ubiquitin) and their interaction with DNA. *Eur. J. Biochem.* **112,** 577–580.
134. Rabbani, A., Goodwin, G. H., Walker, J. M., Brown, E., and Johns, E. W. (1980). Trout liver high mobility group non-histone chromosomal proteins. *FEBS Lett.* **109,** 294–298.
135. Rechsteiner, M., and Kuehl, L. (1979). Microinjection of the non-histone chromosomal protein HMG 1 into bovine fibroblasts and HeLa cells. *Cell* **16,** 901–908.
136. Reeves, R., and Candido, E. P. M. (1980). Partial inhibition of histone deacetylase in active chromatin by HMG 14 and HMG 17. *Nucleic Acids Res.* **8,** 1947–1963.
137. Rhodes, D. (1979). Nucleosome cores reconstructed from poly(dA-dT) and the octamer of histones. *Nucleic Acids Res.* **6,** 1805–1816.
138. Riggs, M. G., Whittaker, R. G., Neumann, J., and Ingram, V. M. (1977). *n*-Butyrate causes histone modifications in HeLa and friend erythroleukemia cells. *Nature (London)* **268,** 462–464.
139. Rodriguez, A. C., Rudkin, G. T., and Cohen, L. H. (1980). Isolation properties and cellular distribution of D1, a chromosomal protein of drosophila. *Chromosoma* **78,** 1–31.
140. Romani, M., Rodman, T. C., Vidali, G., and Bustin, M. (1979). Serological analysis of species specificity in the high mobility group chromosomal proteins. *J. Biol. Chem.* **254,** 2918–2922.
141. Romani, M., Vidali, G., Tahourdin, C. S., and Bustin, M. (1980). Solid phase radioimmunoassay for chromosomal components. *J. Biol. Chem.* **255,** 468–474.
142. Rouviere-Yanif, J., and Gros, F. (1975). Characterization of a novel, low-molecular weight DNA-binding protein from *Escherichia coli. Proc. Natl. Acad. Sci. U.S.A.* **72,** 3428–3432.
143. Rouviere-Yanif, J., Gros, F., Haselkorn, R., and Reiss, C. (1977). "The Organization and Expression oF the Eukaryotic Genome" (E. M. Bradbury and K. Javaherian, eds.), pp. 211–231. Academic Press, New York.
144. Russev, G., Vassilev, L., and Tsanev, R. (1980). Histone exchange in chromatin of hydroxyurea-blocked Ehrlich ascites tumor cells. *Nature (London)* **285,** 584–586.
145. Saffer, J. D., and Glazer, R. I. (1980). The phosphorylation of high mobility group proteins 14 and 17 from Ehrlich ascites and 1210 *in vitro. Biochem. Biophys. Res. Commun.* **93,** 1280–1285.
146. Sandeen, G., Wood, W. I., and Felsenfeld, G. (1980). The interaction of high mobility proteins HMG 14 and 17 with nucleosomes. *Nucleic Acids Res.* **8,** 3757–3778.
147. Sanders, C., and Johns, E. W. (1974). A method for the large-scale preparation of two chromatin proteins. *Biochem. Soc. Trans.* **2,** 547–550.
148. Sanders, C. (1977). A method for the fractionation of the high-mobility group non-histone chromosomal proteins. *Biochem. Biophys. Res. Commun.* **78,** 1034–1042.
149. Sanders, L. A., Schechter, N. M., and McCarty, K. S. (1973). A comparative study of histone acetylation, histone deacetylation, and ribonucleic acid synthesis in avian reticulocytes and erythrocytes. *Biochemistry* **12,** 783–791.
150. Sealy, L., and Chalkley, R. (1978). DNA associated with hyperacetylated histone is preferentially digested by DNase I. *Nucleic Acids Res.* **5,** 1863–1876.
151. Searcy, D. G. (1975). Histone-like protein in the prokaryote thermoplasma acidophilum. *Biochim. Biophys. Acta* **395,** 535–547.
152. Seyedin, S. M., and Kistler, W. S. (1979). Levels of chromosomal protein high

mobility group 2 parallel the proliferative activity of testis, skeletal muscle, and other organs. *J. Biol. Chem.* **254,** 11264–11271.

153. Simpson, R. T. (1978). Structure of chromatin containing extensively acetylated H3 and H4. *Cell* **13,** 691–699.
154. Simpson, R. T. (1978). Structure of the chromatosome, a chromatin core particle containing 160 base pairs of DNA and all the histones. *Biochemistry* **17,** 5524–5531.
155. Simpson, R. T., and Kunzler, P. (1979). Chromatin and core particles formed from the inner histones and synthetic polydeoxyribonucleotides of defined sequence. *Nucleic Acids Res.* **6,** 1387–1415.
156. Smith, B. J., Robertson, D., Birbeck, M. S., Goodwin, G. H., and Johns, E. W. (1978). Immunochemical studies of high mobility group non-histone chromatin proteins HMG 1 and HMG 2. *Exp. Cell Res.* **115,** 420–423.
157. Smith, J. A., and Stocken, L. A. (1973). The characterization of a non-histone protein isolated from histone F1 preparations. *Biochem. J.* **131,** 859–861.
158. Sollner-Webb, B., and Felsenfeld, G. (1975). A comparison of the digestion of nuclei and chromatin by staphylococcal nuclease. *Biochemistry* **14,** 2915–2920.
159. Sommer, A. (1978). Yeast chromatin: Search for histone H1. *Mol. Gen. Genet.* **161,** 323–331.
160. Spiker, S., Mardian, J. K., and Isenberg, I. (1978). Chromosomal HMG proteins occur in three eukaryotic kingdoms. *Biochem. Biophys. Res. Commun.* **82,** 129–135.
161. Steinmetz, M., Streeck, R. E., and Zachau, H. G. (1978). Reconstituted histone–DNA complexes. *Philos. Trans. R. Soc. London* **283,** 259–268.
162. Sterner, R., Boffa, L. C., and Vidali, G. (1978). Comparative structural analysis of high mobility group proteins from a variety of sources. Evidence for a high mobility group protein unique to avian erythrocyte nuclei. *J. Biol. Chem.* **253,** 3830–3836.
163. Stenrner, R., Vidali, G., and Allfrey, V. G. (1979). Discrete proteolytic cleavage of high mobility group proteins. *Biochem. Biophys. Res. Commun.* **89,** 129–133.
164. Sterner, R., Vidali, G., and Allfrey, V. G. (1979). Studies of acetylation and deacetylation in high mobility group proteins. Identification of the sites of acetylation in HMG-1. *J. Biol. Chem.* **254,** 11577–11583.
165. Sterner, R., Vidali, G., Heinrikson, R. L., and Allfrey, V. G. (1978). Postsynthetic modification of high mobility group proteins. Evidence that high mobility group proteins are acetylated. *J. Biol. Chem.* **253,** 7601–7604.
166. Sures, I., and Gallwitz, D. (1980). Histone-specific acetyltransferase from calf thymus. Isolation, properties and substrate specificity of three different enzymes. *Biochemistry* **19,** 943–951.
167. Tahourdin, C. S., and Bustin, M. (1980). Chromatin subunits elicit species-specific antibodies against nucleoprotein antigenic determinants. *Biochemistry* **19,** 4387–4394.
168. Tatchell, K., and Van Holde, K. E. (1978). Compact oligomers and nucleosome phasing. *Proc. Natl. Acad. Sci. U.S.A.* **75,** 3583–3587.
169. Teng, C. S., Andrews, G. K., and Teng, C. T. (1979). Studies on the high-mobility-group non-histone proteins from hen oviduct. *Biochem. J.* **181,** 585–591.
170. Teng, C. S., Gallagher, K., and Teng, C. T. (1978). Isolation of a high-molecular-weight high-mobility-group-type non-histone protein from hen oviduct. *Biochem. J.* **176,** 1003–1006.
171. Teng, C. T., and Teng, C. S. (1980). Immuno-biochemical studies of a non-histone chromosomal protein in *embryonic* and mature chick oviduct. *Biochem. J.* **185,** 169–175.
172. Thoma, F., and Koller, T. (1977). Influence of histone H1 on chromatin structure. *Cell* **12,** 101–107.

173. Thomas, J. O., and Khabaza, J. A. (1980). Cross-linking of histone H1 in chromatin. *Eur. J. Biochem.* **112,** 501–511.
174. Vidali, G., Boffa, L. C., and Allfrey, V. G. (1977). Selective release of chromosomal proteins during limited DNASE 1 digestion of avian erythrocyte chromatin. *Cell* **12,** 409–415.
175. Walker, J. M., Brown, E., Goodwin, G. H., Stearn, C., and Johns, E. W. (1980). Studies on the structures of some HMG-like non-histone chromosomal proteins from trout and chicken tissues. Comparison with calf thymus proteins HMG 14 and 17. *FEBS Lett.* **113,** 253–257.
176. Walker, J. M., Gooderham, K., Hastings, J. R., and Johns, E. W. (1978). An unusual structural feature of non-histone chromosomal high-mobility-group protein 1. *Biochem. Soc. Trans.* **6,** 242.
177. Walker, J. M., Gooderham, K., Hastings, J. R. B., Mayes, E., and Johns, E. W. (1980). The primary structures of the NHC proteins HMG 1 and 2. *FEBS Lett.* **122,** 264–270.
178. Walker, J. M., Gooderham, K., and Johns, E. W. (1979). The isolation, characterization and partial sequence of a peptide rich in glutamic acid and aspartic acid (HGA-2 peptide) from calf thymus non-histone chromosomal protein HMG 2. Comparison with a similar peptide (HGA-1 peptide) from calf thymus non-histone chromosomal protein HMG 1. *Biochem. J.* **179,** 253–254.
179. Walker, J. M., Gooderham, K., and Johns, E. W. (1979). The isolation and partial sequence of peptides produced by cyanogen bromide cleavage of calf thymus non-histone chromosomal high-mobility-group protein 2. Sequence homology with non-histone chromosomal high-mobility-group protein 1. *Biochem. J.* **181,** 659–665.
180. Walker, J. M., Goodwin, G. H., and Johns, E. W. (1978). Chromosomal proteins. The amino terminal sequence of high-mobility group non-histone chromosomal protein HMG 14. Showing sequence homologies with two other chromosomal proteins. *Int. J. Pept. Protein Res.* **11,** 301–304.
181. Walker, J. M., Goodwin, G. H., and Johns, E. W. (1978). The isolation and identification of ubiquitin from the high-mobility group (HMG) non-histone protein fraction. *FEBS Lett.* **90,** 327–330.
182. Walker, J. M., Goodwin, G. H., and Johns, E. W. (1979). The primary structure of the nucleosome-associated chromosomal protein HMG 14. *FEBS Lett.* **100,** 394–398.
183. Walker, J. M., and Johns, E. W. (1980). The isolation, characterization and partial sequences of the chicken erythrocyte non-histone chromosomal proteins HMG 14 and HMG 17. Comparison with the homologous calf thymus proteins. *Biochem. J.* **185,** 383–386.
184. Walker, J. M., Hastings, J. R. B., and Johns, E. W. (1977). The primary structure of a non-histone chromosomal protein. *Eur. J. Biochem.* **76,** 461–468.
185. Walker, J. M., Parker, B. M., and Johns, E. W. (1978). Isolation and partial sequence of the cyanogen bromide peptides from calf thymus non-histone chromosomal protein HMG 1. *Int. J. Pept. Protein Res.* **12,** 269–276.
186. Walker, J. M., Stearn, C., and Johns, E. W. (1980). The primary structure of non-histone chromosomal protein HMG 17 from chicken erythrocyte nuclei. *FEBS Lett.* **112,** 207–210.
187. Waalace, R. B., Dube, S. K., and Bonner, J. (1977). Localization of globin gene in the template active fraction of friend leukemic cells. *Science* **198,** 1166–1168.
188. Wasylyk, B., and Chambon, P. (1980). Studies on the mechanism of transcription of nucleosomal complexes. *Eur. J. Biochem.* **103,** 219–226.
189. Wasylyk, B., Oudet, P., and Chambon, P. (1979). Preferential *in vitro* assembly of

nucleosome cores on some A-T-rich regions of SV-40 DNA. *Nucleic Acids Res.* **7,** 705–715.

190. Watson, D. C., Levy, W. B., and Dixon, G. H. (1978). Free ubiquitin is a non-histone protein of trout testis chromatin. *Nature* (*London*) **276,** 196–199.
191. Watson, D. C., Peters, E. H., and Dixon, G. H. (1977). The purification, characterization and partial sequence determination of a trout testis non-histone protein, HMG-T. *Eur. J. Biochem.* **74,** 53–60.
192. Watson, D. C., Wong, N. C., and Dixon, G. H. (1979). The complete amino-acid sequence of a trout-testis non-histone protein, H6, localized in a subset of nucleosomes and its similarity to calf-thymus non-histone proteins HMG-14 and HMG-17. *Eur. J. Biochem.* **95,** 193–202.
193. Weaver, R. F., Blatti, S. P., and Rutter, W. J. (1971). Molecular structures of DNA-dependent RNA polymerases (II) from calf thymus and rat liver. *Proc. Natl. Acad. Sci. U.S.A.* **68,** 2994–2999.
194. Weber, S., and Isenberg, I. (1980). High mobility group proteins of saccharomyces cervisiae. *Biochemistry* **19,** 2236–2240.
195. Weintraub, H., and Groudine, M. (1976). Chromosomal subunits in active genes have an altered conformation. *Science* **93,** 848–858.
196. Weisbrod, S., Groudine, M., and Weintraub, H. (1980). Interaction of HMG 14 and 17 with actively transcribed genes. *Cell* **19,** 289–301.
197. Weisbrod, S., and Weintraub, H. (1979). Isolation of a subclass of nuclear proteins responsible for conferring a DNase I-sensitive structure on globin chromatin. *Proc. Natl. Acad. Sci. U.S.A.* **76,** 630–634.
198. Weisbrod, S., and Weintraub, H. (1981). Isolation of HMG 14 and 17 and an analysis of δ-globin chromatin. *Cell* **23,** 391–400.
199. Whittaker, R. H. (1969). New concepts of kingdoms of organisms. *Science* **163,** 150–160.
200. Wigle, D. T., and Dixon, G. H. (1971). A new histone from trout testis. *J. Biol. Chem.* **246,** 5636–5614.
201. Wilhelm, J. A., and McCarty, K. S., Sr. (1970). Partial characterization of the histone and histone acetylation in cell cultures. *Cancer Res.* **30,** 409.
202. Woodcock, C. L. F. (1977). Reconstitution of chromatin subunits. *Science* **195,** 1350–1352.
203. Yu, S. H., Li, H. J., Goodwin, G. H., and Johns, E. W. (1977). Interaction of non-histone chromosomal proteins HMG 1 and HMG 2 with DNA. *Eur. J. Biochem.* **78,** 497–502.
204. Yu, S. H., and Spring, T. G. (1977). The interaction of non-histone chromosomal proteins HMG 1 and HMG 2 with subfractions of H1 histone immobilized on agarose. *Biochim. Biophys. Acta* **492,** 20–28.
205. Zweidler, A. F. (1978). Resolution of histones by polyacrylamide gel electrophoresis in presence of non-ionic detergents. *Methods Cell Biol.* **17,** 223–234.

4

RNA Content and Chromatin Structure in Cycling and Noncycling Cell Populations Studied by Flow Cytometry

ZBIGNIEW DARZYNKIEWICZ AND FRANK TRAGANOS

I. Introduction

With the advent of flow cytometry the possibility of rapid and accurate measurements of individual cells in large populations became a reality. The method offered the opportunity to perform biochemical measurements *in situ,* either supravitally or in relatively intact permealized or fixed cells to extract information on the quality and often on the conformation of particular cell constituents. Among the variety of different types of macromolecules that could be analyzed, the most frequently studied was the content of DNA per cell, i.e., the cell feature that

GENETIC EXPRESSION IN THE CELL CYCLE

ISBN 0-12-543720-X

positions cells in the cell cycle. As a consequence, an abundance of new information contributing to the better understanding of the cell cycle has recently become available. This chapter concentrates mainly on these data and does not cover subjects related to the general biology of the cell cycle, which may be found in numerous extensive reviews (*4,6, 35,38,39,48,51,55,56*) and elsewhere in this volume.

Bulk biochemical measurements of whole cell populations often provided essential information at the molecular level that could not be obtained by classical cytochemical methods. These measurements, however, could not reveal important details related to individual cells. In the absence of single cell measurements, information on heterogeneous cell populations was reduced to a single value. Correlation of specific cell features on a cell-to-cell basis was impossible and any information inherent to the *in situ* structure of various cell constituents was destroyed. There are situations, however, when information on individual cells is of great importance. For example, large variations in the duration of the cell cycle of individual cells are observed in many cell systems. This phenomenon, responsible for rapid decay of synchrony of presynchronized populations, complicates any therapeutic approach to the eradication of tumor cells based on cell synchronization and the use of cell cycle-specific drugs. Flow cytometric analysis of individual cells may provide critical information necessary for understanding regulatory mechanisms of cell proliferation in such situations. The ability to measure biochemical features of individual cells, as offered by flow cytometry, may also close the gap in our knowledge between cellular metabolic events at the molecular level and the behavior of cell populations. The purpose of this chapter is to present and discuss the data from recent studies on RNA content and *in situ* chromatin structure in individual cells in cycling and quiescent populations with special emphasis on the heterogeneity of cell populations. Attention is also given to changes in RNA content and in chromatin structure occurring when cells progress through the cycle, and the correlation between the rate of cell progression through the cycle and cellular RNA content.

II. RNA Content

A. Methodology

Two quantitative techniques have been developed that are applicable to the measurement of RNA in individual cells. The first technique, scanning microspectrophotometry, introduced by Casperson and his collaborators (*12,13*) is based on the uv absorption measurement at 265 nm. After subtraction of the "DNA-attributed" absorption from the total absorption value, the RNA content per cell can be estimated. Killander and Zetterberg (*42–44*) used this technique in combination with time-lapse cinematography to measure changes in RNA content during progression of cells through the cycle. They reported that a continuous and rather constant rate of accumulation of RNA occurs in cells traversing the various cell cycle phases. They also observed that RNA (and proteins, as represented by the dry mass of the cell) are unequally distributed into the sister cells during mitosis. The latter observation is of great importance as it directly supports the "metabolic model" of the cell cycle and has been proposed as an explanation of the heterogeneity of cell cycle times of individual cells. This subject will be discussed in more detail later in this chapter.

The second technique employed to quantitate RNA in the cell is based on the use of the metachromatic fluorochrome, acridine orange (AO). AO was introduced in our laboratory (*22,23,65*) as a stain for flow cytometry to measure simultaneously and differentially the content of cellular DNA and RNA. After selective denaturation of double-stranded RNA, which may be induced at certain AO concentrations in the presence of EDTA and/or citrate (*22*), interaction of the dye with DNA results in green fluorescence with maximum emission of 530 nm (F_{530}) while interaction with RNA, gives red metachromasia at 640 nm ($F_{>600}$) [for reviews, see (*17,58*)]. A stoichiometric relationship between the intensity of green fluorescence and DNA content per cell (*16*) as well as the intensity of red fluorescence and RNA content (*7*), have been demonstrated. Considering the existing data on the respective proportions of different RNA species within the cell indicating that about 80% of total cellular RNA is rRNA, one may assume that the RNA-specific $F_{>600}$ after staining with AO represents, for the most part, the stainability of rRNA, i.e., is related to the quantity of ribosomes. Application of AO as a dye in flow cytometry is limited, however, to cells that do not contain large amounts of glycosaminoglycans and requires separate enzymatic digestion of DNA and RNA to evaluate the specificity of staining. Under optimal conditions the technique allows for the rapid estimation of RNA content in relation to the position of cells in the cycle.

It should be emphasized that there are significant, consistent differences in RNA content between different tissues or cell types, unrelated to the cell cycle. This tissue-specific RNA content thus provides a baseline above which cell cycle-specific differences in RNA are measured

(*1,30*). Further discussion on RNA content in this chapter is limited to the intercellular variations within the same cell types, i.e., it refers to the cell cycle-related differences.

B. Cycling Populations

The typical distribution of DNA and RNA values of individual cells from exponentially growing populations is illustrated in Fig. 1. This pattern of stainability was consistently observed in all cell types (*31,32*). Taking into account the DNA content, the cells may be classified into G_1, S, and G_2 + M phases, as shown. Analysis of the RNA content

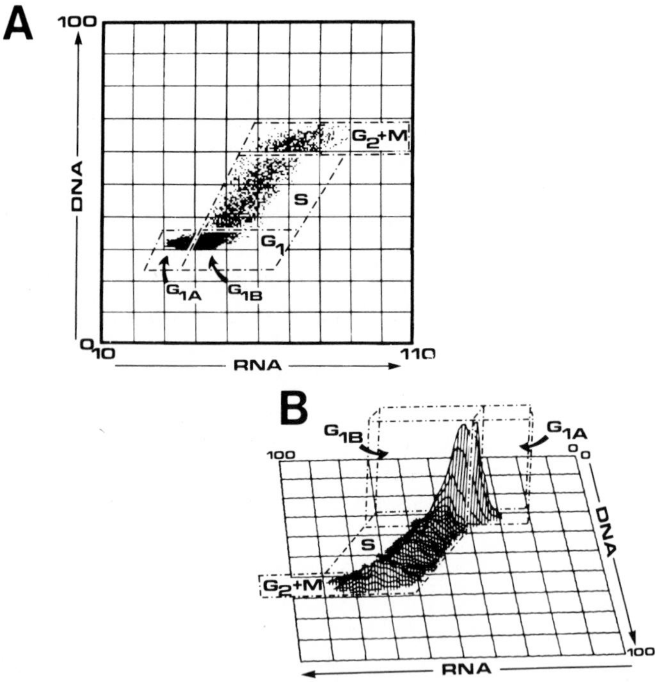

Fig. 1. Scattergram (A) and two-parameter (RNA versus DNA) frequency histogram (B) representing RNA and DNA values of individual cells from an exponentially growing L1210 cell culture, as measured by flow cytometry (see text). Differences in DNA content make it possible to classify cells in G_1, S, and G_2 + M phases. A progressive increase in RNA content during the cell cycle is observed, and the G_2 + M cells have, on average, twice as much RNA as G_1 cells. The G_1 cell population is heterogeneous with respect to RNA content. As evident from the continuity of the cell clusters, only those G_1 cells that have RNA content above the indicated threshold (G_{1B} cells) can enter S phase. The remaining G_1 cells (G_{1A}) prior to entrance to S have to progress through G_{1B}.

indicates that a progressive increase in RNA occurs during interphase so that G_2 + M cells have nearly twice as much RNA as G_1 cells.

Cell heterogeneity with respect to RNA content is a characteristic feature of the G_1 population (Fig. 1). It is evident from the scattergram (A) that only cells on the right side of the G_1 cluster (i.e., cells with RNA values above the indicated threshold) enter S phase directly. Thus, two distinct compartments G_{1A} and G_{1B} may be distinguished within the G_1 population. Cells from the G_{1A} compartment to enter S phase must first progress through the G_{1B} compartment as shown by the continuity of the clusters on the scattergram (A) or ridges on the frequency histogram (B). The threshold dividing these compartments represents the minimal RNA content of the S-phase population. Experiments on synchronized cultures indicated that the low-RNA G_{1A} compartment represents post-mitotic, early G_1 cells whereas the G_{1B} compartment represents late G_1 cells prior to their entrance into S phase *(29,32)*.

The data discussed above, repeated on numerous cell types of both normal and neoplastic origin, clearly demonstrate that a threshold amount of RNA is required for G_1 cells before they are able to enter S phase. So far the single exception to this pattern was seen in the case of AF8 cells infected with adenovirus 2. Adenovirus 2-infected cells moved into the S phase without the obligatory rise in RNA during G_1 (*53*).

Several lines of evidence indicate that G_{1A} and G_{1B} are functionally distinct compartments of the G_1 phase. Thus, the proportion of cells in G_{1A} versus G_{1B} changes specifically with changes in growth rate. Suppression of cell growth by serum deprivation or by addition of *n*-butyrate results in a specific block in G_{1A}. A dramatic shift in cell number from G_{1B} to G_{1A} is also evident in cultures at the plateau phase of growth (*32,34*). Kinetically inactive human leukemia cells have been found to be preferentially arrested in G_{1A} (*54*). In contrast, hydroxyurea- or thymidine-blocked cells are predominantly in G_{1B} or at the G_{1B}/S boundary.

The most interesting evidence in support of the notion that G_{1A} and G_{1B} are functionally distinct compartments comes from the kinetic studies in which the transit times of individual cells through G_{1A} and G_{1B} were analyzed (*32–34*). In those studies, the reentrance of cells into G_1 following mitosis was precluded by addition of Vinblastine or Colcemid so that the rate of cell exit from G_{1A} or G_1 (transit through G_{1B}) could then be measured. Cell exit from the G_1 phase was found to be represented by a biphasic curve. The second phase of cell exit from G_1 was always characterized by an exponentially declining slope. Cell exit from G_{1A}, on the other hand, was represented by a single exponentially declining slope, evident from the onset of stathmokinesis. These data indicate that transit times of individual cells through G_{1A} are exponentially distributed (of indeterminate duration) in contrast to the cell residence times in G_{1B}, which are of rather constant length. Discussion of these data in terms of the "probabilistic" or "metabolic" cell cycle models, as well as further evidence that G_{1A} and G_{1B} compartments represent functionally different portions of G_1 phase, will be given later in this chapter.

Extensive experiments have also been performed on synchronized cultures to correlate the rate of cell progression through the cycle with their RNA content (*28,29*). In the case of CHO cells synchronized by selective detachment at mitosis, a good correlation between RNA content and the transit times through G_1 and S was observed; cells with abundant RNA traversed these phases faster than their counterparts with low RNA content. In the case of cycling lymphocytes synchronized at the G_1/S boundary by hydroxyurea (or thymidine) and then released, their rate of progression through S was also seen to be highly correlated with the degree of RNA accumulation. These data demonstrate that the metabolic activity of individual cells, at least as reflected by RNA (mostly rRNA) content, influences the rate of their progression through the cycle. It is interesting to note that not only was the duration of G_1 correlated with RNA content but also the rate of traverse through S phase was also highly dependent on RNA.

C. Noncycling Populations

Changes in RNA content during the transition of cells from quiescence into the cycle (or vice versa) depended on the cell type and circumstances at which the quiescence was induced. Figure 2 illustrates changes in RNA and DNA occurring during stimulation of lymphocytes. Nonstimulated lymphocytes obtained from peripheral blood are cells in deep quiescence, characterized by 2C DNA content and minimal RNA content (G_0 or G_{1Q} cells). In cultures, following stimulation by mitogens or antigens, lymphocytes enter a long inductive (transitional) phase during which a marked increase in RNA content per cell occurs. Then the lymphocytes undergo a few rounds of replication and return back to quiescence [for review, see (*46*)]. The maximum increase in RNA content coincides in time with maximal proliferation in those cultures (*23*).

Not all lymphocytes undergo stimulation when challenged with mitogens. Noncommitted lymphocytes, i.e., cells which do not respond to particular types of mitogens or antigens, remain quiescent in cultures at all times. The degree of stimulation may thus be assayed by calculating the number of cells with increased RNA content, even before cells enter the phase of DNA replication. This approach has been used by us (*23,29,66*) and others (*9, 57,64*) to discriminate between cycling (stim-

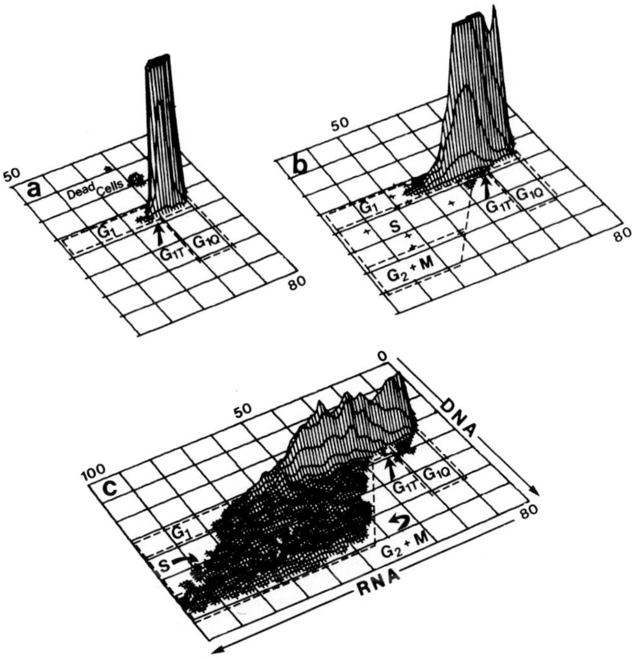

Fig. 2. Two-parameter (RNA versus DNA) frequency histograms illustrating changes in cellular RNA and DNA during progression of lymphocytes through the cell cycle after stimulation by phytohemagglutinin (PHA). (a) Control, nonstimulated cultures. Quiescent cells (G_{1Q}) in these cultures have a 2C DNA content and minimal RNA content. (b) Lymphocytes from 1-day-old PHA-stimulated culture. Nearly 50% of the cells have RNA above the G_{1Q} level. Only a few cells are in the S and G_2 + M phases. Numerous cells are in transition (G_{1T}) between quiescence and the cell cycle. (c) Lymphocytes from a 3-day-old PHA culture. Over 80% of the cells are seen in the cell cycle with elevated RNA values. Many cells are in S and G_2 + M phase.

ulated) and noncycling lymphocytes and to analyze the kinetics of lymphocyte stimulation *in vitro* and *in vivo* (*15,50*). Thus, in summary, in the lymphocyte cell system the noncycling, quiescent cells have distinctly lower RNA content than their cycling counterparts.

During lymphocyte stimulation, prior to their entrance into the cycle, as well as when cells return to quiescence, lymphocytes pass through a transition phase (G_{1T}). The cells often remain in this phase for a considerable period of time. The transition phase, referred to sometimes as the induction, "genome activation" or prereplicative phase, is characterized by a variety of unique biochemical events, some of them preparatory to DNA replication [see reviews (*4,46*)]. Distinctive features of this phase as evidenced by flow cytometry were recently recognized by

Richman (*57*). Lymphocytes in transition (G_{1T} cells) are characterized by 2C DNA content and intermediate (between G_{1Q} and G_{1A}) values of RNA.

Macrophages represent a cell system that is somewhat similar to lymphocytes. Namely, the noncycling, nonstimulated macrophages have a 2C DNA content and minimal RNA content. Upon stimulation, their RNA content increases and the stimulated cells can be distinguished from the quiescent ones by flow cytometry (*63*).

The 3T3 cells kept at confluency in cultures for an extended period of time represent still another cell system in which quiescence correlates with a significant decrease in RNA content. Under these conditions 3T3 cells lose about half of their RNA and become arrested in G_1 (*32*). Likewise, quiescent and nonclonogenic EMT6 cells described by Watson and Chambers (*70,71*) and HeLa cells from crowded and unfed cultures, as shown by Bauer and Dethlefsen (*8*), also contained markedly diminished RNA content in comparison with their cycling controls. The term "G_{1Q}" cells was proposed to characterize and classify the situations described above in which quiescent cells had 2C DNA values and markedly decreased RNA content such that there was no overlap in RNA values between individual cells in cycling and noncycling populations (*31*).

It should be emphasized, however, that there are also situations when cells do not progress through the cycle for a considerable period of time and yet their RNA content is not markedly diminished. As discussed before, 3T3 cells maintained in the presence of 0.5% serum remain quiescent with RNA values typical of those of G_{1A} cells. Also, kinetically inactive leukemic lymphocytes from the peripheral blood of patients were shown to be characterized by G_{1A} rather than G_{1Q} RNA values. Analyzing these cases it is tempting to postulate that the "depth" of quiescence (*3,59*) relates to RNA content, and thus noncycling cells with RNA values similar to those of G_{1A} or G_{1T} cells did not reach the depth of quiescence typical of G_{1Q} cells. Further studies, however, are needed to support this notion. In these studies the depth of quiescence, as measured by other criteria (i.e., reversibility and time required for cells to enter the S phase after release from quiescence) should be correlated with RNA content.

So far the discussion of quiescent cells was restricted to the instances when cells were blocked in G_1. Indeed, these are the most common situations. However, there is growing evidence that cells with S or G_2 DNA values may also remain in quiescence. Nearly two decades ago Gelfant, in his elegant and convincing experiments, demonstrated that epidermal cells could remain quiescent for an extended period of time while in G_2 phase [for reviews, see (*38,39*)]. The presence of quiescent cells with an S DNA content is still regarded as controversial. Such cells, however, were observed by us in three different cell systems. In the first system, quiescent S cells with low RNA values were seen in cultures of leukemic cells the growth of which was suppressed by macrophages (*45*). Despite having an S phase DNA content, these leukemic cells did not incorporate tritiated thymidine. Their block in S phase was reversible; after removal of macrophages they rapidly entered the cycle. In another system, induction of quiescence (paralleled by cell differentiation) was achieved by growth of Friend leukemia cells in the presence of DMSO (*67*). Although most cells were blocked in the G_1 (G_{1Q}) phase, some cells remained quiescent having S or G_2 DNA values.

Chronic myeloid leukemia during blastic crisis represents perhaps the most interesting cell system in which quiescent S phase cells were seen (*18*). Namely, in the peripheral blood and bone marrow of several patients, blast cells with low RNA content and S-phase DNA values were observed. These S-quiescent cells, in contrast to S-phase cells with higher RNA values, were refractory to the patients' treatment with a cell cycle-specific drug. When transferred into tissue culture these cells doubled their RNA content within 2–4 hr and then entered the cell cycle.

To classify quiescent cells that have S or G_2 DNA content and a distinctly lower RNA content than their cycling counterparts in S or G_2 phases, the terms "S_Q" and "G_{2Q}" were proposed (*31*). The respective positions of quiescent versus cycling cell populations, based on differences in their DNA and RNA values, are schematically illustrated in Fig. 3. Further discussion on these cells with respect to their chromatin structure is given later in the chapter.

D. Unbalanced Cell Growth

Progression of cells through the cell cycle may be interrupted under two different sets of circumstances. In the first one, inhibition of cell proliferation is paralleled by an overall decrease in other metabolic functions. This is the case of cell quiescence, i.e., as occurring when lymphocytes after completion of 3–4 rounds of cell division following stimulation by mitogens return to quiescence or when cell growth in cultures is suppressed by serum deprivation, cell crowding, chalones, etc. On the other hand, cells may be stopped in the cycle by a specific agent interfering with DNA replication; the rate of other metabolic functions of such cells may still remain high for a considerable period of time (*68*).

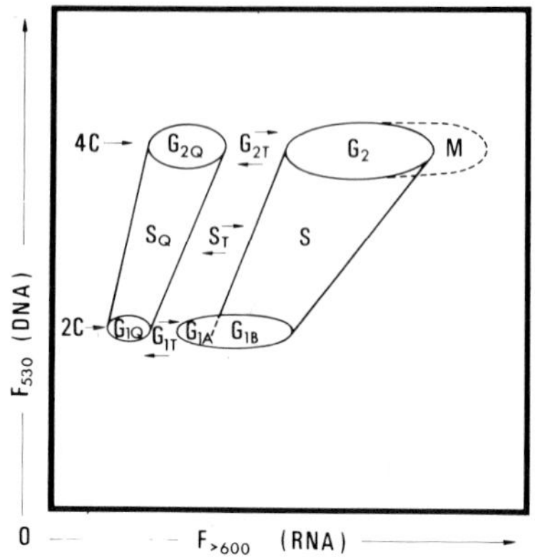

Fig. 3. Cell cycle compartments that may be recognized by flow cytometry following simultaneous staining of RNA and DNA (*30*). Outlined are the respective positions of cell clusters (compartments) on DNA–RNA scattergrams; the arrows indicate the possible transition points. Cells in quiescence have minimal RNA content. Though in most cases quiescent cells have a 2C DNA content (G_{1Q}), in some cell systems noncycling cells also have S or G_2 DNA values (S_Q, G_{2Q} cells). The transition of quiescent cells into the cycle may occur at any DNA level (G_{1T}, S_T, or G_{2T}) and is accompanied by an increase in RNA. The pattern of DNA versus RNA stainability of cycling cells, as seen at the right, is described in the legend to Fig. 1. The prominence of the G_{1A} compartment, or the respective proportions of G_{1A} versus G_{1B} cells vary from cell type to cell type and correlate with cell growth conditions. Under adverse conditions (e.g., serum starvation or cell crowding) cells are blocked in G_{1A} (*30*).

In such situations, in the absence of cell division, cells grow in size and accumulate large quantities of proteins and RNA. This unbalanced cell growth, if it continues for a prolonged period of time, leads to cell death.

Simultaneous measurement of cellular RNA and DNA content offers the possibility to easily distinguish these two situations from normal growth. As discussed before, cell quiescence is either correlated with markedly decreased RNA content or, at least, with a shift of cells from the G_{1B} to the G_{1A} compartment, i.e., with a slight decrease in RNA content. Thus, during quiescence the RNA/DNA ratio is diminished. In contrast, unbalanced growth is characterized by a shift in RNA content to generally higher but occasionally lower values. For instance, cells blocked at the G_1/S interphase by hydroxyurea or thymidine for approximately one generation contain nearly double the RNA content of normal cells in G_1 or S phase (*29*). In contrast, a class II anthracycline such as aclacinomycin causes a shift of all cycling cells to lower RNA content (*69*). Thus, a change in the RNA/DNA ratio is an indication of unbalanced growth and the ratio may be used as a measure of the degree of unbalance.

E. Conclusions

1. During exponential cell growth a continuous increase in RNA content per cell takes place in interphase. Cells in G_2 + M phase have nearly twice as much RNA as early G_1 cells.

2. A critical RNA threshold exists for G_1-phase cells. The cells with RNA values below this threshold (classified as cells in the G_{1A} compartment) cannot enter S phase regardless of the time spent in G_1.

3. The residence times of cells in the G_{1A} compartment are of indeterminate duration (exponentially distributed).

4. G_1-phase cells with RNA values above a critical threshold are classified as cells in the G_{1B} compartment; their transit times through G_{1B} are of relatively constant duration.

5. The rate of progression of individual cells not only through G_1 (G_{1A}) but also through S phase is correlated with their RNA content.

6. In certain situations (i.e., growth at low serum concentration, treatment by *n*-butyrate) cells become arrested in the G_{1A} compartment.

7. Cells in a deep quiescence (Q) (i.e., nonstimulated lymphocytes, 3T3 cells maintained at confluency) have a distinctly lower RNA content than G_{1A} cells. In general, such cells contain a G_1 content of DNA (G_{1Q} or G_0 cells).

8. In certain cell systems, cells may enter quiescence having an S or G_2 DNA content. Such cells (S_Q, G_{2Q}) are characterized by minimal RNA content similar to that of G_{1Q} cells.

9. Cells in transition (T) from quiescence to the cycle, and vice versa, may be distinguished as having intermediate RNA values. During transition such cells may have G_1 (G_{1T}), S (S_T), or G_2 (G_{2T}) DNA content.

10. During unbalanced cell growth, in the absence of cell division, the RNA/DNA ratio changes. This ratio is a sensitive parameter allowing detection and quantitation of the degree of unbalanced cell growth.

III. Chromatin Structure

A. Methodology

Extensive literature exists regarding chromatin changes that occur during progression of cells through the cell cycle or during transition from quiescence to the cycle [see reviews (*4,5,51*)]. At the molecular level the best characterized are cell cycle-related modifications by histones, i.e., acetylation of inner histones and phosphorylation of histone H1 (*47,61*), (Chapter 2). These changes appear to be associated with regulation of genome transcription and organization of chromatin structure at the supranucleosomal level, as reflected by chromatin condensation or packing into metaphase chromosomes. Nearly all studies on this subject were done using biochemical methods on synchronized cell populations, i.e., under conditions where information about individual cells and intercellular variability could not be preserved. The present review is focused on cytochemical studies of nuclear chromatin *in situ* performed on large cell populations with the use of fluorescent probes and flow cytometry.

One of the most common approaches to the analysis of chromatin structure involves studying the accessibility of DNA *in situ* to an intercalating, fluorescent probe. It is well established that the binding of such probes to nuclear DNA in isolated chromatin or *in situ* is restricted by chromosomal proteins. The extent of the unmasked DNA that is accessible for intercalation of the ligand depends on the size of the probe and the involvement of the major or minor groove of the DNA helix. There is also evidence that in certain cell systems the binding correlates with changes in chromatin structure that occur during cell differentiation or quiescence. In the case of small intercalating probes such as AO or ethidium bromide, a change in the extent of binding was reproducibly demonstrated only during cell differentiation in the following cell systems: (1) normal erythroid differentiation in bone marrow (*41*); (2) erythroid differentiation of Friend leukemia cells (*24,67*); and (3) spermiogenesis (*37,40*). In other cell systems, especially in studies of cycling versus noncycling cells, no change in binding of such probes could be observed [for review, see (*17*)].

In contrast to small intercalating probes, binding of larger intercalators, with bulky chains protruding into the grooves of the DNA helix (e.g., actinomycin D) was shown to be lowered in noncycling (*19*) and mitotic (*52*) cells. Also, binding of quinacrine mustard was described recently (*49*) to vary during the cell cycle and be correlated with chromatin condensation (see also Chapter 5). To date, binding of those later probes has not been investigated on large cell populations by flow cytometry.

Stability of DNA *in situ* to heat- or acid-induced denaturation was investigated in a variety of cell types. This DNA feature was found to provide a useful parameter correlated with the extent of chromatin condensation, which in turn appears to change during cell transition to quiescence. DNA *in situ* in chromatin is locally stabilized by positively charged macromolecules that provide counterions for DNA phosphates. The strength and extent of interactions between DNA and these local counterions, as well as any crosslinking which additionally stabilizes the double helix, may be evaluated from the analysis of the patterns (''profiles'') of DNA denaturation. It is expected that histone modifications such as acetylation or phosphorylation, which weaken interactions between histones and DNA, will be reflected as altered stability of DNA *in situ*. The assay of DNA stability is simple and is also based on the metachromatic properties of AO, which in this case is used to differentially stain native versus denatured DNA (*20,21*) In this technique either isolated nuclei or prefixed cells are first incubated with RNase and then heated or treated with acid buffers (*25*) to denature DNA. Subsequent staining with AO reveals the extent of denatured DNA, which stains metachromatically red ($F_{>600}$). In contrast, interactions of AO with native DNA result in green fluorescence (F_{530}). Under appropriate staining conditions and calibration of photomultiplier sensitivities the ratio of red fluorescence to total cell fluorescence [α_t index $= F_{>600}/(F_{>600} + F_{530})$] represents the fraction of denatured DNA and the total cell fluorescence, i.e., the sum of the red and green fluorescence ($F_{>600} + F_{530}$) is proportional to total DNA per cell (*21*). This technique has been widely used to analyze differences between cells at various phases of the cycle and to differentiate between cycling and noncycling cells. The discussion that follows is based mostly on the data resulting from this technique, i.e., relates to the intercellular differences in sensitivity of DNA *in situ* to acid denaturation.

B. Cycling Populations

A characteristic pattern of cell stainability with AO following extraction of RNA and partial DNA denaturation, is shown in Fig. 4. As described in detail before (*25–27*), based on differences between cells in both green and red fluorescence, it is possible to recognize cells in G_1, S, G_2, and M in an exponentially growing population. In addition, the G_1 subpopulation may be further subdivided into two qualitatively different compartments, G_{1A} and G_{1B}. Cells with higher α_t values (G_{1A}) do not enter S phase directly. On the other hand, G_1 cells with lower α_t values (G_{1B}) are in continuity with S-phase cells. The situation thus resembles the subdivision of G_1 phase based on the differences in RNA content except that in this case the cells differ in their α_t values. As shown before, the α_t index correlates with the degree of chromatin condensation (*25,26*). The G_{1A} cells, therefore, have more condensed chromatin than G_{1B} cells. Of all cells, however, cells in mitosis have the highest α_t value.

It is evident from the histogram (Fig. 4) that progression of cells from G_{1A} to S phase involves their transit through the G_{1B} compartment. G_{1B} cells have the lowest α_t values of all cells. After reaching the minimum α_t at G_{1B} and early S phase (the lowest degree of chromatin condensation) further cell progression through S phase is paralleled by a stepwise

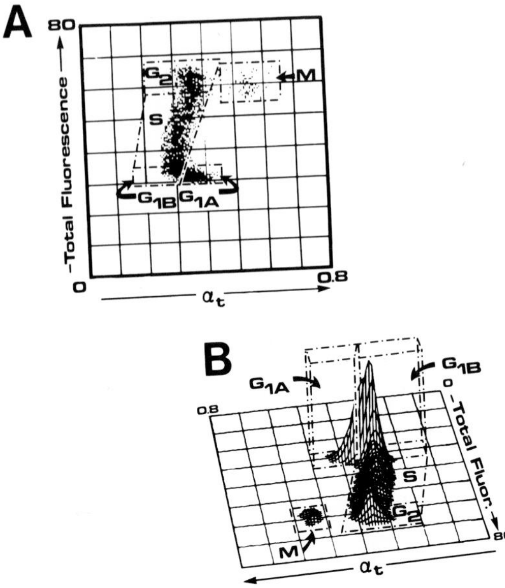

Fig. 4. Scattergram (A) and two-parameter (total fluorescence, i.e., $F_{530} + F_{>600}$ versus α_t value) frequency histogram (B) representing the stainability of chromatin of individual L1210 cells from exponentially growing cultures. Total fluorescence is proportional to DNA content per cell, whereas the α_t value represents the susceptibility of DNA *in situ* to denaturation and correlates with the degree of chromatin condensation (see text). Based on differences in total fluorescence, cells may be classified into G_1, S, and G_2 + M phases. Cells in mitosis have the highest α_t value and form a distinct cluster separated from G_2 cells. Recognition of the G_{1A} and G_{1B} compartments within G_1 phase is possible based on the α_t threshold value, which represents the maximal α_t value of early S phase cells. Thus, progression of G_{1A} cells into S involves their transit through G_{1B}. The degree of chromatin condensation (α_t value) of G_{1B} cells is similar to that of early S-phase cells, whereas G_{1A} cells have α_t values within similar range as G_2 cells.

increase in the α_t value. This is evidenced by the nonvertical position of the S cluster with respect to the α_t axis. The mean α_t value of G_2 cells is similar to that of G_{1A} cells. Experiments on synchronized cultures have confirmed that the G_{1A} population represents early G_1, postmitotic cells, while the late G_1 cells are mostly within the G_{1B} cluster (*33,34*). In summary, both RNA content and chromatin structure (sensitivity of DNA to acid denaturation) may be used independently to distinguish qualitatively different compartments within the G_1 phase. Cells in the first compartment (G_{1A}) increase their RNA content and decrease α_t up to a certain threshold. This threshold represents the minimum RNA content per cell and maximum α_t, below which cells cannot enter S phase. After the threshold is passed the cells still remain in G_1 phase

(G_{1B} compartment) but now can enter the S phase without additional changes in RNA content or chromatin stainability.

It should be emphasized that cell residence times in G_{1A}, regardless of whether this compartment is distinguished by RNA content or chromatin sensitivity to acids, are of indeterminate duration (exponentially distributed). However, the proportions of G_{1A} cells are somewhat lower when the distinction is based on the chromatin stainability pattern rather than on RNA content (*33*). Thus, although in both staining techniques the G_{1A} versus G_{1B} populations represent the early G_1 versus late G_1 cells, the threshold dividing these subpopulations is located somewhat earlier in the cycle when chromatin stainability is used as the method of discrimination.

C. Noncycling Populations

Numerous biochemical observations point out that there are differences in chromatin structure between noncycling cells and cells in the cell cycle [for reviews, see (*4,51,55,56*)]. The flow cytometric technique, based on differential sensitivity of DNA *in situ* to denaturation, was widely applied in a variety of cell systems to discriminate between cycling and noncycling cells (*25–27,36*). With this approach it was consistently observed that the extent of DNA denaturation (α_t value) correlated with the degree of chromatin condensation, which in turn correlated with cell quiescence. In the case of normal lymphocytes, quiescent cells had very condensed chromatin (high α_t values) and a 2C DNA content (*25,31*). Transition of these G_{1Q} cells into the cycle triggered by mitogenic stimulation was paralleled by a substantial decrease in α_t. The decrease in α_t was observed prior to cell entrance into S phase. Due to a large difference in α_t, the G_{1Q} cells could be easily discriminated from cycling cells in G_1 phase. The discrimination of cells in transition (G_{1T}) as well as between cycling lymphocytes in the G_{1A} and G_{1B} compartments was also possible (*31*).

In certain cell systems the noncycling cells had S or G_2 values of DNA. These cells were also characterized by condensed chromatin (high α_t). The presence of these noncycling cells arrested in S or G_2 phase (S_Q, G_{2Q} cells) and having condensed chromatin was detected in chronic myeloid leukemia during blastic crisis (*27*) and in some solid tumors (*36*). The respective positions of cell clusters of cycling and noncycling populations discriminated based on differences in total cellular fluorescence and in α_t values are schematically presented in Fig. 5.

It should be mentioned that in most situations when quiescent cells were characterized by more condensed chromatin, regardless of whether

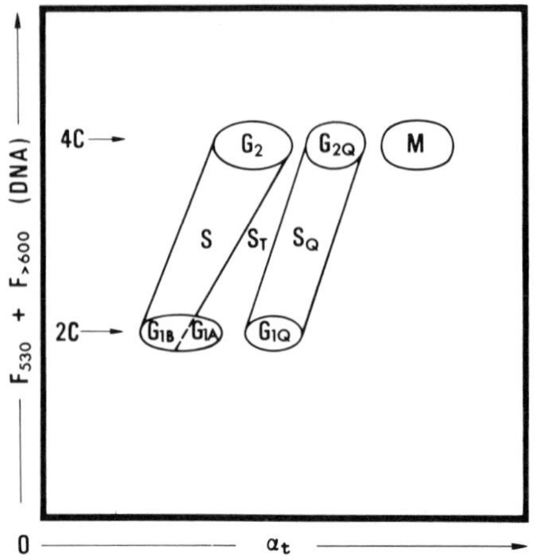

Fig. 5. Cell cycle compartments distinguished by flow cytometry after partial denaturation of DNA *in situ* and differential staining of native versus denatured DNA with AO (*27*). The axes, as in Fig. 4, represent total cell fluorescence, which is proportional to DNA content per cells, and α_t index. The respective positions of the cell clusters (compartments) of cycling and quiescent populations differing in DNA and α_t are schematically outlined. Cells from cycling populations have lower α_t values than noncycling cells. The outline of cycling cells (left) may be compared with the actual, raw data shown in Fig. 4. In some cell systems, quiescent cells with high α_t values, having S and G_2 DNA content (S_Q and G_{2Q} cells) may be detected (see text).

they had 2C or higher DNA content, these cells also exhibited decreased RNA content in comparison with their cycling counterparts (*31,33,34*). Sometimes, however (e.g., L1210 cells brought to quiescence by *n*-butyrate treatment, or 3T3 cells maintained at low serum concentration), cells were arrested specifically in the G_{1A} compartment as recognized by their RNA/DNA stainability following a rather modest decrease in RNA. Based on chromatin sensitivity to acid denaturation, these cells could also be classified as arrested in G_{1A} rather than in G_{1Q} because their chromatin was less condensed (lower α_t) than typical G_{1Q} cells such as nonstimulated lymphocytes.

Noncycling cells with condensed chromatin have smaller nuclei than cycling cells. Since the size of cell nuclei is easily measured by flow cytometry based on the width (duration) of the fluorescence pulse, detection of noncycling cells by this parameter is also possible. Based on this principle, noncycling lymphocytes were distinguished from mitogen-stimulated lymphocytes (*25*). The ratio of the pulse height to pulse width is an especially sensitive parameter to recognize cells with small, condensed nuclei, typical of quiescent cells.

D. Conclusions

1. Changes in chromatin structure occurring during the cell cycle can be conveniently monitored by flow cytometry as alterations in the sensitivity of DNA *in situ* to heat- or acid-induced denaturation. These alterations parallel changes in chromatin condensation.

2. Two distinct compartments, G_{1A} and G_{1B}, are discriminated in the G_1 phase of cycling cells, based on differences in chromatin structure. Functionally and kinetically these compartments resemble the G_{1A} and G_{1B} subphases detected by differences in RNA content.

3. In numerous cell systems noncycling cells are characterized by more condensed chromatin than that of cycling cells in the same tissue. Thus, these cells may be recognized by flow cytometry based on differences in the sensitivity of DNA to denaturation. In addition, quiescent cells having smaller nuclei differ from cycling cells in their nuclear pulse width value, or value of the pulse height/pulse width ratio.

4. In most cell systems noncycling cells have a G_1 DNA content and condensed chromatin (G_{1Q} cells). In some cases, however, quiescent cells with both condensed chromatin and S or G_2 DNA content could be detected.

IV. Detection of the Discrete Cell Cycle Compartments Based on Differences in RNA Content and Chromatin Structure

Flow cytometric measurements as described in an earlier part of this chapter (Sections II,B,C and III,B,C), made it possible to discriminate between different functional states ("compartments") of the cell cycle. The continuity of these compartments and possible cell transitions between them are schematically illustrated in Fig. 6. The nature of the compartments and conditions under which the transitions occur are discussed below.

Relatively little is known at present about the G_{1A} and G_{1B} compartments of the G_1 phase. Experiments on synchronized cultures provided evidence that following mitosis cells have minimal RNA content, i.e., most cells reside in the low-RNA, G_{1A} compartment (*2,29*). During G_1 phase, an increase in RNA content occurs as cells undergo the transition into the G_{1B} compartment. Thus, in a variety of cell systems, cells in the G_{1A} compartment are characterized by lower RNA content and more condensed chromatin in comparison with G_{1B} cells. During exponential cell growth, and when cell reentrance into G_1 is prevented by mitotic blockers, cell exit from G_{1A} is characterized by an exponentially declining

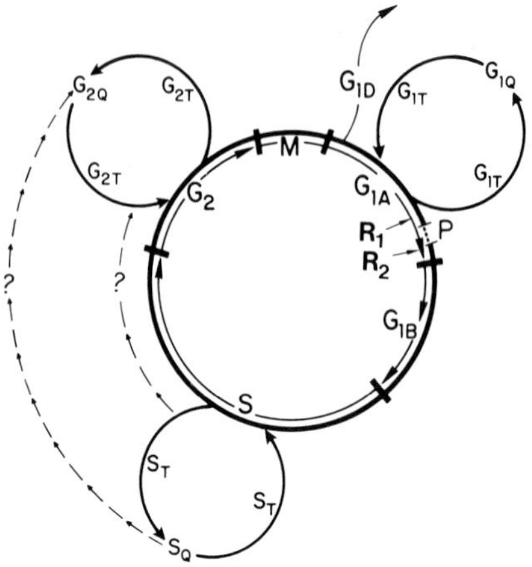

Fig. 6. Various compartments of the cell cycle and various states of quiescence as indicated in this diagram may be recognized by flow cytometry; the kinetics of cell transition between those compartments may thus be analyzed. Following mitosis (M), cells enter the G_{1A} compartment from which they can enter the differentiation pathway (G_{1D}) or quiescence (G_{1Q} via G_{1T} = transition phase to quiescence). The cells' further progression through the cycle may involve a probabilistic event (P) that can explain the indeterminate duration of G_{1A}. Alternatively, the G_{1A} subphase may be an equalization period during which cells accumulate the critical threshold quantitives of some essential constituent(s). The restriction points (R_1 and R_2) (*51*) may be related to these critical thresholds. For instance, the actinomycin D restriction point may indicate accumulation of the threshold quantity of RNA, i.e., G_{1A} to G_{1B} transition. Cell transit through G_{1B} is of rather constant duration. The similarity in the sensitivity of chromatin of G_{1B} and early S cells to denaturation suggests that the G_{1B} subphase may be functionally related to DNA replication. In certain systems, cells in the S phase may undergo a transition (S_T) to quiescence (S_Q). Likewise, G_2 cells may leave the cycle and become quiescent (G_{2Q} via G_{2T}). It is possible that S_T or even S_Q cells having low RNA content and relatively condensed chromatin replicate DNA very slowly, progressing towards G_2.

slope evident from the very onset of stathmokinesis and extending for nearly two decades (Fig. 7), indicating that cell residence times in G_{1A} are of indeterminate duration (exponentially distributed). In contrast, cell transit times through G_{1B} appear to be of rather constant duration (*30*).

Two alternative models may account for the exponential distribution of G_{1A} residence times. The first one is the transition probability model introduced by Smith and his colleagues to explain the exponential distribution of cell generation times (*11,62*). According to their model there is an indeterminate state in G_1 ("A" state) from which cells exit at random. The random transition from the indeterminate ("A") state into the deterministic portion of the cycle ("B") is responsible for the exponential distribution of generation times. Considering the exponential

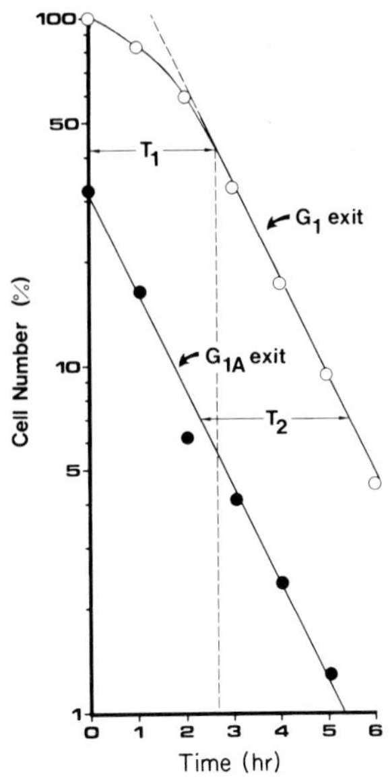

Fig. 7. Kinetics of cell exit from the G_1 and G_{1A} compartments during stathmokinesis. Exponentially growing L1210 cultures were treated at time 0 with Vinblastine and samples were then withdrawn at hourly intervals (abscissa). The number of cells remaining in G_1 phase (○) and in the G_{1A} compartment (●) were then estimated for each time point from the histograms as shown in Fig. 4. A straight exponentially declining curve characterizes exit from G_{1A}, indicating that cells in G_{1A} have exponentially distributed residence times. The half-time of cell residence in G_{1A} estimated from the slope of the G_{1A} exit curve is about 1 hr. The biphasic curve represents the decrease in the number of cells residing in G_1. The second portion of this curve declines exponentially and has a slope similar to the G_{1A} exit curve. The onset of this portion becomes apparent 2.5–3 hr after addition of Vinblastine (T_1). The data may be interpreted as indicating that cell residence times in G_{1A} are of indeterminate duration, whereas their further progression through G_1 (G_{1B}) is of rather constant length (T_2) approximately 2.5–3 hr (*30*).

character of cell exit from the G_{1A} compartment as observed, it is clear that if the probabilistic model is accepted than G_{1A} would be the locus of the indeterminate state. Thus, the hypothetical random event triggering progression of cells through the cycle would be located in G_{1A}.

The alternative model that explains the heterogeneity of generation times of individual cells in populations assumes an inequality of metabolic constituents in daughter cells as a result of their uneven distribution during mitosis [for review, see (*55,56*)]. Thus, in this "metabolic" model, the G_1 period (or part of it) is of varying duration to allow for the equalization of some critical cell constituents prior to their entrance into the DNA pre- or replicative phase. The data of Killander and Zetterberg

(*43,44*) demonstrate unambiguously that the distribution of RNA and proteins into the daughter cells during mitosis is unequal. These authors have also shown that cells with a paucity of RNA or with lower protein content (dry mass) remain in G_1 for longer periods of time than cells with high RNA content and that critical cell mass is required prior to the entrance of cell into S phase. Our data, described in the early part of this review (Section II,B), clearly indicate that cells must attain a critical RNA content (number of ribosomes) before they can enter S phase.

Because mean values of cell size, RNA or protein content in cell populations remain constant over generations, an efficient mechanism of cell "equalization" must occur during the cell cycle. Could it be then that cell equalization takes place specifically in G_{1A}? In the case of equalization with respect to RNA content, the answer to this question, as discussed above, is affirmative. Indeed, in support of this conclusion, we observed that cell heterogeneity, expressed as a coefficient of variation of the mean value of RNA content, is lowest in G_{1B} and in early S phase and then progressively increases during S and G_2 phase to become the highest for the G_{1A} subpopulations (manuscript in preparation). Thus, the intercellular variability is reduced every time the cells undergo the transition from the G_{1A} to G_{1B} compartment. It remains to be seen whether cellular variability of other essential constituents follows the same pattern as RNA. If this is the case, the G_{1A} compartment may be recognized as the specific equalization subphase of the cell cycle during which cells accumulate the essential constituents up to some critical threshold. The mechanism sensing these thresholds should operate within the cell during G_1 and be associated with signal(s) triggering cell replication. Only the cells that have synthesized the threshold quantities of those constituents could then progress into the prereplicative, G_{1B} subphase.

The question may be asked as to whether the exponential distribution of cell residence times in G_{1A} may be explained by the equalization mechanisms discussed above. Several lines of evidence indicate that indeed that may be the case. Cell (cytoplasm) cleavage during mitosis has some elements of randomness; the daughter cells are unequal immediately following cell division (*42–44*). It is likely that their heterogeneity with regard to the quantity of RNA or other essential constituents is reflected in a heterogeneity of metabolic activity, which in turn is proportional to the rate of progression through G_1 phase. Such a correlation was observed with respect to RNA content (*28,29*). Thus, the rates of progression of individual cells through early G_1 phase may be exponentially, or quasiexponentially, distributed. As a consequence, the cell residence times in the equalization subphase may be expected to be also exponentially or quasiexponentially distributed. A mechanism of this type was postulated by Castor (*14*) in his "G_1 rate" model of the cell cycle based on computer simulations of data. This model, which explains exponential distribution of generation times in cell populations, is consistent with the observed differences in cell metabolism rather than the assumption that a transition occurs at an inherently unpredictable time point.

The wide, exponential distribution of the residence times of cells in the G_{1A} compartment conforms well with observations of Roti Roti and his colleagues (*60*). To account for variability of the transit times of G_1 cells in a mathematical analysis of the cell cycle, those authors have introduced a "compressed G_1" estimate. The compression parameter reflected the particular quasiexponential skewness of the distribution of cell cycle transit times. More recently Blair and Roti Roti (*10*) have demonstrated that the early portion of the G_1 phase, prior to the actinomycin D restriction point, is the main locus of the variability in cell generation times. Because RNA accumulation (RNA synthesis) is required for cells to exit G_{1A} (Fig. 1), it is evident that the G_{1A} compartment is located prior to the actinomycin D restriction point (*51*). Thus, the variability measured by Blair and Roti Roti relates most likely to the G_{1A} subphase.

In summary, all the evidence discussed above suggests that the early G_1 period (G_{1A} subphase) is characterized by highly heterogeneous transit times. Such a distribution is not incompatible with the assumption that the G_{1A} subphase is the metabolic equalization period during which an increase in RNA content (and perhaps in the quantity of other essential constituents) up to a critical threshold takes place.

G_{1B} cells are in transition from G_{1A} to S phase, and the G_{1B} period appears to be of relatively constant duration. Considering that cells in G_{1B} have an RNA content similar to early S-phase cells, one may assume that they are located in the cell cycle beyond the actinomycin D restriction point (*51*). Cells in the G_{1B} compartment are also characterized by having the lowest α_t values (lowest DNA sensitivity towards denaturation) when compared to cells in all other phases of the cycle. It is possible that the G_{1B} compartment represents cells that have already initiated certain DNA-prereplicative functions but have not yet begun to synthesize DNA. Taking into account that the G_{1B} period is of rather constant duration and accepting the transition probability model of the cell cycle, one may characterize the G_{1B} compartment as the deterministic subphase of G_1.

In most cases, quiescent noncycling cells contain a G_1 DNA content. The G_{1Q} cell population represents cells in "deep" quiescence (*3,59*). Their characteristic feature is minimal RNA content and highly condensed chromatin. G_{1Q} cells are distinctly different from their cycling counterparts since they do not overlap at all in RNA or α_t values with the latter. Nonstimulated lymphocytes are the most typical example of G_{1Q} cells. Normal cells or 3T3 cells in primary cultures maintained at confluency for long periods of time also approach the G_{1Q} state *(31,32)*, as do low-RNA, nonclonogenic EMT6 cells, described by Watson and Chambers (*70*). When G_{1Q} cells are stimulated to proliferate their transition phase preceding DNA replication is long, varying between 24 and 48 hr.

Diploid cell lines subjected to serum deprivation do arrest in a state, which judged from RNA content or chromatin structure resembles G_{1A} more than G_{1Q}. Kinetically inactive leukemic cells from the peripheral blood or bone marrow also appear to be arrested in G_{1A}. Likewise, cells exposed to *n*-butyrate cease proliferation and become preferentially blocked in G_{1A} (*33*). Stimulation of G_{1A}-arrested cells (as of G_{1Q} cells), results in a rather long transition phase prior to cell entrance into S phase (*2*).

S_Q- and G_{2Q}-phase cells either do not cycle or cycle very slowly and are characterized by an S and G_2 DNA content, respectively. They have very low RNA content and highly condensed chromatin, features that discriminate them from cycling cells. Detection of such cells in solid tumors was recently reported by Dethlefsen *et al.* (*36*). It will be interesting to define conditions under which these cycling cells undergo transition into quiescence (retaining S or G_2 DNA content) and characterize those cells in more detail. In practical terms the presence of S_Q or G_{2Q} cells complicates the analysis of the cell cycle based on measurements of DNA alone and may be the cause of disagreement between the [^{3}H]TdR autoradiography and flow cytometry data.

The differences in cellular RNA content or in chromatin structure, as described, are indicators of the different functional states of the cell. These cell features, therefore, may be used as parameters in cell sorting to separate cell populations representative of these functional states, for biochemical analysis. This approach will undoubtly increase our knowledge of the cell cycle and its regulatory mechanisms. Other cell features such as cell volume, content of specific proteins, activity of particular enzymes, and immunochemistry of the cell membrane will be of further help toward the better characterization of the cell cycle.

Acknowledgments

Supported by PHS Grants CA 23296, CA 28704, and CA 14134 awarded by the National Cancer Institute, DMHS. The skillful assistance of Miss Robin Nager in the preparation of the manuscript is greatly appreciated.

References

1. Andreeff, M., Darzynkiewicz, Z., Sharpless, T., Clarkson, B., and Melamed, M. R. (1980). Discrimination of human leukemia subtypes by the cytometric analysis of cellular DNA and RNA. *Blood* **55,** 282–293.
2. Ashihara, T., Traganos, F., Baserga, R., and Darzynkiewicz, Z. (1978). A comparison of cell cycle related changes in post mitotic and quiescent AF8 cells as measured by flow cytometry after acridine orange staining. *Cancer Res.* **38,** 2514–2518.
3. Augenlicht, L. H., and Baserga, R. (1974). Changes in the G_0 state of WI-38 fibroblasts at different times after confluence. *Exp. Cell Res.* **89,** 255–262.
4. Baserga, R. (1976). "Multiplication and Division in Mammalian Cells." Dekker, New York.
5. Baserga, R. (1978). Resting cells and the G_1 phase of the cell cycle. *J. Cell. Physiol.* **95,** 377–386.
6. Baserga, R. (1981). The cell cycle. *N. Engl. J. Med.* **304,** 453–459.
7. Bauer, K. D., and Dethlefsen, L. A. (1980). Total cellular RNA content: Correlations between flow cytometry and ultraviolet spectroscopy. *J. Histochem. Cytochem.* **28,** 493–498.
8. Bauer, K. D., and Dethlefsen, L. A. (1981). Control of proliferation in HeLa-S3 suspension cultures. I. Simultaneous determination of cellular RNA and DNA content by acridine orange staining and flow cytometry. *J. Cell. Physiol.* **108,** 99–112.
9. Betel, I., Martijmse, J., and Vander Westen, G. (1979). Mitogenis activation and proliferation of mouse thymocytes. *Exp. Cell Res.* **124,** 329–337.
10. Blair, O. C., and Roti Roti, J. L. (1981). Variation in G_1 transit time relative to the cycloheximide and actinomycin D drug restriction points. *Cell Tissue Kinet.* **14,** 91–101.
11. Brooks, R. F., Bennett, D. C., and Smith, J. A. (1980). Mammalian cell cycles need two random transitions. *Cell* **19,** 493–504.
12. Caspersson, T. (1940). Methods for the determination of the absorption spectra of cell structures. *J. R. Microsc. Soc.* [3] **60,** 8–25.
13. Caspersson, T., Jacobson, F., Lomakka, G., Svenson, G., and Safstrom, R. (1953). A high resolution ultra microspectrophotometer for large scale biological work. *Exp. Cell Res.* **5,** 560–563.
14. Castor, L. N. (1980). A G_1 rate model accounts for cell cycle kinetics attributed to "transition probability." *Nature (London)* **287,** 857–859.
15. Collste, L. G., Darzynkiewicz, Z., Traganos, F., Sharpless, T., Whitmore, W. F., and Melamed, M. R. (1979). Regional lymph node reactivity in explanted bladder cancer of mice as measured by flow cytometry. *Cancer Res.* **39,** 2120–2124.
16. Coulson, P. B., Bishop, A. O., and Lenarduzzi, R. (1977). Quantitation of cellular deoxyribonucleic acid by flow cytometry. *J. Histochem. Cytochem.* **25,** 1147–1153.
17. Darzynkiewicz, Z. (1979). Acridine orange as a molecular probe in studies of nucleic acids *in situ*. *In* "Flow Cytometry and Sorting" (M. R. Melamed, M. Mendelsohn, and P. Mullaney, eds.), pp. 283–316. Wiley, New York.
18. Darzynkiewicz, Z. (1979). Drug effects on cell cycle: Discussion. *In* "Effects of Drugs on Cell Nucleus" (H. Bush, S. T. Crooke, and Y. Daskal, eds.), pp. 470–473. Academic Press, New York.
19. Darzynkiewicz, Z., Bolund, L., and Ringertz, N. R. (1969). Actinomycin binding of normal and phytohaemagglutinin stimulated lymphocytes. *Exp. Cell Res.* **55,** 120–123.
20. Darzynkiewicz, Z., Traganos, F., Sharpless, T., and Melamed, M. R. (1974). Thermally-induced changes in chromatin of isolated nuclei and of intact cells as revealed by acridine orange staining. *Biochem. Biophys. Res. Commun.* **59,** 392–399.

21. Darzynkiewicz, Z., Traganos, F., Sharpless, T., and Melamed, M. R. (1975). Thermal denaturation of DNA *in situ* as studied by acridine orange staining and automated cytofluorometry. *Exp. Cell Res.* **90,** 411–428.
22. Darzynkiewicz, Z., Traganos, F., Sharpless, T., and Melamed, M. R. (1975). Conformation of RNA *in situ* as studied by acridine orange staining and automated cytofluorometry. *Exp. Cell Res.* **95,** 143–153.
23. Darzynkiewicz, Z., Traganos, F., Sharpless, T., and Melamed, M. R. (1976). Lymphocyte stimulation: A rapid multiparameter analysis. *Proc. Natl. Acad. Sci. U.S.A.* **73,** 2881–2886.
24. Darzynkiewicz, Z., Traganos, F., Sharpless, T., Friend, C., and Melamed, M. R. (1976). Nuclear chromatin changes during erythroid differentiation of Friend virus induced leukemic cells. *Exp. Cell Res.* **99,** 301–309.
25. Darzynkiewicz, Z., Traganos, F., Sharpless, T., and Melamed, M. R. (1977). Cell cycle related changes in nuclear chromatin of stimulated lymphocytes as measured by flow cytometry. *Cancer Res.* **37,** 4635–4640.
26. Darzynkiewicz, Z., Traganos, F., Sharpless, T., and Melamed, M. R. (1977). Different sensitivity of DNA *in situ* in interphase and metaphase chromatin to heat denaturation. *J. Cell Biol.* **73,** 128–138.
27. Darzynkiewicz, Z., Traganos, F., Andreeff, M., Sharpless, T., and Melamed, M. R. (1979). Different sensitivity of chromatin to acid denaturation in quiescent and cycling cells as revealed by flow cytometry. *J. Histochem. Cytochem.* **27,** 478–485.
28. Darzynkiewicz, Z., Evenson, D. P., Staiano-Coico, L., Sharpless, T., and Melamed, M. R. (1979). Correlation between cell cycle duration and RNA content. *J. Cell. Physiol.* **100,** 425–438.
29. Darzynkiewicz, Z., Evenson, D., Staiano-Coico, L., Sharpless, T., and Melamed, M. R. (1979). Relationship between RNA content and progression of lymphocyte stimulation through the S phase of the cell cycle. *Proc. Natl. Acad. Sci. U.S.A.* **76,** 358–362.
30. Darzynkiewicz, Z., Andreeff, M., Traganos, F., and Melamed, M. R. (1980). Proliferation and differentiation of normal and leukemic lymphocytes as analysed by flow cytometry. *In* "Flow Cytometry IV" (O. D. Laerum, T. Lindmo, and E. Thorud, eds.), pp. 392–397. Universitetsforlaget, Bergen.
31. Darzynkiewicz, Z., Traganos, F., and Melamed, M. R. (1980). New cell cycle compartments identified by flow cytometry. *Cytometry* **1,** 98–108.
32. Darzynkiewicz, Z., Sharpless, T., Staiano-Coico, L., and Melamed, M. R. (1980). Subcompartments of the G_1 phase of cell cycle detected by flow cytometry. *Proc. Natl. Acad. Sci. U.S.A.* **77,** 6696–6699.
33. Darzynkiewicz, Z., Traganos, F., Xue, S., Staiano-Coico, L., and Melamed, M. R. (1981). Rapid analysis of drug effects on the cell cycle. *Cytometry* **1,** 279–286.
34. Darzynkiewicz, Z., Traganos, F., Xue, S., and Melamed, M. R. (1981). Effect of *n*-butyrate on cell cycle progression and *in situ* chromatin structure of L1210 cells. *Exp. Cell Res.* **136,** 279–293.
35. Dethlefsen, L. A. (1979). In quest of the quaint quiescent cells. *In* "Radiation Biology in Cancer Research" (R. F. Meyn and H. R. Withers, eds.), pp. 415–435. Raven, New York.
36. Dethlefsen, L. A., Bauer, K. D., and Rigler, R. M. (1980). Analytical cytometric approaches to heterogeneous cell populations in solid tumors. *Cytometry* **1,** 89–97.
37. Evenson, D. P., Darzynkiewicz, Z., and Melamed, M. R. (1980). Comparison of human and mouse sperm chromatin structure by flow cytometry. *Chromosoma* **78,** 225–238.
38. Gelfant, S. (1977). Cycling–noncycling cell transition in tissue, aging, immunological surveillance, transformation and tumor growth. *Int. Rev. Cytol.* **70,** 1–25.

39. Gelfant, S. (1977). A new concept of tissue and tumor cell proliferation. *Cancer Res.* **37,** 3845–3862.
40. Gledhill, B. L., Gledhill, M. P., Rigler, R., and Ringertz, N. R. (1966). Changes in deoxyribonucleoprotein during spermatogenesis in the bull. *Exp. Cell Res.* **41,** 632–641.
41. Kernell, A. M., Bolund, K., and Ringertz, N. R. (1971). Chromatin changes during erythropoiesis. *Exp. Cell Res.* **65,** 1–10.
42. Killander, D. (1965). Intercellular variation in generation time and amounts of DNA, RNA and mass in a mouse leukemia population *in vitro*. *Exp. Cell Res.* **40,** 21–31.
43. Killander, D., and Zetterberg, A. (1965). Quantitative cytochemical studies on interphase growth. *Exp. Cell Res.* **38,** 272–284.
44. Killander, D., and Zetterberg, A. (1965). A quantitative cytochemical investigation of the relationship between cell mass and initiation of DNA synthesis in mouse fibroblasts *in vitro*. *Exp. Cell Res.* **40,** 12–20.
45. Kurland, J., Traganos, F., Darzynkiewicz, Z., and Moore, M. (1978). Macrophage mediated cytostasis of neoplastic hemopoietic cells. Cytofluorometric analysis of cell block. *Cell. Immunol.* **38,** 318–330.
46. Ling, M. R., and Kay, J. E. (1975). "Lymphocyte Stimulation." Elsevier/North-Holland, New York.
47. Matthews, H. R., and Bradbury, E. M. (1978). The role of H1 histone phosphorylation in the cell cycle. *Exp. Cell Res.* **111,** 343–351.
48. Mitchison, J. M. (1971). "The Biology of the Cell Cycle." Cambridge Univ. Press, London and New York.
49. Moser, G. C., Fallon, R. J., and Meiss, H. K. (1981). Fluorimetric measurements and chromatin condensation patterns of nuclei from 3T3 cells throughout G_1. *J. Cell. Physiol.* **106,** 293–301.
50. Noronha, A. B. C., Richman, D. P., and Arnason, B. G. W. (1980). Detection of *in vivo* stimulated cerebrospinal fluid lymphocytes by flow cytometry in patients with multiple sclerosis. *N. Engl. J. Med.* **303,** 713–717.
51. Pardee, A. B., Dubrow, R., Hamlin, J. L., and Kletzien, R. A. (1978). Animal cell cycle. *Annu. Rev. Biochem.* **47,** 715–750.
52. Pederson, T. (1972). Chromatin structure and the cell cycle. *Proc. Natl. Acad. Sci. U.S.A.* **69,** 2224–2228.
53. Pochron, S., Rossini, M., Darzynkiewicz, Z., Traganos, F., and Baserga, R. (1980). Failure of accumulation of cellular RNA in hamster cells stimulated to synthesize DNA by infection with adenovirus 2. *J. Biol. Chem.* **255,** 4411–4413.
54. Preisler, H., and Darzynkiewicz, Z. (1981). Flow cytometric analysis of proliferating and quiescent human leukemia cells. *Leuk. Res.* (in press).
55. Prescott, D. M. (1976). "Reproduction of Eukaryotic Cells." Academic Press, New York.
56. Prescott, D. M. (1976). The cell cycle and the control of cellular reproduction. *Adv. Genet.* **18,** 99–177.
57. Richman, P. D. (1980). Lymphocyte cell-cycle analysis by flow cytometry. Evidence for a specific postmitotic phase before return to G_0. *J. Cell Biol.* **85,** 459–465.
58. Rigler, R. (1966). Microfluorometric characterization of intracellular nucleic acids and nucleoproteins by acridine orange. *Acta Physiol. Scand.* **67,** Suppl., 1–122.
59. Rossini, M., Lin, J. C., and Baserga, R. (1975). Effects of prolonged quiescence on nucleic and chromatin of WI-38 fibroblasts. *J. Cell. Physiol.* **88,** 1–12.
60. Roti Roti, J. L., Bohling, V., and Dethlefsen, L. A. (1978). Kinetic models of C3H mouse mammary tumor growth: Implications regarding tumor cell loss. *Cell Tissue Kinet.* **11,** 1–21.

61. Ruiz-Carrillo, A., Wangh, L. J., and Allfrey, V. G. (1975). Processing of newly synthesized histone molecules. *Science* **190,** 117–128.
62. Smith, J. A., and Martin, L. (1973). Do cells cycle? *Proc. Natl. Acad. Sci. U.S.A.* **70,** 1263–1267.
63. Stadler, B. M., and DeWeck, A. L. (1980). Flow cytometric analysis of mouse peritoneal macrophages. *Cell. Immunol.* **54,** 36–48.
64. Stadler, B. M., Kristensen, F., and DeWeck, A. L. (1980). Thymocyte activation by cytokines: Direct assessment of G_0–G_1 transition by flow cytometry. *Cell. Immunol.* **55,** 436–443.
65. Traganos, F., Darzynkiewicz, Z., Sharpless, T., and Melamed, M. R. (1977). Simultaneous staining of ribonucleic and deoxyribonucleic acids in unfixed cells using acridine orange in a flow cytofluorometric system. *J. Histochem. Cytochem.* **25,** 46–56.
66. Traganos, F., Gorski, A. J., Darzynkiewicz, Z., Sharpless, T., and Melamed, M. R. (1977). Rapid multiparameter analysis of cell stimulation in mixed lymphocyte culture (MLC) reaction. *J. Histochem. Cytochem.* **25,** 881–887.
67. Traganos, F., Darzynkiewicz, Z., Sharpless, T., and Melamed, M. R. (1979). Erythroid differentiation of Friend leukemia cells as studied by acridine orange staining and flow cytometry. *J. Histochem. Cytochem.* **27,** 382–389.
68. Traganos, F., Evenson, D. P., Staiano-Coico, L., Darzynkiewicz, Z., and Melamed, M. R. (1980). Action of dihydroanthraquinone on cell cycle progression and survival of a variety of cultured mammalian cells. *Cancer Res.* **40,** 671–681.
69. Traganos, F., Staiano-Coico, L., Darzynkiewicz, Z., and Melamed, M. R. (1981). Effects of aclacinomycin on cell survival and cell cycle progression of cultured mammalian cells. *Cancer Res.* **41,** 2728–2737.
70. Watson, J. V., and Chambers, S. H. (1977). Fluorescence discrimination between diploid cells on their RNA content: A possible distinction between clonogenic and non-clonogenic cells. *Br. J. Cancer* **36,** 592–600.
71. Watson, J. V., and Chambers, S. H. (1978). Nucleic acid profile of the EMT6 cell cycle *in vitro. Cell Tissue Kinet.* **11,** 415–422.

5

Nuclear Fluorescence and Chromatin Condensation of Mammalian Cells during the Cell Cycle with Special Reference to the G_1 Phase

GERTRUDE C. MOSER AND HARRIET K. MEISS

I. Introduction

The cell division cycle is conventionally described by subdividing the time between successive mitoses into G_1, S, and G_2 phases (*21*). These terms are based on the periods of DNA replication and intervening periods in relation to mitosis. In parallel with the DNA cycle, the chromatin seems to undergo a decondensation–condensation process between the successive mitoses. The existence of such a chromatin cycle was previously proposed by Mazia (*29*), who stated that decondensation of the chromatin may take place throughout G_1 up to a critical point, at which

GENETIC EXPRESSION IN THE CELL CYCLE

ISBN 0-12-543720-X

time the chromosomes can replicate. Following completion of replication of any part of the chromatin, recondensation would begin and continue to the next mitosis. Staining nuclei with the fluorescent dye quinacrine dihydrochloride (QDH) has enabled us to monitor specific aspects of such a chromatin cycle cytologically. The nuclear fluorescence intensity of cells stained with this dye decreases as the cells pass from mitosis to S phase and increases during the rest of the cycle (*33*). Differences in fluorescence intensity of nuclei from various stages within G_1 were measured with a microfluorometer (*35*). Quinacrine is one of the acridine DNA intercalating dyes, originally used for chromosome staining by Casperson and Zech, who discovered that specific fluorescent metaphase banding patterns result in both plant (*4*) and in human cells (*5*). However, the mechanism of dye–chromatin interaction and fluorescence emission is not yet clearly understood.

The phenomenon of premature chromosome condensation (PCC) described by Johnson and Rao (*22*) has also been useful in determining the chromatin condensation state of cells during interphase. Fusion of mitotic cells with interphase cells results in PCC patterns characteristic of the state of chromatin of the interphase cells at the time of fusion. Thus, G_1 cells yield single-stranded PCC chromatids, S-phase cells "pulverized"-appearing chromatin, and G_2-phase cells long, double-stranded chromatids (*40*). Moreover, as with quinacrine staining, it is possible to position cells within G_1, since the chromatin strands of the PCC figures become considerably longer and thinner between the end of mitosis and the initiation of DNA synthesis (*17*). Since measurement of DNA content cannot be used to discriminate time points within G_1, QDH-staining as well as PCC methods are particularly useful for the analysis of the G_1 interval. Darzynkiewicz *et al.* (*8*) have shown that it is possible to distinguish quiescent lymphocytes from stimulated ones by flow cytometric analysis after acid denaturation of the DNA. It is during the G_1 phase that the critical functions such as cell differention (*1,12,46*) or preparation for cell cycle traverse (*2,36,40*) take place, except in those cells where no G_1 exists (*7,26*).

In this chapter the following findings are reported: the technique of staining with QDH and the nuclear fluorescence patterns obtained throughout the cell cycle for cells derived from several species, results of fluorometric measurements of nuclei from synchronized 3T3 cells as they traverse G_1 as well as for those from cells arrested in G_1 by applying restrictive growth conditions, comparisons between the QDH fluorescence intensities and PCC patterns, and applications of QDH staining for the screening and study of temperature-sensitive (*ts*) mutants of the fibroblast cell line of baby hamster kidney (BHK). The possible mechanism(s) of dye–chromatin interaction in relation to the state of condensation is (are) discussed.

II. The QDH Staining Method and Fluorescent Nuclear Patterns

Monolayer cultures grown on coverslips or slides are fixed in methanol: acetic acid 3:1 and air-dried. The preparations are then exposed to 0.3 N HCl for 1 min, rinsed in distilled H_2O, placed in MacIlvaines buffer (pH 4.1) for 4 min, and then stained in 0.1% solution of QDH. After several rinses in buffer the preparations are dehydrated in an alcohol series and air dried. The coverslips are mounted cell side up and are observed through an epifluorescent microscope (excitor = 436 nm, barrier = 490 nm) with an oil-immersion objective (*23*). The oil is directly placed on the coverslip after it is firmly mounted. When a randomly growing population of cells such as 3T3 is fixed and stained with QDH, a variety of nuclear fluorescence patterns results. As seen in Fig. 1, there are nuclei which are large with a low fluorescence intensity and others varying in size with higher intensities. Bright mitotic figures are easily recognized. Each pattern is characteristic for a cell at a specific point in the cell cycle. The cell cycle position of each nucleus has been established by two procedures for a number of cell types such as HeLa cells, human embryonic fibroblasts (*33*), BHK cells (*32*), and mouse embryonic fibroblast cells (*34*). In the case of HeLa cells, synchronization by mitotic detachment was used and cell samples were fixed and stained at different intervals after mitosis as shown in Fig. 2 (*33*). The second method applied to fibroblasts was as follows: A series of asynchronous monolayer cultures were exposed to [^{3}H]thymidine for varying lengths of time just prior to fixation. The cells were stained with QDH and nuclei with specific fluorescence patterns were photographed and localized on the slide. Then these slides were processed for autoradiography and the cell nuclei previously examined for QDH fluorescence and photographed were localized and evaluated with respect to their [^{3}H]thymidine uptake pattern. Knowing the cell cycle times, it is possible to assume that cells exposed to [^{3}H]thymidine, for example, for 4 hr and free of silver grains were in G_1 at the time of fixation. Nuclei in early S phase can be distinguished from those in late S because of the grain patterns of the clumped heterochromatin, as well as the inactive X chromatin in human female cells, which replicates at the end of the S period (*42*). In contrast,

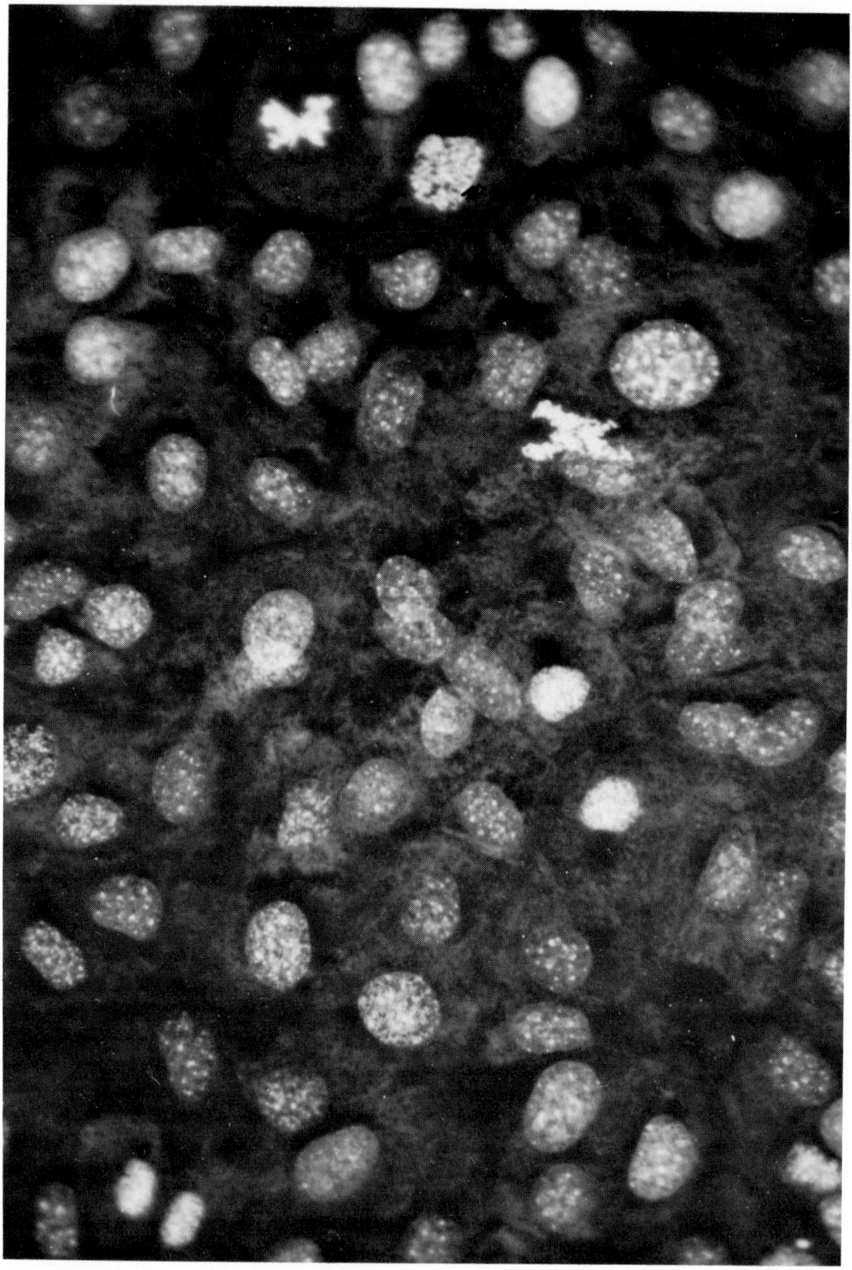

Fig. 1. Micrograph of a monolayer of asynchronous cultured 3T3 cells, stained with QDH. Bar, 20 μm.

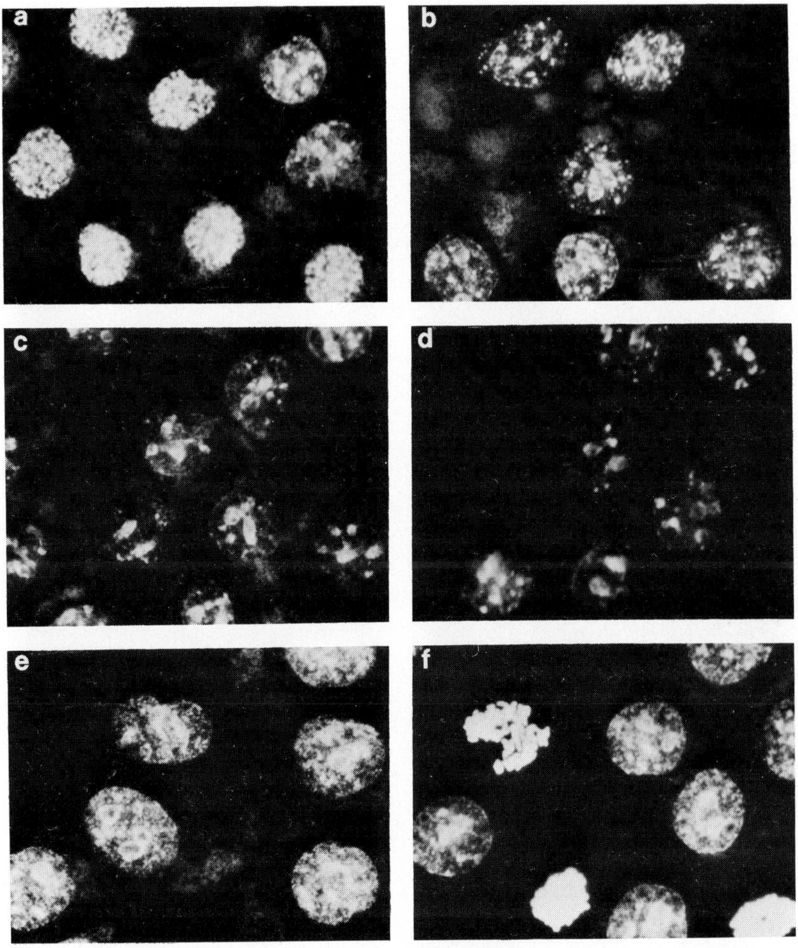

Fig. 2. Montage of synchronized HeLa cells as they procede through the cell cycle. (a) Early G_1, 1.5 hr; (b) early G_1, 2.5 hr; (c) G_1, 6 hr; (d) $G_1 \rightarrow S$, 8 hr; (e) late $S \rightarrow G_2$, 14 hr; and (f) $G_2 \rightarrow M$, 18 hr. Early and late indicate positions within the respective phase; arrows in figure legend indicate that morphological changes do not occur exactly at time of transition from one phase to the next. The times indicated relate to hr after mitosis. [From Moser *et al.* (*33*) with permission.] Bar, 10 μm.

euchromatin, which is more evenly distributed, replicates earlier during the S phase. If a given nucleus has silver grains with a "spotty" distribution covering the heterochromatic regions, it can be concluded that this nucleus was in late S phase at the time of addition of the [^{3}H]thymidine. Correlatively, cells with a low nuclear grain count and unlabeled heterochromatin can be designated as being earlier in S. Figure 3 illustrates

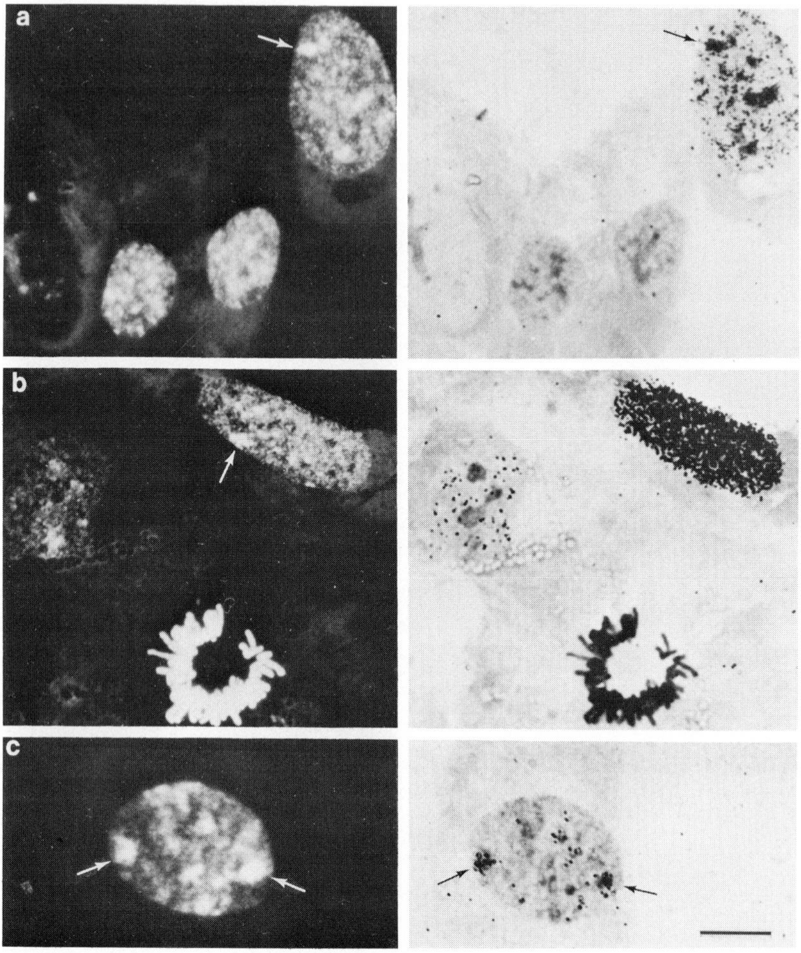

Fig. 3. Micrographs of samples of embryonic human fibroblast nuclei: left, stained with QDH; right, autoradiographs. (a) and (b) are taken from the same slide, and (c) from another slide. Arrow, X-chromatin. [From Moser *et al.* (*33*), with permission.]

corresponding autoradiographic and fluorescence patterns of culture samples of human embryonic fibroblasts that were exposed to [^{3}H]thymidine for 2 hr. The small, paired, brightly fluorescent, unlabeled nuclei in Fig. 3a are examples of the early G_1 pattern. The unlabeled weakly fluorescent nucleus in Fig. 3a is an example of a late G_1 nucleus. In other experiments (not shown) late G_1 cell types were identified by their lack of label after a longer [^{3}H]thymidine pulse. The spottily labeled nucleus in Fig. 3a was in late S or G_2 at the time of fixation. In Fig. 3b examples of early and late S nuclei are shown with their corresponding labeling patterns. Figure 3c represents a nucleus of an embryonic fibroblast line with a 47, XXX chromosome constitution. In this late S or G_2 nucleus, the two X chromatin bodies show preferential uptake of [^{3}H]thymidine. Figure 4 is a montage of photographs which represent a cell cycle sequence of mouse embryonic fibroblasts (*34*). To summarize, early G_1 nuclei are small and usually paired and show a brilliant fluorescence of the chromatin; as the cells progress through G_1, the nuclei appear larger and the fluorescence intensity is reduced; later in G_1 and at the G_1–S boundary, the chromatin shows very low fluorescence except for some bright spots representing

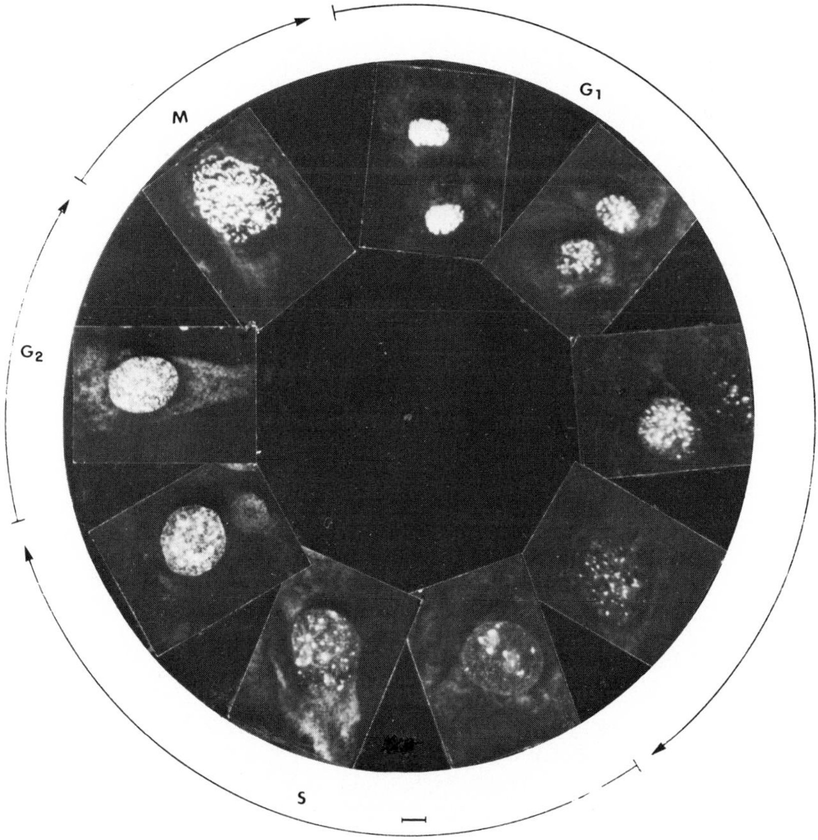

Fig. 4. Micrographs of fluorescence patterns of representative nuclei, stained with QDH, arranged according to cell cycle position determined by autoradiography in an asynchronous mouse embryonic fibroblast culture. Bar, 10 μm.

the mouse heterochromatin. In early S the nuclei are still dark but the intensity increases during middle to late S and continues to increase into G_2, where the chromatin shows a grainy fluorescent appearance. In late G_2 one can begin to see chromatin strands foreshadowing prophase. At metaphase the chromatin is maximally bright.

III. Fluorometric Measurements of QDH-Stained Nuclei from Synchronized 3T3 Cells

We were especially interested in exploring the fluorescence patterns and intensities exhibited in G_1 nuclei. As evident from Fig. 4, the decrease in fluorescence intensity as cells pass from mitosis to S is accompanied by an increase in nuclear size. We therefore considered it important to verify this visual observation with fluorometric measurements to demonstrate that during G_1 there was an actual decrease in total fluorescence and not just a dispersion of the chromatin into a larger nuclear volume.

A. Mitotic Detachment

We measured total nuclear fluorescence using a microfluorometer (Leitz MPV compact) of 3T3 cell samples synchronized by mitotic detachment as they passed through G_1 at 2, 4, and 6.5 hr, at which time 9% of cells were found to be in the S phase. The projected "area" occupied by the fixed nuclei was calculated from diameter measurements with an ocular micrometer. From the micrographs in Fig. 5 it can be seen that the fluorescence of the nuclei decreased with time from 2 hr to 6.5 hr after mitosis. The histograms in Fig. 5 and the data in Table I give the results of these measurements. As seen in Table I there was a small but significant decrease of 19% (from 32 to 26 relative units) in total fluorescence between 2 hr and 4 hr after mitosis. Since the nuclear "area" at 4 hr is 1.7 times higher than that of 2 hr, the somewhat darker appearance of the 4-hr nuclei may be attributed for the most part to dispersion of fluorescence into a larger area. However, during the interval from 4 to 6.5 hr the total fluorescence decreased 58% (from 26 to 11 units), while nuclear "area" was 1.9 times larger at 6.5 hr than at 4 hr. Thus, in this case the dark appearance of the nuclei is primarily a reflection of a decrease in total fluorescence. As can be seen from the calculations in Table I, the rate of decrease in fluorescence intensity and the concurrent increase in nuclear "area" are considerably greater for the 4–6.5 hr interval than for the 2–4 hr interval. The significance of this finding is not known. However, on the whole, our results confirm

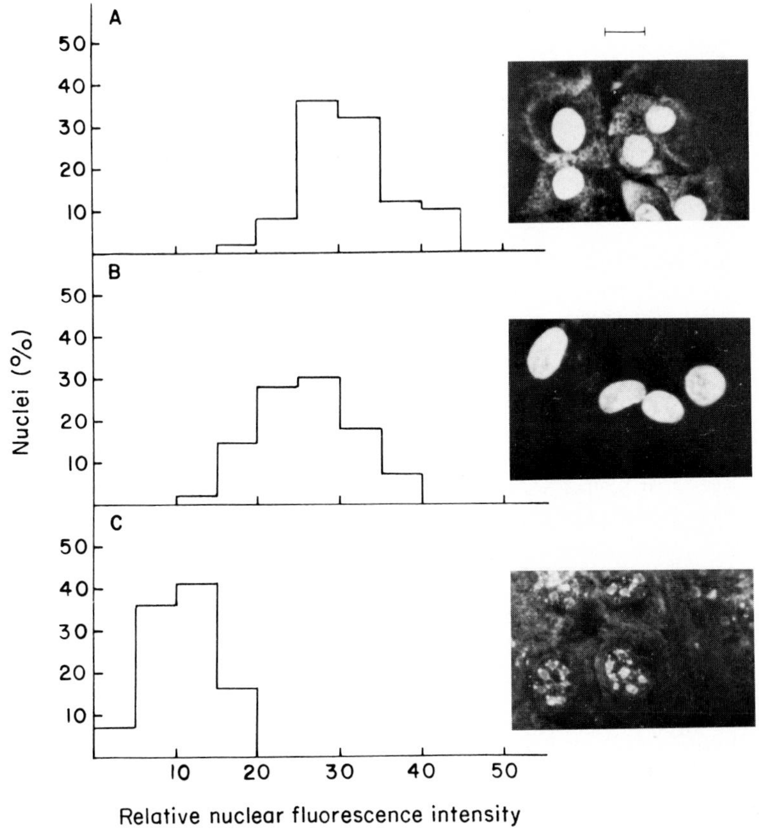

Fig. 5. Histograms of fluorometric measurements of QDH-stained 3T3 nuclei from G_1 cells synchronized by mitotic detachment. (A) 2 hr after mitosis, (B) 4 hr after mitosis, and (C) 6.5 hr after mitosis. The micrograph insets are of the corresponding QDH-stained nuclei. Bar, 20 μm. [From Moser *et al.* (*35*), with permission.]

the decrease in fluorescence intensity reflects a true reduction in total nuclear fluorescence rather than dispersion of chromatin into a larger nuclear area, and the QDH staining patterns reflect the process of chromatin decondensation which occurs during the G_1 phase. It is probable that the process of decondensation of the chromatin and the increase in nuclear size are interrelated. Yen and Pardee (*50*), using the flow fluorometry and light-scatter analysis of serum-starved quiescent 3T3 cells, found a positive correlation between nuclear volume and the rate of entry into the S phase. It is reasonable to assume that these changes in the morphology of the nucleus are preparatory for cells to enter S.

TABLE I

Relative Nuclear Fluorescence Intensity of Cells Synchronized in G_1 by Mitotic Detachment[a]

Time after mitosis (hr)	Mean total nuclear fluorescence (relative units)	Rate of decrease of fluoroescence (relative units/hr)	Nuclear "area" (μm^2)	Rate of increase in "area" (μm^2/hr)
2	32.0 ± 0.75[b]	2.9	145 ± 8.9[b]	51
4	26.2 ± 0.74	6.0	252 ± 9.1	91
6.5	11.1 ± 0.50		480 ± 9.2	

[a] Each time point represents measurements of 60 nuclei from two independent cell samples. Nuclear "area" was calculated from measurements of nuclear diameter. [From Moser *et al.* (*35*), with permission.]

[b] Average value ± SEM.

B. Blocking Conditions

Suboptimal culture conditions, such as contact inhibition and serum starvation (*47*) or isoleucine deprivation (*25*), result in arresting the growth of many normal mammalian cells somewhere during the G_1 phase. Exposure to hydroxyurea (HU) stops cells at the G_1/S boundary (*39*). When 3T3 cells are subjected to any one of these conditions and stained with QDH, the nuclei exhibit relatively uniform fluorescence patterns. Fluorometric measurements of nuclei from cells blocked by serum starvation, contact inhibition, isoleucine deprivation, and HU treatment are reported in Fig. 6 and Table II. Micrographs of the nuclear fluorescence patterns are displayed in Fig. 6. The 28.9 relative units of fluorescence intensity obtained with nuclei of serum-starved cells were not significantly different from the 27.6 relative units obtained for contact-inhibited

TABLE II

Relative Nuclear Fluorescence Intensity of Cells Arrested in G_1 by a Variety of Conditions[a]

Culture condition	Total mean nuclear fluorescence (relative units)	Mean nuclear "area" (μm^2)
Serum starvation	28.9 ± 1.1[b]	343 ± 8.3[b]
Contact inhibition	27.6 ± 1.0	306 ± 8.7
Isoleucine deprivation	13.6 ± 0.6	445 ± 11.2
HU exposure	8.7 ± 0.4	526 ± 7.0

[a] Values are based on measurements of 50 nuclei for each condition pooled from two independent cell samples. [From Moser *et al.* (*35*), with permission.]

[b] Average value ± SEM.

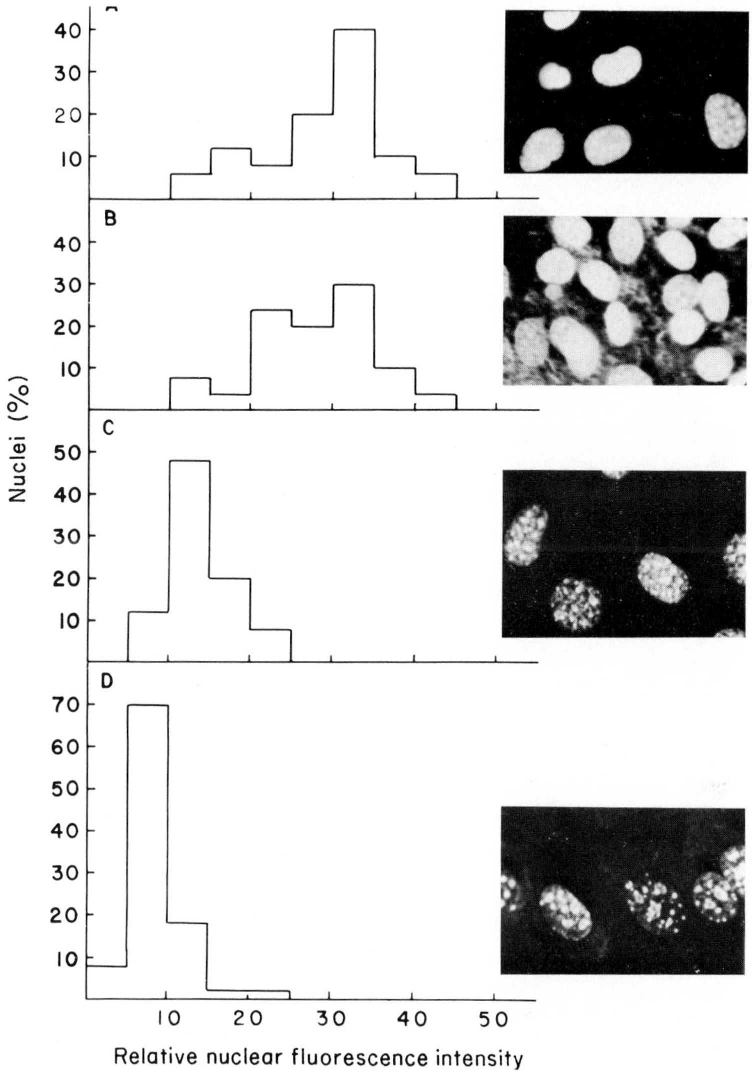

Fig. 6. Histograms of fluorometric measurements of QDH-stained 3T3 nuclei arrested by various G_1 blocking conditions. (A) serum starvation, (B) contact inhibition, (C) isoleucine deprivation, and (D) HU treatment. The micrograph insets are of the corresponding QDH-stained nuclei. Bar, 20 μm. [From Moser *et al.* (*35*), with permission.]

cells. Frequent medium changes assured that these cells were not starved for serum. The average fluorescence intensity from isoleucine-deprived cells was 13.8 and that from HU-treated cells was 8.7. Thus, with this technique it is possible to order these four G_1-arrested cell populations with respect to transit from mitosis to S. Serum-starved and contact-inhibited cells precede isoleucine-deprived ones, and HU-treated cells are blocked latest, at the G_1/S border. Here again as seen in the synchronization experiments above, a decrease in the fluorescence intensity was accompanied by an increase in nuclear "area." In this respect, the nuclei from the blocked cells generally are larger than those of normally growing cells traversing G_1 (see Tables I and II). These treatments also perturb normal nuclear morphology.

Our results on the relative positions of the serum and isoleucine blocks in 3T3 cells within G_1 agrees with Burstin *et al.* (*3*). They concluded from biochemical experiments with the cell cycle BHK mutant, *ts* AF8, that the functions provided by serum precede those provided by isoleucine and that the *ts* function occurs in between (see Section V). Other workers, employing a variety of methods, also showed that the cells come to rest at different states during G_1 (*44,49*) and that the quiescent state is not a single unique state as had earlier been proposed by Pardee (*36*). It is clear from the results of QDH staining that the states of the chromatin are different depending upon the method used to arrest the cells, although that of serum-starved and contact-inhibited cells are similar.

IV. Fluorescence Patterns Resulting upon Release from Serum Block

The method of releasing serum-starved "quiescent" cells by replacement of serum factors has been used in many investigations (*36,41,45*). It is well known that the length of time until the onset of DNA synthesis is much longer than the G_1 period in unperturbed growing cells. In order to probe the sequence of events during this "lag" period, we examined the QDH fluorescence staining patterns of 3T3 nuclei after the serum-blocked cells are released by serum addition. The blocked 3T3 cells are characterized by relatively bright nuclear chromatin patterns (see Fig. 6). No change was observed in nuclear fluorescence for 6 to 8 hr after addition of serum; only by 10 hr, at the time when the cells were just beginning to enter S phase, did 50% of the nuclei appear darker, as would be expected for cells late in G_1 or early S. Thus, during the lag period the chromatin does not decondense, which may signify that the cells remain arrested and are not slowly traversing the cycle. However, when the cells begin to cycle they appear to do so with a normal transit time. Therefore, kinetic studies in combination with QDH studies can give an estimate of the length of time required for cells to reenter G_1 from a quiescent state. These results show that kinetic studies which position cells in G_1 according to the length of time it takes for the cells to enter S may not be reliable.

V. Correlation of PCC Morphology with QDH Staining Patterns

Rao *et al.* (*40*) have shown that fusion of mitotic cells with those in interphase results in PCC patterns characteristic of the state of the chromatin of the interphase cells at the time of fusion. As noted earlier, Hittelman and Rao (*18*) demonstrated that it is possible to position cells in G_1 by classifying PCC morphological patterns in terms of lengths and thickness of the chromosomes. The PCC morphology varies from short, stubby chromosomes, early in G_1 to long, thin, late in G_1, and pulverized ones at the G_1/S boundary. We undertook the experiments described below, comparing results obtained with QDH staining with those obtained with PCC in order to gain further insight into the mechanism of QDH staining (*35*). We examined with both techniques 3T3 cells blocked in G_1 by the culture conditions discussed above. Serum-starved and contact-inhibited cell nuclei had the highest fluorescence intensity and had the shortest, thickest prematurely condensed chromosomes. HU-treated ones which had the lowest intensity had the longest, thinnest, and sometimes pulverized chromosomes, whereas the intensity and PCC pattern for isoleucine-deprived cells were intermediate. Therefore the same relative order of G_1 positions was obtained with both methods, which most likely monitor the same event, namely the state of chromatin condensation (*35*). The fluorescent QDH method monitors the cell nuclei *in situ* after fixation and staining and is therefore a simpler and more rapid procedure than that required for PCC analysis. However with PCC more information can be obtained since the chromosomes can be visualized.

VI. Applications of Cytologic Methods to the Analysis of Blocks Caused by Temperature-Sensitive Mutations

Many *ts* mutant cell clones are selected for their inability to grow at the nonpermissive temperature (39°C), making the characterization of the specific lesion difficult. As a first step in the classification of some of our BHK *ts* growth mutants, we have made use of these cytological techniques. Those which were cell cycle mutants had a relatively uniform fluorescence pattern and PCC morphology. The arrest point in the cycle could likewise be determined from the QDH pattern.

In the case of the BHK *ts* mutant, *ts* AF8, the arrest point was found to be located in G_1 between the blocks induced by serum starvation and isoleucine deprivation in biochemical experiments (*3*). These cells, if incubated in low serum medium at the permissive temperature (33°C) and then shifted to complete medium at 39°C, were not able to enter S, whereas those incubated in isoleucine-deficient medium at 33°C and then shifted to complete medium at 39°C were able to enter S. In cytological studies with fluorescent QDH staining it was confirmed that indeed *ts* AF8 cells exhibit uniform nuclear patterns consistent with a cell cycle position late in G_1 (*32*). More recently we measured by microfluorometry the fluorescence intensity and the PCC patterns of *ts* AF8 cells at 39°C. We demonstrated that the fluorescence intensity of the cell nuclei at 39°C is lower than that of serum-starved ones at 33°C and higher than that of isoleucine-deprived *ts* AF8 cells at 33°C. Likewise, at 39°C, PCC patterns are more condensed than under conditions of isoleucine deprivation and less condensed than under conditions of serum starvation. Therefore, the physiological and cytological data are in full agreement, and a relatively simple staining method has yielded the same information as that of a more complicated biochemical technique.

Meiss *et al.* (*30*) described a group of noncomplementing cell cycle mutant clones of BHK that were mapped to the X chromosome (*43*). One of these, *ts* A45, was examined by us (*31*) in a manner similar to *ts* AF8. The *ts* A45 cells, in contrast to those from *ts* AF8, cannot enter S upon shift up to 39°C after both serum starvation and isoleucine deprivation, but can enter S if blocked at the G_1/S boundary by HU. These results confirm those of Talavera and Basilico (*45*) who studied *ts* 13, a BHK mutant that does not complement with *ts* A45 (*30*). Analysis of *ts* A45 at 39°C by QDH revealed a relatively uniform G_1 pattern. We measured QDH fluorescence intensities for the *ts* A45 cells under the variety of blocking conditions as in the case of *ts* AF8. The pattern for isoleucine-arrested *ts* A45 cells at 33°C was abnormally bright, i.e., only slightly lower than that of serum-arrested *ts* A45 cells at 33°C or serum-arrested wild-type BHK. Exposure to 39°C resulted in a pattern of fluorescence which was slightly lower than that of isoleucine-deprived cells. These results, also confirmed by the method of PCC, agree with the relative order of the biochemically determined arrest point that placed the *ts* block after that of isoleucine block. However, it seems that the *ts* mutation also affects chromatin condensation of the cells at 33°C under conditions of deprivation of isoleucine. Whether these two phenotypic effects were caused by one identical mutation is now being investigated. It is interesting that the cytological studies yielded information which was not revealed in the kinetic studies.

VII. Discussion

A simple and rapid technique of QDH fluorescent staining has proved useful and reliable in monitoring the position of cells throughout the cell cycle. The observed decrease in fluorescence intensity of cells as they progress from mitosis to S phase and the increase in fluorescence intensity as they progress from S phase to mitosis most probably reflect changes in chromatin organization. Our results with QDH correlate with those obtained by studies of PCC, which likewise monitors the chromatin decondensation–condensation cycle. Both these methods give visual support to evidence of other investigators who have reported changes in the chromatin structure of cycling cells. Pederson and Robbins (*38*) found a gradual increase in actinomycin binding to nuclei of synchronized HeLa cells between early G_1 and S and decreased binding in late G_2; similarly, Pederson (*37*) observed increased sensitivity of the nuclear DNA to DNase in late G_1 cells. Hoechst dye 33258 inhibits chromatin condensation, especially of the centromeric heterochromatin (c-chromatin) regions in mouse chromosomes (*16*). This fact has enabled Marcus and Sperling (*27*), using PCC in the presence of Hoechst, to show that the heterochromatin regions decondense late in G_1 and recondense in G_2. The actual mechanism responsible for condensation processes of chromatin has been the subject of many investigations but as yet is not fully understood. For example, the studies of Gurley *et al.* (*13*) and Hildebrand *et al.* (*15*) have provided evidence that a variety of phosphorylations of histone H1 in Chinese hamster ovary cells affect condensation processes during interphase. The release of the protein A24 from histone H2A during late G_2 may be important for chromatin condensation during mitosis (*28*).

The biochemical and physical reactions underlying the differences in nuclear fluorescence intensity patterns exhibited during the cycle are likewise not well understood. QDH is one of the acridine dyes which binds to DNA by intercalation; however, the nature of its binding to chromatin has yet to be elucidated (*10,24*). Investigations into the nature of Q-banding of metaphase chromosomes have shown that A-T rich DNA sequences enhance fluorescence intensity and that G-C sequences have a quenching effect (*6,48*). Other factors, such as chromosomal proteins (*6,11*) and the state of chromatin condensation (*20*), also seem to play a major role in modifying the degree of fluorescence emission. Hatfield *et al.* (*14*) showed that the brilliant fluorescence of the long arm of the Y chromosome bound an amount of tritiated QDH equivalent to the autosomes. Likewise, the binding of another such dye, acridine orange, to chromatin has been shown to be constant and independent of the state of chromatin condensation (*9*). Thus, it appears that changes in the amount of bound QDH do not dramatically alter the fluorescence intensity, and the reduction in fluorescence intensity observed in G_1 is most likely due to the structural changes which take place in the chromatin as the cells traverse G_1.

The procedures for using QDH staining to evaluate cells in suspension, either by cytocentrifugation or by dropping cells subjected to hypotonic medium and fixation directly into slides are now being worked out. QDH analysis can be applied to many other biological systems in which it would be useful to determine the state of chromatin of cells as well as their cell cycle position. For example, it would be interesting to determine the state of chromatin in cells prior to and after differentiation or to study the effects of specific drugs on the state of chromatin and cell cycle transit. The phenomenon of PCC has now been applied clinically to predict the future course of the disease in leukemic patients during chemotherapy and to the forecasting of relapse. The bone marrow cells of patients who will relapse show an accumulation of cells in late G_1 phase instead of early G_1 (*18,19*). The application of the technique of QDH staining for such clinical studies would be relatively simple and therefore more efficient for routine diagnosis in clinical laboratories.

Acknowledgments

The authors would like to thank the E. Leitz Co., Inc. (Rockleigh, N.J.) who so kindly loaned us the MPV compact fluorometer which was used for the fluorometric measurements reported.

Furthermore, we thank Ms. Roni Rossman for her excellent technical assistance and Dr. Menashe Marcus for his helpful suggestions with the manuscript.

Our research has been supported by NIH grant CA-16631. H.K.M. is a recipient of a Career Award, NIH–NCI 00020.

References

1. Buckley, P. A., and Konigsberg, I. R. (1974). Myogenic fusion and the duration of the post-mitotic gap (G_1). *Dev. Biol.* **37**, 193–212.

2. Burns, F. J., and Tannock, I. F. (1970). On the existence of a G_0-phase in the cell cycle. *Cell Tissue Kinet.* **3,** 321–324.
3. Burstin, S. J., Meiss, H. K., and Basilico, C. (1974). A temperature sensitive cell cycle mutant of the BHK cell line. *J. Cell. Physiol.* **84,** 397–408.
4. Casperson, T., Zech, L., Modest, E. J., Foley, G. E., Wagh, U., and Simonsson, E. (1969). Chemical differentiation with fluorescent alkylating agents in *Vicia faba* metaphase chromosomes. *Exp. Cell Res.* **58,** 123–140.
5. Casperson, T., Zech, L., and Johansson, C. (1970). Differential binding of alkylating fluorochromes in human chromosomes. *Exp. Cell Res.* **60,** 315–319.
6. Comings, D. E., Kovacs, B. W., Avelino, E., and Harris, D. C. (1975). Mechanisms of chromosome banding. V. Quinacrine banding. *Chromosoma* **50,** 111–145.
7. Cooper, S. (1979). A unifying model for the G_1 period in prokaryotes and eukaryotes. *Nature (London)* **280,** 17–19.
8. Darzynkiewicz, Z., Traganos, F., Andreeff, M., Sharpless, T., and Melamed, M. R. (1979). Different sensitivity of chromatin to acid denaturationin quiescent and cycling cells as revealed by flow cytometry. *J. Histochem. Cytochem.* **27,** 478–485.
9. Darzynkiewicz, Z. (1979). Acridine orange as a molecular probe in studies of nucleic acids in situ. *In* "Flow Cytometry and Sorting" (M. R. Melamed, M. Mendelsohn, and P. Mullaney, eds.), pp. 285–316. Wiley, New York.
10. Gabby, J. E., and Wilson, W. D. (1978). Intercalating agents as probes of chromatin structure. *Methods Cell Biol.* **18,** 351–384.
11. Gatti, M., Pimpinelli, S., and Santini, G. (1976). Characterization of *Drosophila* heterochromatin. Staining and decondensation with Hoechst 33258 and quinacrine. *Chromosoma* **57,** 351–375.
12. Green, H., and Meuth, M. (1974). An established pre-adipose cell and its differentiation in culture. *Cell* **3,** 127–133.
13. Gurley, L. R., Tobey, R. A., Walters, R. A., Hildebrand, C. E., Hohmann, P. G., D'Anna, J. A., Barham, S. S., and Deaven, L. L. (1978). Histone phosphorylation and chromatin structure in synchronized mammalian cells. *In* "Cell Cycle Regulation" (J. R. Jeter, I. L. Cameron, G. M. Padilla, and A. M. Zimmerman, eds.), Cell Biol. Monogr. Ser., pp. 37–60. Academic Press, New York.
14. Hatfield, J. M. R., Pleden, K. W. C., and West, R. M. (1975). Binding of quinacrine to the human Y-chromosome. *Chromosoma* **52,** 67–71.
15. Hildebrand, C. E., Tobey, R. A., Gurley, J. R., and Walters, R. A. (1978). Action of heparin on mammalian cell nuclei. II. Cell-cycle specific changes in chromatin organization correlate temporally with histone-I, H1, phosphorylation. *Biochim. Biophys. Acta* **517,** 486–499.
16. Hilwig, I., and Gropp, A. (1973). Decondensation of constitutive heterochromatin in L cell chromosomes by a benzimidazole compound (33258 Hoechst). *Exp. Cell Res.* **81,** 474–477.
17. Hittelman, W. N., and Rao, P. N. (1978). Mapping G_1 phase by structural morphology of the premature condensed chromosomes. *J. Cell. Physiol.* **95,** 333–341.
18. Hittelman, W. N., and Rao, P. N. (1978). Predicting response of progression of human leukemia by premature chromosome condensation of bone marrow cells. *Cancer Res.* **38,** 416–423.
19. Hittelman, W. N., Broussard, L. C., Dosik, G., and McCredie, C. H. B. (1980). Predicting relapse of human leukemia by means of premature chromosome condensation. *N. Engl. J. Med.* **303,** 479–484.
20. Holmquist, G. (1975). Hoechst 33258 fluorescent straining of *Drosophila* chromosomes. *Chromosoma* **49,** 333–356.

21. Howard, A., and Pelc, S. R. (1953). The synthesis of deoxyribonucleic acid in normal and irradiated cells and its relation to chromosome breakage. *Heredity, Suppl.* **6,** 261–273.
22. Johnson, R. T., and Rao, P. N. (1970). Mammalian cell fusion. II. Induction of premature chromosome condensation in interphase nuclei. *Nature (London)* **226,** 717–722.
23. Klinger, H. P., and Moser, G. C. (1972). Improved chromatin-fluorescence technique. *Lancet* **2,** 1366.
24. Latt, S. A., Brodie, S., and Munroe, S. H. (1974). Optical studies of complexes of quinacrine with DNA and chromatin: Implications for the fluorescence of cytological chromosomes preparations. *Chromosoma* **49,** 17–40.
25. Ley, K. D., and Tobey, R. A. (1970). Regulation of DNA synthesis in Chinese hamster cells. II. Induction of DNA synthesis and cell division by isoleucine and glutamine in G_1-arrested cells in suspension culture. *J. Cell Biol.* **47,** 453–459.
26. Liskay, R. M. (1978). Genetic analysis of a Chinese hamster cell line lacking a G_1 phase. *Exp. Cell Res.* **114,** 69–77.
27. Marcus, M., and Sperling, K. (1979). Condensation-inhibition by 33258-Hoechst of centromeric heterochromatin in prematurely condensed mouse chromosomes. *Exp. Cell Res.* **123,** 406–411.
28. Matsui, S., Seon, B. K., and Sandberg, A. A. (1979). Disappearance of a structural chromatin protein A 24 in mitosis: Implications for molecular basis of chromatin condensation. *Proc. Natl. Acad. Sci. U.S.A.* **76,** 6386–6390.
29. Mazia, D. (1963). Synthetic activities leading to mitosis. *J. Cell. Comp. Physiol.* **62,** Suppl. I, 123–140.
30. Meiss, H. K., Talavera, A., and Nishimoto, T. (1978). A recurring temperature-sensitive mutant class of BHK-21 cells. *Somatic Cell Genet.* **4,** 125–130.
31. Meiss, H. K., and Moser, G. C. (1980). Cytochemical studies confirm the cell cycle positions of two temperature sensitive mutants of BHK cells. *Eur. J. Cell Biol.* **22,** 600.
32. Moser, G. C., and Meiss, H. K. (1977). A cytological procedure to screen mammalian temperature sensitive mutants for cell cycle related defects. *Somatic Cell Genet.* **3,** 449–456.
33. Moser, G. C., Müller, H., and Robbins, E. (1975). Differential nuclear fluorescence during the cell cycle. *Exp. Cell Res.* **91,** 73–78.
34. Moser, G. C., and Müller, H. J. (1979). Cell cycle dependent changes of chromosomes in mouse fibroblasts. *Eur. J. Cell Biol.* **19,** 116–119.
35. Moser, G. C., Fallon, R. J., and Meiss, H. K. (1981). Fluorometric measurements and chromatin condensation patterns of nuclei from 3T3 cells throughout G_1. *J. Cell. Physiol.* **106,** 293–301.
36. Pardee, A. B. (1974). A restriction point for control of normal animal cell proliferation. *Proc. Natl. Acad. Sci. U.S.A.* **71,** 1286–1290.
37. Pederson, T. (1972). Chromatin structure and the cell cycle. *Proc. Natl. Acad. Sci. U.S.A.* **69,** 2224–2228.
38. Pederson, T., and Robbins, E. (1972). Chromatin structure and the cell division cycle. Actinomycin binding in synchronized HeLa cells. *J. Cell Biol.* **55,** 322–327.
39. Pfeiffer, S. E., and Tolmach, L. J. (1967). Inhibition of DNA synthesis in HeLa cells. *J. Cell Biol.* **55,** 322–327.
40. Rao, P. N., Sunkara, P. S., and Wilson, B. A. (1977). Premature chromosome condensation and cell cycle analysis. *J. Cell. Physiol.* **91,** 131–141.
41. Riddle, V. G. H., Dubrow, R., and Pardee, A. B. (1979). Changes in the synthesis of actin and other cell proteins after stimulation of serum-arrested cells. *Proc. Natl. Acad. Sci. U.S.A.* **76,** 1298–1302.

42. Schmidt, W. (1963). DNA replication patterns of human chromosomes. *Cytogenetics* **2,** 175–193.
43. Schwartz, H. E., Moser, G. C., Holmes, S., and Meiss, H. K. (1979). Assignment of temperature-sensitive mutations of BHK cells to the X chromosome. *Somatic Cell Genet.* **5,** 217–227.
44. Stiles, C. D., Isberg, W. J., Pledger, Antomiades, H. N., and Scher, C. D. (1979). Control of the BALB/c 3T3 cell cycle nutrients and serum factors: Analysis using platelet-derived growth factor and platelet-poor plasma. *J. Cell. Physiol.* **99,** 395–406.
45. Talavera, A., and Basilico, C. (1977). Temperature sensitive mutants of BHK cells affected in cell cycle progression. *J. Cell. Physiol.* **92,** 425–436.
46. Terada, M., Fried, J., Nuder, U., Rifkind, R. A., and Marks, P. A. (1977). Transient inhibition of initiation of S-phase associated with dimethyl sulfoxide induction of murineerythroleukemic cells to erythroid differentiation. *Proc. Natl. Acad. Sci. U.S.A.* **74,** 248–252.
47. Todaro, G. J., Lazar, G. K., and Green, H. (1965). The initiation of cell division in a contact-inhibited mammalian cell line. *J. Cell. Comp. Physiol.* **66,** 325–337.
48. Weisblum, B., and de Haseth, P. L. (1972). Quinacrine, a chromosome stain specific for deoxyadenylate-deoxythymidylate-rich region in DNA. *Proc. Natl. Acad. Sci. U.S.A.* **69,** 629–632.
49. Yen, A., and Pardee, A. B. (1978). Exponential 3T3 cells escape in mid-G_1 from their high serum requirement. *Exp. Cell Res.* **116,** 103–113.
50. Yen, A., and Pardee, A. B. (1979). Role of nuclear size in cell growth initiation. *Science* **204,** 1315–1317.

II

Genetic Expression and Posttranscriptional Modifications

6

Stimulation of Transcription in Isolated Mammalian Nuclei by Specific Small Nuclear RNAs

MARGARIDA O. KRAUSE AND MAURICE J. RINGUETTE

I. Introduction

The basic molecular mechanisms of procaryotic gene expression are now reasonably well understood. In bacteria, details of various elements controlling expression of specific genes have been worked out and found to involve protein molecules, acting as repressors or inducers, synthesized or activated in response to environmental changes and the nutrient requirements of the cell (*11*). By contrast, the internal environment of multicellular eukaryotes is much more stable, with each cell carrying out specialized tissue functions. Only under special situations, such as regeneration or oncogenesis, is there a partial deprogramming of specialized function.

Although eukaryotic organisms are likely to retain some of the fundamental protein mechanisms for gene regulation, the onset and maintenance of stable differentiated phenotypes in the various tissues of

GENETIC EXPRESSION IN THE CELL CYCLE

ISBN 0-12-543720-X

multicellular organisms may require a higher degree of gene selectivity than can be provided by proteins alone. If RNA molecules were involved, their ability to base-pair to DNA would offer greater precision in recognizing specific promoters and maintaining internal stability.

Eukaryotic cells contain a special class of nuclear RNAs, with sedimentation coefficients ranging from 4 to 9 S, referred to as small nuclear RNA (SnRNA). They were first discovered by Knight and Darnell in HeLa cells (*37*) and subsequently have been found in all eukaryotic organisms examined, representing approximately 0.5% of the total cellular RNAs (*25,29,88*). Although they were first found in the nuclei of mammalian cells (*54,70,87*), several species were later found in the cytoplasm (*13,14,57,86*). It appears that most SnRNA species are concentrated in the nucleus; however, some of these RNAs have been shown to shuttle between nucleus and cytoplasm (*24*).

Discrete species of viral-coded SnRNA have also been isolated from adenovirus 2-infected KB and HeLa cells (*55,65,89*), from SV40-infected African green monkey cells (*1*), and from vaccinia virus DNA transcribed *in vitro* (*59*). In addition, RNA tumor virions such as Rous sarcoma virus (*75*), murine leukemia virus (*63*), and mammary tumor virus (*62*) have been found to contain discrete classes of SnRNA of host origin.

The universality of SnRNA in eukaryotes suggests that they may have an important role in the cell. However, until recently, their function was totally unknown. Indirect evidence now implicates certain classes of SnRNA in such roles as DNA replication (*27*), chromatin architecture (*58*), splicing of mRNA (*2,48,72*), transport of mRNA to ribosomes (*28*), and translation regulation (*5*). The possible role of SnRNA in gene expression at the transcriptional level has also been considered by a number of investigators. Bonner and Widholm (*6*) were the first to hypothesize that RNA might function in the regulation of specific genes. Since then several models have been put forward which involve small RNAs as gene derepressors (*7,20,22,38*).

More direct evidence for a role of SnRNA in gene transcription was provided simultaneously by three laboratories. Kanehisa *et al.* (*35*) reported that nuclear 4.5 S RNA could increase the number of binding sites for RNA polymerase to initiate transcription in calf thymus chromatin, whereas Deshpande *et al.* (*10*) found that a small RNA from chick embryonic heart cells could induce the expression of several functions characteristic of differentiated heart when added to postnodal heart cells. From our laboratory we reported that a small RNA extracted from the chromatin of SV40-transformed WI38 human fibroblasts could stimulate transcription of normal WI38 nuclei to a level undistinguishable from that of transformed nuclei, using either endogenous or *E. coli* RNA polymerase (*45*). Homologous WI38 SnRNA had no effect under the same assay conditions.

These results suggested that some classes of SnRNA participate in regulation of transcription and may be involved in the maintenance of the differentiated or transformed phenotypes. It remained to be seen whether they are in fact tissue- and species-specific and whether they conform to any one of the predictions of the various gene derepression models.

II. The Use of Isolated Nuclei for Assay of Regulatory Elements in Transcription

An essential prerequisite to investigating the role of chromosomal components in eukaryotic gene regulation at the transcriptional level is the availability of a cell-free system capable of the high-fidelity transcription occurring in intact cells. Transcriptional systems using purified enzymes and naked DNA or chromatin as templates were limited by the complexity of the eukaryotic genome and its transcribing enzymes. Many factors crucial to faithful transcription have yet to be defined; therefore, one has no choice but to resort to unpurified cellular extracts.

Significant progress in elucidating some of the regulatory processes operating in the control of eukaryotic gene expression has been achieved by using intact isolated nuclei for transcription assays. Intact nuclei have been found to be the system most capable of duplicating transcriptional events occurring in intact cells. Isolated nuclei, unlike intact cells, are permeable to modulators of transcription and large molecules, making possible further elucidation of transcriptional events as well as detection and characterization of components which may play a critical role in gene regulation.

All three classes of RNA polymerases have been shown to be able to transcribe their specific genes in isolated nuclei. Transcription of ribosomal genes by RNA polymerase I was demonstrated in HeLa cells (*92*) and in *Xenopus laevis* nuclei (*66*). Synthesis of heterogenous nuclear RNAs (HnRNA) and by RNA polymerase II was shown in HeLa cells (*92*), mouse myeloma (*52*), chick oviduct (*15*), and slime mold (*33*) nuclei. Preferential transcription of the class II ovalbumin gene has been shown in oviduct nuclei (*58*). Transcription of tRNA and 5 S RNA genes by RNA polymerase III has been demonstrated in mouse plasmacytoma (*79*), and *Xenopus laevis* and *X. borealis* nuclei (*12,53,56,89*).

For this reason, our initial studies utilized intact isolated nuclei under conditions optimized for RNA polymerase II transcription, thus avoiding enzyme purification or removal from its template (*45*). However, the efficiency of our early system was unsatisfactory. Due to the leaky nature of the enzyme during nuclear isolation procedures, the incorporation of [^{3}H]UTP into RNA was found to be tenfold lower than when heterologous *E. coli* RNA polymerase was added to the nuclei at a ratio of 4 μg of DNA per unit of enzyme. This ratio was selected because it was shown to reflect the differences in template activity found *in vitro* between normal and SV40-transformed WI38 human fibroblasts (*45*). Higher amounts of enzyme resulted in higher levels of transcription but abolished the differences between the two cell types. Since the bacterial polymerase appeared to be an effective probe for template structure, and provided a more efficient transcription system, it was used for many subsequent studies investigating the effect of chromosomal components on transcription *in vitro*. The endogenous system was utilized only as a control for possible artifacts introduced by the heterologous enzyme and to investigate possible RNA polymerase-specific effects.

For studies on the mode of action of SnRNA, however, the use of the bacterial enzyme had to be discontinued; therefore, it became imperative to develop an endogenous assay which could be shown to retain a higher RNA polymerase II activity and to be capable of reinitiation *in vitro*.

Our early work used a method of nuclear isolation similar to those reported by Ernest *et al.* (*15*) and Marzluff *et al.* (*51*). These authors employed isotonic sucrose-lysing media containing Triton X-100 as a detergent to solubilize membranes and divalent cations to minimize both leakage of chromosomal proteins and changes in chromatin architecture. They also used heavy sucrose cushions (2 *M*) for separation of intact nuclei from cellular debris.

Even though they obtained quantitative yields of transcriptionally active nuclei, some leakage of weakly bound RNA polymerase II was found to occur. Experiments using endogenous RNA polymerase II often require supplements of purified exogenous enzyme in order to obtain yields of RNA transcripts sufficient for analysis.

The use of heavy sucrose cushions for nuclear purification as well as glycerol concentrations higher than 20% for resuspension of nuclei has been found to be undesirable. Glycerol inhibits RNA polymerase II activity at concentrations greater than 15% (*81*), while higher concentrations of polyhydroxyl compounds in general can act as detergents, weakening hydrophobic bonding within chromatin (*8*).

We have therefore modified the nuclear isolation procedure, avoiding the use of sucrose cushions and higher concentrations of glycerol during transcription. After cell lysis, one centrifugation through 20 volumes of transcription buffer containing 10% glycerol is sufficient to yield a nuclear pellet relatively free of detergent and cytoplasmic debris.

The model we selected to assess the fidelity of *in vitro* transcription in isolated intact nuclei made use of the transforming capacity of simian virus 40 (SV40) in human and mouse cells. SV40 is one of the smallest DNA viruses known to transform human cells (*85*). Its genome is well characterized and has now been completely sequenced (*17*). It is also known that the maintenance of the transformed phenotype requires the expression of only part of the SV40 genome, the early region or gene *A* (*9*). This makes SV40 transformation a very appealing system for molecular studies on the role of cellular and viral products in the onset and maintenance of the transformed state.

The first experiments were designed to determine whether or not differences in transcriptional activity found in intact cells were retained in isolated nuclei. Transformed cells have a greater rate of proliferation than their normal counterparts and synthesize two to three times as much RNA per cell (*64*). The template properties of chromatin in nuclei isolated from log-phase normal and SV40-transformed WI38 human fibroblasts were compared using either endogenous or *E. coli* RNA polymerase (*45*). Under both assay conditions the same template differences observed *in vivo* were maintained *in vitro*. It was apparent, therefore, that the *in vitro* system using nuclei isolated from normal cells could be used to assay chromosomal components from SV40-transformed cells in order to establish their function in transcription and to characterize presumptive gene regulatory molecules.

III. Role of Loosely Bound Non-Histone Chromosomal Proteins and SnRNAs

Evidence accumulated in many laboratories suggests that non-histone chromosomal proteins play a role in stimulating transcription of specific genes, hence acting as positive control elements. Synthesis of certain non-histones was found to be under the influence of hormones which stimulate RNA synthesis in target cells (*78*). The subcellular localization of non-histones and their accumulation in regions of chromatin active in RNA synthesis (*19,50,77*) were consistent with the hypothesis that they may be associated with genetic activation. Moreover, their phosphorylation rates increase at times of gene activation (*36*). More directly, chromatin reconstitution experiments involving various combinations of DNA, histones, and non-histones extracted from different tissues or from cells at different stages of activity have shown that the resulting chromatins exhibit a template activity characteristic of the tissue from which the non-histones were extracted (*3,4,32,60,82*).

Studies in our laboratory were initially dedicated to comparison of both histone and non-histone chromosomal proteins of cultured mammalian cells at different stages of the cell cycle as well as between normal and SV40-transformed cells. These studies detected significant differences in histones, namely the degree of phosphorylation and acetylation, which correlates with tightness of binding to the chromatin complex (*40,41,46,47*). The results suggested that histones are involved in conformational changes of nucleosomes and their packaging in chromatin fibers which occur concomitantly with genetic activation. Likewise, consistent differences in the composition, rates of synthesis, turnover, and phosphorylation of non-histone proteins were found in mammalian cells transformed by SV40 virus (*42–44*). The meaning of these differences, however, remained obscure.

Among non-histone chromosomal proteins, the loosely bound fraction (extractable with 0.35 *M* NaCl) has received a great deal of attention. Some of these proteins have been found to be tissue- and species-specific, are very heterogeneous in size, and are in transient association with chromatin, making them attractive candidates for a role in selective gene regulation (*39,59*). Addition of the 0.35 *M* NaCl extract of Ehrlich ascites hyperdiploid cells was found to restore the transcriptional template activity of the complex to a level similar to that of native chromatin (*39*). It is now well known that this extract contains the high mobility group (HMG) proteins and that HMG 14 and 17 are associated with nucleosome core particles in transcribable regions of the chromatin (*90*).

Research in our laboratory was initially focused on the properties of the 0.35 *M* NaCl extract from the chromatin of SV40-transformed human WI38 when added to nuclei from untransformed WI38 cells. Such an extract, however, is known to contain not only the loosely bound chromosomal proteins (HMG and LMG) but also a sizable proportion of the SnRNAs (*30,54,87*). Thus, we decided to separate the protein and RNA in the extract in order to investigate the effect of the individual components on transcription *in vitro*. Unexpectedly, we found that the RNA and not the protein was the active element stimulating transcription in WI38 cells (*45*). We did ascertain, however, that the two occurred in tight association with each other and that the proteins must be denatured in either SDS or urea in order to release the active RNA species, which is otherwise lost into the phenol phase during RNA purification. The stimulatory effect of these RNAs was found to be dose dependent as well as RNase and NaOH sensitive, indicating that RNA was indeed responsible for the effect rather than some accidental protein or DNA contaminant in the preparation. We conducted similar experiments with SV40-transformed human (SV-WI38) and mouse 3T3 cells (SVT2) and found that, in analogy with the human system, SVT2 SnRNA could stimulate the transcriptional activity of mouse 3T3 nuclei, whereas homologous 3T3 SnRNA had little or no activity (Table I). Furthermore, when assayed under conditions utilizing endogenous polymerase, the stimulatory effect was shown to be totally dependent on RNA polymerase II, since it was completely eliminated by addition of 1 μg/ml α-amanitin to the reaction mixture (Fig. 1).

Since the properties of SnRNAs from SVT2 were very similar to those from SV-WI38 cells, and the former are easier and more economical to cultivate, more detailed experiments were conducted with the mouse system.

IV. Tissue and Species Specificity of SnRNAs

The obvious question stemming from the above results is whether the active SnRNA is a unique characteristic of viral transformation or

TABLE I

Effect of SnRNA from SV40-Transformed Human (WI38) and Mouse (3T3) Cells Using the Improved Transcription Assay Optimized for RNA Polymerase II Activity[a]

	$[^3H]$UTP incorporated (pmole/100 μg DNA)	Stimulation (treated/control)
WI38 nuclei (control)	20.3	—
+SV-WI38 SnRNA	33.5	1.65
+SV-WI38 SnRNA (Na0H-digested)	23.7	1.05
3T3 nuclei (control)	23	—
+SVT2 SnRNA	35	1.52
+SVT2 SnRNA (Na0H-digested)	23.8	1.03
3T3 SnRNA	25.4	1.10

[a] Nuclei were prepared under mild conditions as described in the text and incubated at 25°C for 30 min. Endogenous assay conditions were modified as compared to those previously reported (*45*) to include higher concentrations of nucleoside triphosphates (0.8 m*M*), particularly ATP (1.2 m*M*), plus 4 m*M* phosphoenolpyrurate (*58*). SnRNAs were added at 0.01:1 ratio to DNA. Na0H hydrolyses were carried out either at 60°C for 1 hr or at 37°C for 16 hr in 0.3 *M* Na0H, with no apparent differences. Stimulation was calculated relative to either the human or the mouse control, as appropriate.

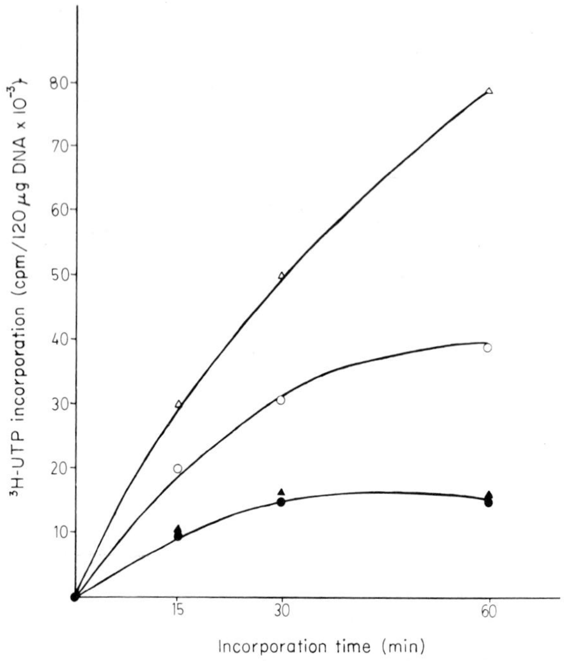

Fig. 1. Effect of SnRNA extracted from the chromatin of SV40-transformed mouse 3T3 (SVT2) cells on transcription of 3T3 cell nuclei incubated under conditions optimized for endogenous RNA polymerse II activity as described in the text and Table I. (○) 3T3 nuclei, control; (△) with SVT2 SnRNA at 0.01:1 w/w proportion to DNA: (●) 3T3 nuclei plus 1 μg/ml α-amanitin; (▲) with SVT2 SnRNA plus 1 μg/ml α-amanitin.

whether there are other regulatory RNAs which may be involved in gene expression in eukaryotes in general. If the latter is the case, one should be able to find tissue-specific SnRNAs in normal cells. The reported activity of embryonic small RNAs in tissue differentiation (*10*) supports such a concept. However, the data presented in this paper were obtained using an RNA preparation which bound to oligo(dT)-cellulose and, therefore, it is not possible to conclude whether some of the RNA acts as messenger for the synthesis of new tissue-specific proteins or whether it has a direct effect on transcription of chromatin into tissue-specific messengers. Since the author found that the active RNA not only was nontranslatable but it also caused inhibition of globin and rat liver mRNA translation *in vitro,* the first alternative appears improbable. In order to test the latter possibility, SnRNA was extracted from the chromatin of two human tissues (WI38 fibroblasts and placenta) as well as from monkey cells (vero) and tested for its activity in transcription of nuclei from homologous and heterologus tissue (see Table II). It is apparent that SnRNA from normal cells has only a small stimulatory effect on their own homologous nuclei. However, if one interchanges the SnRNA between two tissues of the same species (e.g., human fibroblast and placenta cells), a much more marked stimulation can be observed. The transcriptional activity of placenta nuclei, with no SnRNA added, is twofold higher that of WI38. Yet SnRNA from WI38 cells does stimulate transcription of placenta nuclei almost as much as SnRNA from transformed cells, whereas homologous SnRNA has a much smaller effect. Similarly, SnRNA from placenta cells has much greater stimulatory activity on WI38 than on placenta nuclei. However, when SnRNA from normal human cells is added to nuclei of a different species there appears to be a slight inhibition. On the other hand, SnRNA from transformed human cells (SV-WI38) can be seen to stimulate transcription in both human and monkey nuclei, although the stimulation observed in human nuclei is considerably greater.

These results indicate that regulatory SnRNAs are not unique to viral transformation and are tissue and species specific. The finding that WI38 SnRNA shows considerable stimulation of transcription in placenta nuclei, and vice versa, could be explained on the basis of induction of extra genes in the heterologous tissue of the same species. Placenta is a very active tissue. It functions simultaneously as fetal lung, liver, gut, and

TABLE II

Activity of Unfractionated SnRNA from Human and Monkey Cells on Homologous and Heterologous Transcription[a]

	WI38 nuclei	SV-WI38 nuclei	Placenta nuclei	Vero nuclei	No nuclei
No addition	1650	3670	3450	1810	50
+SVWI38 SnRNA	6950(4.2 ×)	6310(2 ×)	9382(2.7 ×)	3076 (2 ×)	100
+WI38 SnRNA	2584(1.6 ×)	—	8970(2.6 ×)	1578(0.87 ×)	110
+Vero SnRNA	1445(0.87 ×)	—	—	2639 (1.5 ×)	90
+Placenta SnRNA	4450(2.7 ×)	—	5458(1.6 ×)	—	—

[a] Data in pmole [^{3}H]UTP incorporated/mg DNA (30 min). SnRNAs were added to nuclei at a concentration of 0.1:1 w/w ratio to DNA. Nuclei were incubated at 25°C for 30 min using *E. coli* RNA polymerase at 4.3 μg DNA/unit of enzyme (*49,69*). Numbers within parentheses represent the multiplier factor relative to the control (minus SnRNA). [From (*69*), with permission.]

kidney (*21*); therefore, it must contain active genes which are normally turned off in WI38 fibroblasts. On the other hand, cycling fibroblasts must contain cell cycle genes and other fibroblast-specific genes that are turned off in placenta. It is likely, therefore, that their regulatory RNAs will recognize different regions of the chromatin and induce transcription in those regions. The results obtained by interchanging SnRNAs between two different primate species could be explained on the basis that sequence homology between the RNAs and their recognition sites in DNA being higher in a homologous than in a heterologous, albeit related, species. However, SV-WI38 "active" RNAs appear different from their normal counterpart in that they show considerably higher stimulatory activity on transcription of WI38 nuclei and are the only ones capable of activating nuclei from both human and monkey cells. Since SV40 is a monkey virus, this latter property could perhaps be explained on the basis of an SV40-coded component. It is also possible that SV-WI38 SnRNAs have a pleiotropic effect on a greater number of sites in chromatin.

V. Effect on RNA Polymerase II: Initiation and Sizing of RNA Transcripts

Most of the transcription assays described in Sections III and IV were carried out using *E. coli* RNA polymerase at low enzyme-to-DNA ratios. Although, at these low ratios, the bacterial enzyme was shown to be an effective probe for chromatin conformational changes occurring as a result of gene activation, it can shed little light on the mode of action of SnRNA. For these studies, it was necessary to use our improved endogenous transcription system as described in Section II.

Two of the essential prerequisites for the use of an endogenous transcription assay for testing the mode of action of SnRNA are that the nuclei must be shown to have a high RNA polymerase II activity and to support reinitiation *in vitro*. The first prerequisite was easily demonstrated (see Fig. 1). The second, however, was more difficult to establish. Most previous attempts by other researchers to demonstrate *in vitro* reinitiation have utilized γ-^{32}P-labeled nucleoside triphosphates in order to label the 5′-ends of newly initiated transcripts. Since only newly initiated transcripts contain the γ-phosphate at their 5′-end, the quantity of ^{32}P incorporated should reflect the degree of initiation *in vitro*. However, isolated nuclei contain other enzymatic activities such as endogenous kinases and phosphatases which compete for the labeled γ-phosphate. In order to correct for this, one must therefore estimate initiation indirectly via a selective inhibitor of RNA polymerase II initiation.

In our first approach, we utilized 5,6-dichloro-β-D-ribosylbenzimidazole (DRB), thought to be an inhibitor of RNA polymerase II chain initiation (*76,84*), together with [γ-^{32}P]ATP and [^{3}H]UTP in order to estimate relative amounts of initiation and elongation of RNA transcripts. One can deduce from the results illustrated in Fig. 2 that initiation takes place in WI38 nuclei, since DRB reduces [γ-^{32}P]ATP incorporation by about 25% with no corresponding effect on [^{3}H]UTP incorporation. This is not unexpected, since inhibition of initiation is likely to be compensated for by increased elongation. However, we were not entirely satisfied with this approach, since more recent evidence indicates that DRB may cause premature chain termination and have little effect on initiation *per se* (*18*).

Consequently, *in vitro* initiation was demonstrated by a second approach involving the use of 5′-(γ-S)ATP. It has been shown that when 5′-(γ-S)ATP is used for measuring initiation in isolated nuclei, the sulfur is present at the γ position of the 5′-end of newly synthesized RNAs (*26*). With the thiol analogue at the 5′-end and [^{3}H]UMP at internal positions of the chains, it is possible to measure both initiation and elongation of RNA transcripts. Sulfur-terminated sequences can be retained in mercury–agarose columns, whereas chains preinitiated *in vivo*

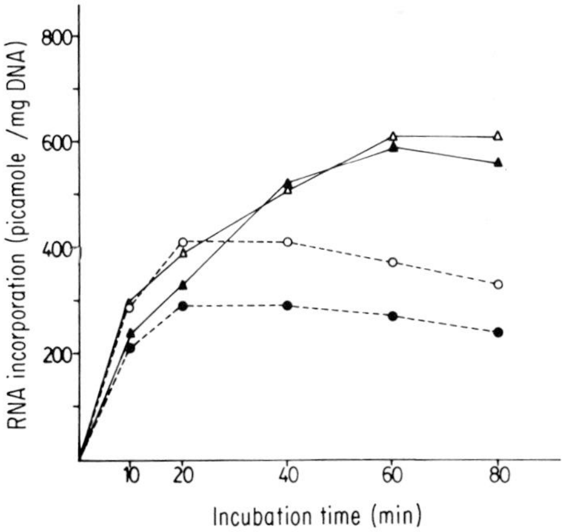

Fig. 2. Inhibition by 5,6-dichloro-1-β-D-ribosylbenzimidazole (DRB) of RNA synthesis *in vitro* in WI38 nuclei. Transcription was optimized for endogenous RNA polymerase II and assayed using both [^{32}P]ATP and [^{3}H]UTP. Incubations were carried out at 25°C with and without 100 μ*M* DRB: (△) [^{3}H]UTP and (○) [^{32}P]ATP incorporation (control); (▲) [^{3}H]UTP and (●) [^{32}P]ATP incorporation in the presence of DRB.

lack the sulfur and will pass through the column (*67,80*). The column-bound chains can then be released by adding dithiothreitol (DTT) to the elution buffer.

In preliminary experiments with 5′-(γ-S)ATP using either the human WI38 or the mouse 3T3 endogenous transcription system, we observed no increase in initiation in the presence of "active" SnRNA from SV40-transformed cells, a result we attributed to the presence of spermine in the transcription assay. Spermine, used as a nuclease inhibitor during SnRNA isolation and purification, was found to coprecipitate with RNA in ethanol and to inhibit initiation by about twofold. This inhibition has also been observed in other laboratories (*61*). We decided, therefore, to purify the SnRNA samples using Sephadex G-50 chromatography after ethanol precipitation. Purification was carried out in the presence and absence of another nuclease inhibitor, polyvinyl sulfate (PVS), in order to protect the SnRNA; therefore, we tested the effect of the same amount of PVS (25 μg/ml) in the transcription assay. As seen in Table III, 5′-(γ-S)ATP is utilized in this system even more so than unmodified ATP, presumably because it is not as good a substrate for phosphatases and/or kinases, therefore decreasing the competition and increasing the amount of triphosphate available for the initiation reaction. Addition of PVS can be seen to decrease initiation almost by half, a result analogous to that obtained with spermine. When SVT2 SnRNA minus PVS is assayed, less than twofold increase in transcription can be observed in the unbound fraction, but there is a twofold stimulation of initiation. When SnRNA is tested in the presence of PVS, a very high increase in transcriptional activity can still be observed in spite of the sizable decrease

TABLE III

Effect of SVT2 SnRNA and PVS on Initiation of Transcription in 3T3 Nuclei[a]

Treatment	[^{3}H]UTP incorporation (cpm) Unbound	Bound	Initiation (Percentage of control)
ATP	12,200	0	—
[γ-S]ATP	18,207	1165	6.4
[γ-S]ATP + SnRNA[b]	22,590	2417	13.3
[γ-S]ATP + PVS[c]	35,000	595	3.3
[γ-S]ATP + SnRNA + PVS	72,838	1020	5.6

[a] Hg-Sepharose chromatography using (γS)ATP.
[b] SnRNA was added at 0.05:1 proportion to DNA.
[c] PVS was added to SnRNA throughout isolation steps at 25 μg/ml. An equivalent amount of PVS alone was added to nuclear incubation as control.

in initiation. It appears, therefore, that PVS has a marked effect in protecting the nascent transcripts, yet it must be left out whenever assaying for initiation. It must be noted, however, that the estimates of transcription initiation are only relative. Since nuclease activity in the system cannot be eliminated, absolute values are likely to be considerably higher.

These experiments still give no answer as to whether the active SnRNA has any effect on elongation of RNA transcripts. In order to answer this question, we developed a fractionation procedure that would allow qualitative as well as quantitative analysis of RNA molecules synthesized *in vitro*. This was achieved by fractionation of the RNA in composite 2.5% polyacrylamide–1% agarose slab gels. As illustrated in Fig. 3, the composite gel system is capable of resolving a broad spectrum of chain lengths. The two heavier bands on the top half of the gel in lane 1 are 28 and 18 S ribosomal RNAs. On the bottom half, the small molecular weight RNAs are still well resolved. Discrimination between newly synthesized and preexisting RNAs was obtained by slicing gel tracks and quantitating [^{3}H]UTP incorporation by liquid scintillation counting. Figure 4 illustrates the results of such an analysis, including the effect of SVT2 SnRNA on the quantity and size of the RNA transcripts synthesized *in vitro* in 3T3 nuclei. The counts obtained under similar conditions in the presence of 1 μg/ml α-amanitin were subtracted from the total counts so as to record only transcription by RNA polymerase II. Here we can see that, in the presence of PVS alone, there is a significant increase in the size of the transcripts, confirming PVS's role in protecting newly synthesized RNA against nuclease attack. The effect of SVT2 SnRNA can be seen here to be strictly quantitative, since no size shift is apparent when PVS is present in both treated and control samples. On the basis of the relative values for initiation in the presence and absence of SVT2 SnRNA as seen in Table III, we conclude that its stimulatory effect can be explained by the twofold increase in initiated chains.

The alternative interpretation that SnRNA might function as a nuclease inhibitor appears most unlikely for the following reasons: (1) The amounts of active SnRNA are extremely small, since its activity can be detected at a total SnRNA ratio as low as 0.01:1 w/w to DNA and, as we will see in the following section, only a very small fraction of the added SnRNA has any activity at all; (2) the observed tissue and species specificity of the active RNA is hardly compatible with a simple inhibitor of nucleases; and (3) if nuclease inhibition were the primary mechanism of the observed stimulatory activity, one would expect to see a definite increase in the size of the transcripts synthesized in the presence of

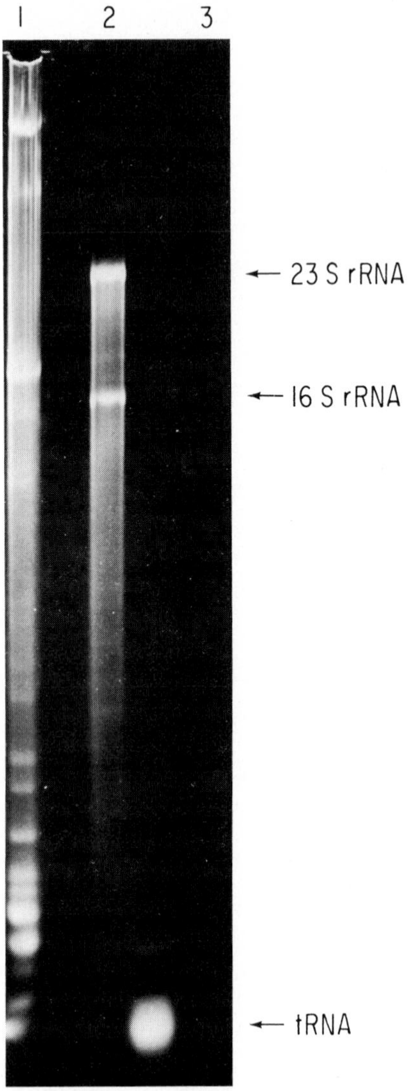

Fig. 3. Composite gel electrophoresis of total nuclear RNA from SV-WI38 cells after transcription *in vitro*. Nuclei were incubated with [^{3}H]UTP for 45 min at 25°C under endogenous RNA polymerase assay conditions. Total RNA was recovered from proteinase K digested nuclei and further deproteinized by repeated SDS–phenol–chloroform–isoamyl alcohol extractions as described (*69*). RNA samples were denatured in 85% formamide prior to loading onto the 1.5% polyacrylamide–1% agarose slab gel and electrophoresed at 160 V until the bromphenol blue marker was 5 cm from the bottom. The gels were stained with ethidium bromide and photographed under uv: lane 1, total nuclear RNA; lane 2, markers 23 S and 16 S rRNA; lane 3, tRNA. [From (*69*), with permission.]

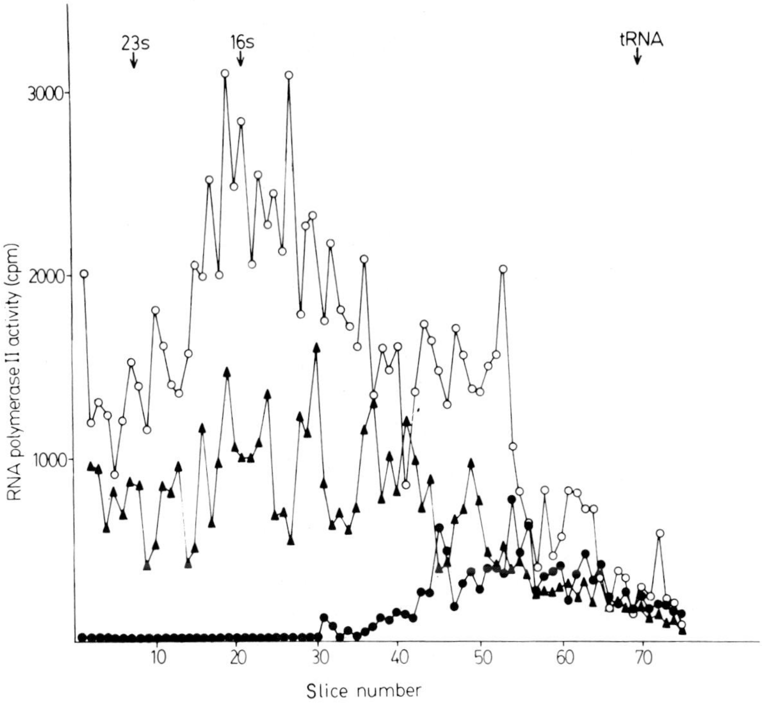

Fig. 4. Analysis of RNA transcribed by endogenous RNA polymerase II in 3T3 nuclei in the presence and absence of SVT2 SnRNA and polyvinyl sulfate (PVS). Transcription assays and gel electrophoresis were performed as in Fig. 3. Gel tracks were sliced and counted in a liquid scintillation counter. The radioactivities obtained under the same conditions in the presence of 1 μg/ml α-amanitin were subtracted from all points, therefore, the values shown represent only RNA polymerase II activity. (●) 3T3 nuclei (control); (▲) with 25 μg/ml PVS and (○) with SVT2 SnRNA at 0.01/1 w/w proportion to DNA and 25 μg/ml PVS.

active RNA. Such a shift was only detected by addition of PVS or spermine. In properly controlled experiments, we could only detect quantitative increases in transcription with no size shift, compatible with the interpretation of its role in promoting new chain initiations.

We propose that the active SnRNA, acting in conjunction with nuclear proteins, may help to destabilize the double helix in controlling regions of the DNA, thereby facilitating entry of RNA polymerase II for the formation of an initiation complex. Our data do not support the hypothesis that the active SnRNA itself may serve as a primer as postulated by Kolodny (*38*). If such were the case, we would not be able to detect any increase in the number of 5′-γ-substituted chains. Moreover, current studies in our laboratory have shown that when SVT2 SnRNA is oxidized at the 3′-end with $NaIO_4$, its activity does not change. Both overall incorporation and initiation of new transcripts are stimulated to the same extent as with untreated SnRNA.

VI. The Search for the Active SnRNA Subfraction

All of the above studies utilized unfractionated SnRNA preparations. Such preparations, however, are known to contain several subspecies (*25,29,37,54,70,87,88*). It is important, therefore, to determine whether one or more of these species acts as transcription regulator(s). If one succeeded in purifying one such species, its sequence might provide invaluable information on the structure of eukaryotic promoters. For best resolution we decided to fractionate SnRNAs by electrophoresis in 5–15% gradient slab gels. Figure 5 represents a combined autoradiograph and ethidium bromide-stained gel of ^{32}P-labeled SnRNA extracted from SV-WI38 cells.

Panel A is an autoradiograph, whereas panel B shows the ethidium bromide-stained gel comparing an untreated SnRNA sample (lane 2) with a sample denatured in 7 *M* urea prior to loading on the gel (lane 1) and another sample predigested with RNase (lane 3). 5 S and tRNA markers were run in lane 4. Letters without parentheses designate bands as described by Penman (*91*), whereas those within parentheses correspond to the nomenclature of Ro-Choi and Busch (*70*). It is apparent that the band pattern does not change as a result of urea treatment but that all visible bands are RNase-sensitive since none can be detected in the RNase-treated sample track. Labeling *in vivo* with ^{32}P does seem to improve detection of some minor bands.

The transcriptional activity and estimated size of SV-WI38 extracted from each of the gel regions numbered in Fig. 5 is shown in Table IV. The size estimates are based on the relative mobilities of tRNA, 5 S

Fig. 5. A combined autoradiograph and ethidium bromide stained gel of SnRNA purified from the chromatin of SV-WI38 cells. SnRNA samples were fractionated by electrophoresis on a 5–15% polyacrylamide gradient slab gel and electrophoresis at 150 volts using 5 m*M* Mg $(OAc)_2$, 40 m*M* Tris-HCl pH 7.2, 1 m*M* EDTA as tray buffer. Cells were labeled *in vivo* with ^{32}P for 24 hrs and the SnRNAs detected by autoradiography on Kodak X-OMAT X-ray film (Panel A) or under uv after staining the gel with ethidium bromide (*69*). Panel B. Lane 1, SnRNA was denatured in 7 *M* urea prior to electrophoresis; lane 2, untreated SnRNA; lane 3, RNase-treated SnRNA; and lane 4, markers 5 S and tRNA. Numbers indicate the areas of the gel which were later excised, the RNA extracted and tested for its effect on transcription of WI38 nuclei. Letters identify the bands according to the nomenclature of Penman (*90*) and between parentheses that of Busch (*70*).

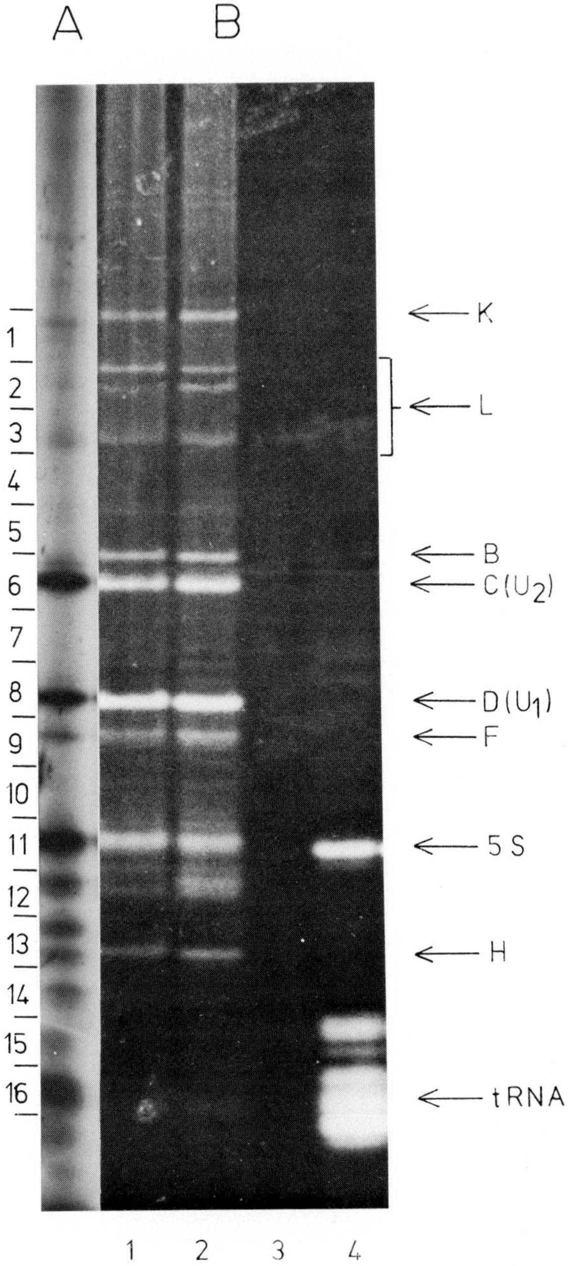
A
B
1
2
3
4
5
6
7
8
9
10
11
12
13
14
15
16
K
L
B
C(U2)
D(U1)
F
5 S
H
tRNA
1 2 3 4

TABLE IV

Effect of SV-WI38 SnRNA Purified by Polyacrylamide Gel Electrophoresis on Transcription of WI38 Nuclei with *E. coli* RNA Polymerase

SnRNA[a] slice number	Approximate nucleotide intervals	DNA-dependent $[^3H]UTP$ incorporation[b] (pmole/mg DNA in 30 min)		Inhibition by actinomycin D (untreated RNAs) (%)
		Urea-treated	Untreated	
No addition		1713		
1	290–320	1576	7571	95
2	265–290	3265	3597	85
3	250–265	3611	3590	87
4	225–250	3176	3426	90
5	208–225	2227	3207	92
6	190–208	2059	2398	89
7	175–190	2501	1370	87
8	160 175	10278	4908	90
9	140–160	3665	2141	93
10	124–140	3802	1833	87
11	115–125	3939	1413	80
12	108–115	1025	866	75
13	102–108	1107	904	65
14	92–102	965	942	82
15	78–92	1306	1268	80
16	68–78	1025	1086	76
Unfractionated at 0.1:1 (w/w)		5930		

[a] Slice numbers correspond to gel areas illustrated in Fig. 5.

[b] SnRNAs were extracted from gel slices indicated. Recovered RNA was divided into two lots and tested in the presence and absence of actinomycin D. To the gel, 10 μg SnRNA were applied, with 85–95% recovery per band. Values shown for AMD-sensitive incorporation were corrected by subtracting pmole UMP incorporated in the presence of 100 μg/ml AMD. Results were obtained by averaging data from two separate experiments.

RNA, and major SnRNA species (U_1 and U_2), which have been completely sequenced (*70,71*). Transcription assays in this case utilized normal WI38 nuclei as templates and *E. coli* RNA polymerase as enzyme. Actinomycin D (AMD) controls were run in parallel in order to test for possible RNA-dependent RNA polymerase activity of the *E. coli* enzyme (*23*). The relative inhibition by 100 μg/ml AMD is indicated, and the incorporation of $[^3H]UTP$ into RNA is corrected by subtracting the incorporation obtained in the presence of AMD; therefore, the counts per minute shown represent only DNA-dependent RNA synthesis. Corresponding extracts from the RNase-treated gel track gave no stimulation (data not shown).

Although the pattern of the major bands does not appear to change, there is an activity shift toward a smaller size when SnRNA preparations are pretreated with 7 *M* urea. Untreated RNA shows the highest activity in region 1 (290–320 nucleotides), whereas urea-treated RNA shows the highest activity in region 8 (160–175 nucleotides).

When we fractionated SnRNAs from normal WI38 cells and placenta, we found identical band patterns as obtained with SV-WI38 cells. An autoradiograph of ^{32}P-labeled SnRNAs from human and monkey cells is illustrated in Fig. 6. Lanes 1 and 2 show SnRNAs from monkey (vero) cells. The sample in lane 2 was previously denatured in urea. Lane 3 has SnRNA from SV-WI38, and lanes 4 and 5 have SnRNA from normal WI38 cells. The only apparent difference in band pattern between monkey and human cells is in the region of the L bands; other differences appear merely quantitative. However, when we extracted the RNA from the different regions of the gels and tested it in transcription of homologous versus heterologous nuclei, again in all cases only one gel region, estimated to be 160–175 nucleotides in length, showed true activity, discriminating between homologous and heterologous nuclei as seen with unfractionated SnRNAs in Table II above. Although in these cases we did not detect activity in the upper region (300–320 nucleotides), we believe now that this is likely due to nicking of the active RNA around the middle of the molecule. Evidence for this interpretation came out during subsequent studies using normal and SV40-transformed mouse 3T3 cells.

Since the active SVT2 SnRNA was found to have a gene depression potential similar to that of SV-WI38 SnRNA, it was decided to investigate whether SVT2 SnRNA contains active subspecies analogous to those found within SV-WI38 SnRNA samples fractionated by gel electrophoresis.

A comparison of the gel banding pattern of SV-WI38, 3T3, and SVT2 SnRNAs is shown in Fig. 7, in an ethidium bromide-stained 5–15% polyacrylamide gradient slab gel containing in lanes A SVT2, lanes B 3T3, and lanes C SV-WI38 SnRNAs. Letters designate bands as described in Fig. 5. Here all samples were denatured in 85% formamide prior to loading on the gel. Both quantitative and qualitative differences in band mobilities can be seen. Bands B and C are better resolved from each other in human than in mouse samples; band D has a lower mobility in mouse than in human; band K is not apparent in transformed mouse cells, whereas a new band is evident below species D. The difference observed with band K, however, may not be real. We considered the possibility that transformed cells might contain more nucleases than normal cells. As a result, band K, visible only in 3T3 SnRNA tracks, might also be present in transformed cells; however, it may be more easily nicked into half molecules migrating below the D band, as seen in this

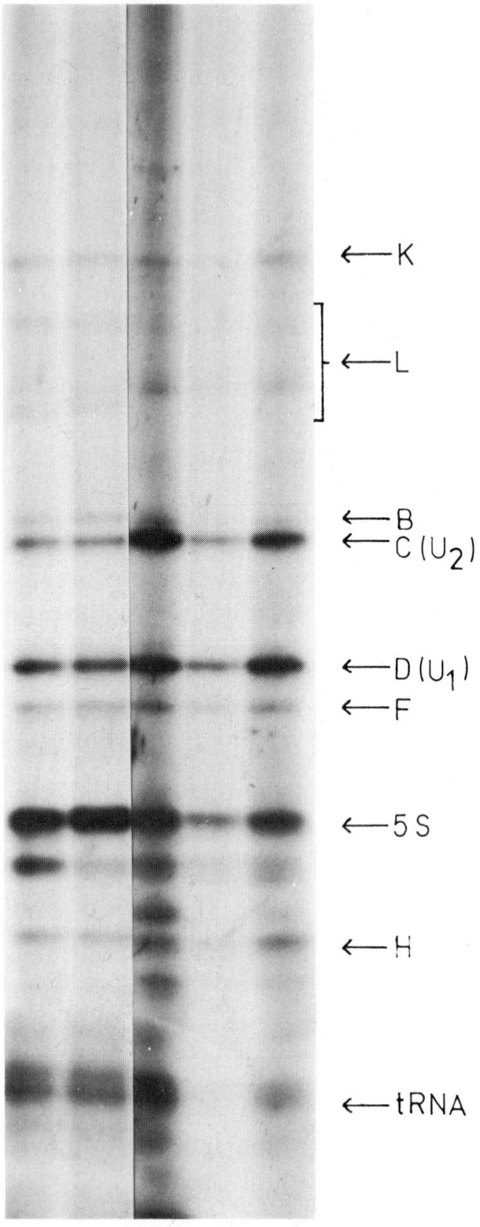

Fig. 6. Autoradiograph of a gradient acrylamide slab gel identical to the one described in Fig. 5, but comparing ^{32}P-labeled SnRNAs from monkey (Vero) cells (lanes 1 and 2), human SV-WI38 (lane 3), and normal WI38 (lanes 4 and 5). Vero SnRNAs in lane 2 were predenatured in 7 *M* urea.

gel. The use of nuclease inhibitors such as combined spermine and PVS, together with greater speed during preparation of SnRNA, did in fact restore band K to its proper position in the gel in later preparations. The gel regions (indicated by arabic numerals) were excised from the SVT2 tracks and the RNAs extracted and assayed for their effect on transcription of 3T3 nuclei under endogenous assay conditions. As presented in Table V, stimulation was observed only in regions 2 and 8, with mean nucleotide lengths estimated at 350 and 165, respectively, in analogy with the results obtained by fractionation of SV-WI38 SnRNA. The similarity of the results obtained under both endogenous and exogenous assay conditions supports the validity of the heterologous assay carried out with *E. coli* RNA polymerase.

The two-to-one relationship in the size of the active RNAs suggests that the smaller species is a breakdown product of the larger one, nicked at about the middle of the molecule. This could be explained if the active RNA had a hairpin configuration, with a nuclease-sensitive single-stranded loop. Thus, following denaturation, intact molecules would stay in the larger 300–350 nucleotide region, whereas the nicked half molecules would migrate in the 160–175 nucleotide region. This hypothesis is supported by the finding that the relative amounts of active species recovered from the larger size region of the gel increased when both nuclease inhibitors, PVS and spermine, were added to all solutions after cell lysis for isolation of SnRNA. Perhaps this shift in the size of the active species can be correlated with the shift in the K band observed in SnRNA from SVT2 cells. No such shift was observed in normal 3T3 or in human or monkey cell SnRNA. In the case of human and monkey cells, this could be due to the fact that any partial or total shift of the K band to the one-half molecular weight region would not be visible as a new band but would instead overlap with band D, which migrates in the same region. While there may be a structural similarity between band K and the active SnRNA species, we think it is unlikely that the active molecules can be totally identified with it, since we have observed activity shifts in the absence of any apparent differences in band patterns. Moreover, as seen in Fig. 7 and Table V, no K band is visible in the highest activity region (slice 2). This makes the job of isolating a pure species for sequence analysis extremely difficult. Furthermore, it is likely that these regulatory RNAs not only exist in small quantities in the cell but may be heterogeneous, having only the common feature of a unique size and possible hairpin configuration. Sequence heterogeneity is to be expected if in fact these RNAs are capable of recognizing regulatory regions of different gene families in different tissues, as suggested by their tissue and species specificity.

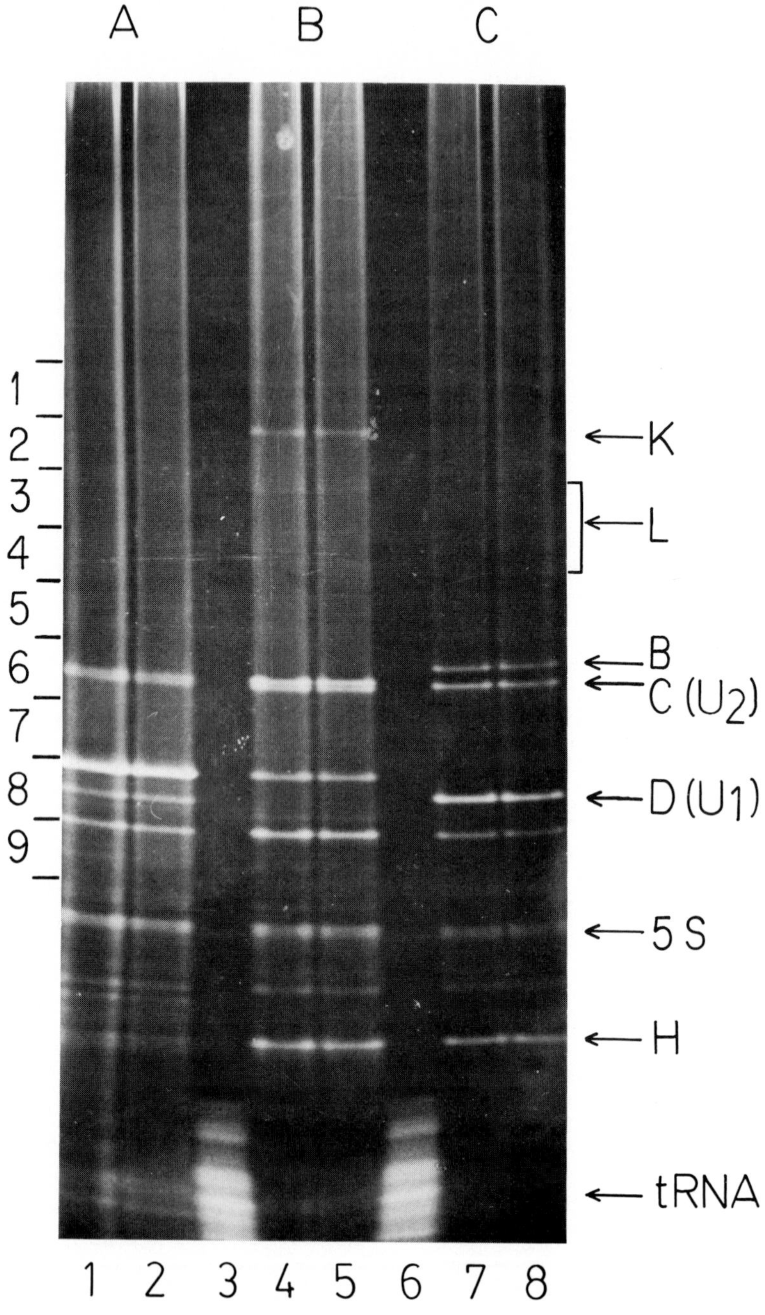

A
B
C
1
2
3
4
5
6
7
8
9
K
L
B
C (U2)
D (U1)
5 S
H
tRNA
1 2 3 4 5 6 7 8

TABLE V

Effect of SVT2 SnRNA Fractionated by Polyacrylamide Gel Electrophoresis on Transcription of 3T3 Nuclei[a] under Conditions Optimized for Endogenous RNA Polymerase II Activity

Slice number[b]	Mean nucleotide length	[^{3}H]UTP incorporation cpm/10 μg DNA	Activity (treated/control)
1	420	1,626	1.1
2	350	11,512	7.5
3	310	2,590	1.7
4	260	1,168	0.8
5	230	2,052	1.3
6	210	1,922	1.2
7	195	1,326	0.9
8	165	8,990	5.8
9	145	1,172	0.8
Control	—	1,544	

[a] 3T3 nuclei were incubated for 30 min at 25°C under the same assay conditions as described in Table 1.
[b] Slice numbers correspond to gel areas indicated in Fig. 7.

As to their origin, both mammalian and viral genomes are known to contain inverted sequences which could give rise to hairpin configurations in RNAs. Inverted sequences of about 300 nucleotides in length have been found in SV40 (*31,83*). Ubiquitous, interspersed repeated sequences of about the same size have been found in mammalian DNAs (*34,73*) as well as in pre-mRNAs (*16,74*). It would be interesting to discover whether a relationship exists between these DNA regions and the active SnRNAs.

VII. Implications and Prospects

From all of the above results it appears that active SnRNA, which coextracts with loosely bound non-histone chromosomal proteins, is involved in gene regulation in eukaryotic cells. Even though some non-histone proteins may, by themselves, be able to recognize promoters of

Fig. 7. Comparison of gel banding patterns of SnRNAs purified from normal mouse 3T3, SV40-transformed 3T3 (SVT2) and human SV-WI38 cells. Gel preparation and electrophoresis were carried out as in Figs. 5 and 6. Samples were denatured in 85% formamide prior to loading. Numbers on the left correspond to gel slices excised from lanes A and assayed for activity (Table V). The gels were stained with ethidium bromide and photographed under uv. Lanes A, SVT2 SnRNA; lanes B, 3T3 SnRNA, and lanes C, SV-WI38 SnRNA.

unique genes, RNA has a greater structural potential, via base pairing, to recognize unique promoter sequences. Thus, working in conjunction with each other, chromosomal proteins and SnRNA could ensure accurate and selective expression as required to maintain phenotypic differentiation in eukaryotic cells.

A eukaryotic cell may contain on the order of 50,000 to 100,000 structural genes, of which only a fraction are selectively expressed at any one time within a particular cell type.

The finding of a tight association of active SnRNA with some proteins in the loosely bound fraction of chromatin could be critical to both their functions, since the proteins, in the absence of SnRNA, had no stimulatory effect on transcription. This also suggests that when active SnRNAs are added to isolated nuclei, they probably interact with nuclear proteins, possibly HMGs, in order to stimulate transcription. If such a cooperative situation exists, one can envisage a mechanism whereby the proteins would unfold chromatin at promoter regions to allow the active RNA to base-pair to DNA, thus facilitating entrance of RNA polymerase.

The possible double-strandedness of the active molecules and the uniform size in all species and tissues so far examined could also be an important feature of their regulatory role. Perhaps RNA–DNA recognition occurs at the level of the nucleosome, since the size of each half of the active SnRNA molecule is similar to that of DNA associated with the nucleosome (*40*). If the two halves are indeed self-complementary, they could base-pair to both strands of DNA upstream from the RNA polymerase initiation site. The observation of a doubling in the number of RNA chains initiated in the presence of the active SnRNA can be interpreted to mean that active SnRNA does not act as a primer as postulated by Kolodny (*38*). If such were the case, its 3′-end could serve as the initiation point and no additional 5′ triphosphate-ended chains would be detected. It seems reasonable to conclude, therefore, that the active SnRNA stimulates transcription by promoting new chain initiations.

The next challenge—chemical characterization of the active SnRNA in normal and transformed cells—appears difficult to achieve at the present time. Up to now we have been unable to isolate a pure species suitable for sequence analysis. Our current strategy is directed toward selecting out an SV40-coded species, if such does exist in transformed cells, by preparative hybridization with various restriction fragments from the SV40 genome. Considering the very small amount of active SnRNA present in the cell, this search is not going to be easy. We are also working on the development of a cDNA probe suitable for qualitative analysis of RNA transcripts produced *in vitro* in nuclei of normal and transformed cells. This probe could then be applied to a qualitative analysis of transcription in normal nuclei in the presence and absence of SnRNA from their transformed counterpart. We hope that these and other studies, which include specific associations of SnRNA with chromosomal proteins, will contribute greatly to our understanding of the mechanism of gene regulation in eukaryotes in general and to the mechanism of malignant transformation in particular.

References

1. Alwine, J. C., Dhar, R., and Khoury, G. (1980). A small RNA induced late in Simian virus 40 infection can associate with early viral mRNAs. *Proc. Natl. Acad. Sci. U.S.A.* **77,** 1379–1383.
2. Avvedimento, V. E., Vogeli, G., Yoshihiko, Y., Maizel, J. V., Pastan, I., and de Crombrugghe, B. (1980). Correlation between splicing sites within an intron and their sequence complementarity with U1 RNA. *Cell* **21,** 689–696.
3. Barret, T., Maryanka, D., Humlyn, R. H., and Gould, H. J. (1974). Nonhistone proteins control gene expression in reconstituted chromatin. *Proc. Natl. Acad. Sci. U.S.A.* **71,** 5057–5061.
4. Bekhor, I., and Samal, B. (1977). Nonhistone chromosomal protein interaction with DNA/histone complex. 1. Transcription. *Arch. Biochem. Biophys.* **179,** 537–543.
5. Bester, A. J., Kennedy, D. S., and Heywood, S. M. (1975). Two classes of translation control RNA: Their role in the regulation of protein synthesis. *Proc. Natl. Acad. Sci. U.S.A.* **72,** 1523–1527.
6. Bonner, J., and Widholm, J. (1967). Molecular complementarity between nuclear DNA and organ-specific chromosomal RNA. *Proc. Natl. Acad. Sci. U.S.A.* **57,** 1279–1385.
7. Britten, R. J., and Davidson, E. H. (1969). Gene regulation for higher cells: A theory. *Science* **165,** 349–357.
8. Buss, W. C., and Stalter, K. (1978). Stimulation of eukaryotic transcription by glycerol and polyhydroxylic compounds. *Biochemistry* **17,** 4825–4832.
9. Crawford, L. V., Cole, C. N., Smith, A. E., Paucha, E., Tegtmeyer, P., Rundell, K., and Berg, P. (1978). Organization and expression of early genes of simian virus 40. *Proc. Natl. Acad. Sci. U.S.A.* **75,** 117–121.
10. Desphande, A. K., Kakowlew, S. B., Arnol, H. H., Crawford, P. A., and Siddiqui, M. A. Q. (1977). A novel RNA affecting embryonic gene functions in early chick blastoderm. *J. Biol. Chem.* **252,** 6521–6527.
11. Dickson, R. C., Abelson, J., Barnes, W. M., and Reznikoff, W. S. (1975). Genetic regulation: The lac control region. *Science* **187,** 27–34.
12. Doering, J. (1977). The structure of *X. borealis* oocyte and somatic DNAs. *Year Book—Carnegie Inst. Washington* **75,** 102–105.
13. Eliceiri, G. L. (1974). Short-lived, small RNAs in the cytoplasm of HeLa cells. *Cell* **3,** 11–14.
14. Eliceiri, G. L. (1979). Sensitivity of low molecular weight RNA synthesis to UV radiation. *Nature (London)* **279,** 80–81.
15. Ernest, M. J., Schutz, G., and Feigelson, P. (1976). RNA synthesis in isolated hen oviduct nuclei. *Biochemistry* **15,** 824–829.
16. Federoff, N., Wellauer, P. K., and Wall, R. (1977). Intermolecular duplexes in heterogeneous nuclear RNA from HeLa Cells. *Cell* **10,** 597–610.
17. Fiers, W., Contreras, R., Haegeman, G., Rogiers, R., Van de Voorde, A., Van Heu-

verswyn, H., Van Herreweghe, J., Volckaert, G., and Ysebaert, M. (1978). Complete nucleotide sequence of SV40 DNA. *Nature (London)* **273,** 113–120.

18. Fraser, N. W., Sehgal, P. B., and Darnell, J. E. (1979). DRB-induced premature termination of late adenovirus transcription. *Nature (London)* **272,** 590–593.
19. Frenster, J. H. (1965). Nuclear polyanions as de-repressor of synthesis of ribonucleic acid. *Nature (London)* **206,** 680–683.
20. Frenster, J. H. (1976). Selective control of DNA helix openings during gene regulation. *Cancer Res.* **36,** 3394–3398.
21. Friesen, H. (1973). Placenta protein and polypeptide hormones. *Endocrinology* **2,** 295–309.
22. Georgiev, G. P. (1969). On the structural organization of operon and the regulation of RNA synthesis in animal cells. *J. Theor. Biol.* **25,** 475–490.
23. Giesecke, K., Sippel, E., Nguyen-Huu, M. C., Groner, B., Hynes, N. E., Wartz, T., and Schutz, G. (1977). A RNA-dependent RNA polymerase activity: Implications for chromatin transcription experiments. *Nucleic Acids Res.* **4,** 3943–3958.
24. Goldstein, L., and Ko, C. (1974). Electrophoretic characterization of shuttling and nonshuttling SnRNAs. *Cell* **2,** 259–266.
25. Goldstein, L., and Trescott, O. H. (1970). Characterization of RNAs that do not migrate between cytoplasm and nucleus. *Proc. Natl. Acad. Sci. U.S.A.* **67,** 1367–1374.
26. Gross, R. H., and Ringler, J. (1979). Ribonucleic acid synthesis in isolated Drosophila nuclei. *Biochemistry* **18,** 4923–4927.
27. Harada, F., and Ikawa, Y. (1979). A new series of RNAs associated with the genome of spleen focus forming (SFFV) and poly (A)-containing RNA from SFFV-infected cells. *Nucleic Acids Res.* **7,** 895–908.
28. Harada, F., and Kato, W. (1980). Nucleotide sequences of 4.5S RNAs associated with poly (A)-containing RNAs of mouse and hamster cells. *Nucleic Acids Res.* **8,** 1273–1285.
29. Hellung-Larsen, P., and Frederiksen, S. (1972). Small molecular weight RNA components in Ehlich Ascites tumor cells. *Biochim. Biophys. Acta* **262,** 290–307.
30. Holmes, D. S., Mayfield, J. E., and Bonner, J. (1974). Sequence composition of rat ascites chromosomal ribonucleic acid. *Biochemistry* **13,** 849–855.
31. Hsu, M. T., and Jelinek, W. R. (1977). Maping of inverted repeated DNA sequences within the genome of SV40. *Proc. Natl. Acad. Sci. U.S.A.* **74,** 1631–1634.
32. Huang, R. C. C., and Huang, P. C. (1969). Effect of protein-bound RNA associated with chick embryo chromatin on template specificity of the chromatin. *J. Mol. Biol.* **39,** 351–365.
33. Jacobson, A., Firtel, R. A., and Lodish, H. F. (1974). Synthesis of messenger and ribosomal RNA precursors in isolated nuclei of the cellular slime mold *Dictyostelium discoideum. J. Mol. Biol.* **82,** 213–230.
34. Jelinek, W. R., Toomey, T. P., Leinwand, L., Duncan, C. H., Biro, P. A., Choudary, P. V., Weissman, S. M., Rubin, C. M., Houck, C. M., Deininger, P. L., and Schmid, C. W. (1980). Ubiquitous, interspersed repeated sequences in mammalian genomes. *Proc. Natl. Acad. Sci. U.S.A.* **77,** 1398–1402.
35. Kanehisa, T., Kitazume, Y., Ikuta, K., and Tanaka, Y. (1977). Release of template restriction in chromatin by nuclear 4–5 S RNA. *Biochim. Biophys. Acta* **475,** 501–513.
36. Kleinsmith, L. J., Allfrey, V. G., and Mirsky, A. E. (1966). Phosphorylation of nuclear protein early in the course of gene activation in lymphocytes. *Science* **154,** 780–781.
37. Knight, E., Jr., and Darnell, J. E. (1967). Distribution of 5S RNA in HeLa cells. *J. Mol. Biol.* **28,** 491–502.
38. Kolodny, G. M. (1975). The regulation of gene expression in eukaryotic cells. *Med. Hypotheses* **1,** 1–8.

39. Kostraba, N. C., Montagna, R. A., and Wang, T. Y. (1975). Study of the loosely bound non-histone chromatin proteins. *J. Biol. Chem.* **250,** 1548–1555.
40. Krause, M. O. (1978). The binding of histones in mammalian chromatin: Cell-cycle induced and SV40-induced changes. *In* "Cell Cycle Regulation" (J. R. Jeter, I. L. Cameron, G. M. Padilla, and A. M. Zimmerman, eds.) pp. 61–74.
41. Krause, M. O., and Inasi, B. (1974). Histones from exponential and stationary L-cells; evidence for metabolic heterogeneity of histone fractions retained after isolation of nuclei. *Arch. Biochem. Biophys.* **164,** 179–184.
42. Krause, M. O., Kleinsmith, L. J., and Stein, G. S. (1975). Properties of the genome in normal and SV40-transformed WI38 human diploid fibroblasts. II. Turnover of nonhistone chromosomal proteins and their phosphate groups. *Life Sci.* **16,** 1047–1058.
43. Krause, M. O., Kleinsmith, L. J., and Stein, G. S. (1975). Properties of the genome in normal and SV40-transformed WI38 human diploid fibroblasts. I. Composition and metabolism of nonhistone chromosomal protein. *Exp. Cell Res.* **92,** 164–174.
44. Krause, M. O., Noonan, K. D., Kleinsmith, L. J., and Stein, G. S. (1976). The effect of SV40 transformation on the chromosomal proteins of 3T3 mouse embryo fibroblasts. *Cell Differ.* **5,** 83–96.
45. Krause, M. O., and Ringuette, M. J. (1977). Low molecular weight nuclear RNA from SV40-transformed WI38 cells; effect on transcription of WI38 chromatin *in vitro. Biochem. Biophys. Res. Commun.* **76,** 796–803.
46. Krause, M. O., and Stein, G. S. (1975). Properties of the genome in normal and SV40-transformed WI38 human diploid fibroblasts. *Exp. Cell Res.* **92,** 175–190.
47. Krause, M. O., Yoo, B. Y., and MacBeath, L. (1974). Histones from exponential and stationery cells; evidence for differential binding of lysine-rich and arginine-rich fractions in chromatin. *Arch. Biochem. Biophys.* **164,** 172–178.
48. Lerner, M. R., Boyle, J. A., Mount, S. M., Wolin, S. L., and Steitz, J. A. (1980). Are SnRNP's involved in splicing? *Nature (London)* **283,** 220–224.
49. Liu, W. C., Godbout, R., Jay, E., Yu, K. K.-Y., and Krause, M. O. (1981). Tissue and species specific effects of small molecular weight nuclear RNA's on transcription in isolated mammalian nuclei. *Can. J. Biochem.* **59,** 343–352.
50. Marushige, K., and Dixon, H. (1969). Developmental changes in chromosomal composition and template activity during spermatogenesis in trout testes. *Dev. Biol.* **19,** 397–414.
51. Marzluff, W. F. (1978). Transcription of RNA in isolated nuclei. *Methods Cell Biol.* **19,** 317–332.
52. Marzluff, W. F., Murphy, E. C., and Huang, R. C. C. (1973). Transcription of ribonucleic acid in isolated mouse myeloma nuclei. *Biochemistry* **12,** 3440–3446.
53. Marzluff, W. F., Murphy, E. G., and Huang, R. C. C. (1974). Transcription of the genes for 5S ribosomal RNA and transfer RNA in isolated mouse myeloma cell nuclei. *Biochemistry* **13,** 3689–3696.
54. Marzluff, W. F., White, E. L., Benjamin, R., and Huang, R. C. C. (1975). Low molecular weight RNA species from chromatin. *Biochemistry* **14,** 3715–3724.
55. Mathews, M. B., and Petterson, U. (1978). The low molecular weight RNA of adenovirus 2-infected cells. *J. Mol. Biol.* **119,** 293–328.
56. McReynolds, L., and Penman, S. (1974). Pre-4S RNA made in isolated HeLa cell nuclei terminates with U. *Cell* **3,** 185–188.
57. Miller, T. E., Huang, C. Y., and Pogo, O. (1978). Rat liver nuclear skeleton and small molecular weight RNA species. *J. Cell Biol.* **76,** 692–704.
58. Nguyen-Huu, M. C., Sippel, A. A., Hynes, N. E., Groner, B., and Sehutz, G. (1978). Preferential transcription of the ovalbumin gene in isolated hen oviduct nuclei by RNA polymerase B. *Proc. Natl. Acad. Sci. U.S.A.* **75,** 686–690.

59. Paoletti, E., Lipinskas, B. R., and Panicali, D. (1980). Capped and polyadenylated low-molecular weight RNA synthesized by Vaccinia virus *in vitro*. *J. Virol.* **33,** 208–219.
60. Paul, J., and Gilmour, R. S. (1968). Organ-specific restriction of transcription in mammalian chromatin. *J. Mol. Biol.* **34,** 305–314.
61. Pays, E., Donaldson, D., and Gilmour, R. S. (1978). Specificity of chromatin transcription *in vitro*. Anomalies due to RNA-dependent RNA synthesis. *Biochim. Biophys. Acta* **562,** 112–130.
62. Peters, G., and Glover, C. (1980). tRNAs and priming of RNA-directed DNA synthesis in mouse mammary tumor virus. *J. Virol.* **35,** 31–40.
63. Peters, G. F., Harada, J. E., Dahlberg, A., Panet, W., Haseltine, A., and Baltimore, D. (1977). Low molecular weight RNAs of moloney murine leukemia virus: Identification of the primer for RNA-directed RNA synthesis. *J. Virol.* **21,** 1031–1042.
64. Rapp, F., and Westmoreland, D. (1976). Cell transformation by DNA-containing viruses. *Biochim. Biophys. Acta* **458,** 167–211.
65. Raska, K., Sehulster, L. M., and Varricchio, F. (1976). Three new virus-specific low molecular weight RNAs in adenovirus type 2 infected KB cells. *Biochem. Biophys. Res. Commun.* **69,** 79–84.
66. Reeder, R. H., and Roeder, R. G. (1972). Ribosomal RNA synthesis in isolated nuclei. *J. Mol. Biol.* **67,** 433–441.
67. Reeve, A., Smith, M. M., Pigiet, V., and Huang, R. C. C. (1977). Incorporation of purine nucleoside 5′ (-S) triphosphates as affinity probes for initiation of RNA synthesis *in vitro*. *Biochemistry* **16,** 4464–4469.
68. Ringuette, M. J., Gordon, K., and Krause, M. O. (1982). Specific small nuclear RNAs from SV40-transformed cells stimulate transcription initiation in nontransformed isolated nuclei. *Can. J. Biochem.* (in press).
69. Ringuette, M. J., Liu, W. C., Jay, E., Yu, K. K.-Y., and Krause, M. O. (1980). Stimulation of transcription of chromatin by specific small nuclear RNAs. *Gene* **8,** 211–224.
70. Ro-Choi, T. S., and Busch, H. (1974). Low molecular weight nuclear RNAs. *In* "The Cell Nucleus" (H. Busch, ed.), Vol. 3, pp. 151–208. Academic Press, New York.
71. Ro-Choi, T. S., and Henning, D. (1977). Sequence of 5′ oligonucleotide of U1 RNA from Novikoff hepatoma cells. *J. Biol. Chem.* **252,** 3818–3820.
72. Rogers, J., and Wall, R. (1980). A mechanism for RNA splicing. *Proc. Natl. Acad. Sci. U.S.A.* **77,** 1877–1879.
73. Rubin, C. M., Houck, C. M., Deininger, P. L., Friedman, T., and Schmidt, C. W. (1980). Partial nucleotide sequence of the 300-nucleotide interspersed repeated human DNA sequence. *Nature (London)* **284,** 372–374.
74. Ryskov, A. P., Saunders, G. F., Farashyan, G. P., and Georgiev, G. P. (1973). Double-helical regions in nuclear precursor of mRNA (pre-mRNA). *Biochim. Biophys. Acta* **312,** 152–164.
75. Sawyer, R. C., and Dallberg, J. E. (1973). Small RNAs of Rous sarcoma virus: Characterization by two-dimensional polyacrylamide gel electrophoresis and fingerprint analysis. *J. Virol.* **12,** 1226–1237.
76. Sehgal, P. B., Derman, E., Molloy, G. R., Tamm, I., and Darnell, J. E. (1976). 5,6-Dichloro-1-β-D-ribofuranosyl benzimidazole inhibits initiation of nuclear heterogeneous RNA chains in HeLa cells. *Science* **194,** 431–433.
77. Seligy, V., and Miyagi, M. (1969). Studies of template activity of chromatin isolated from metabolically active and inactive cells. *Exp. Cell Res.* **58,** 27–34.
78. Shelton, K. S., and Allfrey, V. G. (1970). Selective synthesis of a nuclear acid protein in liver cells stimulated by cortisol. *Nature (London)* **288,** 132–134.

79. Sluyterman, L. A. A. E., and Wijdenes, J. (1970). An agarose mercurial column for the separation of mercaptopapain and nonmercaptopapain. *Biochim. Biophys. Acta* **200,** 595.
80. Smith, M. M., Reeve, A. E., and Huang, R. C. C. (1978). Transcription of bacteriophage DNA *in vitro* using purine nucleoside 5′(γ-S) triphosphates as affinity probes for RNA chain initiation. *Biochemistry* **17,** 493–500.
81. Smith, R. J., and Duerksen, J. D. (1975). Glycerol inhibition of purified and chromatin-associated mouse liver hepatoma RNA polymerase II activity. *Biochem. Biophys. Res. Commun.* **67,** 916–923.
82. Stein, G., Park, W., Thrall, C., Mans, R., and Stein, J. (1975). Regulation of cell cycle stage-specific transcription of histone genes from chromatin by non-histone chromosomal proteins. *Nature (London)* **257,** 764–767.
83. Subramanian, K. W., Reddy, V. B., and Weissman, S. M. (1977). Occurrence of reiterated sequences in an untranslated region of Simian Virus 40 DNA determined by nucleotide sequence analysis. *Cell* **10,** 497–507.
84. Tamm, I., Hand, R., and Caliguiri, A. (1976). Action of dichlorobenzimidazole riboside on RNA synthesis in L-929 and HeLa cells. *J. Cell Biol.* **69,** 229–240.
85. Tooze, J., ed. (1973). "The Molecular Biology of Tumor Viruses," pp. 352–403. Cold Spring Harbor Lab., Cold Spring Harbor, New York.
86. Walker, T. A., Pace, N. H., Frikson, R. I., Frikson, T., and Bahr, F. (1974). The 7S RNA common to oncornaviruses and normal cells is associated with polysomes. *Proc. Natl. Acad. Sci. U.S.A.* **71,** 3390–3394.
87. Weinberg, R. A., and Penman, S. (1968). Small molecular weight monodisperse nuclear RNA. *J. Mol. Biol.* **38,** 289–304.
88. Weinberg, R. A., and Penman, S. (1969). Metabolism of small molecular weight monodisperse nuclear RNA. *Biochim. Biophys. Acta* **190,** 10–29.
89. Weinmann, R., Brender, T. G., Raskas, H. J., and Roeder, R. G. (1976). Low molecular weight viral RNAs transcribed by RNA polymerase III during adenovirus 2 infection. *Cell* **7,** 557–566.
90. Weisbrod, S., and Weintraub, H. (1979). Isolation of a subclass of nuclear proteins responsible for conferring a DNase I-sensitive structure on globin chromatin. *Proc. Natl. Acad. Sci. U.S.A.* **76,** 630–634.
91. Zieve, G., and Penman, S. (1976). Small RNA species of the HeLa cell: Metabolism and subcellular localization. *Cell* **8,** 19–31.
92. Zylbler, E. A., and Penman, S. (1971). Products of RNA polymerases in Hela cell nuclei. *Proc. Natl. Acad. Sci. U.S.A.* **68,** 2861–2865.

7

Transcription of rRNA Genes and Cell Cycle Regulation in the Yeast *Saccharomyces cerevisiae*

R. A. SINGER AND G. C. JOHNSTON

I. Introduction

In order to reproduce, cells must have the capacity to grow (increase in mass) and to divide (replicate genomic DNA and segregate replicated DNA, along with cellular constituents, to progeny cells). Although these two processes are coordinated so that cells normally do not continue division in the absence of growth, these activities of growth and division occur independently [for review, see (*16*)]. This independence has been

GENETIC EXPRESSION IN THE CELL CYCLE

ISBN 0-12-543720-X

formalized by Mitchison (*26*), who described the cell cycle as consisting of two independent cycles: the "growth cycle" and the "DNA-division cycle." The growth cycle refers to those processes which go to make up the bulk of new cytoplasm, and the DNA-division cycle or "DNA-division sequence" refers to those periodic, sequential events involved in the replication and segregation of DNA (see also Chapter 4).

We have been interested in the mechanisms by which cells normally coordinate the processes of growth and cell division. More specifically, we are interested in how metabolic processes (aspects of the growth cycle) affect the normal regulation of cell division (the DNA-division sequence). For our investigations we have employed the unicellular yeast *Saccharomyces cerevisiae* as a model system because it has proved advantageous for study of the eukaryotic cell cycle (*12*).

II. Yeast as a Model Eukaryote

Since this volume may be more directed toward the researcher involved with animal cells, we feel it would be useful to describe the yeast *S. cerevisiae*, particularly with respect to both execution and regulation of its cell cycle. Before discussing the various aspects of yeast cell cycle regulation, we must digress briefly into a more general description of the yeast system.

In general, the biochemistry of yeast appears to be that of a typical eukaryote. For example, its nuclear chromosomes are distributed with the aid of a mitotic spindle and exhibit classical meiotic and mitotic chromosome segregation (*27*). One or more of the multiple DNA polymerase enzymes (*5*) replicates nuclear DNA within a specific period of the cell cycle (S phase), and multiple RNA polymerase species transcribe the DNA into RNA (*1*). Ribosomal RNA genes are clustered in tandemly repeated units on one chromosome (*30*) and are transcribed into high molecular weight precursor ribosomal RNA (pre-rRNA), which in turn is cleaved to produce mature ribosomal RNA (rRNA) (*31*). Like other eukaryotic cells, yeast also utilizes RNA splicing to mature the precursors to some transfer RNA (tRNA) (*24*) and messenger RNA (mRNA) (*28*) species.

The *S. cerevisiae* system has a number of useful technical properties. Yeast can be cultured in either the haploid or the diploid state; hence, mutants can be isolated in the stable haploid condition and complementation tests can be carried out in the diploid. In addition to a large number of metabolic mutants available in yeast, there is also available a large collection of mutants defective in specific processes of cell division (the so-called "cell division cycle" or *cdc* mutants) (*12*).

A further advantage is morphological. Yeast reproduce through the production of buds; the initiation of a new bud roughly corresponds to the transition from the G_1 to the S phase of the cell cycle (*21,40*). Thus, the presence or absence of buds (easily determined by direct microscopic examination) reveals if cells are in the S, G_2, or M phases (budded) or in the G_1 portion (unbudded) of the cell cycle.

III. Regulation of the Yeast Cell Cycle

A. G_1 Regulation

As found for most mammalian cells, the yeast cell cycle is divided into four major periods, with control of cell division exerted within the G_1 period. Starved cells arrest preferentially in G_1 and do not reinitiate division until the required nutrient is resupplied (*16*). Haploid yeast cells of opposite mating type can fuse and form a diploid only when both cells are arrested in G_1 by the action of mating pheromones (*12*). To accomplish this synchronized arrest for mating, cells of each mating type of yeast (*a* or α mating type) are arrested in G_1 by the action of a small polypeptide secreted by cells of the opposite mating type. Thus, *a* cells are arrested by what is referred to as α-factor; likewise, α cells are arrested by *a* factor. The G_1 arrest brought about by the application of mating factors has proved a useful landmark for cell cycle analysis [see (*13*)].

B. Growth and the Critical Size Requirement

Yeast cells (and indeed most cells in culture) display a rather narrow and characteristic range of cell sizes [for review, see (*16*)]. Thus, some mechanism must exist to coordinate growth and cell division. A model to explain this coordination has been proposed by several workers (*14,16*). This model suggests that growth, rather than progression through the DNA-division sequence, is the rate-limiting activity for cell division. That is, cells can normally complete the DNA-division sequence faster than they can double in cell mass. There is experimental support for this feature of the model, since abnormally large cells blocked at some point in the DNA-division sequence can, upon release from the block, rapidly "divide themselves back to the normal cell size range" while normal growth continues (*16*).

A second aspect of this model is that sufficient growth, to what is mgasured as a critical size, is required for the completion of at least one step in the DNA-division sequence. We have shown that abnormally small cells (produced by nitrogen starvation) grow to a certain cell size prior to completion of this step and the initiation of a bud (*16*). This requirement for cell size, or more reasonably some property related to size, can also be demonstrated during steady state growth and division (*19*). Because of the mode of division of yeast cells, the bud (the incipient daughter cell) usually does not enlarge to equal the size of the mother cell prior to cytokinesis (*14*). These normally small daughter cells produced under steady state growth conditions must subsequently grow to a critical size before they can initiate a bud. Because the larger mother cells need not go through an extensive period of growth before initiating another bud, consistent differences in cell cycle times between mother cells and daughter cells are observed. These differences can be accounted for in the longer G_1 periods of daughter cells required for growth to the critical size. Thus, G_1 is also the period in which coordination of growth and the DNA-division sequence occurs.

C. The Concept of the Start Event

A point of regulation in G_1 has been referred to by Hartwell and others as *start* (*12,14,16*). This is the earliest known event in the yeast cell cycle and is defined as the *cdc*28 gene-mediated step. Start is also the step sensitive to the presence of mating pheromone. Starved cells, and cells below the critical size, are arrested at or before the start event. Once start is completed, cells generally complete the mitotic cycle regardless of growth conditions (*12,16,17*). It appears that at start a large number of internal and external inputs are integrated such that the cell ultimately undergoes a mitotic cycle, or arrests cell division, or mates (in the case of haploids), or sporulates (in the case of diploids). An understanding of the molecular nature of start and the molecular signals that affect start is one of our major goals and is a major concern of this chapter.

D. Involvement of Macromolecular Metabolism in Cell Cycle Regulation

Once the start event has been completed, cells can then complete the balance of the DNA-division sequence with little or no net increase in mass (*16,17*). Thus, it would appear that cell cycle regulation is responsive to cell size or mass only at, or prior to, the start event. Several workers have investigated the effect of macromolecular metabolism on the ability of cells to complete the start event and initiate a new cell cycle (*14,18*). Changes in the nutritional environment so that growth rates were limited either by carbon source (*40*) or by the presence of the protein synthesis inhibitor cycloheximide (*14*) caused the rates of cell number increase to be slowed. These rate-limiting effects on growth did not seem to affect the time required to complete the budded period of the cell cycle (roughly that period between the completion of start and mitosis). Instead, only the length of the G_1 period was markedly affected. Thus, longer generation times result in more time spent in G_1 (perhaps as cells require longer periods to grow to the critical size).

An enlightening example of the cell cycle effects caused by altered macromolecular metabolism was provided by Unger and Hartwell (*43*). They demonstrated that for a mutant strain defective in methionyl-tRNA synthetase (EC 6.1.1.10) slowing the rate of tRNA aminoacylation (and thus the rate of polypeptide chain elongation) caused cells to accumulate in G_1 at or before the start event. Their experinents allowed them to conclude that a metabolic signal acting to affect cell cycle regulation must be generated by the general process of protein synthesis at or after the step of polypeptide chain elongation. This conclusion effectively explains how the many metabolic perturbations which slow the rate of protein synthesis all make cells less able to execute the start event.

In the following discussion we describe another experimental approach to identify other aspects of macromolecular metabolism to which the cell cycle regulatory machinery is responsive.

IV. Experimental Approach

A. Use of Inhibitors in the Study of Cell Cycle Regulation

Although several conditional mutations have been identified which cause yeast cells at the nonpermissive temperature to arrest at several positions in the cell cycle, including in G_1 (*12*), experiments with these mutations have shed little light on the molecular basis for cell cycle control. The major reason behind this disappointing realization is that for most of these mutations we do not know the specific molecular defects involved. An alternative approach, then, has been to identify treatments which specifically cause actively dividing cells to be unable to execute start, but still be able to complete ongoing cell cycles. The metabolic effects of these treatments have then been determined, with the hope of identifying correlations between cell cycle regulation and particular aspects of macromolecular metabolism (*4,18,36,37*).

1. *Cellular Effects of G_1 Arresting Compounds*

We have identified a limited but diverse collection of compounds, all of which cause cells to arrest permanently or transiently in the G_1 period of the cell cycle. These compounds are listed in Table I. When we examined the cellular effects of these compounds we found that they all, when added at appropriate concentrations, caused cells, within one cell division cycle, to arrest in the G_1 period (the unbudded portion) of the cell cycle. The point in the cell cycle at which these cells were G_1 arrested was in each case more precisely determined by the use of the technique of order-of-function mapping. In this procedure, the ability to complete the cell cycle is assessed after arrested cells are shifted from an initial arresting condition to a second arresting condition (see Fig. 1). From the results of this and the converse reciprocal shift experiment (arrested cells shifted from the second arresting condition to the first arresting condition), the functional relationships of the two induced arrest conditions may be determined [e.g., see (*37*)]. In this way, and using α-factor exposure as the second arresting treatment, we found that almost all of these particular compounds caused cells to arrest at the start event. [Results for nalidixic acid (*36*) could not be this clear, since this compound causes only a transient G_1 arrest; thus, we were unable to perform a complete order-of-function experiment. Nevertheless, nalidixic acid does arrest cells transiently at or before the start event.]

2. *Metabolic Effects of G_1 Arresting Compounds*

Having identified a number of compounds which cause cells to arrest in G_1, we then examined the effects of these G_1 arresting treatments on macromolecular metabolism. Initially we determined the effects of these

TABLE I

List of Compounds Causing G_1 Arrest

Compound	Reference
o-Phenanthroline	Johnston and Singer (*18*)
8-Hydroxyquinoline	Johnston and Singer (*18*)
L-Ethionine	Singer *et al.* (*37*)
β-2-DL-Thienylalanine	Bedard *et al.* (*4*)
Nalidixic acid	Singer and Johnston (*36*)

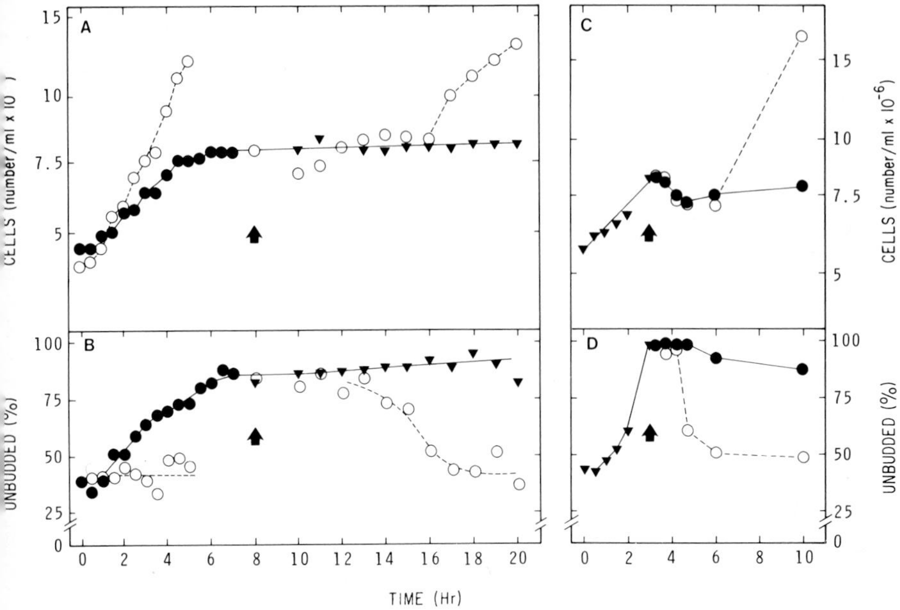

Fig. 1. Order-of-function mapping of L-ethionine- and α-factor-sensitive steps in the yeast cell division cycle. Panels A and B: cells were arrested in G_1 by treatment with L-ethionine. At the time indicated by the arrows, cells were transferred to media with or without α-factor. Panels C and D: cells were arrested in G_1 by treatment with α-factor. At the time indicated by the arrows, cells were transferred to media with or without L-ethionine. (○) cells in the absence of treatments causing G_1 arrest; (●) cells in the presence of L-ethionine; (▼) cells in the presence of α-factor. [From (*37*).]

compounds on the rates of protein and RNA synthesis. This was done by a pulse-labeling protocol, with label incorporation rates corrected for changes in precursor pool specific activities. We routinely found little or no effect on the rates of protein synthesis under conditions causing G_1 arrest. However, we always found a significant effect on the rates of RNA production. For the purpose of illustration we will refer in detail to the effects of specific compounds causing G_1 arrest; in general, results for any one treatment were similar to those for any other.

As shown in Fig. 2, the major effect of treatment of cells was on the production of RNA. Yeast has three RNA polymerase species responsible for the synthesis of three major classes of RNA: rRNA, mRNA, and the low molecular weight RNAs, tRNA and 5 S rRNA (*1*). When we examined the effect of one of these G_1-arresting agents on the production of poly(A)-containing RNA (mRNA), we found essentially no effect during treatment leading to G_1 arrest (*18*). Likewise, there was

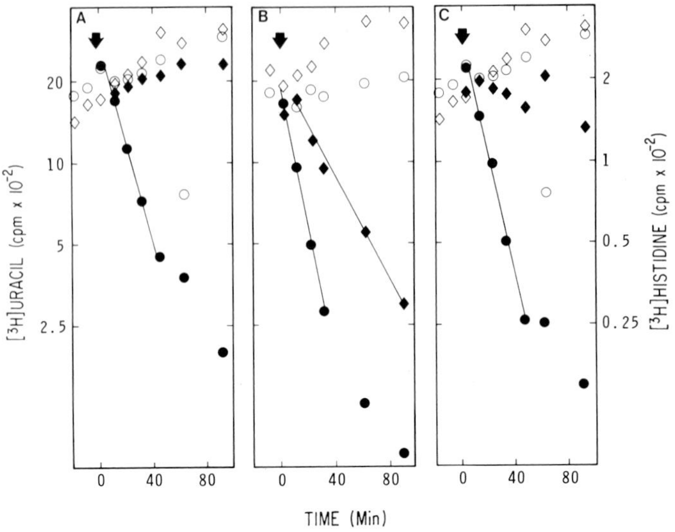

Fig. 2. Rates of uracil and histidine incorporation during treatment with methionine analogues. Samples of treated and untreated cells were removed to tubes containing either [^{3}H]histidine (10 μCi/ml) or [^{3}H]uracil (10 μCi/ml). After a further 5-min incubation, incorporation of precursors was stopped by the addition of an equal volume of 10% trichloroacetic acid. Panel A: effect of ethionine. Panel B: effect of trifluoromethionine. Panel C: effect of selenomethionine. Symbols: (○) uracil incorporation in the absence of analogues; (●) uracil incorporation in the presence of the analogue; (◇) histidine incorporation in the absence of analogues; (◆) histidine incorporation in the presence of the analogue. [From (*37*).]

little effect when we measured by gel electrophoretic techniques the production of 4 S RNA. However, when we examined the production of rRNA, we found a significant decrease in the rate of production of high molecular weight pre-rRNA and on its subsequent processing to yield mature rRNA (Fig. 3).

Thus, in all cases examined we found a striking correlation between G_1 arrest at start and the decreased abilities to produce and process pre-rRNA.

B. Transcription of pre-rRNA and Cell Cycle Control

Temperature shift experiments have been employed to eliminate from cell cycle considerations those aspects of RNA metabolism not correlated with cell cycle regulation (*20*). Upon shift from 23° to 36°C, a growing population of yeast displays transient decreases in the rates of synthesis of pre-rRNA (*45*) and of mRNA for ribosomal proteins (*10*). We found similar results; although protein synthesis continued at rates comparable

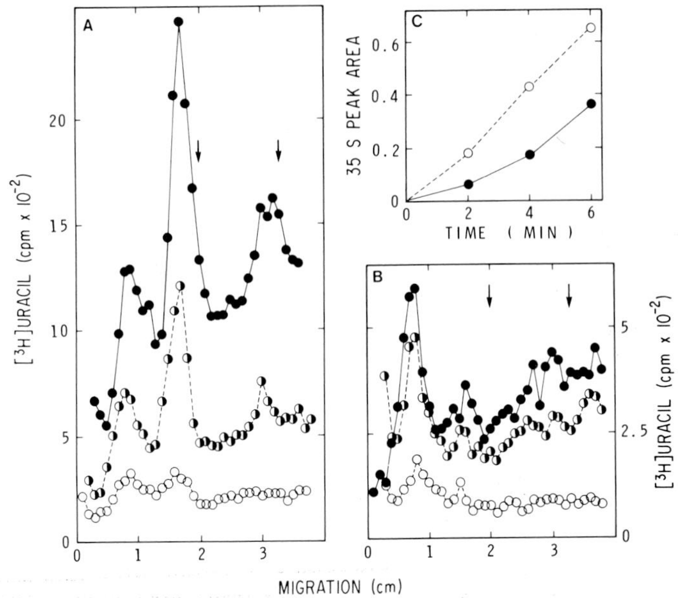

Fig. 3. Production of precursor ribosomal RNA (pre-rRNA) during treatment with L-ethionine. A [^{14}C]uracil prelabeled culture was divided; to one-half (panel A) no additions were made, and to the other half (panel B) L-ethionine was added. After a further 30-min incubation, cells were labeled for periods of 2 (○--○), 4 (◑--◑), and 6 (●--●) min with [^{3}H]uracil. RNA was extracted and resolved by polyacrylamide gel electrophoresis. Arrows indicate the positions of ^{14}C-labeled 25 S and 18 S rRNA. Panel C. Accumulation of label in the 35 S peak for ethionine-treated (●—●) and untreated (○--○) cells, normalized to ^{14}C-labeled 25 S RNA. [From (*37*).]

to those of untreated cells, RNA synthesis (after correction of label incorporation rates for altered specific activities of nucleoside triphosphate pools) was decreased. These observations led us to look for cell cycle effects. As expected from the described effects on pre-rRNA production, upon temperature shift these cells also displayed a transient arrest in G_1 as unbudded cells. Cells transiently arrested in the unbudded period of the cell cycle were found by α-factor sensitivity experiments to be at or before the start event. Thus, temperature shift is one more instance in which decreased pre-rRNA production is correlated with specific G_1 arrest.

The use of particular mutant strains allowed us to conclude that cell cycle behavior is correlated with effects on pre-rRNA production but not with effects on pre-rRNA processing. Strains carrying certain of the *rna* mutations are like wild type in the transcription of rRNA genes (*34*), but, unlike wild type, become permanently defective in the synthesis of ribosomal proteins at the nonpermissive temperature of 36°C (*9;* C. Gorenstein, personal communication). In the absence of such ribosomal protein synthesis, proper processing of the high molecular weight pre-rRNA transcript does not occur, and instead the pre-rRNA is degraded (*45*). Thus, these mutant cells at the nonpermissive temperature can transcribe rRNA genes but can produce neither mature rRNA nor ribosomes. Significantly, upon temperature shift of strains carrying *rna* mutations, they too, like wild-type strains, exhibited only a transient G_1 arrest. Reinitiation of new cell cycles occurred as pre-rRNA production returned to normal, even when mutant cells had lost the ability to synthesize ribosomal proteins and to produce mature rRNA. Thus, the ability of a cell to execute the start event must be responsive to some aspect of pre-rRNA production itself.

C. Mechanism of Action of G_1-Arresting Treatments

Although we can detail the cellular and macromolecular effects of the G_1-arresting compounds, their modes of action remain unknown. To this end, we have examined the nature of the active form of one class of compound, exemplified by the chelating agents *o*-phenanthroline (OP) and 8-hydroxyquinoline (HQ) (Fig. 4a and g, respectively). We explored the modes of action of these compounds by looking for reversal of the G_1-arresting effects. The addition to OP- or HQ-arrested cells of either Fe^{2+} (our unpublished observations) or Zn^{2+} caused a resumption of cell cycle activity. No other divalent cations tested reversed this G_1 arrest. Upon addition of Fe^{2+} or Zn^{2+} salts, cells began to bud with kinetics similar to those found after removal of OP or HQ from the growth

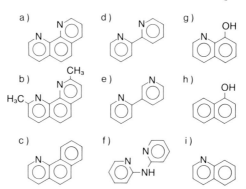

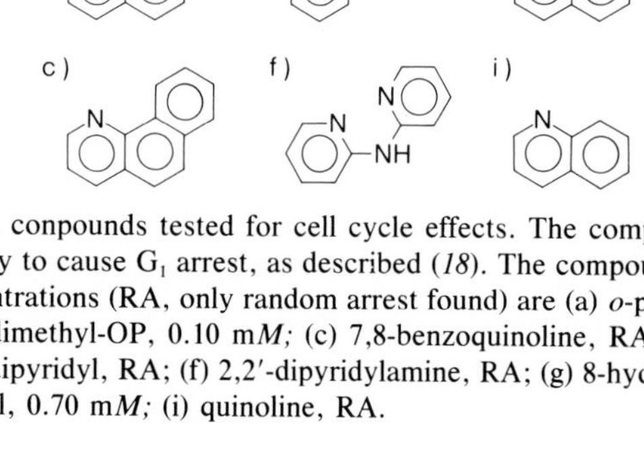

Fig. 4. Related conpounds tested for cell cycle effects. The compounds shown were tested for the ability to cause G_1 arrest, as described (*18*). The compounds, along with the G_1-arresting concentrations (RA, only random arrest found) are (a) *o*-phenanthroline (OP), 0.10 m*M;* (b) 2,9-dimethyl-OP, 0.10 m*M;* (c) 7,8-benzoquinoline, RA; (d) 2,2′-dipyridyl, 0.20 m*M*; (e) 2,3′-dipyridyl, RA; (f) 2,2′-dipyridylamine, RA; (g) 8-hydroxyquinoline, 0.14 m*M;* (h) 1-naphthol, 0.70 m*M;* (i) quinoline, RA.

medium. Apparently these G_1-arresting compounds do not affect pre-rRNA production simply by chelating intracellular Zn^{2+} or Fe^{2+}, as shown by the effects of the compound 2,9-dimethyl-OP (Fig. 4b). This analogue of OP has an affinity for Zn^{2+} decreased by 10^3 (*35*). Nevertheless, this *nonchelating* analogue, added at concentrations similar to those of OP giving cell cycle activity, caused similar G_1 arrest but one which could not be reversed by divalent cations (J. Knowles, unpublished data). From these results, we could conclude that the metabolically active configuration of these G_1-arresting compounds is the unchelated form and that Zn^{2+} or Fe^{2+} reverses the cell cycle effects by forming inactive metal-chelated complexes.

Both OP and HQ have similar arrangements of aromatic rings and of N or O atoms. Therefore, to determine if these common structural components are related to cell cycle effects, other aromatic molecules with similar alignments were checked for cell cycle activity. Some of these related compounds we tested are shown in Fig. 4. As indicated, some could indeed cause G_1 arrest (J. Knowles, unpublished data). Since at least one of these compounds (1-naphthol, Fig. 4h) does not bind divalent cations, this result supports our conclusion that with these aromatic compounds cation binding per se is not involved in cell cycle arrest. (This interpretation depends upon the reasonable assumption that these related compounds have the same mode of action as OP or HQ.) In addition, the inability of many of these compounds to effect G_1 arrest, specifically those compounds with different configurations of N atoms and aromatic rings, suggests the importance of these features in the activity of this group of compounds.

One unusual observation shown by the dose–response curve for HQ bears further comment. As noted earlier, at concentrations of HQ lower than those causing G_1 arrest there was immediate inhibition of cell cycle activity, so that populations of cells were arrested randomly throughout the cell cycle (*18*). It is difficult for us to imagine how increasing the concentration of this agent can relieve what appears to be a nonspecific cell cycle inhibition in favor of a more specific effect on pre-rRNA production. Perhaps further study of this anomalous concentration effect may help to elucidate the mode of action of HQ and the other, similar compounds mentioned here.

The mechanism by which the amino acid analogue L-ethionine causes decreased pre-rRNA production also has not been resolved. Of the three methionine analogues that were studied, only L-ethionine caused G_1 arrest. (The D-ethionine isomer had no effect on cell kinetics.) The other two analogues, selenomethionine and trifluoromethionine, were found by bud morphology analysis to inhibit further cell division at various points throughout the cell cycle (*37*). The reasons for this differential cell cycle response became clear when macromolecular metabolism was assessed. When cells were treated with the lowest concentration of each analogue still giving cell cycle effects, all these analogues were found to significantly decrease RNA labeling. However, addition of selenomethionine or trifluoromethionine also led to large decreases in rates of amino acid incorporation. We suspect that, upon this type of inhibition of protein synthesis, cells rapidly become unable to complete DNA division sequence events. If so, then regardless of effects on RNA synthesis, cells treated with selenomethionine or trifluoromethionine lose the ability to G_1 arrest solely because of these effects on protein synthesis.

The ability of the amino acid analogue β-2-DL-thienylalanine to cause G_1 arrest (*4*), as well as the stereochemical relationship of this molecule to methionine, particularly in the location of the S atom, make it possible that this compound may act as an analogue of methionine. However, the similar analogue β-3-DL-thienylalanine also caused G_1 arrest, although at concentrations threefold higher than required for β-2-DL-thienylalanine (R. A. Singer and G. C. Johnston, unpublished observations). Since in β-3-DL-thienylalanine the S atom is sterically positioned differently than it is in methionine, this result makes it less likely that β-3-DL-thienylalanine and β-2-DL-thienylalanine act as analogues of methionine.

The mechanism of action of nalidixic acid in causing decreases in pre-rRNA production and transient G_1 arrest may be suggested by results from another system. For the bacterium *E. coli* and some of its viruses, nalidixic acid (an inhibitor of DNA gyrase) inhibits transcription of some, but not all, operons (*33*). Among the transcriptional units differentially sensitive to DNA gyrase inhibitors are those for rRNA. This sensitivity, and that of these same operons to inhibition by coumermycin and its derivatives (*46*), implies that promoter-specific transcription of these bacterial genes requires the functioning of both subunits of DNA gyrase. A DNA gyrase-like activity has not yet been found in *S. cerevisiae;* also, treatment of this yeast with coumermycin does not cause even transient G_1 arrest (unpublished observations). Nevertheless, from the results cited here it may be reasonable to imagine that treatment of yeast with nalidixic acid inhibits pre-rRNA production, and thereby causes G_1 arrest, through promoter-specific inhibition of the transcription of rRNA genes.

V. Discussion

A. Cellular Response to Changes in pre-rRNA Production: A General Mechanism for Probing the Environment?

In the introduction we presented the observations that the cell cycle behavior of yeast cells is responsive to certain perturbations, such as the absence of a required nutrient or the presence of mutations or inhibitors which affect the rate of protein synthesis. In Section IV we showed that the cell cycle is also responsive to changes in pre-rRNA production. Since the only known regulatory point in the yeast cell cycle is start, it has been concluded that the start event is sensitive to large number of metabolic and environmental perturbations (*12*). One could hypothesize that start itself is directly responsive to each of these perturbations causing G_1 arrest; however, we have suggested an alternative way that cells may be responsive to these perturbations, by what we have termed an integrative sequence of responses (*22*). We propose that such an integrative sequence (see Fig. 5) consists of at least three definable steps. The first step is some aspect of the process of protein synthesis, because one type of perturbation in macromolecular metabolism leading to cells accumulated at start is that affecting rates of protein synthesis [see (*43*)]. A second step in this integrative sequence is the production of pre-rRNA, because affecting this process also leads to G_1 arrest. For the integrative sequence, the essential linkage between the rates of protein synthesis and of pre-rRNA production is provided by the *stringent* response of cells, recently shown to operate in yeast (*44*). The stringent response, which has been well characterized in bacteria, is a metabolic coordination in which decreased rates of protein synthesis rapidly cause decreased rates of pre-rRNA production. Although the mechanism of this response in yeast may be different from that found in bacteria (*29*), the end result is the same. Thus, through the stringent response, decreased rates of protein synthesis have indirect effects on the production of pre-rRNA and thereby generate cell cycle responses; in contrast, affecting pre-rRNA production more directly, as we have described here, can affect cell cycle regulation without affecting protein synthesis rates. A third step in this sequence is start itself, which is sensitive to mating pheromone. When macromolecular metabolism was examined in cells of *a* mating type treated with the α-factor mating pheromone, there was no effect of this G_1 arrest treatment on rates of protein synthesis or RNA synthesis (*42;* also our unpublished data). Thus, we suppose that G_1 arrest brought about by mating pheromone

Amino acid starvation Cycloheximide *mes*1 Mutation	Heat shock *o*-Phenanthroline L-Ethionine	α-Factor *cdc*28 Mutation
↓	↓	↓

Protein synthesis → → pre-rRNA synthesis → → → "Start"

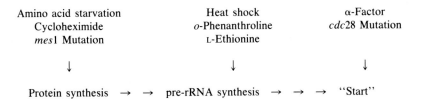

Fig. 5. The integrative sequence of events affecting completion of the start event.

acts more directly on the start event, without affecting the above-mentioned aspects of general metabolism.

We therefore suggest that the start event itself is responsive only to the penultimate step in an integrative sequence of metabolic responses and that perturbations causing cell cycle responses only affect processes "downstream" in the sequence. Events in the sequence are sufficiently sensitive to the metabolic status of the cell to ensure that a large number of perturbations will be detected and the resulting information integrated by the cell for cell cycle control.

One prediction of the integrative sequence model is that metabolic alterations not directly or even indirectly perturbing steps in the integrative sequence will not lead to G_1 arrest. There are many situations in which this prediction is fulfilled. Strains bearing any one of the *rna* mutations 2 through 11 seem to be defective in the ability to produce mRNA for ribosomal proteins (*9;* C. Gorenstein, personal communication). Nevertheless, such strains continue to transcribe rRNA genes (*34*). Thus, changes in ribosomal protein synthesis do not directly affect any step in the integrative pathway and, as shown in Section IV, do not lead to G_1 arrest. Alterations in membrane and cell wall metabolism also fail to affect the integrative pathway and do not result in G_1 arrest. For example, starvation of inositol-requiring strains for that essential precursor of membrane biosynthesis causes cells to halt randomly in the cell cycle (*15*). Similarly, defects in cell wall organization also have no cell cycle effects. Sloat and Pringle (*41*) have described the effects of a conditional mutation (*cdc*24), which at the nonpermissive temperature causes a failure to lay down chiton in the appropriate fashion. Although cell wall metabolism is disturbed by this mutation, cells continue to initiate nuclear DNA-division cycles.

Starvation of certain strains for deoxythymidine monophosphate (*25*), mutational alterations in thymidylate synthetase (*8*) or DNA ligase (*23*), or treatment of cells with the DNA synthesis inhibitor hydroxyurea (*39*), even in limiting concentrations (*38*), all fail to cause cells to G_1 arrest, even though each situation has effects on DNA metabolism. Thus, manipulations which specifically block DNA synthesis are also unable to affect the integrative sequence.

B. Implications for Regulation of Cell Size in Yeast

The mechanism used by yeast to coordinate growth with cell division involves the requirement for what is seen as growth to some critical size prior to completion of the start event in G_1 (*16*). But cells may not actually measure cell size per se. It is more likely that what is seen operationally as a requirement for the attainment of a certain cell size really reflects the assessment by the cell of one or a few monitored components, whose production or intracellular concentration is normally proportional to cell size. Our findings that decreased rates of pre-rRNA production are correlated with G_1 arrest have led us to suggest that one such monitored component may be related to ribosomes. A measure of the number of new ribosomes may serve as an estimate of cell mass. Since decreases in rates of ribosome biosynthesis or mature rRNA production do not themselves lead to G_1 arrest, but decreased rates of pre-rRNA production do so (*20*), the production of pre-rRNA itself may be the monitored event correlated with cell size in yeast.

Our suggestion that pre-rRNA metabolism is involved in cell cycle regulation and in estimation of cell size is not new. Indeed, Prescott (*32*) has suggested that the initiation of a new cell cycle (as represented by DNA synthesis) "is not governed by the attainment of a given cell mass . . . but rather by some relatively specific component of growth (for example, ribosome accumulation) that increases in parallel with cell mass." Various aspects of metabolism of rRNA have been implicated as components of cell cycle regulation in animal cells. Early work by Baserga and co-workers (*2,3*) showed that low concentrations of actinomycin D, thought to inhibit RNA production selectively, caused Erlich ascites cells growing asynchronously in the peritoneal cavity of mice to arrest in G_1. They reported a major effect of this treatment on the production of rRNA. More recently, Darzynkiewicz and co-workers (Chapter 4; *6*) found an inverse relationship for animals cells between RNA content and time spent in G_1; cells with higher RNA contents displayed shorter G_1 periods. They, too, suggested that for individual cells the rate of passage through G_1 is correlated with the number of ribosomes per cell. It must be pointed out that others have shown that ribosome accumulation per se is not necessary for exit from the G_1 phase of the cell cycle (*11*). Our work with yeast leads us to agree with this assertion. Even though our experiments have ruled out ribosome accumulation as the governing factor for cell cycle initiation in yeast, we do feel that it is some aspect of pre-rRNA production that is correlated with the ability to initiate a new cell cycle.

Many studies of animal cells have dealt with the requirements for initiation of new cell cycles through the stimulation of nonproliferating, G_1-arrested cells to enter S phase. These types of studies using nonproliferating cells are fundamentally different from work summarized here, concerning cell cycle arrest of actively dividing cells. Nevertheless, upon stimulation of nonproliferating animal cells, increases in rRNA production are almost always observed prior to initiation of S phase [but see

(7)]. These findings are consistent with our conclusion that pre-rRNA production is involved in yeast cell cycle regulation.

Acknowledgments

Work reported here has been supported by grants to R. A. S. and G. C. J. from the Medical Research Council of Canada and the National Cancer Institute of Canada. One of us (R. A. S.) was also supported by a grant from the Department of Medicine Research Foundation. The authors also wish to thank David Carruthers for expert technical assistance and Dr. D. Bedard for helpful discussions. Kris Calhoun provided invaluable assistance in the preparation of this manuscript.

References

1. Adman, R., Schultz, L. D., and Hall, B. D. (1972). Transcription in yeast: Separation and properties of multiple RNA polymerases. *Proc. Natl. Acad. Sci. U.S.A.* **69,** 1702–1706.
2. Baserga, R., Estensen, R. D., Peterson, R. O., and Layde, J. P. (1965). Inhibition of DNA synthesis in Ehrlich ascites cells by actinomycin D. I. Delayed inhibition by low doses. *Proc. Natl. Acad. Sci. U.S.A.* **54,** 745–751.
3. Baserga, R., Estensen, R. D., and Peterson, R. O. (1965). Inhibition of DNA synthesis in Ehrlich ascites cells by actinomycin D. II. The presynthetic block in the cell cycle. *Proc. Natl. Acad. Sci. U.S.A.* **54,** 1141–1148.
4. Bedard, D. P., Singer, R. A., and Johnston, G. C. (1980). Transient cell cycle arrest of the yeast *Saccharomyces cerevisiae* by the amino acid analog β-2-DL-thienylalanine. *J. Bacteriol.* **141,** 100–105.
5. Chang, L. M. S. (1977). DNA polymerases from Baker's yeast. *J. Biol. Chem.* **252,** 1873–1880.
6. Darzynkiewicz, Z., Evenson, D. P. Staiano-Coies, L., Sharpless, T. K., and Melamed, M. L. (1979). Correlation between cell cycle duration and RNA content. *J. Cell. Physiol.* **100,** 425–438.
7. Galanti, N., Jonak, G. J., Soprano, K. J., Floros, J., Kaczmarek, L., Weissman, S., Reddy, V. B., Tilghman, S. M., and Baserga, R. (1981). Characterization and biological activity of cloned simian virus 40 DNA fragments. *J. Biol. Chem.* **256,** 6469–6474.
8. Game, J. C. (1976). Yeast cell-cycle mutant *cdc*21 is a temperature-sensitive thymidylate auxotroph. *Mol. Gen. Genet.* **146,** 313–315.
9. Gorenstein, C., and Warner, J. R. (1976). Coordinate regulation of the synthesis of eukaryotic ribosomal proteins. *Proc. Natl. Acad. Sci. U.S.A.* **73,** 1547–1555.
10. Gorenstein, C., and Warner, J. R. (1977). Coordinate regulation of the synthesis of yeast ribosomal proteins. *ICN–UCLA Symp. Mol. Cell. Biol.* **8,** 203–211.
11. Grummt, F., Grummt, I., and Mayer, E. (1979). Ribosome biosynthesis is not necessary for initiation of DNA replication. *Eur. J. Biochem.* **97,** 37–42.
12. Hartwell, L. H. (1974). *Saccharomyces cerevisiae* cell cycle. *Bacteriol. Rev.* **38,** 164–198.
13. Hartwell, L. H. (1976). Sequential function of gene products relative to the DNA synthesis in the yeast cell cycle. *J. Mol. Biol.* **104,** 803–817.

14. Hartwell, L. H., and Unger, M. W. (1977). Unequal division in *Saccharomyces cerevisiae* and its implications for the control of cell division. *J. Cell Biol.* **75,** 422–435.
15. Henry, S. A., Atkinson, K. D., Kolat, A. I., and Culbertson, M. R. (1977). Growth and metabolism of inositol-starved *Saccharomyces cerevisiae. J. Bacteriol.* **130,** 472–484.
16. Johnston, G. C., Pringle, J. R., and Hartwell, L. H. (1977). Coordination of growth and cell division in the yeast *Saccharomyces cerevisiae. Exp. Cell Res.* **105,** 79–98.
17. Johnston, G. C., Singer, R. A., and McFarlane, E. S. (1977). Growth and cell division during nitrogen starvation of the yeast *Saccharomyces cerevisiae. J. Bacteriol.* **132,** 723–730.
18. Johnston, G. C., and Singer, R. A. (1978). RNA synthesis and control of cell division in the yeast *Saccharomyces cerevisiae. Cell* **14,** 951–958.
19. Johnston, G. C., Ehrhardt, C. W., Lorincz, A., and Carter, B. L. A. (1979). Regulation of cell size in the yeast *Saccharomyces cerevisiae. J. Bacteriol.* **137,** 1–5.
20. Johnston, G. C., and Singer, R. A. (1980). Ribosomal precursor RNA metabolism and cell division in the yeast *Saccharomyces cerevisiae. Mol. Gen. Genet.* **178,** 357–360.
21. Johnston, G. C., Singer, R. A., Sharrow, S. O., and Slater, M. L. (1980). Cell division in the yeast *Saccharomyces cerevisiae* growing at different rates. *J. Gen. Microbiol.* **118,** 479–484.
22. Johnston, G. C., and Singer, R. A. (1981). A model for cell cycle regulation in the yeast *Saccharomyces cerevisiae. In* "Current Developments in Yeast Research" (G. G. Stewart and I. Russell, eds.), pp. 555–560. Pergamon, Toronto.
23. Johnston, L., and Nasymth, K. A. (1978). *Saccharomyces cerevisiae* cell cycle mutant *cdc*9 is defective in DNA ligase. *Nature (London)* **274,** 891–893.
24. Knapp, G., Beckmann, J. S., Johnson, P. F., Fuhrman, S. A., and Abelson, J. (1978). Transcription and processing of intervening sequences in yeast tRNA genes. *Cell* **14,** 221–236.
25. Little, J. G., and Haynes, R. H. (1979). Isolation and characterization of yeast mutants auxotrophic for 2′-deoxythymidine-5′-monophosphate. *Mol. Gen. Genet.* **168,** 141–151.
26. Mitchison, J. M. (1971). "The Biology of the Cell Cycle." Cambridge Univ. Press, London and New York.
27. Mortimer, R. K., and Hawthorne, D. C. (1969). Yeast genetics. *In* "The Yeasts" (A. H. Rose and J. S. Harrison, eds.), Vol. 1, pp. 386–543. Academic Press, New York.
28. Ng, R., and Abelson, J. (1980). Isolation and sequence of the gene for actin in *S. cerevisiae. Proc. Natl. Acad. Sci. U.S.A.* **77,** 3912–3916.
29. Oliver, S. G., and McLaughlin, C. S. (1977). The regulation of RNA synthesis in yeast. 1. Starvation experiments. *Mol. Gen. Genet.* **154,** 145–153.
30. Petes, T. D. (1979). Yeast ribosomal DNA genes are located on chromosome XII. *Proc. Natl. Acad. Sci. U.S.A.* **76,** 410–414.
31. Planta, R. J., Retel, J., Klootwijk, J., Meyerink, J. H., DeYonge, P., VanKeulen, H., and Brand, R. C. (1977). Synthesis and processing of ribosomal rubonucleic acid in eukaryotes. *Biochem. Soc. Trans.* **5,** 462–466.
32. Prescott, D. M. (1976). "Reproduction of Eukaryotic Cells," p. 40. Academic Press, New York.
33. Sanzey, B. (1979). Modulation of gene expression by drugs affecting deoxyribonucleic acid gyrase. *J. Bacteriol.* **138,** 40–47.
34. Shulman, R. W., and Warner, J. R. (1978). Ribosomal RNA transcription in a mutant of *Saccharomyces cerevisiae* defective in ribosomal protein synthesis. *Mol. Gen. Genet.* **161,** 221–223.

35. Sillén, L. G., and Mantell, A. E. (1964). "Stability Constants of Metal-Ion Complexes," pp. 665, 686. Chemical Society, London.
36. Singer, R. A., and Johnston, G. C. (1979). Nalidixic acid causes a transient G_1 arrest in the yeast *Saccharomyces cerevisiae*. *Mol. Gen. Genet.* **176,** 37–39.
37. Singer, R. A., Johnston, G. C., and Bedard, D. P. (1978). Methionine analogs and cell division regulation in the yeast *Saccharomyces cerevisiae*. *Proc. Natl. Acad. Sci. U.S.A.* **75,** 6083–6087.
38. Singer, R. A., and Johnston, G. C. (1981). Nature of the G_1 phase of the yeast *Saccharomyces cerevisiae*. *Proc. Natl. Acad. Sci. U.S.A.* **78,** 3030–3033.
39. Slater, M. L. (1973). Effect of reversible inhibition of deoxyribonucleic acid synthesis on the yeast cell cycle. *J. Bacteriol.* **113,** 263–270.
40. Slater, M. L., Sharrow, S. O., and Gart, J. J. (1977). Cell cycle of *Saccharomyces cerevisiae* in populations growing at different rates. *Proc. Natl. Acad. Sci. U.S.A.* **74,** 3850–3854.
41. Sloat, B. F., and Pringle, J. R. (1978). A mutant of yeast defective in cellular morphogenesis. *Science* **200,** 1171–1173.
42. Throm, E., and Duntze, W. (1970). Mating-type-dependent inhibition of deoxyribonucleic acid synthesis in *Saccharomyces*. *J. Bacteriol.* **104,** 1388–1390.
43. Unger, M. W., and Hartwell, L. H. (1976). Control of cell division in *Saccharomyces cerevisiae* by methionyl-tRNA. *Proc. Natl. Acad. Sci. U.S.A.* **73,** 1664–1668.
44. Warner, J. R., and Gorenstein, C. (1978). Yeast has a true stringent response. *Nature (London)* **275,** 338–339.
45. Warner, J. R., and Udem, S. A. (1972). Temperature-sensitive mutations affecting ribosome synthesis in *Saccharomyces cerevisiae*. *J. Mol. Biol.* **65,** 243–257.
46. Yang, H. L., Heller, K., Gallert, M., and Zubay, G. (1979). Differential sensitivity of gene expression *in vitro* to inhibitors of DNA gyrase. *Proc. Natl. Acad. Sci. U.S.A.* **76,** 3304–3308.

8

Posttranscriptional Regulation of Expression of the Gene for an Ammonium-Inducible Glutamate Dehydrogenase during the Cell Cycle of the Eukaryote *Chlorella*

ROBERT R. SCHMIDT, KATHERINE J. TURNER,
NEWELL F. BASCOMB, CHRISTOPHER F. THURSTON,
JAMES J. LYNCH, WILLIAM T. MOLIN,
AND ANTHONY T. YEUNG

I. Introduction

In this laboratory, synchronous cultures of the eukaryotic microorganism *Chlorella sorokiniana* are being used as an experimental tool to elucidate the types of regulatory mechanisms that control expression of

GENETIC EXPRESSION IN THE CELL CYCLE

ISBN 0-12-543720-X

genes that code for enzymes involved in metabolism of both inorganic and organic forms of nitrogen (*16,17,28,29,34,38,39*). Recent studies have focused on the molecular mechanisms that regulate the levels of two isozymes of glutamate dehydrogenase (*3,16,17,22,36–39*). This organism contains a constitutive NAD-GDH,* which is localized in the mitochondrion (*20,21*) and synthesized in cells cultured in medium containing either nitrate or ammonia as the sole source of nitrogen (*16,17,34*). *Chlorella* cells also contain a GDH isozyme, which is specific for NADP and is not localized in the mitochondrion. This NADP-GDH only accumulates in cells cultured in ammonium-containing medium (*16,17,34*). In this laboratory, both isozymes have been purified, partially characterized, and shown (*10,20,21,40*) to be physically, chemically, and antigenically distinct from each other (Table I).

Talley *et al.* (*34*) showed that the NADP-GDH was inducible throughout the *Chlorella* cell cycle. In these cell cycle experiments, cells were periodically harvested from a parent synchronous culture growing in the absence of inducer (i.e., ammonia) and then challenged to synthesize the NADP-GDH. At each stage of the cell cycle analyzed, the enzyme was observed to accumulate in a linear manner, following an approximately 30-min induction lag. The rate of enzyme accumulation between 30 and 60 min was taken as a measure of the initial rate of induction, i.e.,

TABLE I

Properties of Glutamate Dehydrogenase Isozymes of *Chlorella*

Properties	NADP-GDH	NAD-GDH
MW Holoenzyme	354,000	180,000
MW Subunit	59,000	45,000
Subunits/holoenzyme	6	4
N-terminal amino acid	Blocked	Lysine
C-terminal amino acid	Alanine	Unknown
Amino acid composition	Different from NAD-GDH	—
Rabbit antiserum	Monospecific	Not reactive with anti-NADP-GDH IgG
pH optimum	7.2	8.0
NADP:NAD activity ratio	Specific for NADP(H)	1:4
Type of regulation	Inducible	Constitutive
Light requirement	Yes	No
Cell cycle pattern	Linear	Step
in Vivo half-life	<90 min	Unknown
Cellular location	Unknown	Mitochondrial

* Key to abbreviations: Nicotinamide adenine dinucleotide: NAD; nicotinamide adenine dinucleotide phosphate: NADP; glutamate dehydrogenase: GDH.

enzyme potential (*28*), at each stage of cell development. When enzyme potentials were compared to the pattern of DNA accumulation in a synchronous culture in which each cell was dividing into four daughter cells, the enzyme potential abruptly increased fourfold within the S phase of the cell cycle. With improved culture conditions and more highly synchronous cells, Turner *et al.* (*37*) showed that enzyme potential also increases continuously in essentially a linear manner during the G_1 phase prior to its abrupt fourfold increase during the S phase. In fact, a close correlation was observed between the increase in enzyme potential and the increase in total cellular protein. The timing of increase in enzyme potential during the S phase was insensitive to large changes (i.e., doubling) in the cellular growth rate. By specific inhibition of DNA synthesis with 2′-deoxyadenosine, the fourfold increase in enzyme potential normally observed during the S phase was blocked. These results taken collectively suggested that the structural gene of this enzyme is continuously available for transcription (even shortly after its replication) during the cell cycle and that the abrupt increase in NADP-GDH potential during the S phase is dependent on DNA replication within that cell cycle. The increase in enzyme potential prior to the S phase remained unexplained in these experiments.

Israel *et al.* (*16,17*) used a different experimental approach to study the cell cycle regulation of the NADP-GDH. They wanted to determine whether the regulatory strategy of inducible gene expression would change when preinduced cells were cultured in the continuous presence of an inducer for an entire cell cycle. When preinduced synchronous cells were cultured in the continuous presence of inducer, under conditions to give a fourfold increase in cell number, NADP-GDH activity accumulated in a linear manner throughout the cell cycle with a positive rate change observed within the S phase. However, when the growth rate of the organism was doubled, the positive rate change was displaced from the S phase in the first cycle to the fourth or fifth hour of the G_1 phase of the subsequent cell cycle. One possible explanation was that the replication of the NADP-GDH gene was displaced outside of the major S phase at certain cellular growth rates. However, since the addition of 2′-deoxyadenosine to the cells early in the cell cycle had no effect on the positive rate change in catalytic activity, the early-replication gene model was tentatively discarded.

Many eukaryotic cells have detectable or high basal levels of inducible (or adaptable) enzymes (*2,11,27,35*). In contrast, catalytic activity of the *Chlorella* NADP-GDH cannot be detected in uninduced cells. Turner *et al.* (*36*) proposed that the absence of detectable basal activity in uninduced cells of the NADP-GDH might be due to the presence of the inactivation system that Israel *et al.* (*16,17*) observed to become operational upon removal of ammonia from induced *Chlorella* cells. Thus, the absence of a basal level of NADP-GDH activity should not be assumed to reflect the absence of enzyme synthesis. For example, Funkhouser and Ramadoss (*7*) and Funkhouser *et al.* (*8*) have recently shown that an inactive protein precursor of *Chlorella* nitrate reductase was actively synthesized in ammonium-cultured cells in which nitrate reductase activity was at extremely low levels. When cells were transferred from nitrate medium (inducing conditions) to ammonium medium, nitrate reductase was shown to undergo rapid inactivation by covalent modification (*31–33*).

Because only the catalytic activity of the NADP-GDH was measured in all of the aforementioned studies, the possibility existed that catalytic activity might not reflect the actual pattern of accumulation of NADP-GDH antigen. Therefore, as a specific probe for use in measurements of enzyme antigen levels, Yeung *et al.* (*40*) prepared rabbit antiserum against the purified NADP-GDH. The anti-NADP-GDH IgG was purified from the crude antiserum by use of a stable, reusable antigen-affinity column. The affinity column was prepared by coupling the NADP-GDH holoenzyme to CNBr-activated Sepharose and linking the subunits within the enzyme together with a chemical cross-linking agent. With the purified anti-NADP-GDH IgG, two highly specific immunochemical procedures (i.e., indirect immunoprecipitation and immunoadsorption) were used to recover radioactive NADP-GDH antigen from extracts of ^{35}S-labeled cells. By use of the purified antibody and highly specific indirect immunoprecipitation and immunoadsorption procedures, coupled with pulse-chase experiments with ^{35}S-labeled sulfate, it has been possible to gain insight into the molecular mechanism(s) regulating the induction of NADP-GDH activity by ammonia and the loss in enzyme activity after removal of the inducer. In this chapter, we will review our experimental findings, which show that induction of the NADP-GDH is probably regulated at the posttranscriptional level.

II. Ammonia and Light Requirement for Induction of NADP-GDH Antigen

Molin *et al.* (*22*) showed that the *Chlorella* NADP-GDH required light both for its induction by ammonia in uninduced synchronous daughter cells and for its continuous accumulation in fully induced cells. These workers used a rocket immunoelectrophoresis procedure (*3,10,40*) to show that the addition of ammonia to uninduced cells in continuous light

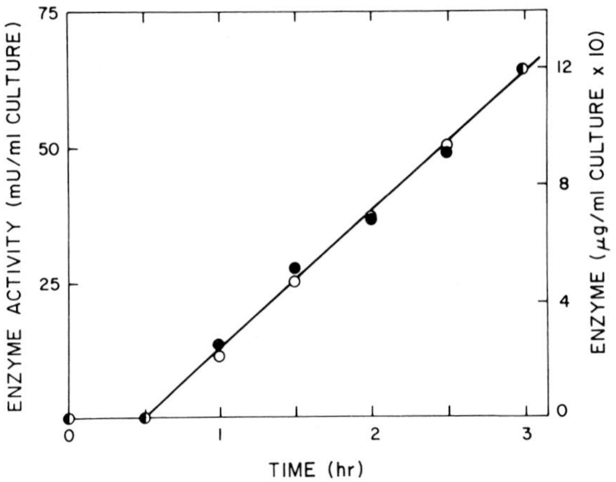

Fig. 1. Comparison of catalytic activity and antigen of *Chlorella* NADP-GDH during induction in synchronous daughter cells in ammonium medium (30 m*M*) in continuous light. Zero time was the beginning of the cell cycle and the time of addition of ammonia. NADP-GDH antigen was measured by a rocket immunoelectrophoresis/activity stain procedure. Pure NADP-GDH was used to establish a standard curve of "rocket" heights versus enzyme concentration. mu, milliunits of enzyme activity. Catalytic activity (●); enzyme antigen (○). [From Molin *et al.* (*22*), with permission.]

results in the linear and coincident accumulation of NADP-GDH activity and antigen following a 30 min induction lag (Fig. 1). In contrast, when uninduced cells were placed in the dark, NADP-GDH activity was not induced even after a 3-hr period in ammonium medium. However, the transfer of these cells to the light resulted in an immediate increase in NADP-GDH activity (Fig. 2).

To determine if light is also required for the continuous accumulation of NADP-GDH activity in induced cells, uninduced synchronous daughter cells were induced in ammonium medium for 3 hr and then transferred to the dark. The NADP-GDH activity and total cellular protein ceased to accumulate very shortly after termination of the light period (Fig. 3). Ammonia continued to be absorbed by the cells in the dark. After 1.3 hr into the dark period, the activity of the enzyme began to decrease rapidly. The loss in NADP-GDH activity was assumed to be related to the concentration of ammonia in the medium rather than to the duration of the dark period per se. In either the light or dark, the activity of the NADP-GDH was observed to decay very rapidly if the concentration of ammonia in the medium decreased below 0.5 m*M*. Although the accumulation of the NADP-GDH isozyme was shown to be light dependent, Molin *et al.* (*22*) observed that the accumulation of the constitutive NAD-GDH was light independent in synchronous cells cultured in either nitrate

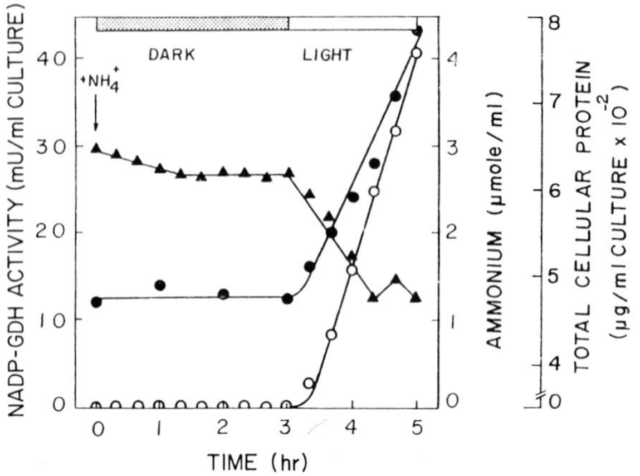

Fig. 2. Light requirement for induction of *Chlorella* NADP-GDH by ammonia (3 m*M*) in uninduced synchronous daughter cells. NADP-GDH antigen also was measured, as described in Fig. 1, and observed to increase coincident with catalytic activity at the onset of the light period at the third hour (data not plotted). Zero time was explained in Fig. 1. mu, milliunits of enzyme activity. Catalytic activity (○); total cellular protein (●); ammonium uptake (▲). [From Molin *et al.* (*22*), with permission.]

or amnonium medium. This observation lends further support to the mounting evidence that these two GDH isozymes are regulated independently.

The reason for the lack of induction of NADP-GDH activity in the dark is unclear. The same amounts of ammonia adsorbed by cells in the light and dark resulted in induction of NADP-GDH activity only in lighted cells (Fig. 2). The failure to observe (Figs. 2 and 3) a net increase in total cellular protein in the dark does not indicate that enzyme synthesis cannot occur under these conditions. In this laboratory, it was previously shown (*2*) that synthesis of isocitrate lyase occurs within 15–20 min after transfer of cells of this species of *Chlorella* to the dark in the absence of exogenous organic substrates at any time during the cell cycle. Moreover, the constitutive NAD-GDH isozyme was synthesized during the dark in either nitrate or ammonium medium.

In higher plants, there is quite an extensive literature (*5,41*) dealing with the light induction of a number of enzymes and other proteins of which some are localized within and others outside of the chloroplast. Some of the induced activities are related to activation by light-dependent redox systems (*5*), whereas others are undoubtedly related to *de novo* enzyme synthesis (*41*), as appears to be the case for *Chlorella* NADP-GDH antigen (Fig. 1). However, the present studies with the *Chlorella* NADP-GDH have not ruled out the possibility that an inactive nonan-

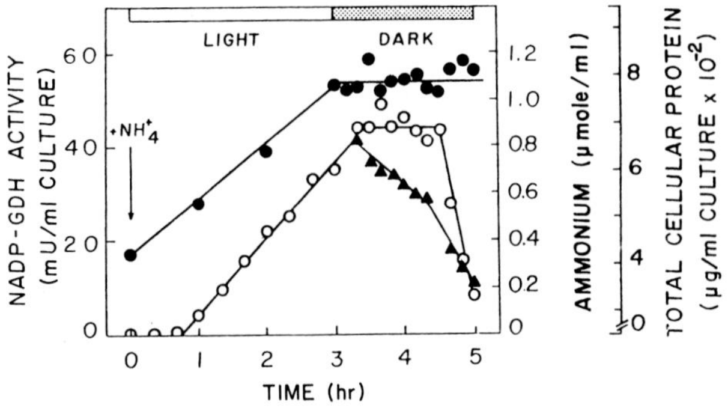

Fig. 3. Light requirement for the continuous accumulation of *Chlorella* NADP-GDH in induced synchronous cells growing in ammonium medium (2 m*M*). The cells were placed into the dark at the third hour of the cell cycle. Zero time was explained in Fig. 1. mU, milliunits of enzyme activity. Catalytic activity (○); total cellular protein (●); ammonium uptake (▲). [From Molin *et al.* (*22*), with permission.]

tigenic precursor of the enzyme is converted to an active antigenic form by a light-dependent redox reaction (*5*) or by some phytochrome-linked reaction (*41*). Alternatively, the synthesis of the NADP-GDH subunits (*36*) or their assembly into holoenzyme might be closely coupled to photophosphorylation (*30*). However, unless cellular compartmentalization is altering the accessibility or rate of flow of ATP within the cells in the dark, the ATP level per se is unlikely to be rate-limiting the synthesis of the NADP-GDH. The total cellular levels of ATP have been shown to decrease only momentarily and then to increase to equal or higher levels in *Chlorella* cells transferred from light to darkness (*4*). Although the constitutive NAD-GDH has been shown (*20*) to be situated in the mictochondrion of *Chlorella,* the intracellular location of the NADP-GDH is currently unknown. In certain higher plants (*18*) and in another unicellular green alga (*9*), which contain more than one GDH isozyme, the NADP-GDH isozyme has been shown to be localized in the chloroplast. Thus, since the activity of the mitochondrial NAD-GDH increases in the dark, whereas the accumulation of the NADP-GDH is light dependent (Figs. 2 and 3) in *Chlorella,* an important step in the elucidation of the mechanism of the light-dependent induction of the *Chlorella* NADP-GDH will be to determine whether it is a chloroplast enzyme.

In an attempt to determine the minimum concentration of ammonia required to induce the NADP-GDH, Molin and Schmidt (*23*) placed uninduced synchronous daughter cells in culture media containing a range of ammonia concentrations (i.e., 0.1 to 30 m*M*). At concentrations of 1.0 m*M* and above, the initial rates of induction of the NADP-GDH appeared to be equal (Fig. 4). The enzyme did not appear to be inducible below 0.5 m*M*. Interpretation of the results of this experiment was complicated by the rapid loss in NADP-GDH activity (i.e., deinduction) that was observed when the uptake of ammonia decreased its concentration to approximately 0.5 m*M* in the culture medium. To avoid the problem of a changing inducer concentration during the induction period, ammonia was added at 5-min intervals. By this experimental approach, it was possible to show that the enzyme was inducible, although not maximally at ammonia concentrations down to 0.15 m*M* (Fig. 5). The enzyme was not inducible at amnonia levels below 0.1 m*M*. A maximal rate of accumulation of total cellular protein was observed at concentrations of ammonia below those which supported a maximal rate of NADP-GDH

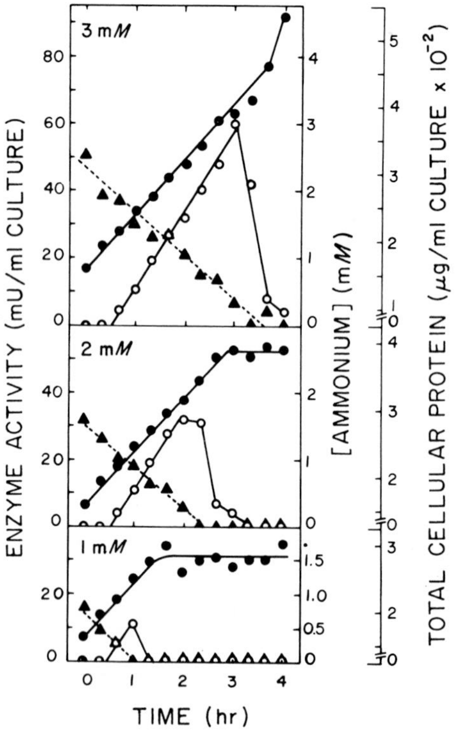

Fig. 4. Induction of *Chlorella* NADP-GDH activity by addition of different concentrations of ammonia to the culture medium. The ammonia was added at zero time to previously uninduced cells. mu, milliunits of enzyme activity. Catalytic activity (○—○); total cellular protein (●—●); ammonium uptake (▲----▲). [From Molin and Schmidt (*23*), with permission.]

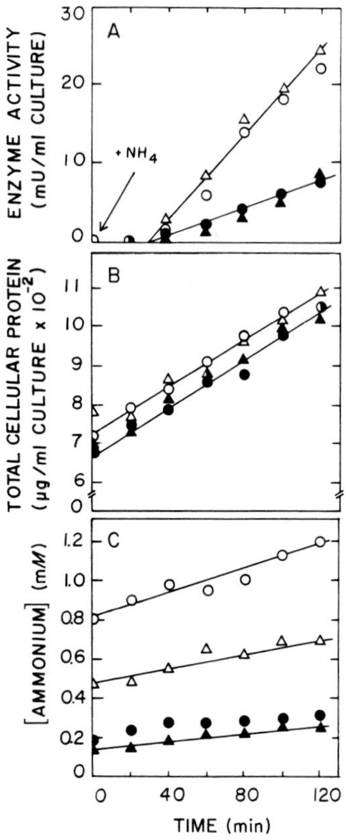

Fig. 5. Induction of *Chlorella* NADP-GDH activity in cultures in which ammonia was added continuously in an attempt to maintain constant low levels of this inducer in the culture medium. (A) Catalytic activity (mU, milliunits of enzyme activity); (B) total cellular protein; and (C) concentration of ammonia in the culture medium. The initial concentrations of ammonia were (▲) 0.15, (●) 0.2, (△) 0.5, and (○) 0.8 m*M*, respectively. [From Molin and Schmidt (*23*), with permission.]

induction. Therefore, it is unlikely that this enzyme is a rate-limiting step in assimilation of anmonia for protein synthesis in this organism.

III. Turnover of NADP-GDH during Induction and its Rapid Inactivation by Covalent Modification during Deinduction Period

One of the striking features of the regulation of the ammonium-inducible NADP-GDH is its very rapid loss ($t_{1/2}$ = 5–10 min) of catalytic activity in cells transferred from ammonium medium to either nitrate or nitrogen-free medium (*16,17*). As discussed in the previous section, the same deinduction phenomenon was observed when the ammonia concentration decreased below 0.5 m*M* in growing cells (Fig. 4). Similar losses of catalytic activity have been observed for GDH isozymes in other eukaryotic microorganisms during transition between various nutritional conditions (*12–14,19,26*). Some of these losses of activity are associated with apparently reversible covalent modifications in enzyme structure which result in rapid enzyme inactivation (*13,14*), and others are associated with enzyme degradation (*12,19*). Thus, it was important to know whether the rapid loss in catalytic activity of the *Chlorella* NADP-GDH was accompanied by (a) a similar rate of enzyme degradation (i.e., loss of enzyme antigen) or (b) some type of covalent modification of the enzyme that results in its very rapid inactivation before degradation.

The first experimental approach involved the use of a rocket immunoelectrophoresis/activity stain procedure (*3,10,39,40*) to show whether the loss in NADP-GDH activity is accompanied by a loss in enzyme antigen(icity) after removal of ammonia from fully induced *Chlorella* cells. However, by this procedure, it is not possible to distinguish between the loss in NADP-GDH antigenicity due to enzyme modification or to loss of enzyme antigen due to degradation.

A control experiment was performed to show that this procedure can detect inactive form(s) of the NADP-GDH which retain their antigenicity. Fully induced cells in ammonium medium were transferred from 38.5° to 55°C, and enzyme inactivation was measured over a 15-min time course. Under these conditions, NADP-GDH activity decayed very rapidly ($t_{1/2}$ = 6 min), but enzyme antigenicity (i.e., rocket height) remained almost constant. Thus, the anti-NADP-GDH IgG had essentially the same affinity for the heat-inactivated form as for the active form of the enzyme.

Since the rocket immunoelectrophoresis/activity stain procedure can detect at least one type of inactive form of NADP-GDH, it was used to analyze extracts that were prepared from cells taken during a deinduction period. When ammonia was removed from a culture, enzyme activity decreased with a half-life of 8 min, and NADP-GDH antigen decreased with a half-life of 12.5 min (Fig. 6). These experimental results indicated that either NADP-GDH was structurally modified so that its antigenicity was significantly changed, or the enzyme was rapidly degraded during the deinduction period. The faster loss in enzyme activity than in antigenicity (8 versus 12.5 min) suggested that the loss of NADP-GDH activity might involve a modification step prior to enzyme degradation.

The shift from ammonium medium to nitrogen-free medium resulted

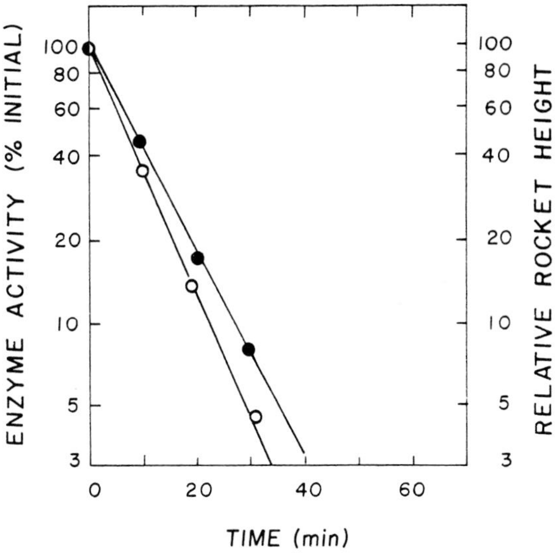

Fig. 6. Comparison of rates of decay in catalytic activity and antigen(icity) of NADP-GDH during a deinduction period after removal of ammonia from *Chlorella* cells. Cells were cultured for 3 hr in ammonium medium and then transferred to nitrogen-free medium. The half-life values for enzyme activity and antigen(icity) were 8 min and 12.5 min, respectively. Catalytic activity (○); rocket heights from immunoelectrophoresis (●). [From Bascomb *et al.* (*3*), with permission.]

in the cessation of accumulation of total cellular protein. Since the level of total protein remained constant during the deinduction period, the loss of NADP-GDH activity and antigen(icity) could not be attributed to a net loss of total cellular proteins.

To determine whether the rapid decrease in NADP-GDH activity during the deinduction period might be due to an increase in rate of NADP-GDH degradation, the half-life of the NADP-GDH antigen was measured before and after removal of ammonia from cultures. The cells were labeled with [^{35}S]sulfate in ammonium medium for 90 min. During the following chase period with nonradioactive sulfate, the cells were in ammonium medium for 90 min, and then transferred to nitrogen-free medium for another 60 min. When ^{35}S-labeled NADP-GDH antigen was quantitatively recovered from radioactive cell extracts by either indirect immunoadsorption or immunoprecipitation, the half-life of the NADP-GDH antigen was shown to be 88 min before and after removal of amnonia from the cultures (Fig. 7A,B). This experiment revealed several important findings. First, the enzyme appears to undergo rapid degradation even in rapidly growing, fully induced cells. Second, since the half-life of NADP-GDH antigen during the deinduction period was com-

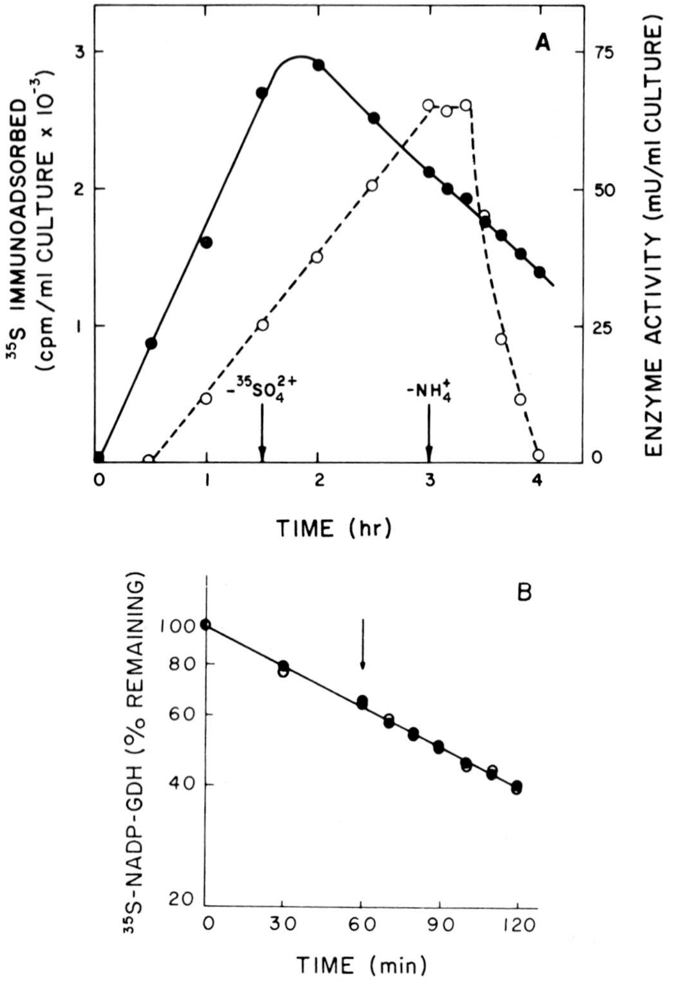

Fig. 7. (A) Pulse-chase experiment in which uninduced *Chlorella* cells were transferred to ammonium medium and labeled for 1.5 hr with [^{35}S]sulfate, chased with nonradioactive sulfate for 1.5 hr in ammonium medium, and then transferred to nitrogen-free medium to begin the deinduction period. Catalytic activity (mU, milliunits of enzyme activity), (○--○); total ^{35}S-labeled NADP-GDH antigen (●—●) recovered from cell extracts by indirect immunoadsorption. (B) Comparison on a semilogarithmic plot of the *in vivo* rates of decay of total ^{35}S-labeled NADP-GDH antigen during the chase period before and after transfer (arrow) of the cells to nitrogen-free medium. The ^{35}S-labeled NADP-GDH antigen was recovered from aliquots of extracts of ^{35}S-labeled cells by use of either indirect immunoadsorption (●) or indirect immunoprecipitation (○). The half-life of total NADP-GDH antigen was 88 min. Zero time in (B) corresponds to the second hour in (A). The half-life of enzyme antigen was also measured in this experiment by the rocket immunoelectrophoresis method and found to be 13 min (data not plotted). [From Bascomb *et al.* (*3*), with permission.]

parable to that observed in fully induced cells, the loss of NADP-GDH activity during the induction period was not due to enhanced degradation. Third, NADP-GDH catalytic activity decreased more rapidly than enzyme antigenicity during the deinduction period (8 versus 88 min), strongly suggesting that the enzyne was inactivated by a chemical or physical modification before its degradation.

As an initial step in elucidation of the type of enzyme modification which might be occurring during the deinduction period, NADP-GDH antigen was immunoadsorbed from extracts prepared from cells in which 65% of the enzyme activity had decayed *in vivo*. The immunoadsorbed protein(s) were dissolved by boiling in a mercaptoethanol/SDS buffer and then subjected to SDS–polyacrylamide gel electrophoresis. The deinduced cells contained two major antigenic proteins, one corresponding in electrophoretic mobility to the NADP-GDH subunit and the other with a much slower mobility (Fig. 8). When the electrophoretic mobilities of these two proteins were compared to protein standards, the two proteins were shown to have molecular weights of 59,000 and 118,000, respectively. Since the molecular weight of the NADP-GDH subunit has been previously shown to be 59,000 (*10*), the higher molecular weight protein will be referred to as the putative dimer of the NADP-subunit.

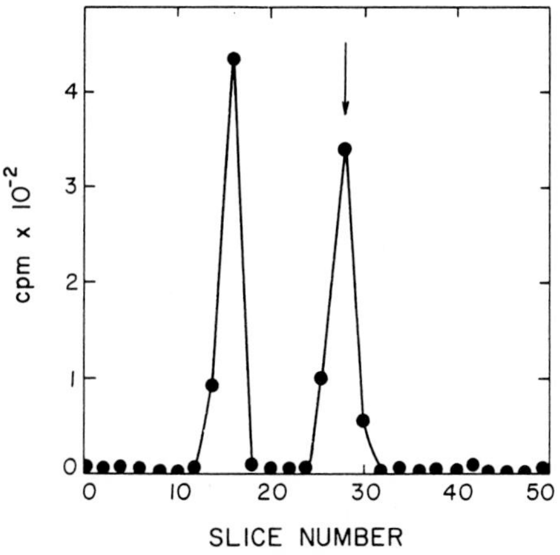

Fig. 8. SDS polyacrylamide gel electrophoresis of the proteins obtained by indirect immunoadsorption with purified anti-NADP-GDH IgG and *Staphylococcus* Protein A Sepharose-4B from extracts of ^{35}S-labeled cells harvested during the deinduction period after 65% of the NADP-GDH catalytic activity had decayed. Comparison of the relative mobilities of these two proteins with those of protein standards (i.e., myosin, β-galactosidase, phosphorylase B, bovine serum albumin, authentic ^{35}S-labeled NADP-GDH, ovalbumin, and chymotrypsinogen A) showed them to have mw of 59,000 and 118,000. The arrow corresponds to the position of the authentic NADP-GDH (MW = 59,000) in a separate gel. [From Bascomb *et al.* (*3*), with permission.]

The dimer was stable during boiling in the presence of high concentrations (10 m*M*) of mercaptoethanol and dithiothreitol. These data strongly suggest that a covalent linkage (other than a disulfide bond) exists between the subunits.

The distribution of radioactivity between the NADP-GDH subunit and the putative dimer was next measured at intervals during the deinduction period (Fig. 9). The reciprocal relationship observed between the loss of radioactivity from subunits and its appearance in the putative dimer strongly supports the inference that the subunit is the precursor of the putative dimer. The half-life of radioactivity in the NADP-GDH subunit was 12.5 min, which is almost identical to the half-life of NADP-GDH antigen measured by rocket immunoelectrophoresis. A summation of the amount of radioactivity associated with subunits and the putative dimer at each sample time showed that their combined radioactivities decreased with a half-life of 88 min (Fig. 9). This half-life is identical to the one measured for total NADP-GDH antigen by either the immunoprecipitation or immunoadsorption procedures during the deinduction period (Fig. 7B).

Because the immunoprecipitates (and immunoadsorbed proteins) were obtained from whole cell extracts and were directly subjected to SDS electrophoresis, it is currently uncertain whether the holoenzyme must

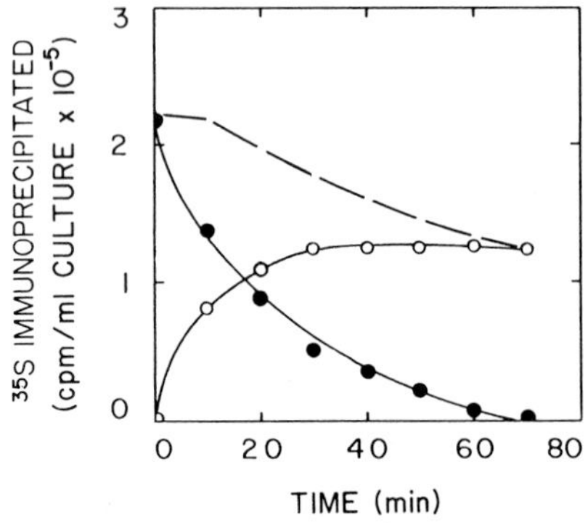

Fig. 9. Disappearance of ^{35}S-labeled NADP-GDH subunits and rapid accumulation of ^{35}S-labeled putative NADP-GDH dimer after the removal of ammonia from the culture medium of ^{35}S-labeled *Chlorella* cells. The subunit and putative dimer were recovered from cell extracts by indirect immunoprecipitation. The immunoprecipitates were subjected to SDS-polyacrylamide gel electrophoresis, and the radioactivity associated with each peak was measured. Radioactivity associated with NADP-GDH subunits (●) and putative dimer (○). The radioactivity of total NADP-GDH antigen (— —) was calculated by addition of radioactivity associated with subunits and dimer. On semilogarithmic plots (data not shown), the half-life of the subunits and the total NADP-GDH antigen were 12.5 min and 88 min, respectively. [From Bascomb *et al.* (*3*), with permission.]

dissociate prior to formation of the putative dimer from the free subunits or whether the subunits can be covalently linked while still situated within the holoenzyme as a first step in NADP-GDH inactivation. Although the half-life of the NADP-GDH subunit was 12.5 min (Fig. 9), which was almost identical to the loss of NADP-GDH antigen measured by rocket immunoelectrophoresis (Fig. 6), NADP-GDH catalytic activity decreased with a half-life of 8 min (Figs. 6 and 7A). This difference in half-lives could be the result of enzyme inactivation by cross-linking of subunits within the NADP-GDH holoenzyme. The holoenzyme is composed of six identical subunits (*40*). The dimerization of a single pair of subunits within the holoenzyme might cause a conformational change sufficient to inactivate the enzyme.

Although the putative dimer rapidly increased and became a major fraction of the total NADP-GDH during the deinduction period, it was observed to be a minor but relatively constant component of the total NADP-GDH antigen during the induction period (Fig. 10). Moreover, in an earlier study (*40*), in which the specific direct and indirect immunoprecipitation and immunoadsorption procedures were tested on extracts of radioactive fully induced *Chlorella* cells, a protein with the same electrophoretic mobility as the putative dimer was observed to be a small but constant fraction of the total NADP-GDH antigen.

IV. Presence of NADP-GDH mRNA on Polysomes of Both Induced and Uninduced Cells

As a first step in the elucidation of the molecular mechanism regulating the induction of the NADP-GDH, the amounts of NADP-GDH mRNA on polysomes of uninduced and induced cells were measured. An isolation procedure was developed which gave high yields of intact *Chlorella* polysomes with a profile similar to those reported for other organisms (*36*).

Labeled antibodies that bind to nascent polypeptides have been used to identify and quantify polysomes involved in the synthesis of specific proteins from different types of cells (*1,24*). Purified rabbit anti-NADP-GDH IgG was radioiodinated by an enzyme-catalyzed reaction and then incubated with polysomes isolated from ammonium-cultured (i.e., induced) and nitrate-cultured (i.e., uninduced) *Chlorella* cells. After fractionation of the ^{125}I-labeled polysomes on sucrose gradients, a peak of radioactivity was observed (*36*) to be present in the polysome region of both types of cells (Fig. 11A,B). The peaks of radioactivity were associated with a class of large polysomes (i.e., greater than 15 ribosomes

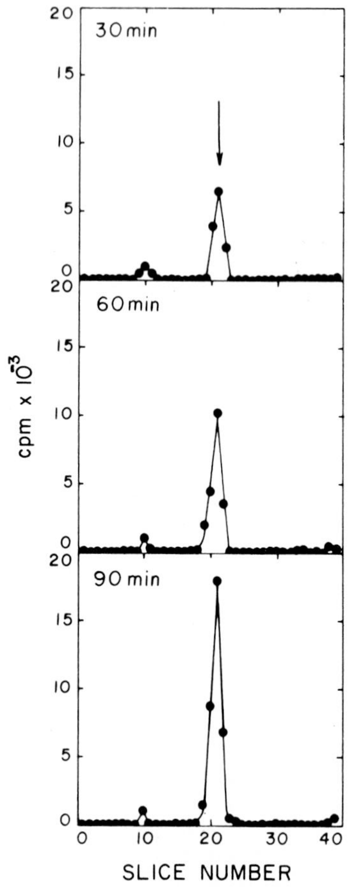

Fig. 10. SDS electrophoresis of the proteins obtained by indirect immunoprecipitation from extracts of ^{35}S-labeled cells during an induction period. The [^{35}S]sulfate and ammonia were added together to the medium of uninduced cells at time zero. The major and minor peaks of radioactivity correspond to positions of proteins with MWs of 59,000 and 118,000, respectively. [From Bascomb *et al.* (*3*), with permission.]

per polysome) which were present in approximately the same region of the two gradients. The position of the radioactive polysomes in the gradient is consistent with polysomes which would be expected to be engaged in synthesis of a protein with a molecular weight similar to that of the NADP-GDH subunit (MW = 59,000). The presence of a specific peak of radioactivity in uninduced cells suggested that even though NADP-GDH catalytic activity was not detectable in these cells, the cells contained functional NADP-GDH mRNA which was being translated *in vivo* (*36*).

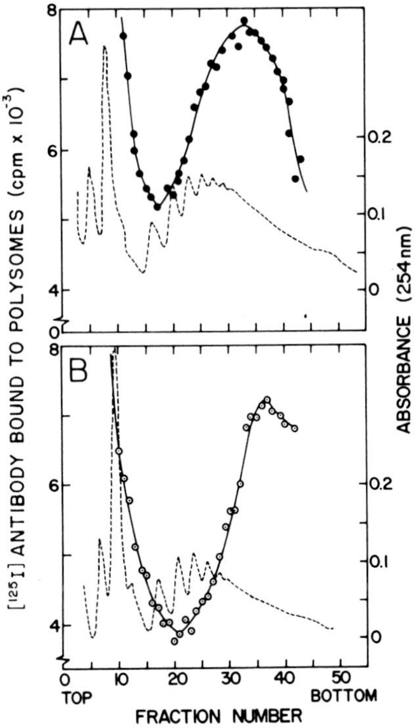

Fig. 11. Binding of ^{125}I-labeled rabbit anti-NADP-GDH IgG to total cellular polysomes from induced and uninduced *Chlorella* cells. Polysome preparations were incubated with 20 μg of ^{125}I-labeled rabbit anti-NADP-GDH IgG in a 3.0 ml volume for 2 hr at 1.5°C. (A) and (B) contained 10 A_{260} units of polysomes isolated from induced and uninduced cells, respectively. The radioactivity was measured in 0.25 ml fractions. (A) Radioactivity (●—●); absorbance (----). (B) Radioactivity (○—○); absorbance (----). [From Turner *et al.* (*36*), with permission.]

Total RNA was extracted from polysomes from both uninduced and induced *Chlorella* cells and was fractionated into non-poly(A)-containing RNA and poly(A)-containing RNA species by use of oligo(dT)-cellulose column chromatography. Total polysomal RNA along with the two species of fractionated RNA were added to a mRNA-dependent cell-free protein-synthesizing system as described by Pelham and Jackson (*25*). The translation of all three *Chlorella* RNA species was observed (*36*) to be a linear function of radioactive amino acid incorporation into total protein synthesized *in vitro* over the concentration range of the RNA species tested (Fig. 12).

The proteins synthesized *in vitro* were analyzed by SDS–gel electrophoresis. The results indicated that the radioactive proteins synthesized

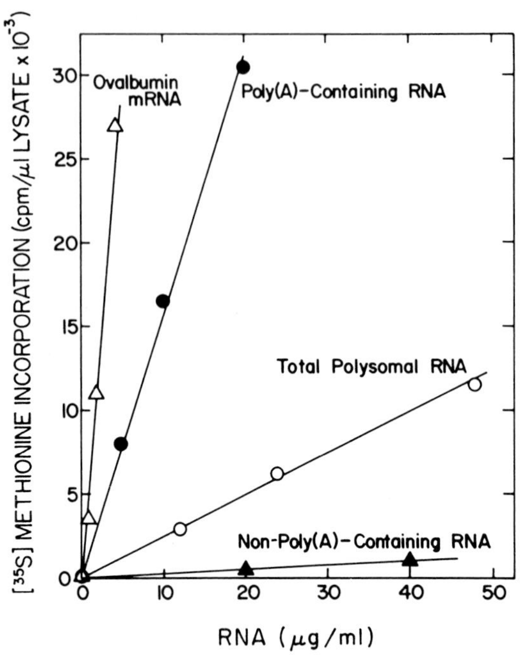

Fig. 12. *In vitro* translation, in an mRNA-dependent rabbit reticulocyte lysate system, of different RNA fractions from *Chlorella* polysomes isolated from ammonium-induced cells. The incorporation of ^{35}S-labeled methionine into total protein was determined in a 2 μl aliquot removed from the lysate reaction mixture after 60 min of protein synthesis. Essentially identical results were obtained with equivalent amounts of these polysomal RNA fractions isolated from uninduced cells. Total protein synthesized from *Chlorella* total polysomal poly(A)-containing RNA (●); total polysomal RNA (○); and total non-poly(A)-containing RNA (▲). Ovalbumin mRNA (△) was purified from hen oviducts by the procedure of Buell *et al.* (*6*) and used as a positive control. [From Turner *et al.* (*36*), with permission.]

in vitro from poly(A)-containing RNA, isolated from total polysomal RNA or from total cellular RNA, were very similar in size and relative number (*36*). The proteins comprised a heterogeneous range of polypeptide sizes, which would be expected if the RNA were undegraded and the translation assay were synthesizing full-length polypeptides.

The amount of radioactive NADP-GDH antigen synthesized *in vitro* was measured after immunoprecipitation from the lysate, and the distribution of radioactive proteins in the solubilized immunoprecipitates was analyzed by SDS–gel electrophoresis. When polysomal poly(A)-containing RNA, isolated from ammonium cells, was incubated with the lysate, a major peak of radioactivity was observed to correspond to the position of the authentic NADP-GDH subunit in SDS gels (Fig. 13A, B).

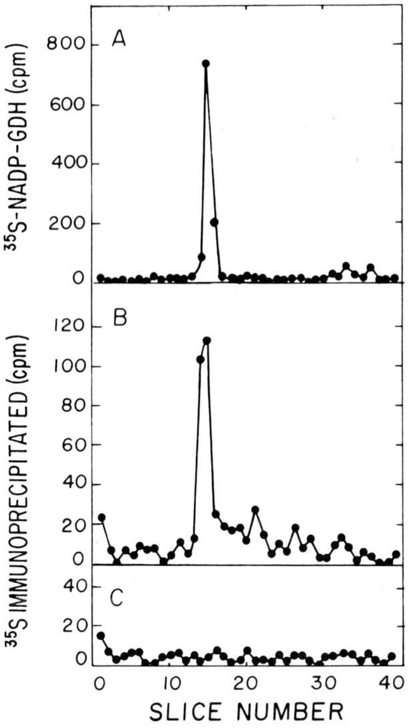

Fig. 13. SDS–polyacrylamide gel electrophoresis of the ^{35}S-labeled NADP-GDH subunit synthesized *in vitro* from poly(A)-containing RNA isolated from polysomes of induced *Chlorella* cells. (A) Standard ^{35}S-labeled NADP-GDH subunit purified from whole cells cultured in ammonium medium with [^{35}S]sulfate. The standard *in vitro* translation assay was performed in the presence (B) and absence (C) of 20 μg/ml of total polysomal poly(A)-containing RNA isolated from induced cells. Direct immunoprecipitation of the *in vitro* synthesized enzyme was performed by addition of 5 μg carrier NADP-GDH and 60 μg of affinity purified rabbit anti-NADP-GDH IgG. The first gel slice contained all of the radioactivity that was present in the whole stacking gel. [From Turner *et al.* (*36*), with permission.]

When no *Chlorella* RNA was added to the lysate, no radioactive immunoprecipitable products were detected in the SDS gel (Fig. 13C). Moreover, based on the amount of RNA added to the translation assay and on the proportion of radioactivity incorporated into the NADP-GDH subunits relative to total amino acid incorporation, less than 5% of the NADP-GDH mRNA was shown to be present in the non-poly(A)-containing RNA fraction extracted from ammonium-cultured cells. Therefore, this result strongly suggests that the mRNA coding for the NADP-GDH contains a poly(A) sequence at its 3′-end.

To estimate the amount of NADP-GDH mRNA present in uninduced cells, equivalent amounts of polysomal poly(A)-containing RNA from uninduced and induced cells were translated *in vitro,* and the radioactive immunoprecipitates were subjected to SDS–gel analysis. Surprisingly, the peak of radioactive NADP-GDH antigen in uninduced cells was observed to be 75% of that of fully induced cells. These data, along with the polysome binding studies with radioactive anti-NADP-GDH, indicated that the NADP-GDH mRNA was at high levels in uninduced cells and that induction of this enzyme might be regulated at the posttranscriptional level (*36*).

V. Synthesis and Rapid Degradation of NADP-GDH Subunits in Uninduced Cells

The presence of the NADP-GDH mRNA and the absence of enzyme activity in uninduced cells suggested that the NADP-GDH subunits might be synthesized and then inactivated and/or degraded in these cells. To test this possibility, a pulse-chase experiment was performed with [^{35}S]sulfate and uninduced cells (*36*). During a 30-min labeling period, two major radioactive proteins could be immunoprecipitated with purified rabbit anti-NADP-GDH IgG from extracts of these cells (Fig. 14A). By SDS electrophoresis, the protein with the faster electrophoretic mobility was shown to have a molecular weight identical to that reported (*10*) for the authentic NADP-GDH subunit (MW = 59,000). Because the other protein was shown to have a molecular weight of 118,000 and to be stable to boiling in high concentrations (10 m*M*) of β-mercaptoethanol or dithiothreitol, it was assumed to be a dimer of covalently linked NADP-GDH subunits (i.e., putative dimer) probably identical to the one observed in the earlier deinduction studies (Fig. 8).

In the present study with uninduced cells, the kinetics of incorporation of radioactivity into the two protein species during the 30-min pulse period clearly indicated that the subunit was synthesized before the putative dimer (Fig. 14A, B). During the initial period of the chase period, an almost reciprocal relationship was observed between the loss of radioactivity from the subunit and its appearance in the dimer (Fig. 14A, B). During the chase period, the radioactivity in the dimer also rapidly decreased. These kinetic data strongly suggest that the NADP-GDH subunits are synthesized and covalently modified to form dimers and then degraded to nonantigenic products in uninduced cells.

The simultaneous addition of ammonia and [^{35}S]sulfate to uninduced cells gave further insight into the mechanism of induction of NADP-

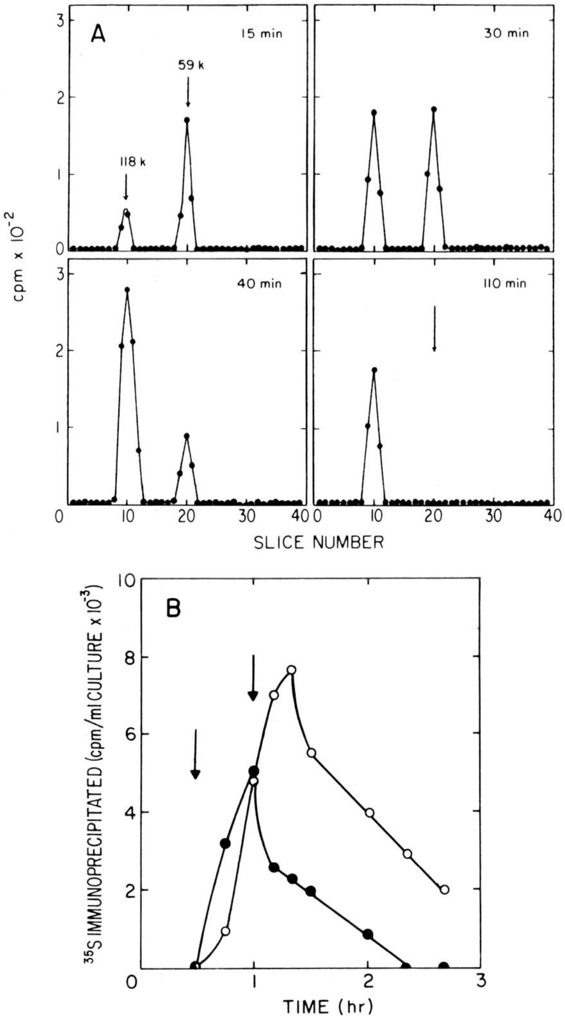

Fig. 14. (A) SDS electrophoresis of proteins obtained by indirect immunoprecipitation from extracts of *Chlorella* during a pulse-chase experiment with [^{35}S]sulfate in uninduced cells growing in nitrate medium. The times indicated in each panel are those at which the cells were harvested after addition of [^{35}S]sulfate to the culture medium. The arrows indicate the positions of authentic NADP-GDH subunit and putative dimer with molecular weights of 59,000 and 118,000, respectively. (B) Precursor–product relationship between the NADP-GDH subunit and the putative dimer was determined by measurement of total radioactivity associated with each radioactive peak in (A). The first and second arrows indicate the times for the onset of the pulse and chase periods, respectively. NADP-GDH subunit (●); putative dimer of NADP-GDH subunits (○). [From Turner *et al.* (*36*), with permission.]

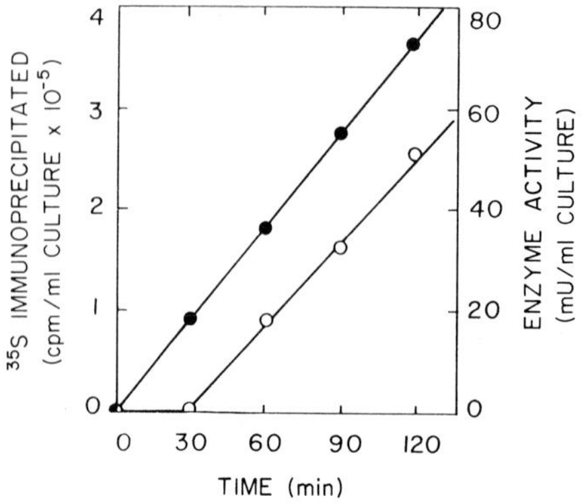

Fig. 15. Comparison of kinetics for the increase in total NADP-GDH antigen and catalytic activity upon the simultaneous addition of ammonia and [^{35}S]sulfate to uninduced *Chlorella* cells. Total NADP-GDH antigen was recovered by indirect immunoadsorption. NADP-GDH antigen (●); NADP-GDH catalytic activity (mu, milliunits of enzyme activity), (○). [From Turner *et al.* (*36*), with permission.]

GDH antigen. The incorporation of radioactivity into total NADP-GDH antigen appeared to occur without an induction lag and proceeded in a linear manner for the duration of the induction experiment (Fig. 15). These data are consistent with the inference that the NADP-GDH mRNA is present at higher than basal levels in uninduced cells and that the induction of NADP-GDH antigen is regulated at least in part at the posttranscriptional level in *Chlorella*.

VI. Posttranscriptional Model for Induction of NADP-GDH Activity

Based on the experimental findings described in the previous sections of this chapter, we propose that expression of the gene for the ammonium-inducible NADP-GDH is regulated primarily at the posttranscriptional level. A posttranscriptional model is presented which is considered to be tentative and will be used to direct further experimentation on the regulation of the level of NADP-GDH antigen and catalytic activity (Fig. 16). In both uninduced and induced cells, the NADP-GDH mRNA is proposed to be synthesized continuously and translated on polysomes to form subunits with a molecular weight (i.e., 59,000) identical to the subunits found in the active holoenzyme. In the absence of inducer, the subunits are proposed to be covalently linked together to form dimers

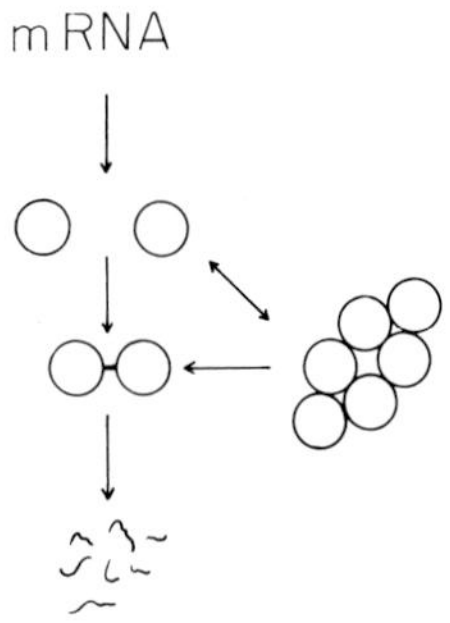

Fig. 16. Posttranscriptional model for induction of the activity of the *Chlorella* NADP-GDH. In both uninduced and induced cells, the NADP-GDH mRNA is proposed to be synthesized continuously and translated on polysomes to form subunits with a molecular weight identical to that of subunits found in the catalytically active holoenzyme. In the absence of inducer, the subunits are proposed to be covalently linked together to form dimers, which are degraded by an endogenous protease to nonantigenic products. The inducer is proposed to slow the rate of dimer formation, which allows the catalytically active holoenzyme to accumulate. It is uncertain whether the dimerization reaction occurs between free subunits or between subunits within the assembled holoenzyme. The molecular weights of the subunit as indicated by large open circle, dimer by two open circles, linked, and holoenzyme by six open circles.

which are degraded by an endogenous protease to nonantigenic products. The inducer is proposed to slow the rate of dimer formation, which allows the catalytically active holoenzyme to accumulate.

Further experimentation is required to determine where the dimerization reaction occurs. Does this reaction occur between two free subunits, between subunits already present within the assembled holoenzyme, or do both reactions proceed simultaneously? This question actually relates to the mechanism by which ammonia induces the accumulation of active enzyme.

If the dimerization reaction occurs between free subunits, the inducer could presumably act in several different ways. The rate of the dimerization reaction per se could be decreased, the rate of association of the subunits to form holoenzyme could be increased, and/or the rate of dissociation of the holoenzyme to free subunits could be decreased. The dimerization and assembly reactions could compete for a pool of free subunits, and the addition of inducer could shift the equilibrium from dimer formation to one in favor of holoenzyme assembly. To account for the rapid inactivation of the holoenzyme, observed during the deinduction period, the holoenzyme would have to undergo rapid dissociation to free subunits in order for the dimerization reaction to occur.

For the other situation, in which the dimerization reaction occurs between subunits within the assembled holoenzyme, the inducer (or one of its metabolites) could function in one of two different ways. First, the inducer could directly inhibit the activity of the dimerizing enzyme(s). Second, the inducer could bind to the NADP-GDH holoenzyme and induce a conformational change to make the holoenzyme have a lower affinity for the dimerizing enzyme(s). The removal of ammonia (i.e., deinduction period) could allow a higher percentage of the holoenzyme molecules to assume a conformation more susceptible to covalent modification. In either case, the close proximity of the subunits within the holoenzyme would seem to favor the rapid inactivation of the enzyme during the deinduction period. The holoenzyme would not have to dissociate before the covalent modification could occur. In fact, the covalent coupling of the subunits into dimers within the holoenzyme could inactivate it and facilitate dissociation of the holoenzyme.

The data are consistent with the inference that the dimer, and not free subunits nor unmodified holoenzyme, is the substrate for a degradation system [i.e., protease(s)]. However, does the degradation system only react on free dimers, or can a dimer situated within the holoenzyme be susceptible to proteolytic cleavage? In other words, if the dimerization reaction occurs between subunits within the holoenzyme, does the dimer have to dissociate from the enzyme before the first proteolytic step can occur?

Research in another laboratory (*15*) has revealed an ATP-dependent reaction in which a low molecular weight polypeptide is conjugated with rabbit reticulocyte proteins as a possible prerequisite step in their degradation. The attachment of this polypeptide is assumed to allow the modified proteins to be recognized by an endogenous protease. The dimerization of identical NADP-GDH subunits might be the covalent modification required to allow recognition by an endogenous protease in *Chlorella*.

The half-life of the total NADP-GDH antigen was shown to be the same in cells during the induction and deinduction periods. Moreover, the dimer was shown to be present at low levels in fully induced cells and to increase rapidly to high levels immediately after removal of inducer from fully induced cells. These data are consistent with a model in which the rate-limiting step in the loss of total NADP-GDH antigen is the reaction in which the dimer is converted to nonantigenic products. In other words, in the presence or absence of inducer, the dimerization reaction appears to be occurring fast enough to keep the proteolytic system saturated with the dimer. Thus, the rate of synthesis of the dimer, not the rate of degradation, appears to be regulated.

The above model does not take into account all of the factors known to affect the induction process. For example, light is required both for the initial induction of the enzyme in previously uninduced cells and for the continuous accumulation of the enzyme in fully induced cells. As discussed earlier, the addition of ammonia to uninduced cells in the dark abolishes the 30-min induction lag when these cells are subsequently returned to the light. What is it that accumulates in the dark and abolishes the induction lag in the light? Since the NADP-GDH antigen has been shown to accumulate from the onset of the time of addition of inducer in the light, what is the nature of the 30-min induction lag in catalytic activity? Does this induction lag indicate the presence of a light-dependent covalent modification step required for the assembly of the holoenzyme from free subunits or for the activation of the holoenzyme? Thus, as stated at the beginning of this section, the construction of a model can help formulate questions that can give a deeper insight into a problem and can help to direct future experimental work.

VII. Accumulation of NADP-GDH mRNA in Uninduced Synchronous Cells: A Possible Explanation for Observed Continuous Increase in Enzyme Potential during the Cell Cycle

In studies on the regulation of inducible enzymes during the bacterial cell cycle, it was shown that the activity of any given inducible enzyme could be induced to accumulate at any time during the cell cycle (*6*). Moreover, enzyme potential, i.e., initial rate of induction, measured at frequent intervals during the cell cycle, was shown to remain constant until the structural gene of the enzyme replicated. At the time of gene replication, enzyme potential doubled (*6*). Thus, a step pattern for enzyme potential was observed during the bacterial cell cycle, suggesting that enzyme potential is proportional to the gene dosage during the bacterial cell cycle.

In contrast to the results obtained with bacteria, enzyme potential for the NADP-GDH was observed to increase continuously in a linear manner during the *Chlorella* cell cycle (*37*). A sharp increase in NADP-GDH enzyme potential was observed during the S phase. This latter increase could be blocked by the DNA synthesis inhibitor 2′-deoxyadenosine; however, the continuous increase in enzyme potential observed before the S phase did not appear to be dependent upon DNA replication. The observation that the NADP-GDH mRNA is present and being translated on polysomes in uninduced cells prompted us to propose a possible explanation for the observed continuous linear increase in NADP-GDH potential before the S phase. If the induction of NADP-GDH activity occurs by the posttranscriptional mechanism proposed earlier (Fig. 16), the initial rate of NADP-GDH induction, i.e., enzyme potential, should be proportional to the amount of NADP-GDH mRNA on the polysomes at any given time during the cell cycle. Thus, if the NADP-GDH structural gene is not replicated before the S phase (i.e., constant gene dosage) and the gene is transcribed continuously to give rise to a relatively stable mRNA which accumulates on the polysomes, it was predicted that the NADP-GDH enzyme potential should increase in parallel with the linear accumulation of NADP-GDH mRNA on the polysomes before the S phase in the cell cycle. After replication of the NADP-GDH structural gene, the increase in dosage of structural genes should support a proportionally higher rate of accumulation of the NADP-GDH mRNA and corresponding increase in enzyme potential.

As an initial step to test these predictions, Turner *et al.* (*38*) extracted total cellular RNA from synchronous cells (Fig. 17) at different times in the cell cycle, and used oligo(dT)-cellulose chromatography to isolate total cellular poly(A)-containing RNA. They observed that the total poly(A)-containing RNA, total cellular RNA, and total cellular protein

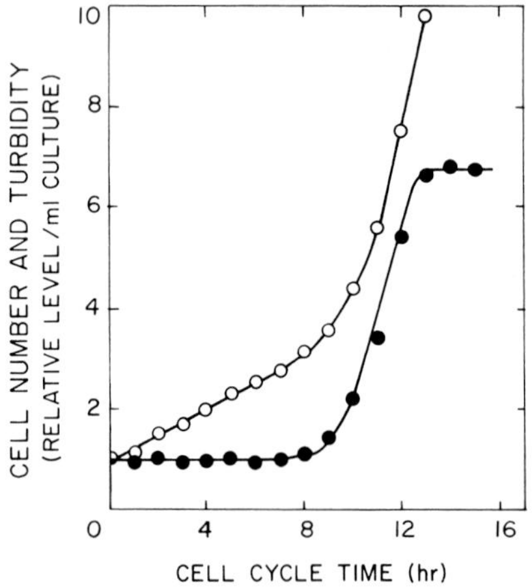

Fig. 17. Patterns of cell number and culture turbidity during the cell cycle of synchronous *Chlorella* cells growing at an average rate of 18% per hr in nitrate-containing medium (i.e., uninduced cells). Cell number (●); culture turbidity (○). [From Turner *et al.* (*38*), with permission.]

increased with coincident patterns during the cell cycle (Fig. 18). To measure the amount of translatable NADP-GDH mRNA in the total poly(A)-containing RNA fraction, the latter RNA was translated *in vitro,* and the amount of newly synthesized NADP-GDH antigen was measured after precipitation with anti-NADP-GDH IgG.

The amount of translatable NADP-GDH mRNA was shown to increase continuously throughout the cell cycle (Fig. 19). Although it is evident that the level of NADP-GDH mRNA increases before the S phase (i.e., DNA replication begins at 7.75 hr), there are insufficient data points to ascertain its type of accumulation pattern (i.e., exponential, linear, etc.). However, the mRNA pattern is consistent with the aforementioned predictions and provides an explanation for the type of enzyme potential pattern observed for the NADP-GDH during the *Chlorella* cell cycle.

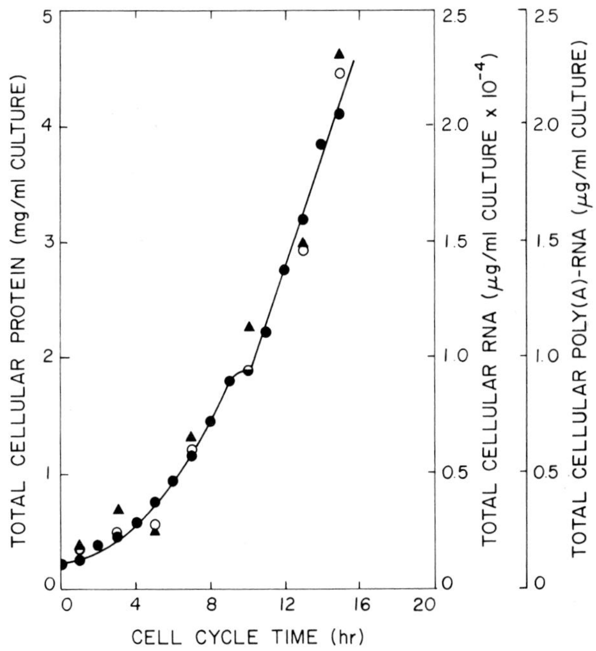

Fig. 18. Patterns of accumulation of total cellular protein, total cellular RNA, and total cellular poly(A)-containing RNA during the cell cycle of synchronous *Chlorella* cells growing at an average rate of 18% per hr in nitrate-containing medium (i.e., uninduced cells). The samples were taken from the synchronous culture described in Fig. 17. Total cellular protein (●—●); total cellular RNA (○); total cellular poly(A)-containing RNA (▲). [From Turner *et al.* (*38*), with permission.]

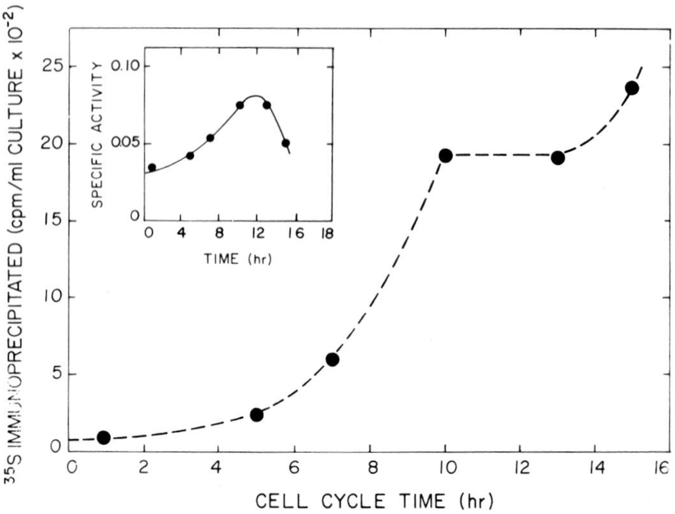

Fig. 19. Pattern of accumulation of total cellular, translatable NADP-GDH mRNA per ml of culture during the cell cycle of synchronous *Chlorella* cells growing at an average rate of 18% per hr in nitrate-containing medium (i.e., uninduced cells). The samples were taken from the synchronous culture described in Fig. 17. Inset: The amount of NADP-GDH synthesized in an *in vitro* translation assay directed by 1 μg of total cellular poly(A)-containing RNA isolated from the synchronous cells. [From Turner *et al.* (*38*), with permission.]

Acknowledgments

Thanks are extended to Dr. Robert T. Schimke and his research group at Stanford University for teaching various immunochemical and nucleic acid fractionation procedures to Dr. Robert R. Schmidt during his research leave in their laboratory. The authors appreciate the drawing of the figures by Mrs. Waltraud Dunn and the typing of the manuscript by Ms. Tonie Henry.

This research was supported by U.S. Public Health Grants 5 RO1 GM 19871 and GM29733-01 from the National Institutes of General Medical Sciences, National Institutes of Health. The research leave of Dr. Christopher F. Thurston in Dr. Schmidt's laboratory was supported by the award of an EMBO long-term fellowship.

References

1. Alt, F. W., Kellems, R. E., Bertino, J. R., and Schimke, R. T. (1978). Selective multiplication of dihydrofolate reductase genes in methotrexate-resistant variants of cultured murine cells. *J. Biol. Chem.* **253,** 1357–1370.
2. Baechtel, F. S., Hopkins, H. A., and Schmidt, R. R. (1970). Continuous inducibility

of isocitrate lyase during the cell cycle of the eucaryote *Chlorella*. *Biochim. Biophys. Acta* **217**, 216–219.

3. Bascomb, N. F., Yeung, A. T., Turner, K. J., and Schmidt, R. R. (1981). Turnover of an ammonium inducible glutamate dehydrogenase during induction and its rapid inactivation after removal of inducer from *Chlorella* cells. *J. Bacteriol.* **145,** 1266–1272.
4. Bassham, J. A., and Jensen, R. G. (1967). Photosynthesis of carbon compounds. *In* "Harvesting the Sun" (A. San Pietro, F. A. Greer, and T. J. Army, eds.), pp. 79–110. Academic Press, New York.
5. Buchanan, B. B. (1980). Role of light in regulation of chloroplast enzymes. *Annu. Rev. Plant Physiol.* **31,** 341–374.
6. Donachie, W. D., and Masters, M. (1969). Temporal control of gene expression in bacteria. *In* "The Cell Cycle: Gene–Enzyme Interactions" (G. M. Padilla, G. L. Whitson, and I. L. Cameron, eds.), pp. 37–76. Academic Press, New York.
7. Funkhouser, E. A., and Ramadoss, C. S. (1980). Synthesis of nitrate reductase in *Chlorella*. II. Evidence for synthesis in ammonia-grown cells. *Plant Physiol.* **65,** 944–948.
8. Funkhouser, E. A., Shen, T.-C., and Ackerman, R. (1980). Synthesis of nitrate reductase in *Chlorella*. I. Evidence for an inactive protein precursor. *Plant Physiol.* **65,** 939–943.
9. Gayler, K. R., and Morgan, W. R. (1976). An NADP-dependent glutamate dehydrogenase in chloroplasts from the marine green alga *Caulerpa simpliciuscula*. *Plant Physiol.* **58,** 283–287.
10. Gronostajski, R. M., Yeung, A. T., and Schmidt, R. R. (1978). Purification and properties of the inducible nicotinamide adenine dinucleotide phosphate-specific glutamate dehydrogenase from *Chlorella sorokiniana*. *J. Bacteriol.* **134,** 621–628.
11. Halvorson, H. O., Bock, R. M., Tauro, P., Epstein, R., and LaBerge, M. (1966). Periodic enzyme synthesis in synchronous cultures of yeast. *In* "Cell Synchrony: Studies in Biosynthetic Regulation" (I. L. Cameron and G. M. Padilla, eds.), pp. 102–116. Academic Press, New York.
12. Hemmings, B. A. (1978). Evidence for the degradation of nicotinamide adenine dinucleotide phosphate-dependent glutamate dehydrogenase of *Candida utilis* during rapid enzyme inactivation. *J. Bacteriol.* **133,** 876–877.
13. Hemmings, B. A. (1978). Phosphorylation of NAD-dependent glutamate dehydrogenase from yeast. *J. Biol. Chem.* **253,** 5255–5258.
14. Hemmings, B. A., and Sims, A. P. (1977). The regulation of glutamate metabolism in *Candida* utilis. Evidence for two interconvertible forms of NAD-dependent glutamate dehydrogenase. *Eur. J. Biochem.* **80,** 143–151.
15. Hershko, A., Ciechanover, A., Heller, H., Hass, A. L., and Rose, I. A. (1980). Proposed role of ATP in protein breakdown: Conjugation of proteins with multiple chains of the polypeptide of ATP-dependent proteolysis. *Proc. Natl. Acad. Sci. U.S.A.* **77,** 1783–1786.
16. Israel, D. W., Grontostajski, R. M., Yeung, A. T., and Schmidt, R. R. (1977). Regulation of accumulation and turnover of an inducible glutamate dehydrogenase in synchronous cultures of *Chlorella*. *J. Bacteriol.* **130,** 793–804.
17. Israel, D. W., Gronostajski, R. M., Yeung, A. T., and Schmidt, R. R. (1978). Regulation of glutamate dehydrogenase induction and turnover during the cell cycle of the eucaryote *Chlorella*. *In* "Cell Cycle Regulation" (J. R. Jeter, I. L. Cameron, and A. M. Zimmerman, eds.), pp. 185–201. Academic Press, New York.
18. Leech, R. M., and Kirk, P. R. (1968). An NADP-dependent L-glutamate dehydrogenase from chloroplasts of *Vicia faba*. *Biochem. Biophys. Res. Commun.* **32,** 685–690.

19. Mazon, M. J., and Hemmings, B. A. (1979). Regulation of *Saccharomyces cerevisiae* nicotinamide adenine dinucleotide phosphate-dependent glutamate dehydrogenase by proteolysis during carbon starvation. *J. Bacteriol.* **139,** 686–689.
20. Meredith, M. J. (1977). Purification, physical and chemical characterization of the nicotinamide adenine dinucleotide-specific glutamate dehydrogenase from *Chlorella sorokiniana*. Ph.D. Thesis, Virginia Polytechnic Institute and State University, Blacksburg, Virginia.
21. Meredith, M. J., Gronostajski, R. M., and Schmidt, R. R. (1978). Physical and kinetic properties of the nicotinamide adenine dinucleotide-specific glutamate dehydrogenase purified from *Chlorella sorokiniana. Plant Physiol.* **61,** 967–974.
22. Molin, W. T., Cunningham, T. P., Bascomb, N. F., White, L. H., and Schmidt, R. R. (1981). Light requirement for induction and continuous accumulation of an ammonium-inducible NADP-specific glutamate dehydrogenase in *Chlorella. Plant Physiol.* **67,** 1250–1254.
23. Molin, W. T., and Schmidt, R. R. (1982). Effect of different carbon sources and concentrations of ammonia on induction of *Chlorella* NADP-specific glutamate dehydrogenase. *Plant Physiol.* (in preparation).
24. Palacios, R., Palmiter, R. D., and Schimke, R. T. (1972). Identification and isolation of ovalbumin-synthesizing polysomes. *J. Biol. Chem.* **247,** 2316–2321.
25. Pelham, H. R. B., and Jackson, R. J. (1976). An efficient mRNA-dependent translation system from reticulocyte lysates. *Eur. J. Biochem.* **67,** 247–256.
26. Sanchez, F., Compomanes, M., Quinto, C., Hansberg, W., Mora, J., and Palacios, R. (1978). Nitrogen source regulates glutamine synthetase mRNA levels in *Neurospora crassa. J. Bacteriol.* **136,** 880–885.
27. Schimke, R. T., and Doyle, D. (1970). Control of enzyme levels in animal tissues. *Annu. Rev. Biochem.* **39,** 929–976.
28. Schmidt, R. R. (1974). Transcriptional and post-transcriptional control of enzyme levels in eucaryotic microorganisms. *In* "Cell Cycle Control" (G. M. Padilla, I. L. Cameron, and A. M. Zimmerman, eds.), pp. 201–233. Academic Press, New York.
29. Schmidt, R. R. (1974). Continuous dilution culture system for studies on gene–enzyme regulation in synchronous cultures of plant cells. *In Vitro* **10,** 306–320.
30. Simonis, W., and Urbach, W. (1973). Photophosphorylation *in vivo. Annu. Rev. Plant Physiol.* **24,** 89–114.
31. Solomonson, L. P. (1974). Regulation of nitrate reductase activity by NADH and cyanide. *Biochim. Biophys. Acta* **334,** 297–308.
32. Solomonson, L. P. (1978). Algal reduction of nitrate. *In* "Microbiology—1978" (D. Schlessinger, ed.), pp. 315–319. Am. Soc. Microbiol., Washington, D.C.
33. Solomonson, L. P., Lorimer, G. H., Hall, R. L., Borchers, R., and Bailey, J. L. (1975). Reduced nicotinamide adenine dinucleotide-specific nitrate reductase of *Chlorella vulgaris. J. Biol. Chem.* **250,** 4120–4127.
34. Talley, D. J., White, L. H., and Schmidt, R. R. (1972). Evidence for NADH- and NADPH-specific isozymes of glutamate dehydrogenase and the continuous inducibility of the NADPH-specific isozyme throughout the cell cycle of the eucaryote *Chlorella. J. Biol. Chem.* **247,** 7927–7935.
35. Tomkins, G. M., Gelehrter, T. D., Granner, D., Martin, D., Samuels, H. H., and Thompson, E. B. (1969). Control of specific gene expression in higher organisms. *Science* **166,** 1474–1480.
36. Turner, K. J., Bascomb, N. F., Lynch, J. J., Thurston, C. F., Molin, W. T., and Schmidt, R. R. (1981). Evidence for mRNA of an ammonium-inducible glutamate

dehydrogenase, and synthesis, covalent-modification, and degradation of enzyme subunits in uninduced *Chlorella* cells. *J. Bacteriol.* **146,** 578–589.

37. Turner, K. J., Gronostajski, R. M., and Schmidt, R. R. (1978). Regulation of the initial rate of induction of nicotinamide adenine dinucleotide phosphate-specific glutamate dehydrogenase during the cell cycle of synchronous *Chlorella. J. Bacteriol.* **134,** 1013–1019.
38. Turner, K. J., Thurston, C. F., Lynch, J. J., and Schmidt, R. R. (1982). Some properties of the mRNA for ammonium inducible NADP-specific glutamate dehydrogenase and the increase of this mRNA during the cell cycle of non-induced synchronous cells of *Chlorella. Mol. Cell. Biol.* (in preparation).
39. Yeung, A. T., Bascomb, N. F., Turner, K. J., and Schmidt, R. R. (1981). Regulation of accumulation of ammonium-inducible glutamate dehydrogenase catalytic activity and antigen during the cell cycle of fully induced synchronous *Chlorella sorokiniana* cells. *J. Bacteriol.* **146,** 571–577.
40. Yeung, A. T., Turner, K. J., Bascomb, N. F., and Schmidt, R. R. (1981). Purification of an ammonium-inducible glutamate dehydrogenase and the use of its antigen affinity-column purified antibody in specific immunoprecipitation and immunoadsorption procedures. *Anal. Biochem.* **110,** 216–228.
41. Zucker, M. (1972). Light and enzymes. *Annu. Rev. Plant Physiol.* **23,** 133–156.

9

Genes and the Regulation of the Cell Cycle

DIETER E. WAECHTER AND RENATO BASERGA

I. Introduction

Conditional lethal mutants defective in certain steps of the cell division cycle have been isolated from several types of eukaryotic cells. Such mutants have been described in yeast by Hartwell (*14*) and have been used to outline the temporal and functional sequence map of the cell cycle events. On the basis of the number of complementation groups of yeast mutants, Edwards *et al.* (*8*) have calculated that the cell cycle of yeast needs a total of 32 gene products. A number of conditional lethal mutants of the cell cycle have also been isolated from mammalian cells especially by Basilico (*4*). Most of these cell cycle mutants are temperature sensitive (*ts*) for one step in the cell cycle. We can therefore give the following definition: Cell cycle-specific *ts* mutants are operationally defined as mutants that arrest at the nonpermissive temperature in a specific phase of the cell cycle (*5*). Some qualifications are necessary at this point. The definition of a mutant is based on the circumstantial evidence reviewed by Siminovitch (*35*) indicating that in mammalian cells with a stably altered phenotype the defect may have, at least in some cases, a genetic basis. In agreement with him, we are using for these

GENETIC EXPRESSION IN THE CELL CYCLE

ISBN 0-12-543720-X

cells the term "mutant," although the term may turn out to be inappropriate in some cases. A cell cycle-specific mutant must be differentiated from growth mutants. Growth mutants are simply mutants that do not grow at all at the nonpermissive temperature and that will stop under restrictive conditions at any point in the cell cycle, usually at the point at which they find themselves when the temperature is raised. Cell cycle-specific *ts* mutants, instead, continue their progression through the cell cycle even at the nonpermissive temperature until they arrive at that specific phase of the cell cycle in which the *ts* function is required. A number of cell cycle-specific *ts* mutants have been described and are listed in a review by Siminovitch and Thompson (*36*). Other mutants have been subsequently reported, and one can say that there is a variety of *ts* mutants of the cell cycle, some of which will arrest in G_1, others in the S phase and, again, others in G_2 or mitosis. In this review we are particularly interested in G_1 *ts* mutants, i.e., mutants operationally defined as arresting at the nonpermissive temperature in the G_1 phase of the cell cycle but growing normally at the permissive temperature.

The definition of G_1 *ts* mutant implies also that (1) when collected by mitotic detachment and plated at the nonpermissive temperature, the cells do not enter S phase; (2) when made quiescent by serum restriction and subsequently stimulated at the nonpermissive temperature, the cells do not enter S phase; (3) the cells enter S phase at the permissive temperature whether plated after mitotic detachment or stimulated after nutritional deprivation; (4) the parental cell line must be capable of entering S phase at both temperatures; and finally (5) the cells arrest in G_1 even when shifted up to the nonpermissive temperature in other phases of the cell cycle. When a cell line meets these criteria, it is defective in a gene product that is required for the transition of the cell from G_0 or G_1 to S under optimal nutritional conditions.

The fact that the defective gene product is required for the $G_0,G_1 \rightarrow$ S transition does not necessarily mean that the gene is expressed only in G_1. In fact, we have no information whatsoever about how genes are expressed throughout the cell cycle, and there is still the possibility that gene transcription is constant throughout the cell cycle and that the various phases are regulated only at a translational level. Although this possibility still exists, there is increasing evidence, as we shall relate later, that some kind of unique-copy gene transcription is necessary for the $G_0,G_1 \rightarrow$ S transition, whereas it may not be necessary after the cells have reached the S phase.

In addition, the fact that these gene products are needed for the transition of mammalian cells from a resting to a growing stage under optimal nutritional conditions does not necessarily mean that they do control the cell cycle. It may be possible that their function is purely permissive, i.e., that it is necessary for the transition from $G_0,G_1 \rightarrow S$ but does not actually control the proliferation rate of the cell population. The objection can also be raised that the G_1 arrest of these *ts* mutants may not be different from the situation in which certain cells become quiescent in the absence of nutrients or growth factors. For instance, it is well known that when the culture medium is deficient in isoleucine, serine, or other amino acids, the cells have a tendency to arrest in the G_1 period of the cell cycle (*1,19,22*). A mutant incapable of synthesizing isoleucine, in theory, would also arrest in the G_1 phase of the cell cycle and again, in theory, would be indistinguishable from a *ts* mutant of the type described above. However, the arrest of cells in G_1 under certain nutritional deprivations is dependent on the presence of at least a certain amount of the amino acid. For instance, cells arrest in G_1 not when the medium is completely deficient in isoleucine but only when the isoleucine content is strongly reduced. A much more important difference, though, is the fact that, when nutritional conditions bring about an arrest in the G_0–G_1 phase of the cell cycle, macromolecular synthesis is markedly reduced. On the contrary, as we shall see below, in *ts* mutants of G_1, protein and RNA synthesis are unaffected for several hours after the shift up to the nonpermissive temperature.

Perhaps a lot of confusion could be avoided if one defines the cell cycle-specific mutants as mutants whose execution point is in G_1. The execution point is defined by Pringle (*28*) as the point in the cell cycle at which the *ts* function is no longer needed for progression. Again, one should emphasize that this is not the point at which the *ts* function is expressed. The *ts* function could have been expressed at earlier times: What the execution point says is simply that the *ts* function is no longer needed beyond that point.

II. Execution Points

We have determined the execution points of three *ts* mutants of the cell cycle that arrest in G_1 at the nonpermissive temperature. The details of the experimental procedure followed to determine the execution points have been given by Ashihara *et al.* (*2*). Briefly, it consists in shifting up to the nonpermissive temperature cell populations at different times after mitosis or serum stimulation and determining the number of cells that can still enter S phase. The results (Table I) show that, regardless of whether the cells come from mitosis or from G_0, the execution point is located at the same distance from the beginning of S phase. It indicates

TABLE I

Execution Points of Three G_1-Specific *ts* Mutants[a]

Cell line	Execution point (hr before onset of S phase)		Complementing human chromosome
	Postmitotic cells	Cells stimulated from quiescence	
*ts*AF8	9.7	8.8	3
*ts*13	3.3	3.4	4
K12	1.6	1.5	14

[a] The execution point is the point in the cell cycle at which the *ts* function is no longer needed. Data obtained from Ming *et al.* (*23*), Ashihara *et al.* (*3*), Floros *et al.* (*9*), and Ming *et al.* (*24*).

that there must be some orderly arrangement in the sequence of events in G_1, particularly in the time between the execution point and S phase.

Another important observation made by Talavera and Basilico (*40*) is the difference in serum requirements among different *ts* mutants. For instance, when *ts*AF8 cells are stimulated at the nonpermissive temperature and then shifted down to the permissive temperature while the serum is simultaneously reduced to 0.5%, the cells will not enter DNA synthesis. On the contrary, when *ts*13 cells are stimulated at the nonpermissive temperature and then shifted down both in temperature and in serum concentration, they will enter S phase (*39*). These findings indicate that the serum-dependent events have not been carried out in *ts*AF8 cells at the nonpermissive temperature, whereas they have been carried out in *ts*13 cells at the nonpermissive temperature. This is one of the few leads we have in determining what serum-dependent events may take place after stimulation. Incidentally, *ts*AF8, *ts*13, and two other mutants, HJ4 and K12 cells, all complement each other and, presumably, have mutations in different genes. This is confirmed by the fact that *ts*AF8, K12, and *ts*13 are each complemented by different human chromosomes (*23,24*).

III. Informational Content of Cells and of Cytoplasts

A number of experiments have been carried out in our laboratory to determine the informational content of cells in different stages of the cell cycle using temperature-sensitive mutants. These experiments were based on the original observation of Rao and Johnson (*29*) that S-phase cells can induce DNA synthesis in nuclei of G_1 cells when the two types of cells are fused together. We have carried out two sets of experiments, both of which indicate that S-phase cells contain all the necessary information for the transition from $G_0,G_1 \rightarrow S$. In the first of these experiments (*43*), we have shown that S-phase cells of *ts* mutants will reactivate chick erythrocytes in heterokaryon formation even when the heterokaryons are incubated at the temperature nonpermissive for the mammalian cells. In control experiments when G_0 cells were fused with chick erythrocytes and the heterokaryons were incubated at the nonpermissive temperature, neither the mammalian nucleus nor the chick nucleus entered DNA synthesis. Furthermore, we have shown that fusion of S-phase *ts*AF8 cells with G_0 *ts*AF8 cells will induce DNA synthesis in the G_0 nuclei (*10*). This indicates that S-phase *ts*AF8 cells contain all the necessary information for the transition from G_0 to S and that gene expression is no longer needed to effect that transition. This is probably the first example of self-complementation in which a cell in one phase of the cell cycle can complement a defective gene in the same cell in a different phase of the cell cycle.

We have also determined the informational content of cytoplasts of *ts* mutants, and for this purpose we have fused cytoplasts made by the cytochalasin technique of Prescott and co-workers (*44*) with G_0 cells of other complementing *ts* mutants. The results (Table II) show the following: The cytoplasts of S-phase cells contain all the necessary information for reactivating G_0 cells even when the "cybrids" are incubated at the

TABLE II

Informational Content of Cytoplasts from G_0 and S Phase Cells[a]

Recipient whole cell	Donor cytoplasts	Cybridoids entering S phase after stimulation at the nonpermissive temperature (%)
G_0 *ts*13	none	4
G_0 *ts*AF8	none	7
G_0 *ts* HJ4	none	10.8
G_0 *ts*13	G_0 *ts*AF8	19.5
G_0 *ts*13	S *ts*AF8	67.0
G_0 *ts* HJ4	G_0 *ts*AF8	18
G_0 *ts* HJ4	S *ts*AF8	66

[a] From Jonak and Baserga (*16,17*). Whole cells or fusion products were stimulated with 10% serum at the nonpermissive temperature in the presence of [^{3}H]thymidine. The experiments were terminated 24 hr later, and autoradiographs were made by standard techniques. At permissive temperature, cells and fusion products regularly enter S. Heterokaryons between G_0 *ts*AF8 × G_0 *ts*13 or G_0 *ts*AF8 × G_0 HJ4 enter S phase even at the nonpermissive temperature.

nonpermissive temperature (*16*). However, when G_0 cytoplasts are fused with complementing *ts* mutants in G_0, the resultant cybrids do not enter S phase (*17*). The results indicate that the information for the complementation of the *ts* defect in these mutants is not present in the cytoplasts of G_0 cells, although it is present in the cytoplasts of S-phase cells. It further indicates that this information must be generated from the nuclei of G_0 cells at some time between the application of the stimulus to proliferate and the entry of cells into S phase. This is the first demonstration that some of the events necessary for the transition of cells from the resting to the growing stage require some nuclear function.

IV. Nature of the *ts* Mutations

Despite the pessimism of Stanners (*37*) who stated that "efforts toward detailed molecular characterization of *ts* cell cycle mutants are usually a waste of time," a number of *ts* cell cycle mutants have been characterized at the molecular level. For instance, Tsai *et al.* (*42*) have described a *ts* mutant of the cell cycle that is defective in the DNA polymerase α. Marunouchi *et al.* (*20*) have described a mutant of the cell cycle, *ts*85, which is defective in chromosome condensation and whose defect is apparently related to the phosphorylation of histone H1. Another *ts* mutant of the cell cycle, *ts*AF8, which arrests in G_1 at the nonpermissive temperature, has been identified as a mutant of RNA polymerase II. The characterization of this mutant began when Rossini and Baserga (*30*) showed that in *ts*AF8 the RNA polymerase II activity of isolated nuclei disappeared with a half-life of ~10–12 hr when the cells were shifted to the nonpermissive temperature. In subsequent experiments, Rossini *et al.* (*31*) showed that the RNA polymerase II molecule actually disappeared with a half-life of ~10–12 hr when *ts*AF8 cells were shifted to the nonpermissive temperature. This finding was based on the use of α-amanitin, a drug that specifically inhibits RNA polymerase II by binding to the α-amanitin-binding subunit of that molecule (*6*). In the experiments of Rossini *et al.* (*31*) it was shown that the α-amanitin-binding subunit of RNA polymerase II disappeared from *ts*AF8 cells at 40.6°C, whereas it remained constant in the same cells at 34°C or in BHK cells at either 34° or 41°C (BHK cells are the wild type from which *ts*AF8 cells have been derived). RNA polymerase I activity is not affected, and the cells remain viable for at least 60 hr, although RNA polymerase II activity is virtually gone by 24 hr. A rapid survey of protein synthesis indicated that protein synthesis was not affected for at least 30 hr at the nonpermissive temperature (*31*). Furthermore, studies by flow cytophotometry have demonstrated that *ts*AF8 accumulates RNA just as effectively at the nonpermissive temperature as at the permissive temperature (*3*). Thus, in the absence of RNA polymerase II, the cells cannot enter S phase but continue to synthesize rRNA and proteins at normal rates. Further studies have confirmed the conclusion that *ts*AF8 cells are a mutant of RNA polymerase II. Shales *et al.* (*34*) have found that *ts*AF8 does not complement a *ts* mutant of CHO cells that has a point mutation in the α-amanitin-binding subunit of RNA polymerase II.

The implications of these findings are (1) RNA polymerase II is needed for the progression of cells in G_1, which, in turn, is strong evidence that unique-copy genes must be transcribed for entry into S; (2) whereas RNA polymerase II is necessary for the entry of cells into S, it is not required for growth in size (as measured by the accumulation of RNA and proteins); and (3) since it can be shown that α-amanitin-resistant mutants have a point mutation in the α-amanitin-binding subunit of RNA polymerase II, the experiments of Shales *et al.* (*34*) seem to indicate that at least in the case of *ts*AF8 we may also be dealing with a true mutant. The requirement for RNA polymerase II for entry into S can also be demonstrated by microinjecting α-amanitin directly into the nuclei of cells, using the glass capillary technique of Graessmann and Graessmann (*13*). This is shown in Fig. 1, in which groups of quiescent *ts*AF8 cells

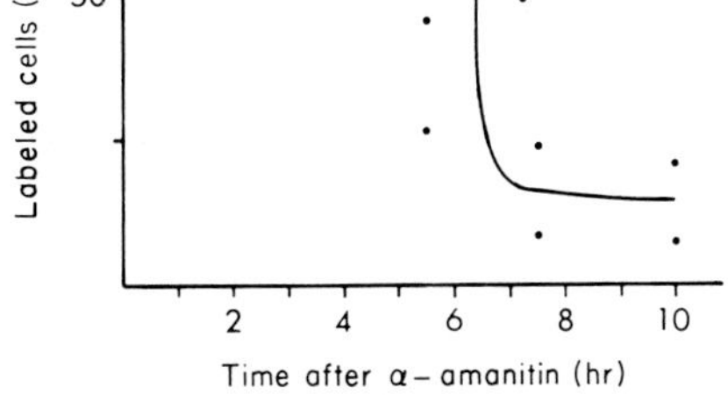

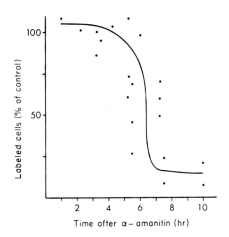

Fig. 1. Determination of the last α-amanitin-sensitive point in the G_1 period of AF8 cells. Quiescent AF8 cells were stimulated with 10% serum. After 24 hr they were microinjected into the nucleus with 50 μg/ml α-amanitin. The cells were then incubated for different lengths of time, pulse-labeled for 30 min with 1 μCi/ml [^{3}H]thymidine, and fixed. The percentage of labeled cells was determined by autoradiography and compared to cells that were not microinjected, which were set at 100%.

were stimulated with serum. The cells were microinjected 24 hr later with α-amanitin and pulse-labeled with [^{3}H]thymidine at various intervals after microinjection. Entry into S is inhibited at 34°C, with a 50% point 8 hr before S, which coincides with the execution point of the *ts* mutation in these cells (see Table I). The amount of α-amanitin to be microinjected (a concentration of 50 μg/ml) was determined by microinjecting simultaneously the drug and a plasmid containing the thymidine kinase gene of Herpes simplex virus [HSV (*tk*)]. The *tk* gene is transcribed by RNA polymerase II, and in this way (Fig. 2) one can determine the concentration of α-amanitin that must be microinjected to inhibit unique-copy gene transcription [When this work had been completed, we found that McKnight and Gavis (*21*) used the same procedure to determine the optimal concentration of α-amanitin to be microinjected into *Xenopus* oocytes]. The percentage of labeled cells, in the absence of α-amanitin, is 151% after 24 hr. This is due to the fact that some of the microinjected cells have divided in the interval.

Another cell cycle *ts* mutant, whose lesion is known, is 422 E, derived, like *ts*AF8 and *ts*13, from BHK cells. This mutant is defective in the

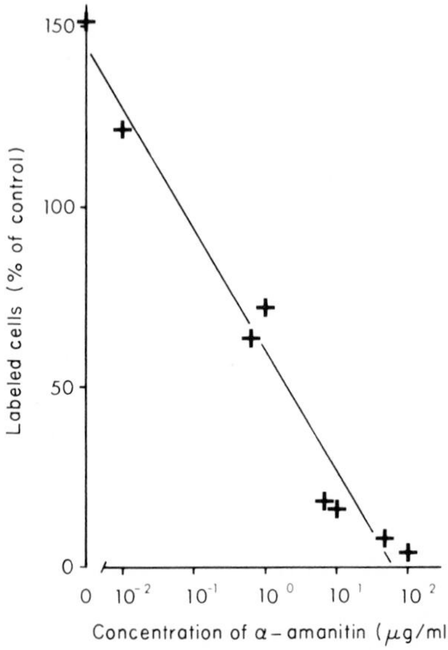

Fig. 2. Determination of the amount of α-amanitin necessary to block gene expression when microinjected into viable animal cells. *ts*13 *tk*$^-$ cells made quiescent by serum deprivation were microinjected with 0.12–0.16 mg/ml HSV *tk* plasmid DNA plus the indicated amounts of α-amanitin. The cells were incubated in presence of 10% serum and labeled with 1 μCi/ml [^{3}H]thymidine. The expression of the *tk* gene was monitored by autoradiography.

processing of 28 S rRNA (*41*). As would be expected, 422 E cells enter S even at the nonpermissive temperature (*25*), indicating that accumulation of rRNA is not necessary for entry into S. Interestingly, at the nonpermissive temperature, 422 E cells do not divide, suggesting a requirement for rRNA for cell division (*25*) (see also Chapter 4).

V. Induction of Cellular DNA Replication in G_1-Specific *ts* Mutants by Viruses

It has been known for several years that small DNA viruses such as SV40 and polyoma- and adenoviruses can induce resting cells to enter the S phase (*7,15,46,47;* for a review, see *45*). A virus like SV40 is known under appropriate conditions to induce, not only cell DNA replication, but also growth in size of the cell and eventually mitosis; for this reason, Weil (*45*) has called these viruses "mitogenic viruses." Because these viruses are known to alter the growth requirements of normal mammalian cells, we thought it would be interesting to determine whether the information contained in the genome of some of these viruses could also alter the requirements for certain cellular functions necessary for the serum-stimulated transition of cells from G_0 to S, i.e., from a resting to a growing stage. Thus far, we have tested two DNA viruses, SV40 and adenovirus 2.

With adenovirus 2, the experiment could be carried out by infecting *ts*AF8 or *ts*13 cells. These cells, being derived from BHK cells, are semipermissive for adenovirus. Adenovirus 2 infection causes stimulation of cell DNA synthesis in serum-deprived *ts*AF8 and *ts*13 cells at 34°C, as does 10% serum (*32*). Furthermore, infection with adenovirus 2 also stimulates cell DNA replication in *ts*AF8 and *ts*13 cells at the nonpermissive temperature, 40.6°C. In the paper by Rossini *et al.* (*32*), we have given the evidence that the stimulation of cell DNA replication by adenovirus 2 in these cells at the nonpermissive temperature as detected by autoradiography is bona fide semiconservative cellular DNA synthesis (*32*). Although the cells are semipermissive for adenovirus, viral DNA is not replicated at the nonpermissive temperature (*27,32*). The results then indicate that adenovirus 2 contains in its genome the necessary information to induce cell DNA replication in *ts*AF8 and *ts*13 cells in the absence of certain G_1 functions that are required by these same cells when stimulated by serum.

With SV40, other techniques have to be used, because the mutant cells, like their parent cell line BHK, are resistant to infection by SV40. It is possible, though, to microinject DNA manually into cells by the technique of Graessmann and collaborators (*12*) as already applied to SV40 DNA by Mueller *et al.* (*26*). In our experiments, we microinjected into the nuclei of *ts*13 and *ts*AF8 cells the DNA from recombinant plasmids in which we had cloned a fragment of the SV40 genome that includes the entire early region (*11*).

Table III shows that SV40 causes induction of cell DNA replication in these cells at the nonpermissive temperature of 40.6° or 39.5°C. The DNA from the recombinant plasmid contained a fragment of SV40 DNA extending from map unit 1 counterclockwise to map unit 0.144, cloned in the pBR322 plasmid. This plasmid, therefore, contains the entire early region, the so-called *A* gene, which extends from map unit 0.67 to map unit 0.17. When this recombinant DNA is manually microinjected into cells, the cells become positive for SV40 T antigen. Approximately 75 to 85% of the microinjected cells become T positive within 24 hr after microinjection. Table III shows that the cells rendered T positive by microinjection also enter DNA synthesis in a much higher number than control cells not microinjected or microinjected with plasmid pBR322 not containing the SV40 information. In the paper by Floros *et al.* (*11*), we have shown that the stimulation of cell DNA replication by microinjected SV40 DNA at the nonpermissive temperature as detected by autoradiography is due, not to DNA repair or to viral DNA replication, but to bona fide semiconservative cellular DNA synthesis. Furthermore, we have been able to induce cell DNA replication in *ts*13 cells at the nonpermissive temperature with microinjection of SV40-specific mRNA, i.e., mRNA isolated by hybridization to the SV40 genome. Finally, we have

TABLE III

Induction of Cell DNA Replication by Microinjection of SV40[a]

Cell line	Temperature (°C)	DNA microinjected	Cells in DNA synthesis (%)
*ts*13	34	none	7.3
	39.6	none	6.0
	34	pBR322	5.0
	34	pSV2G	85.0
	39.6	pSV2G	78.0
*ts*AF8	34	pSV2G	86.0
	40.6	pSV2G	78.0

[a] Quiescent cells were microinjected into the nucleus with the desired DNA, at a concentration of 0.5 mg/ml. The cells were then incubated in conditioned medium (1% serum) at either permissive or nonpermissive temperatures for 24 hr in the presence of [^{3}H]thymidine. Percentage of labeled cells was determined by standard techniques. Experimental details are given in the paper by Floros *et al.* (*11*).

been able to show that the effect of SV40 DNA on cell DNA replication is mediated through the product of the *A* gene, the T antigen, since cells that have been simultaneously microinjected with antibody against T antigen did not enter DNA synthesis. If the cells are microinjected with preimmune IgG fraction, they enter DNA synthesis as cells microinjected only with the recombinant DNA (*11*).

These experiments then show that at least two G_1 cellular functions necessary for the G_0-to-S transition in serum-stimulated cells can be dispensed with at least for one round of DNA synthesis in cells infected with adenovirus 2 or microinjected with the early region of the SV40 genome. Regardless of one's views of the cell cycle, one must say that these G_1 mutants have a *ts* defect in a cellular function that is required for the progression of cells from mitosis or G_0 to S under standard conditions. These functions are no longer needed for entry into S when quiescent cells are provided with information from the early region of the SV40 genome or from the adenovirus 2. This finding raised some interesting questions because it is generally believed that the products of the SV40 *A* gene and of the adenovirus genome act by decreasing the cell's requirement for growth factors. Although this is certainly true (*33,38*), our findings indicate that viral antigens may also act by decreasing the requirements for certain intracellular functions.

VI. Future Directions of Research

The future directions of research include different approaches that can be summarized as follows: The BHK genome is now being cloned in Charon λ phage, and we are in the process of identifying those phages which contain BHK genes that can complement the *ts* defect in either *ts*AF8 or *ts*13 cells. Preliminary experiments have shown that this is feasible and that we should be able in the near future to obtain clones that will correct the *ts* defect when the recombinant DNA is microinjected into the *ts* cells. In the case of *ts*AF8 cells, it is known that the *ts* defect has something to do with the RNA polymerase II molecule (see Section IV). We do not know the temperature-sensitive defect of *ts*13, but one has to start thinking in terms, not of proteins, but of genes. Just as a book of genetics nowadays is more intelligible if one starts from DNA than from the experiments of Mendel, similarly, we should try to look at cell proliferation as a problem of genetics rather than as a problem of biochemistry. In other words, the isolation of genes that are required for certain cellular functions is in itself of importance, regardless of whether we know how the gene product is acting at a biochemical level.

In this respect, the isolation of these genes will allow us to isolate at least two genes whose products are required in serum-stimulated cells for the progression from mitosis or G_0 to S.

The other line of approach along these lines is to determine the sequences within the SV40 *A* gene that specifically stimulate cell DNA replication. The SV40 *A* gene, as pointed out by Lebowitz and Weissman (*18*), codes for a multifunctional protein, and the various functions are located on different domains of the protein. It is conceivable that the ability of the T antigen to induce cell DNA replication is located in a region of the T antigen distinguished from other regions and other functions. For this purpose, deletion mutants of the SV40 *A* gene microinjected into cells will be extremely useful in identifying the sequence of DNA within the SV40 *A* gene that contains the necessary information for the induction of cell DNA replication in resting cells. These experiments will narrow down the boundaries of the information for this important function of cell growth and allow us to elucidate the mechanism by which the product of the viral genome can replace certain cellular functions in the induction of cell proliferation.

Acknowledgment

This work was supported by U.S. Public Health Service Research Grant CA 25898.

References

1. Allen, R. W., and Moskowitz, M. (1978). Arrest of cell growth in the G_1 phase of the cell cycle by serine deprivation. *Exp. Cell Res.* **116,** 127–137.
2. Ashihara, T., Chang, S. D., and Baserga, R. (1978). Constancy of the shift-up point in two temperature-sensitive mammalian cell lines that arrest in G_1. *J. Cell. Physiol.* **96,** 15–22.
3. Ashihara, T., Traganos, F., Baserga, R., and Darzynkiewicz, Z. (1978). A comparison of cell cycle-related changes in postmitotic and quiescent AF8 cells as measured by cytofluorometry after acridine orange staining. *Cancer Res.* **38,** 2514–2518.
4. Basilico, C. (1978). Selective production of cell cycle specific *ts* mutants. *J. Cell. Physiol.* **95,** 367–376.
5. Chu, E. H. Y. (1978). Nature of temperature-sensitive mutants. *J. Cell. Physiol.* **95,** 365–366.
6. Cochet-Meilhac, M., Nuret, P., Courvalin, J. C., and Chambon, P. (1974). Animal DNA-dependent RNA polymerases. 12. Determination of the cellular number of RNA polymerase B molecules. *Biochim. Biophys. Acta* **353,** 185–192.
7. Dulbecco, R., Hartwell, L. H., and Vogt, M. (1965). Induction of cellular DNA synthesis by polyoma virus. *Proc. Natl. Acad. Sci. U.S.A.* **53,** 403–410.
8. Edwards, D. R. W., Taylor, J. B., Wakeling, W. F., Watts, F. Z., and Johnston, I. R. (1978). Studies on the prereplicative phase of the cell cycle in *Saccharomyces cerevisiae. Cold Spring Harbor Symp. Quant. Biol.* **43,** 577–586.

9. Floros, J., Ashihara, T., and Baserga, R. (1978). Characterization of *ts*13 cells. A temperature-sensitive mutant of the G_1 phase of the cell cycle. *Cell Biol. Int. Rep.* **2,** 259–269.
10. Floros, J., and Baserga, R. (1980). Reactivation of G_0 nuclei by S-phase cells. *Cell Biol. Int. Rep.* **4,** 75–82.
11. Floros, J., Jonak, G., Galanti, N., and Baserga, R. (1981). Induction of cell DNA replication in G_1-specific *ts* mutants by microinjection of SV40 DNA. *Exp. Cell Res.* **132,** 215–223.
12. Graessmann, A., Graessmann, M., and Mueller, C. (1977). Regulatory function of simian virus 40 DNA replication for late viral gene expression. *Proc. Natl. Acad. Sci. U.S.A.* **74,** 4831–4834.
13. Graessmann, M., and Graessmann, A. (1976). "Early" simian-virus-40-specific RNA contains information for tumor antigen formation and chromatin replication. *Proc. Natl. Acad. Sci. U.S.A.* **73,** 366–370.
14. Hartwell, L. H. (1978). Cell division from a genetic perspective. *J. Cell Biol.* **77,** 627–637.
15. Henry, P., Black, P. H., Oxman, M. N., and Weissman, S. M. (1966). Stimulation of DNA synthesis in mouse cell line 3T3 by simian virus 40. *Proc. Natl. Acad. Sci. U.S.A.* **56,** 1170–1176.
16. Jonak, G. J., and Baserga, R. (1979). Cytoplasmic regulation of two G_1-specific temperature-sensitive functions. *Cell* **18,** 117–123.
17. Jonak, G. J., and Baserga, R. (1980). The cytoplasmic appearance of three functions expressed during the $G_0,G_1 \rightarrow S$ transition is nucleus-dependent. *J. Cell. Physiol.* **105,** 347–354.
18. Lebowitz, P., and Weissman, S. M. (1979). Organization and transcription of the simian virus 40 genome. *Curr. Top. Microbiol. Immunol.* **87,** 44–172.
19. Ley, K. D., and Tobey, R. A. (1970). Regulation of initiation of DNA synthesis in Chinese hamster cells. *J. Cell Biol.* **47,** 453–459.
20. Marunouchi, T., Yasuda, H., Matsumoto, Y., and Yamada, M. (1980). Disappearance of a basic chromosomal protein from cells of a mouse temperature-sensitive mutant defective in histone phosphorylation. *Biochem. Biophys. Res. Commun.* **95,** 126–131.
21. McKnight, S. L., and Gavis, E. R. (1980). Expression of the herpes thymidine kinase gene in *Xenopus laevis* oocytes: An assay for the study of deletion mutants constructed *in vitro*. *Nucleic Acids Res.* **8,** 5931–5948.
22. Melvin, W. T., Burke, J. F., Slater, A. A., and Keir, H. M. (1979). Effect of amino acid deprivation on DNA synthesis in BHK-21/C13 cells. *J. Cell. Physiol.* **98,** 73–80.
23. Ming, P. M. L., Chang, H. L., and Baserga, R. (1976). Release by human chromosome 3 of the block at G_1 of the cell cycle, in hybrids between *ts*AF8 hamster and human cells. *Proc. Natl. Acad. Sci. U.S.A.* **73,** 2052–2055.
24. Ming, P. M. L., Lang, B., and Kit, S. (1979). Association of human chromosome 14 with a *ts* defect in G_1 of Chinese hamster K12 cells. *Cell Biol. Int. Rep.* **3,** 169–178.
25. Mora, M., Darzynkiewicz, Z., and Baserga, R. (1980). DNA synthesis and cell division in a mammalian cell mutant temperature sensitive for the processing of ribosomal RNA. *Exp. Cell Res.* **125,** 241–249.
26. Mueller, C., Graessmann, A., and Graessmann, M. (1978). Mapping of early SV40-specific functions by microinjection of different early viral DNA fragments. *Cell* **15,** 579–585.
27. Nishimoto, T., Raskas, H. J., and Basilico, C. (1975). Temperature-sensitive cell mutations that inhibit adenovirus 2 replication. *Proc. Natl. Acad. Sci. U.S.A.* **72,** 328–332.
28. Pringle, J. R. (1978). The use of conditional lethal cell cycle mutants for temporal and

functional sequence mapping of cell cycle events. *J. Cell. Physiol.* **95,** 393–406.
29. Rao, P. N., and Johnson, R. T. (1970). Mammalian cell fusion: Studies on the regulation of DNA synthesis and mitosis. *Nature (London)* **225,** 159–164.
30. Rossini, M., and Baserga, R. (1978). RNA synthesis in a cell cycle-specific temperature sensitive mutant from a hamster cell line. *Biochemistry* **17,** 858–863.
31. Rossini, M., Baserga, S., Huang, C. H., Ingles, C. J., and Baserga, R. (1980). Changes in RNA polymerase II in cell cycle-specific temperature-sensitive mutant of hamster cells. *J. Cell. Physiol.* **103,** 97–103.
32. Rossini, M., Weinmann, R., and Baserga, R. (1979). DNA synthesis in temperature-sensitive mutants of the cell cycle infected by polyoma virus and adenovirus. *Proc. Natl. Acad. Sci. U.S.A.* **76,** 4441–4445.
33. Scher, C. D., Pledger, W. J., Martin, P., Antoniades, H., and Stiles, C. D. (1978). Transforming viruses directly reduce the cellular growth requirement for a platelet derived growth factor. *J. Cell. Physiol.* **97,** 371–380.
34. Shales, M., Bergsagel, J., and Ingles, C. J. (1980). Defective RNA polymerase II in the G_1-specific temperature-sensitive hamster cell mutant *ts*AF8. *J. Cell. Physiol.* **105,** 527–532.
35. Siminovitch, L. (1976). On the nature of hereditable variation in cultured somatic cells. *Cell* **7,** 1–11.
36. Siminovitch, L., and Thompson, L. H. (1978). The nature of conditionally lethal temperature-sensitive mutations in somatic cells. *J. Cell. Physiol.* **95,** 361–366.
37. Stanners, C. P. (1978). Characterization of temperature-sensitive mutants of animal cells. *J. Cell. Physiol.* **95,** 407–409.
38. Stiles, C. D., Isberg, R. R., Pledger, W. J., Antoniades, H. N., and Scher, C. D. (1979). Control of the BALB/c-3T3 cell cycle by nutrients and serum factors: Analysis using platelet-derived growth factor and platelet-poor plasma. *J. Cell. Physiol.* **99,** 395–406.
39. Talavera, A., and Basilico, C. (1977). Temperature sensitive mutants of BHK cells affected in cell cycle progression. *J. Cell. Physiol.* **92,** 425–436.
40. Talavera, A., and Basilico, C. (1978). Requirements of BHK cells for the exit from different quiescent states. *J. Cell. Physiol.* **97,** 429–440.
41. Toniolo, D., Meiss, H. K., and Basilico, C. (1973). A temperature-sensitive mutation affecting 28 S ribosomal RNA production in mammalian cells. *Proc. Natl. Acad. Sci. U.S.A.* **70,** 1273–1277.
42. Tsai, Y. J., Hanaoka, F., Nakano, M. M., and Yamada, M. (1979). A mammalian DNA^- mutant decreasing nuclear DNA polymerase α activity at nonpermissive temperature. *Biochem. Biophys. Res. Commun.* **91,** 1190–1195.
43. Tsutsui, Y., Chang, S. D., and Baserga, R. (1978). Failure of reactivation of chick erythrocytes after fusion with temperature-sensitive mutants of mammalian cells arrested in G_1. *Exp. Cell Res.* **113,** 359–367.
44. Veomet, G., Prescott, D. M., Shay, J., and Porter, K. R. (1974). Reconstruction of mammalian cells from nuclear and cytoplasmic components separated by treatment with cytochalasin B. *Proc. Natl. Acad. Sci. U.S.A.* **71,** 1999–2002.
45. Weil, R. (1978). Viral 'tumor antigens': A novel type of mammalian regulator protein. *Biochim. Biophys. Acta* **516,** 301–388.
46. Weil, R., Salomon, C., May, E., and May, P. (1975). A simplifying concept in tumor virology: Virus-specific "pleiotropic effectors." *Cold Spring Harbor Symp. Quant. Biol.* **39,** 381–395.
47. Zimmerman, J. E., Jr., and Raska, K. (1972). Inhibition of adenovirus type 12 induced DNA synthesis in G_1-arrested BHK21 cells by dibutyryl adenosine cyclic 3′:5′-monophosphate. *Nature (London), New Biol.* **239,** 145–147.

10
The Nature of G_0 in Yeast

D. P. BEDARD, R. A. SINGER, AND G. C. JOHNSTON

I. Introduction: The Question of the G_0 State

Nonproliferating cells are generally considered to be in a unique physiological state, when compared to actively dividing cells. This unique state has been designated the G_0 state. Cells arrested in the cell cycle after mitosis but before the initiation of DNA synthesis, in the G_0 state, exhibit many biochemical, structural, and physiological properties different from those of actively dividing cells progressing through G_1 (*1,12*). A vast literature exists that details these differences in animal cells; an equally compelling, although more limited, body of work describes these differences in other eukaryotes including yeast.

In addition to this accepted view of G_0 as an altered physiological state, G_0 may represent a unique developmental state, qualitatively different from the G_1 of actively dividing cells (Fig. 1). In this view, the attainment of, and exit from, the G_0 state is dependent on *unique* gene functions not required during active growth and cell division; a particularly thoughtful treatment of this possibility has been presented by

GENETIC EXPRESSION IN THE CELL CYCLE

ISBN 0-12-543720-X

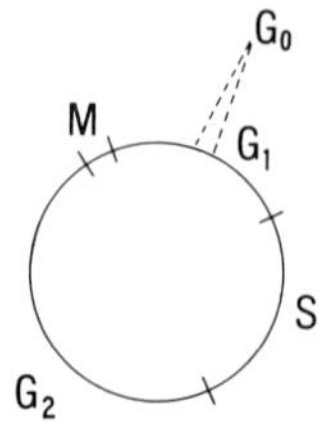

Fig. 1. Diagram of the cell cycle showing a developmentally unique G_0 state.

Baserga (*2*). It is this aspect of the G_0 concept that we have addressed using the yeast *Saccharomyces cerevisiae*.

Before describing our work we will first describe, with a few specific examples, the types of approaches used to investigate the nature of G_0 in both animal cells and yeast.

A. Studies with Animal Cells

1. Kinetic Analyses

One type of approach in the study of G_0 has relied on kinetic analyses of populations of quiescent animal cells after stimulation to proliferate. Resting cells that have been stimulated to divide usually have a prereplicative period, between growth stimulation and the initiation of DNA synthesis, that is longer than the normal G_1 period of actively dividing cells (*27*). In many cases, the resting phase to S phase prereplicative interval is as long as the generation time. This type of observation served as one basis for suggestions that quiescent cells are in a special state; upon stimulation to enter S phase, they may complete unique events in addition to those also completed by actively cycling cells. However, the nature of the differences between nonproliferating and proliferating cells may be only quantitative. A nonproliferating state is generally produced by starvation for some required nutrient or growth factor, or by manipulation which renders cells functionally deprived (*46*). Therefore, the extended prereplicative phase of quiescent cells may simply reflect the time required to resynthesize those components that are depleted during the period of extensive turnover that usually accompanies starvation.

2. Biochemical and Physiological Studies

A second line of investigation bearing on the nature of G_0 has been the examination of biochemical and physiological differences between quiescent cells and actively dividing cells (*1*). Quantitative differences between resting and dividing cells have been well documented. For example, Becker *et al.* (*4*) reported that resting cells have only 70% the ribosome content of actively dividing cells; similarly, Darzynkiewicz *et al.* (*10*, and Chapter 5) demonstrated that resting cells have reduced contents of total RNA compared to actively dividing cells. These types of comparisons between quiescent and dividing cells do indeed show that quiescent cells are physiologically different from actively dividing cells. Unfortunately, because they do not identify unique cell cycle events, these studies cannot clearly determine whether quiescent cells are in a unique developmental state that is dependent upon, and not simply coincident with, other metabolic changes (this is the latter view of G_0 that we initially presented).

3. *Search for Unique Gene Products*

A third and quite different approach has involved attempts to identify gene products that are associated uniquely with either resting cells or proliferating cells. There are gene products unique to nonproliferating cells. Certain soluble proteins (*3*), DNA-binding proteins (*39*), and phosphoproteins (*25*) are found only in extracts of quiescent cells. On this basis, it has been suggested that resting cells are qualitatively different from proliferating cells. Still, there remains the fundamental question of whether these unique gene products are required for entry into or exit from the G_0 state or whether, instead, the appearance of these gene products is only coincident with, but not functionally related to, cell cycle arrest.

4. *Prospects*

Although there is ample evidence that quiescent cells differ from proliferating cells, it remains to be demonstrated that quiescent cells *must* complete unique cell cycle events normally bypassed by actively dividing cells. Indeed, there is some evidence that quiescent cells are simply blocked at a step normally completed by proliferating cells traversing G_1. Rubin and Steiner (*38*) have reported that for cultures of resting cells incubated in the presence of tritiated thymidine, up to 96% of the cells incorporate radioactivity after 120 hr of incubation. These results and those of others indicate that few cells, if any, are truly "withdrawn" from the cycle. Rubin and Steiner (*38*) suggest that cells in resting cultures go through the same sequence of events preceding DNA synthesis as do actively dividing cells, except that the former complete these events in a more protracted fashion. In this view, there are no qualitatively different cell cycle gene functions distinguishing resting from dividing cells.

One resolution of this issue would be through the isolation of mutants conditionally defective only for resumption of growth from the quiescent state, but unaffected for active growth and division (*2*). This kind of genetic evidence would establish the existence of genes whose functions are required uniquely for exit from the G_0 state, but which are not required for execution of the cell cycle. Thus far, no mutant of this type has been reported, even in an organism such as the yeast *S. cerevisiae,* whose metabolism and cell cycle have been well studied genetically.

B. Studies with Yeast

The budding yeast *Saccharomyces cerevisiae* has been a good eukaryotic model system for analysis of the cell cycle. This is in large part due to ease of genetic and physiological manipulation (*15,16*). More importantly for this study, the yeast cell cycle is so much like that of animal cells, both in execution and regulation, that several workers have described cellular responses interpreted to indicate an off-cycle resting state (G_0). For example, upon starvation, cells of *S. cerevisiae* enter a resting state in which the vast majority of cells (>90%) are in the G_1 (or G_0) portion of the cell cycle (cells in G_1 are devoid of a bud, and thus the proportion of cells in G_1 can be readily estimated by direct microscopic examination) (*15*). After the required nutrient is replenished, the prereplicative lag period is often as long as an entire cell cycle. This situation is analogous to that found on stimulation of quiescent mammalian cells.

Physiological and structural differences between quiescent cells (stationary phase cells in yeast literature) and dividing cells, similar to that discussed above for animal cells, have also been reported for *S. cerevisiae*. In Sections I,B,1 and 2, we briefly summarize these findings.

1. Physiological Differences

One of the differences between resting and dividing cells is in sensitivity to high temperatures (*31,40,48*). Yeast cells in exponentially growing populations rapidly lose viability when incubated at 57°C (*48*); resting cells are relatively resistant to this treatment. This suggests some underlying physiological difference between dividing and resting cells. A second significant difference between resting and dividing cells is sensitivity to glusulase, a crude extract of snail gut that digests the yeast cell wall. Glusulase disrupts and kills growing cells but is much less effective on resting cells (*11,41*). The cell wall of growing cells may be more sensitive because of active cell wall metabolism. Alternatively, the differential sensitivity may reflect structural differences in the walls of growing and dividing cells.

2. *Structural Differences*

Structural differences in the spindle plaque, nuclear membrane, and chromatin have been reported for resting and dividing cells.

a. Spindle Plaque. The spindle pole body of *S. cerevisiae,* called the spindle plaque, is a structure functionally analogous to the centriole in higher eukaryotes (*7*). Electron microscopic examination has revealed structural differences between the spindle plaques of resting and growing cells (*8*). Growing cells blocked at the cell cycle regulatory step in G_1 (called "start," see Section I,B,3) have dense amorphous material, referred to as satellite, associated with the spindle plaque; in contrast, resting cells have only a simple spindle plaque. Absence of the spindle plaque satellite in resting cells suggests that there may be steps required of resting cells prior to the initiation of the cell cycle. However, the presence of simple spindle plaques in cells that have just undergone cell division, and are therefore in early G_1 (*8*), indicates that if the additional step of satellite production does exist, it is not unique to cells leaving the resting phase, since actively dividing cells also traverse a period in the cell cycle in which a simple spindle plaque can be seen.

b. Nuclear Pores. Willison and Johnston (*50*) conducted a systematic study of nuclear pore sizes for yeast under different growth conditions. They noted that, for cells freeze-fractured without the use of cryoprotective agents, nuclear pores were on average larger in resting-phase cells than in dividing cells. The significance of this difference is unknown.

c. Chromatin Structure. Pinon (*32*) reported differences in sedimentation velocity of chromatin from resting-phase cells and dividing cells of *S. cerevisiae*. Chromatin isolated from nitrogen-starved cells behaved differently on sucrose gradients than chromatin isolated from cells growing in nitrogen-replete medium. Two types of chromatin, distinguished by sedimentation velocity centrifugation, were recovered from growing cells, one consisting of G_1 phase chromatin, the other consisting of S and G_2 phase chromatin. Chromatin isolated from nitrogen-starved cells was resolved as a third type, probably due to changes in its tertiary structure. Differences in the complement of chromosomal non-histone proteins were found between resting and dividing cells; this different complement of proteins may be responsible for the alteration in the tertiary structure of resting-phase chromatin. This third type of chromatin was not found in G_1 cells in an actively dividing population. However, this situation may be like that described above for spindle plaque structure; it may be difficult to exclude the possibility that the resting-phase structure normally exists in the G_1 period (*32*), if only for a brief interval.

Direct evidence that the resting-phase chromatin configuration is in fact found during the cell cycle was provided by analysis of conditional cell division cycle (*cdc*) mutant strains incubated at the nonpermissive temperature. These studies showed that the mutation *cdc*28, which causes G_1 arrest at the nonpermissive temperature, prevented formation of the resting-phase chromatin structure, whereas the mutation *cdc*4 prevented formation of the G_1 chromatin configuration (*33*). An important observation was that when cells bearing the *cdc*28 mutation entered a resting state at the permissive temperature and were subsequently shifted to the nonpermissive temperature, the "G_0 folded genome" structure could no longer be found. Therefore, because the resting-state chromatin configuration is dependent on factors, such as the *cdc*28 gene product, that regulate the progression of dividing cells through G_1, it is improbable that the resting-phase structure represents an off-cycle state.

3. Prospects

The first known event after mitosis in the normal cell cycle of *S. cerevisiae* has been termed "start" (*15*), the completion of which is dependent on the products of several genes and sensitive to mating pheromones. Johnston *et al.* (*22*) established that stationary-phase cells are functionally at or before start. When placed in fresh medium under conditions in which start is specifically blocked, stationary-phase cells do not complete any other cell cycle events. However, it has not been possible to identify experimentally any additional steps that resting cells must complete prior to reaching start, or to determine if events unique to the resumption of growth from stationary phase must be completed.

Genetic studies hold the most promise in the attempt to identify events uniquely associated with resumption of growth from stationary phase. A conditional mutant unable to reinitiate division from stationary phase under nonpermissive conditions, but unaffected in cell division if first allowed to initiate a cell cycle under permissive conditions, would be strong evidence for an off-cycle resting state. No such mutant has yet been reported. We therefore undertook a systematic search for *ts* mutations in *S. cerevisiae* that exerted a cell cycle effect in resting cells stimulated to divide, but not in actively dividing cells. This approach was based on the premise that if there were unique processes associated with resumption of growth from the resting phase, they should be identifiable by mutation.

Although other extensive efforts for the isolation of *cdc* mutations have been carried out (*17,35*), the design of those experiments precluded the possibility that cells carrying putative G_0 mutations would be among those mutants recovered. This aspect of the mutant isolation procedure is crucial here, since our aim is to produce evidence bearing on the question of the existence of a genetically defined G_0 state.

In the next section, we discuss the rationale and results of the mutant isolation procedures employed to isolate putative G_0 mutants in *S. cerevisiae*.

II. Mutant Isolation Procedures

A. Rationales

A major goal of this work was to isolate mutants specifically defective for resumption of growth from stationary phase, and to provide genetic evidence bearing both on the existence of a developmentally unique G_0 state and on the nature of G_1 regulation in yeast. A mutation specific to G_0 can be distinguished from a G_1 mutation, because a G_1 mutation would be expressed both during resumption of growth from stationary phase and also during active division. In contrast, a putative G_0 mutation would be expressed only during growth from stationary phase.

Mutations in G_0-specific functions might produce several different phenotypes. Therefore, to increase the possibility that a G_0 mutant would be isolated, we developed several complementary methods for isolating G_0 mutants that, taken together, anticipated a spectrum of phenotypes. In all cases, conditional *ts* mutants were sought, since we anticipated that these mutations may affect processes essential for survival.

Although each of the four methods that were developed was based on a different assumption, each had a common format. That is, in each procedure, resting-phase cells (cells in G_0) were exposed to the nonpermissive temperature under conditions favoring growth. This format allows isolation of mutations either in growth from stationary phase (a G_0 mutant) or in completion of G_1 (a G_1 mutant). One of the procedures, the nonselective screen, was designed to isolate G_0 mutants only. Each selection procedure will be discussed with respect to general rationale, protocol, and yield to mutants.

1. *Inositol-less Death*

The first selection procedure was based on the assumption that a stationary-phase cell bearing a G_0 mutation would be unable to increase in mass when placed in fresh medium under nonpermissive conditions. A selection method was developed that employed a yeast strain (MC-

6A) auxotrophic for inositol (*19*). When supplied with all the requirements for growth except inositol, this strain rapidly loses viability. If increase in cell mass is inhibited, such as by the addition of inhibitors of macromolecular synthesis, then viability is preserved. Thus, if a G_0 mutant were unable to increase in cell mass when suspended in fresh medium at the nonpermissive temperature, it should also be protected from inositol-less death. In this protocol, stationary-phase cells were placed at the nonpermissive temperature in fresh medium without inositol, and, after a suitable period of incubation, plated for survival. The survivors of inositol-less death were tested for temperature sensitivity and for cell cycle defects.

2. *Mating Recovery*

Two of the selection procedures utilized some of the *ts cdc* mutations isolated and characterized by Hartwell *et al.* (*17*). The relevant mutations for this work are *cdc*28, *cdc*4, and *cdc*7. These genes mediate steps in a dependent sequence leading to DNA synthesis, and their order of function in the cell cycle is shown schematically in Fig. 2 (*20*). Under nonpermissive conditions, the *cdc*28 mutation prevents execution of the α-factor-sensitive step, thus blocking cells at start (*20*). The *cdc*28 gene product is required for spindle pole body (SPB) duplication (*7*), the *cdc*4 gene product is required for SPB separation (*7*), and the *cdc*7 gene product is required for DNA synthesis (*20*). The *cdc*7 mutation blocks cells at a point in the cell cycle at which protein synthesis is no longer required for DNA synthesis (*20*). Strains bearing these mutations were used for the following mating procedure and for the *cdc*4 survival procedure.

Reid and Hartwell (*36*) showed that haploid cells blocked at start (for example by the *cdc*28 mutation, Fig. 2) retain the ability to conjugate, whereas cells blocked at other points in the cell cycle are unable to mate. However, the boundaries of the period during which mating is possible have not been precisely defined with respect to all cell cycle events. Therefore, it is possible (see *42*) that cells blocked after mitosis but prior to start have the ability to conjugate. The following procedure was based on the assumption that a G_0 mutant blocked prior to start is at a point in the cycle from which it can mate. For this procedure, we employed a strain bearing the *ts* mutation *cdc*7, which blocks cells prior to the initiation of DNA synthesis, at a step in the cell cycle at which conju-

$$\xrightarrow{cdc28} \quad \xrightarrow{cdc4} \quad \xrightarrow{cdc7} \quad \text{Initiation of DNA synthesis}$$

Fig. 2. Order of execution of gene-mediated steps during the G_1 period of yeast.

gation does not occur (*36*). A new mutation that blocked cells earlier in the cycle, but at a point where mating could still occur, would then suppress the loss of mating ability resulting from the *cdc*7 mutation. For this protocol, stationary-phase haploid cells bearing a *cdc*7 mutation were suspended in fresh medium at the nonpermissive temperature; after an appropriate period of incubation, cells that could still conjugate were selected (by selection of diploids) and examined for new *ts* mutations.

3. cdc4 Survival

Another characteristic which distinguishes *cdc* mutants blocked at start from *cdc* mutants blocked at other cell cycle positions is the rapid loss of viability of cells blocked at steps other than start. The *cdc*4 survival procedure was based on the assumption that new mutations that block cells at or prior to start would protect such cells from the death that occurs for mutants blocked at the *cdc*4 step. This method involved resuspending stationary phase cells bearing the *cdc*4 mutation in fresh medium at the nonpermissive temperature and incubating for a period sufficient to allow 99.9% of the cells to lose viability at the *cdc*4 block. After two such rounds of killing, survivors were examined for the presence of new cell cycle mutations.

4. Nonselective Screen

A fourth enrichment procedure was designed using the fewest assumptions about the phenotype of a G_0 mutant. The only major assumption underlying this method concerned the unique aspects of the G_0 state. That is, a *ts* G_0 mutant should not cause temperature sensitivity during exponential growth, but it should cause temperature sensitivity for the resumption of growth from stationary phase. In this procedure, we tested clones that were able to grow exponentially at the nonpermissive temperature for the ability to divide when stationary-phase cells were resuspended in fresh medium at the nonpermissive temperature. This procedure should allow the isolation of G_0 mutants only.

B. Protocols

1. Inositol-less Death

Stationary-phase cultures of the inositol-auxotrophic strain MC-6A (*19*) were preincubated at 37°C and then suspended at 37°C for 30 hr in fresh medium lacking inositol. After 30 hr, the concentration of viable cells was decreased by four orders of magnitude. (The concentration of viable cells in inositol-less, nitrogen-free medium, in which the cells cannot grow, was essentially unchanged by this incubation.) Five hundred independent cultures, each mutagenized with ethylmethane sulfonate (EMS) (*13*), were processed in this way, and from each culture 100–200 survivors were recovered on solid medium. One *ts* clone was picked from each original culture. The cell cycle behavior of each of these 500 *ts* clones was then examined after a shift to the nonpermissive temperature of 37°C. This procedure should isolate both G_0 and G_1 mutants.

A modification of this procedure was designed to isolate G_0 mutants only. After the 30 hr incubation in inositol-less medium, cells were placed on solid medium at the permissive temperature (23°C), incubated for 15 hr to allow cell cycle initiation, and then shifted to the nonpermissive temperature. Those colonies which grew at 37°C were tested for temperature sensitivity during growth from a stationary-phase condition produced by nitrogen starvation. Since these colonies had grown on solid medium at 37°C, they were not *ts* for exponential growth; thus, if they appeared to be temperature sensitive for growth from resting phase, they were considered to be candidate G_0 mutants.

2. *Mating Protocol*

Two-hundred-sixty independent cultures of strain 18032 (*17*), which carries the *cdc*7 mutation, were mutagenized with EMS and grown up to stationary phase. Each culture was then prewarmed at the nonpermissive temperature (37°C) and resuspended in fresh medium at 37°C. After a 12-hr incubation in fresh medium at the nonpermissive temperature, the efficiency of mating of a test culture was decreased approximately three orders of magnitude. Therefore, in the selection protocol, 12 hr was allowed for most cells to "escape" the mating period and proceed to the *cdc*7 block.

The mating procedure was essentially as described previously (*36*). It involved mixing cells of strain 18032 with an equal number of cells of the opposite mating type, collecting them on 0.45-μm pore size nitrocellulose filters, and placing the filters "cell side up" on solid medium. These mating mixtures were incubated at 35°C for 3 hr; cells were then plated on media selective for diploids on the basis of auxotrophic requirements. From each independent mating mixture, one diploid clone was purified and sporulated; haploid segregants were tested by complementation with strains bearing the *cdc*7 mutation. This procedure allowed the detection of clones harboring new *ts* mutations.

3. *cdc4 Survival*

The *cdc*4 survival-enrichment method was based on the premise that cell cycle mutations that blocked cells at the start event would prevent the loss of viability that occurs at the *cdc*4 block. A strain carrying both mutations *cdc*4 and *cdc*7 was used to reduce the frequency of reversion of the *ts* phenotype. For this enrichment procedure, only one mutagenized culture was used; thus, independent mutational events were not ensured. Loss of viability was induced by resuspension of prewarmed stationary-phase cells in fresh medium at the nonpermissive temperature. After 24 hr incubation, the viability of the culture was three orders of magnitude less than at the time of resuspension in fresh medium. At this time, the cells were resuspended in fresh medium at the 23°C permissive temperature and grown to stationary phase. This killing cycle was repeated once more. The next steps were designed to identify clones that remained viable at the nonpermissive temperature. The 10,000 colonies that grew up on solid medium after the second round of cell killing were transferred sequentially to solid medium at the nonpermissive temperature for 3 days and then to solid medium at the permissive temperature. Those colonies that exhibited rapid growth upon return to the permissive temperature were suspected of harboring new mutations that protected cells from the *cdc*4-mediated death. Therefore, they were mated to a temperature-insensitive strain, and the diploids were sporulated. Haploid segregants were tested by complementation with strains bearing the *cdc*4 and *cdc*7 mutations to determine whether they harbored new *ts* mutations. Segregants that did contain new *ts* mutations were examined for cell cycle defects.

4. *Nonselective Screen*

In this procedure, we simply screened a large number of colonies for the G_0 phenotype. All colonies examined were from the same mutagenized culture of strain GR2 (*23*). Cells of this strain were plated on solid medium to produce more than 2×10^5 isolated colonies, and after an initial 15-hr incubation at the permissive temperature (23°C) to allow cells to "escape" stationary phase, or G_0, the plates were transferred to 37°C. Colonies that grew up at the nonpermissive temperature of 37°C were then tested for temperature sensitivity after stimulation of growth from stationary phase. (In this case, stationary phase was brought on by starvation for uracil.)

III. Mutant Characterization

A. G_1 Mutants

1. *Genetic Analysis*

The mutant-enrichment schemes described above yielded eight cell cycle mutants that define six complementation groups (Table I), which

TABLE I

Strains Harboring New Cell Cycle Mutations Listed by Isolation Procedure[a]

Inositol-less death	Mating recovery	*cdc*4 Survival
ID-1 (*cdc*60)	M7-7A	S7 (*cdc*62)
ID-2 (*cdc*61)		S24 (*cdc*62)
		S33 (*cdc*62)
		S44 (*cdc*63)
		S104 (*cdc*64)

[a] The complementation groups are in parentheses. Isolation procedures are described in the text.

will be described in more detail elsewhere. No G_0 mutants were recovered using these procedures.

All the mutations were recessive, and the *ts* phenotypes segregated as single genes, with the exception of the *ts* phenotype in strain M7-7A (Table II). The meiotic segregation pattern of the *ts* phenotype in this strain suggested the involvement of more than one gene. No linkage was detected between the various new complementation groups.

2. *Cell Cycle Analysis*

The cell cycle mutants were considered to be arrested in the G_1 phase, since most (>90%) of the cells incubated at the nonpermissive temperature were unbudded, contained a single nucleus (as revealed by Giemsa stain), and had a pre-S phase content of DNA (data not shown). Each mutation was tested by order-of-function mapping using α-factor. Whether cells were first blocked with α-factor and then resuspended in fresh medium at the nonpermissive temperature, or first incubated at the non-

TABLE II

Meiotic Segregation of Temperature-Sensitive Mutations in Crosses between Temperature-Sensitive and -Insensitive Strains

	Segregation of mutant × non-mutant alleles		
Complementation group	2:2	1:3	0:4
*cdc*60	47	0	0
*cdc*61	15	0	0
*cdc*62	34	0	0
*cdc*63	37	0	0
*cdc*64	29	0	0
ts Phenotype in strain M7-7A	3	4	9

permissive temperature and then treated with α-factor at the permissive temperature, cell division did not occur during the second incubation (data not shown). These results, taken together, indicate that each of the *ts* defects blocks cells at the α-factor-sensitive step, i.e., start.

3. Macromolecular Metabolism

Cell cycle arrest may be due either to primary defects in cell division events, or to alterations in general metabolism that indirectly affect the cell division process (*18,35*). In light of this, the mutants were also characterized with respect to the effects of temperature on macromolecular synthesis, as estimated by pulse-labeling experiments. The mutants isolated fell into two groups. For mutants in one group, a transfer to the nonpermissive temperature resulted in an immediate and drastic decrease in the rate of protein labeling (Fig. 3), which probably reflects a decrease in the rate of protein synthesis. The completion of start has been shown to be particularly sensitive, relative to other cell cycle events, to changes in the rates of protein synthesis (*18*). Thus, for this class of mutants, the observed G_1 arrest at the nonpermissive temperature probably is a consequence of alterations in general growth processes that

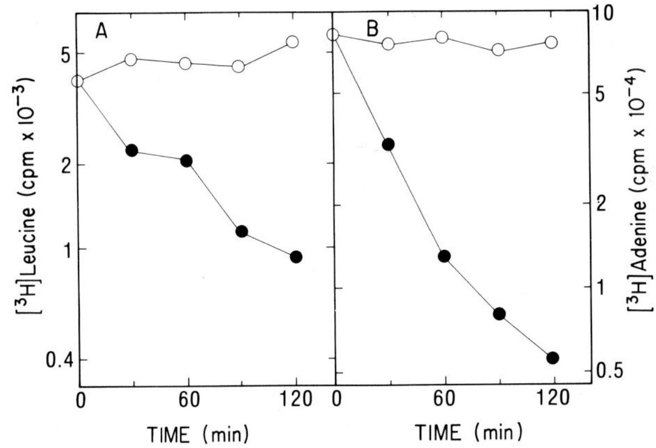

Fig. 3. Effect of temperature on adenine and leucine incorporation into cells bearing the *cdc*61 mutation. A culture was grown at 23°C for several generations in minimal medium (*22*) supplemented with adenine and leucine. At time zero, the culture was divided into two portions; one half was incubated at 36°C, and the other half was incubated at 23°C. At intervals, 1.0-ml portions of each culture were removed and incubated for 5-min periods in the presence of either [^{3}H]adenine or [^{14}C]leucine. Incorporation was terminated by the addition of cold 10% trichloracetic acid (TCA) containing an excess of unlabeled adenine and leucine. Material precipitable by TCA was collected on filter disks, and the radioactivity of the filter disks was determined after washing (*24*). (A) [^{14}C]leucine, 23°C (○); [^{14}C]leucine, 36°C (●); (B) [^{3}H]adenine, 23°C (○); [^{3}H]adenine, 36°C (●).

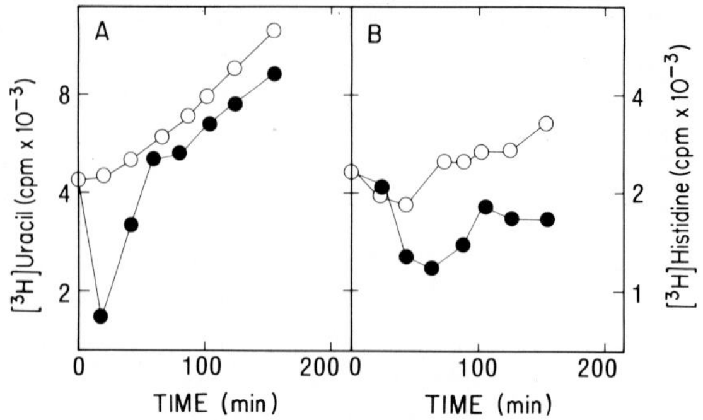

Fig. 4. Effect of temperature on histidine and uracil incorporation into cells of strain M7-7A. A culture was grown at 23°C for several generations in YM-1 (*14*). At zero time, the culture was split into two portions; one portion was incubated at 23°C, and the other portion was incubated at 36°C. Each portion was then treated as described in Fig. 3. (A) [^{3}H]uracil, 23°C (○); [^{3}H]uracil,36°C (●); (B) [^{3}H]histidine, 23°C (○); [^{3}H]histidine, 36°C (●).

indirectly (via normal control mechanisms) (*18*) affect the rate of cell cycle initiation.

For mutants in the second category, a transfer to the nonpermissive temperature resulted in an approximately 50% decrease in the rates of RNA and protein labeling (Fig. 4). The actual magnitudes of the decrease in the rates of macromolecular synthesis are probably small, since precursor pool-specific activities were also reduced by about 40% (data not shown). In this case, it appears that the *ts* defect is in a process more specifically involved in cell cycle regulation, because general metabolism is not affected to any great extent.

B. Mutants Isolated in Nonselective Screen

No G_1 or G_0 mutants were isolated from the nonselective screening procedure. On solid medium, some strains initially appeared to be G_0 mutants, because exponentially growing cells underwent several cell divisions during a period of time in which growth of resting cells after stimulation was negligible. Further testing showed all of these to be division-rate mutants. Nevertheless, the phenotypes of these mutants illuminate the underlying nature of the stationary phase. These *ts* mutants also fell into two classes, based on the kinetics of cell number increase at the nonpermissive temperature.

1. Cell Division Kinetics

Strain NS-R2 is a representative of one class of mutant. Exponentially growing cultures of this strain, when shifted to the nonpermissive temperature, initially divided at a higher rate than did the 23°C culture. After a tenfold increase in cell number, the rate of cell division fell to a very low value (Fig. 5). This decrease in the rate of cell division was independent of cell concentration, since the rate of division fell to the same level at cell concentrations ranging from 2×10^5/ml to 8×10^6/ml. When stationary-phase cells of this strain were preincubated at 37°C and then diluted into fresh medium at 37°C, the rate of cell number increase was similar to the low rate induced by shifting an exponential culture to 37°C (Fig. 6).

Strain NS-C9 is representative of the other class of mutants isolated by this method. An exponentially growing culture of this strain, when shifted to the nonpermissive temperature, immediately began to divide at a new, slower rate compared to the 23°C culture (Fig. 7). When a stationary-phase culture of this strain was preincubated at 37°C and then diluted into fresh medium at 37°C, the lag period before the onset of cell

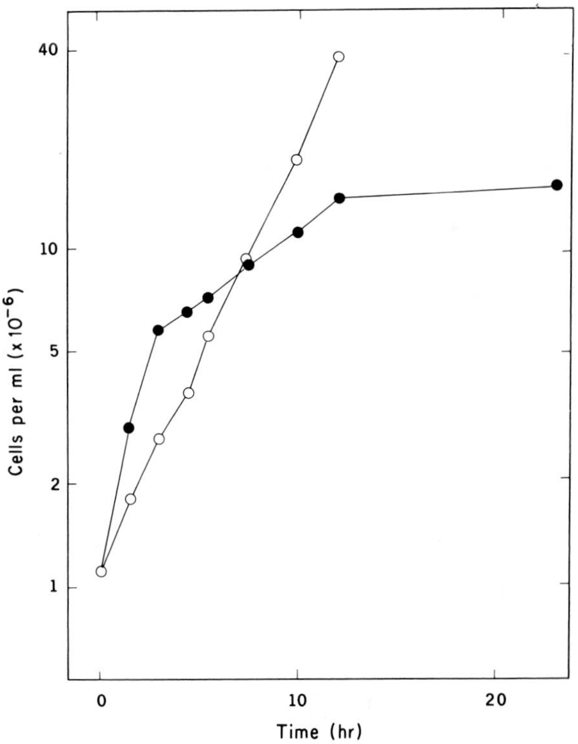

Fig. 5. Kinetics of cell division of strain NS-R2. A culture of exponentially growing cells (23°C) was divided into two portions. One portion (●) was incubated at 37°C; the control (○) was left at 23°C.

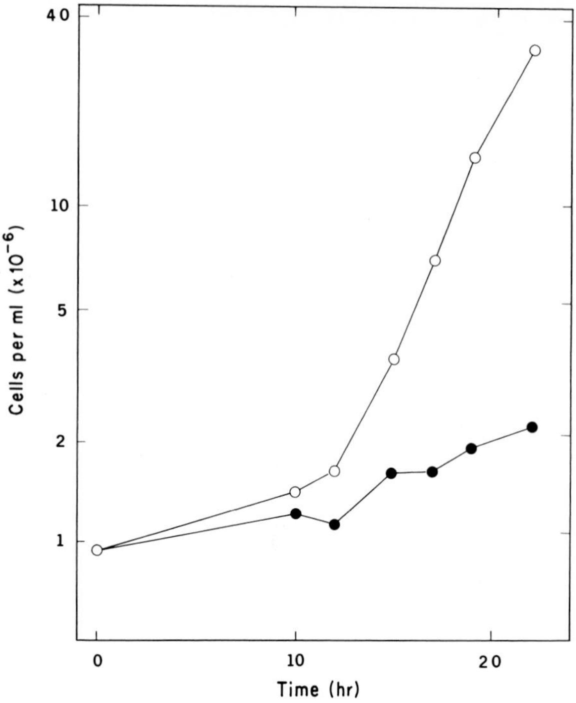

Fig. 6. Kinetics of cell division of strain NS-R2. A stationary-phase culture was pre-warmed to 37°C and then diluted into fresh medium at 23°C (○) or 37°C (●).

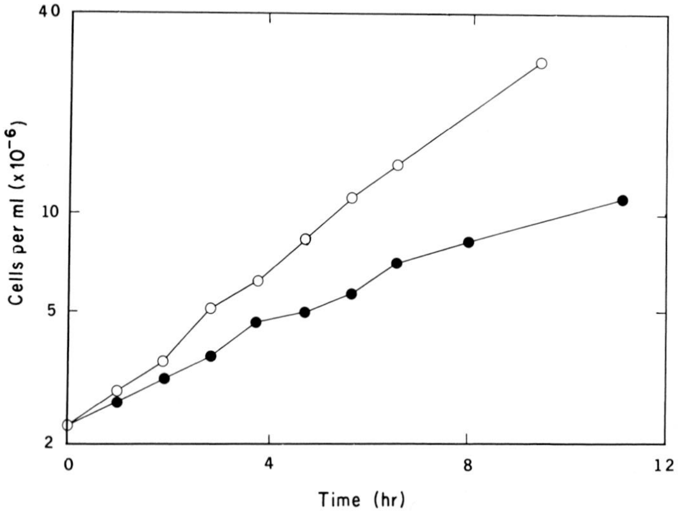

Fig. 7. Kinetics of cell division of strain NS-C9. A culture of exponentially growing cells (23°C) was divided into two portions. One portion (●) was incubated at 37°C; the control (○) was left at 23°C.

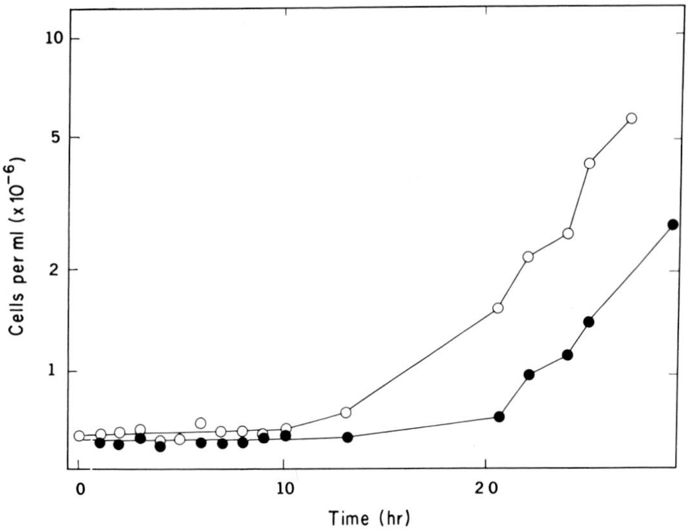

Fig. 8. Kinetics of cell division of strain NS-C9. A stationary-phase culture was prewarmed to 37°C and then diluted into fresh medium at 23°C (○) or 37°C (●).

division was much longer than the lag period exhibited by stationary-phase cells diluted into fresh medium at 23°C (Fig. 8).

2. *Time-Lapse Microscopy*

For mutants like strain NS-C9, most cells in the population were unable to divide at the nonpermissive temperature, resulting in the lack of cell number increase. Conceivably, the eventual increase in cell number may then have been due to a minority of cells not initially blocked in the resting state, which could continue to grow and divide at the nonpermissive temperature. In this situation, the lag period would represent the time required for the subpopulation of cells not blocked in the resting state to overgrow the arrested population.

To assess this possibility, a stationary-phase culture of strain NS-C9 was incubated at 37°C and subsequently plated for single cells on solid medium at 23°C and 37°C. At each temperature, 30 unbudded cells and 10 budded cells were then monitored. This analysis showed that the proportions of both unbudded cells and budded cells from the stationary-phase culture that were able to divide were the same at 23°C and at 37°C. Moreover, at each temperature, buds emerged on the majority of unbudded cells at approximately the same time. These results, then, exclude the possibility that a subpopulation of cells able to divide at the nonpermissive temperature overgrew the original culture.

Thus, for this strain, resting cells transferred to fresh medium do have a much longer lag prior to the resumption of cell division when incubated at 37°C.

IV. Start as the Sole Regulatory Point

Proliferating cells enter a resting state when essential nutrients are depleted (*21*). Cells are thought to enter the resting state from a point in G_1 (*29*), because such resting cells have a G_1 content of DNA and upon stimulation must first replicate DNA before undergoing mitosis (*28*). The term G_0 is used to refer to this resting state.

There are two major reasons resting cells are considered to be in a unique state. First, the prereplicative lag phase (the time required for a resting cell to reach S phase after stimulation) is often longer than the average G_1 period for the particular cell line during active proliferation (*1,30*). Second, a large number of physiological and cytological differences have been described between resting and dividing cells.

We have attempted to provide additional evidence bearing on the nature of the resting phase. Specifically, using the yeast *S. cerevisiae,* we sought to find mutations that were expressed in resting cells stimulated to divide but not in actively dividing cells. This approach was based on the premise that, if there were unique processes specifically required for resumption of growth from the resting phase, they should be identifiable by mutation.

Enrichment procedures were designed to isolate *ts* mutations that blocked the progress of resting-phase cells toward the initiation of DNA synthesis. Potential mutations could fall into one of two categories. One type of mutation would also block proliferating cells in G_1. These mutations would be in genes mediating events common both to the G_1 period of proliferating cells and to the initial growth period of resting cells stimulated to divide. The second type of mutation (the G_0 mutation) would block only *resumption of growth* from stationary phase; *proliferating* cells bearing a G_0 mutation would continue growth and division after being shifted to the nonpermissive temperature.

In attempting to isolate both types of mutations, we sought evidence bearing on two aspects of the genetic control of cell division. One aspect is the complexity of genetic regulation in G_1; the other aspect is the issue of specialized gene functions required only for growth from stationary phase.

Mutations that cause G_1 arrest were recovered using the methods designed to yield both G_0 and G_1 mutations. The phenotypes of these mutations have implications for the genetic complexity of regulation in G_1. As described here and in other work, all conditions that cause G_1 arrest in cells of *S. cerevisiae* also cause cells to arrest at start (see also Chapter 7). These conditions include the presence of mating pheromones (*20,49*), inhibitors of synthetic processes (*18,23*), amino acid analogues (*5,43*), and certain *ts* mutations (*20,35*). That these diverse arresting conditions all block cells at start was determined in each case by order-of-function mapping. Thus, the accumulation of evidence obtained from order-of-function mapping experiments, in conjunction with other physiological data (*22*), strongly suggests that start is the sole regulatory event in the yeast cell cycle.

In addition to being the only apparent regulatory event, start also appears to be the first event in the new cell cycle (after cell separation). One reason for concluding this is that resting phase cells are arrested at or before start (*15*). Another reason is that deprivation of nutrients permits completion of all known cell cycle events except start (*24*). Thus, the completion of start represents a commitment to the initiation of DNA synthesis (and perhaps to the cell cycle as a whole). Although the characteristic of a commitment event does not necessarily establish start as the first event in the cell cycle, no events that occur after cytokinesis and cell separation but prior to start have been identified.

More compelling evidence that start is actually the first event in the cell cycle is found in the results of experiments (*42*) that suggest that most of the G_1 period is devoid of G_1-specific events. By protracting S phase with low concentrations of hydroxyurea (HU) that did not affect the rate of cell number increase, we found that the execution of the start event, which normally occurs at 0.3 of the cycle with respect to cytokinesis, now occurred in these HU-treated cells at 0.0 of the cycle. We interpreted this result to mean that during the extra time taken for completion of the DNA-division cycle in the presence of HU, cells were able to increase in mass to the critical size required for completion of start (*22*). In other words, the G_1 period is simply part of a larger period of preparation for DNA synthesis. Hence, G_1 exists for the most part as a function of the relative rates of the DNA-division cycle and "growth." As the relative rate of progression through the DNA-division cycle decreases, more growth can occur before mitosis, and less time is required after mitosis to attain the critical cell size. If sufficient growth occurs before mitosis, then cells can complete start immediately after completion of cell division.

A similar situation appears to be true in the prereplicative phase of animal cells. In animal cells there appears to be a single point of commitment to DNA synthesis (*45*), called the restriction point (*29*). The existence of certain animal cell types in which the G_1 period is absent during active growth (*26,37*) suggests there need be only one regulatory point in animals as well. Despite the usual G_1-less phenotype of these cells, the restriction-point regulatory mechanism still exists as evidenced by its appearance when growth is limiting. Presumably, the G_1 period results when cells are unable to pass the restriction point immediately after cell division (*34*). Thus, in animal cells, as in yeast, it appears that the commitment to the initiation of DNA synthesis need be the sole regulatory event in the cell cycle.

V. Resting Phase Is Quantitatively Different

The results of the enrichment procedures presented here also have implications for the nature of the stationary phase for *S. cerevisiae*. The nonselective screen, designed to yield only G_0 mutants, resulted only in the isolation of mutations that affected the rate of cell division but not in the isolation of a G_0 mutation. The reason no G_0 mutants were isolated could be any one of several possibilities.

One possible reason no G_0 mutants were isolated is that a sufficient number of clones was not examined. The strain used for the nonselective screening procedure contained a mutation in the *ura*1 gene making this strain auxotrophic for uracil. Reversions to prototrophy were seen in 6 of the 210,000 clones examined. Thus, if a G_0 gene mutated at a rate comparable to the *ura*1 gene in this strain, the chances were good that it would have been detected.

Another possibility is that G_0 mutants may exhibit phenotypes that make their isolation unlikely by the enrichment procedures used here. For example, a particular gene product may be involved in the cell division process during stimulation of growth from the resting phase, but may also be involved in other cellular processes unrelated to the cell cycle. A mutation that affects both functions may be seen as a mutation only in the function unrelated to cell division; in other words, one defective function may be epistatic to the other.

A third possible reason no G_0 mutants were isolated is that G_0, as a unique developmental state, does not exist. Although our negative results are obviously not conclusive, the outcome of all four mutant isolation procedures is certainly consistent with this conclusion.

The nonselective screening procedure is especially important in this respect. No enrichment techniques with associated assumptions of cellular behavior were incorporated into this procedure. The primary assumption upon which this screening procedure was based was that cells bearing a *ts* G_0 mutation would not, by definition, be temperature sensitive during exponential growth. Although some mutants isolated by this method were affected in the rate of division, no mutants with a G_0 phenotype were found.

In light of these results, we favor the concept that resting cells are simply blocked in G_1. In this view, cells in the resting state are more severely limited than are proliferating cells with respect to requirements for initiation of a new cell cycle. All the data proposed as evidence for the existence of a G_0 state (see introduction, Section I) may be incorporated into models of cell division in which the resting state of nonproliferating cells is quantitatively, but not qualitatively, different from the G_1 phase of proliferating cells (*47*). Since the simplest explanation is often the best explanation, we are forced to conclude that resting cells of *S. cerevisiae* are blocked in G_1 at the start event; there is no need to postulate a unique resting state.

The view that the resting phase is only quantitatively different than the normal G_1 phase is compatible with the results presented here and with several models of regulation for animal cells. Cooper (*9*) suggested that there are no G_1-specific events prior to commitment to DNA replication, but rather that the G_1 period is the end of a larger period of preparation for DNA synthesis. The length of G_1 would be dependent on the proportion of preparation that still had to occur after mitosis was completed. In this way, resting cells may be viewed as being less well-prepared than proliferating cells to initiate DNA synthesis. In addition, transition probability models (*6,44*) propose that resting cells simply have a lower probability of entering S phase than do proliferating cells. The work presented in this chapter does not address the issue of whether the commitment event is probabilistic, but our results do agree with the notion that resting cells are not qualitatively different from actively dividing cells.

In summary, the results of the mutant isolation procedures reported here corroborate other physiological and genetic evidence bearing on the nature of the prereplicative phase in the cell cycle of *S. cerevisiae*. The isolation of only G_1 mutants, by procedures designed to isolate both G_1 mutants and G_0 mutants, suggests that G_0 mutants do not exist, although other explanations, as discussed above, are possible. The results here support the concept that start is the sole regulatory event in the yeast cell cycle and that several genes mediate its completion, both directly and indirectly. Finally, the resting phase is described as an extreme case of G_1 arrest, during which no unique cell cycle events occur.

Acknowledgments

This work was supported by grants from the National Cancer Institute of Canada and the Medical Research Council of Canada. The authors thank R. Roy, D. Carruthers, and K. Calhoun for technical assistance.

References

1. Baserga, R. (1976). "Multiplication and Division in Mammalian Cells." Dekker, New York.
2. Baserga, R. (1978). Resting cells and the G_1 phase of the cell cycle. *J. Cell. Physiol.* **95,** 377–386.
3. Becker, H., and Stanners, C. P. (1972). Control of macromolecular synthesis in proliferating and resting Syrian hamster cells in monolayer culture. *J. Cell. Physiol.* **80,** 51–62.
4. Becker, H., Stanners, C. P., and Kudlow, J. E. (1971). Control of macromolecular synthesis in proliferating and resting Syrian hamster cells in monolayer culture. *J. Cell. Physiol.* **77,** 43–50.
5. Bedard, D. P., Singer, R. A., and Johnston, G. C. (1980). Transient cell cycle arrest of *Saccharomyces cerevisiae* by the amino acid analog β-2-DL-thienylalanine. *J. Bacteriol.* **141,** 100–105.
6. Brooks, R. F., Bennent, D. C., and Smith, J. A. (1980). Mammalian cell cycles need two random transitions. *Cell* **19,** 493–504.
7. Byers, B., and Goetsch, L. (1973). Duplication of spindle plaques and integration of the yeast cell cycle. *Cold Spring Harbor Symp. Quant. Biol.* **38,** 123–131.
8. Byers, B., and Goetsch, L. (1975). Behavior of spindles and spindle palques in the cell cycle and conjugation of *Saccharomyces cerevisiae. J. Bacteriol.* **124,** 511–523.
9. Cooper S. (1979). A unifying model for the G_1 period in procaryotes and eucaryotes. *Nature* (*London*) **280,** 17–19.
10. Darzynkiewicz, Z., Traganos, F., Sharpless, T., and Melamed, M. R. (1976). Lymphocyte stimulation: A rapid multiparameter analysis. *Proc. Natl. Acad. Sci. U.S.A.* **73,** 2881–2884.
11. Deutch, C. E., and Parry, J. M. (1974). Sphaeroplast formation in yeast during the transition from exponential phase to stationary phase. *J. Gen. Microbiol.* **80,** 259–268.
12. Epifanova, O. I. (1977). Mechanisms underlying the differential sensitivity of proliferating and resting cells to external factors. *Int. Rev. Cytol., Suppl.* **5,** 303–332.
13. Fink, G. R. (1970). The biochemical genetics of yeast. "Methods in Enzymology" (H. Tabor and C. W. Tabor, eds.), Vol. 17, Part A, pp. 59–78. Academic Press, New York.
14. Hartwell, L. H. (1967). Macromolecule synthesis in temperature-sensitive mutants of yeast. *J. Bacteriol.* **93,** 1662–1670.
15. Hartwell, L. H. (1974). *Saccharomyces cerevisiae* cell cycle. *Bacteriol. Rev.* **38,** 164–198.
16. Hartwell, L. H., Culotti, J., Pringle, J. R., and Reid, B. J. (1974). Genetic control of the cell division cycle in yeast. *Science* **183,** 46–51.
17. Hartwell, L. H., Mortimer, R. K., Culotti, J., and Culotti, M. (1973). Genetic control of the cell division cycle in yeast. V. Genetic analysis of *cdc* mutants. *Genetics* **74,** 267–286.

18. Hartwell, L. H., and Unger, M. W. (1977). Unequal division in *Saccharomyces cerevisiae* and its implication for the control of cell division. *J. Cell Biol.* **75,** 422–435.
19. Henry, S. A., Donahue, T. F., and Culbertson, M. R. (1975). Selection of spontaneous mutants by inositol starvation in yeast. *Mol. Gen. Genet.* **143,** 5–11.
20. Hereford, L. M., and Hartwell, L. H. (1974). Sequential gene function in the initiation of *Saccharomyces cerevisiae* DNA synthesis. *J. Mol. Biol.* **84,** 445–461.
21. Holley, R. W. (1975). Control of growth of mammalian cells in culture. *Nature* (*London*) **258,** 487–490.
22. Johnston, G. C., Pringle, J. R., and Hartwell, L. H. (1977). Coordination of growth and cell division in the yeast *Saccharomyces cerevisiae*. *Exp. Cell Res.* **105,** 79–98.
23. Johnston, G. C., and Singer, R. A. (1978). RNA synthesis and control of cell division in the yeast *Saccharomyces cerevisiae*. *Cell* **14,** 1951–1958.
24. Johnston, G. C., Singer, R. A., and McFarlane, E. S. (1977). Growth and cell division during nitrogen starvation of the yeast *Saccharomyces cerevisae*. *J. Bacteriol.* **132,** 723–730.
25. Kletzien, R. F., Miller, M. R., and Pardee, A. B. (1977). Unique cytoplasmic phosphoproteins are associated with cell growth arrest. *Nature* (*London*) **270,** 57–59.
26. Liskay, R. M. (1978). Genetic analysis of a Chinese hamster cell line lacking a G_1 phase. *Exp. Cell Res.* **114,** 69–77.
27. Martin, R. G., and Stein, S. (1976). Resting state in normal and simian virus 40 transformed Chinese hamster lung cells. *Proc. Natl. Acad. Sci. U.S.A.* **73,** 1655–1659.
28. Nilhausen, K., and Green, H. (1965). Reversible arrest of growth in G_1 of an established fibroblast line (3T3). *Exp. Cell Res.* **40,** 166–168.
29. Pardee, A. B. (1974). A restriction point for control of normal animal cell proliferation. *Proc. Natl. Acad. Sci. U.S.A.* **71,** 1286–1290.
30. Pardee, A. B., and Dubrow, R. (1977). Control of cell proliferation. *Cancer* **39,** 2747–2754.
31. Parry, J. M., Davies, P. J., and Evans, W. E. (1976). The effects of "cell age" upon the lethal effects of physical and chemical mutagens in the yeast *Saccharomyces cerevisiae*. *Mol. Gen. Genet.* **146,** 27–35.
32. Pinon, R. (1978). Folded chromosomes in non-cycling yeast cells. *Chromosoma* **67,** 263–274.
33. Pinon, R. (1979). A probe into nuclear events during the cell cycle of *Saccharomyces cerevisiae*: Studies of folded chromosomes in *cdc* mutants which arrest in G_1. *Chromosoma* **70,** 337–352.
34. Rao, P. N., and Sunkara, P. S. (1980). Correlation between the high rate of protein synthesis during mitosis and the absence of the G_1 period in V79-8 cells. *Exp. Cell Res.* **125,** 507–511.
35. Reed, S. I. (1980). Selection of *Saccharomyces cerevisiae* mutants defective in the start event of cell division. *Genetics* **95,** 561–577.
36. Reid, B. J., and Hartwell, L. H. (1977). Regulation of mating in the cell cycle of *Saccharomyces cerevisiae*. *J. Cell Biol.* **75,** 355–365.
37. Robbins, E., and Scharff, M. D. (1967). The absence of a detectable G_1 phase in a cultured strain of Chinese hamster lung cell. *J. Cell Biol.* **34,** 684–685.
38. Rubin, H., and Steiner, R. (1975). Reversible alterations in the mitotic cycle of chick embryo cells in various states of growth regulation. *J. Cell. Physiol.* **85,** 261–270.
39. Salas, J., and Green, H. (1971). Proteins binding to DNA and their relation to growth in cultured mammalian cells. *Nature* (*London*) *New Biol.* **229,** 165–169.
40. Schenberg-Frascino, A., and Moustacchi, E. (1972). Lethal and mutagenic effects of

elevated temperature on haploid yeast. I. Variations in sensitivity during the cell cycle. *Mol. Gen. Genet.* **115,** 243–257.

41. Shahin, M. M. (1972). Relationship between yield of protoplasts and growth phase in *Saccharomyces. J. Bacteriol.* **110,** 769–771.
42. Singer, R. A., and Johnston, G. C. (1981). Nature of the G_1 phase of the yeast *Saccharomyces cerevisiae. Proc. Natl. Acad. Sci. U.S.A.* **78,** 3030–3033.
43. Singer, R. A., Johnston, G. C., and Bedard, D. P. (1978). Methionine analogs and cell division regulation in the yeast *Saccharomyces cerevisiae. Proc. Natl. Acad. Sci. U.S.A.* **75,** 6083–6087.
44. Smith, J. A., and Martin, L. (1973). Do cells cycle? *Proc. Natl. Acad. Sci. U.S.A.* **70,** 1263–1267.
45. Temin, H. M. (1971). Stimulation by serum of multiplication of stationary chicken cells. *J. Cell. Physiol.* **78,** 161–170.
46. Todaro, G. J., Lazar, G. K., and Green, H. (1965). The initiation of cell division in a contact-inhibited mammalian cell line. *J. Cell. Comp. Physiol.* **66,** 325–334.
47. Van Putten, L. M. (1974). G_0, a useful term? *Biomedicine* **20,** 5–8.
48. Walton, E. F., Carter, B. L. A., and Pringle, J. R. (1979). An enrichment method for temperature-sensitive and auxotrophic mutants of yeast. *Mol. Gen. Genet.* **171,** 111–114.
49. Wilkinson, L. E., and Pringle, J. R. (1974). Transient G_1 arrest of *Saccharomyces cerevisiae* cells of mating type α by a factor produced by cells of mating type *a*. *Exp. Cell Res.* **89,** 175–187.
50. Willison, J. H. M., and Johnston, G. C. (1978). Altered nuclear pore diameters in G_1-arrested cells of the yeast *Saccharomyces cerevisiae. J. Bacteriol.* **136,** 318–323.

11

The Effect of Morphogenetic Hormones on the Cell Cycle of Cultured *Drosophila* Cells

BRYN STEVENS AND JOHN D. O'CONNOR

I. Introduction

In the invertebrate phylum Arthropoda, the external manifestation of growth is the shedding of the articulated exoskeleton and its replacement with a larger one, allowing the animal to increase in size. This periodic shedding of the limiting exoskeleton is accomplished by ecdysis, or molting. Insect development proceeds through a series of stages (instars) separated by molts. Evidence for the hormonal control of ecdysis was first provided by Kopec (*45*) and has been reviewed extensively (*35,42,67*).

Insect development is controlled by the interaction of two classes of endocrine secretions: a steroid, 20-hydroxyecdysone (20-OH-ecdysone), and a sesquiterpenoid, juvenile hormone (JH) (*15*). The insect order Diptera, to which the fruit fly *Drosophila melanogaster* belongs, has a varying number of larval instars, which are followed by an intermediate larval/pupal molt and complete metamorphosis at the pupal/adult molt (*26*). Although ecdysis is initiated by 20-OH-ecdysone, the qualitative nature of the molt (i.e., larval/larval, larval/pupal, or pupal/adult) is de-

GENETIC EXPRESSION IN THE CELL CYCLE

ISBN 0-12-543720-X

termined by the circulating titer of JH (*26*). The initiation of adult metamorphosis is generally attributed to an increase in 20-OH-ecdysone titer and a decrease in JH titer (*67*).

Observations of the dramatic tissue reorganization which occurs at metamorphosis have stimulated numerous investigations focused on the endocrine control of molting. Much of this work has employed surgical techniques uniquely applicable to insect systems (*65*). Surgical ablation and implantation of various neural and endocrine tissues have elucidated many of the hormonal parameters of insect development. However, in attempting to define the cellular responses to developmental hormones, the use of intact animals with intricate endocrine systems presents a complicated picture. Because of the limitations of *in vivo* endocrine experiments, a number of hormonally responsive organ- and tissue-culture systems have been developed.

II. Ecdysteroid Responsive Tissues *in Vitro*

Drosophila melanogaster has been the organism of choice for a number of *in vitro* endocrine studies. In addition to the fact that it has an historically well-known genome, a number of explanted tissues respond specifically to hormonal stimuli *in vitro*. Cells in several *Drosophila* larval tissues undergo repeated rounds of DNA replication without subsequent chromosomal scission or cytokinesis (*6*). The many chromatids formed during such polytenization remain in register, forming the characteristic giant chromosome of *Drosophila*. Polytene chromosomes are easily visualized under the light microscope, allowing direct observation of gene activity. In some cases, transcriptionally active sites can be distinguished from inactive sites by the presence of a puff, a region of laterally extended chromatin (*47*). The polytene chromosome of larval salivary glands has proven to be a valuable model system for the study of ecdysteroid-induced alterations of transcription (*3*). In fact, the first suggestion that steroid hormones directly alter gene activity was made by Clever and Karlson (*20*) in their studies of the *Chironomus tentans* polytene chromosome. Analysis of hormonally induced changes in polytene chromosome puffing patterns *in vitro* has allowed the formulation of a detailed model of ecdysteroid and JH action in *Drosophila* (*7*). Recently, Gronemeyer and Pongs (*38*) have demonstrated the specific localization of 20-OH-ecdysone at an inducible puff. This is the first direct visualization of hormone binding at a locus of induced transcription.

A second *Drosophila* tissue which responds to developmental hormones *in vitro* is the imaginal disk. Imaginal disks are ectodermal derivatives which are maintained throughout the larval instars in a determined but undifferentiated state (*34*). The imaginal disk cell is one of a few cell types which retains the ability to undergo normal cell division rather than polytenization in late embryonic and larval states (*55*). Undifferentiated disks grow and divide during the larval instars. At metamorphosis, imaginal disks cease division, evaginate, and differentiate into adult structures such as wings, legs, eyes, antennae, and external genitalia, replacing many of the necrotic larval structures (*64*).

Explanted imaginal disks reproduce *in vitro,* a normal differentiative response to physiological concentrations of 20-OH-ecdysone (*30*). Evagination occurs within 14 hr of hormone addition, followed by cuticle formation and imaginal differentiation (*32*). The development of techniques for the mass isolation of imaginal disks (*19*) has made possible studies on hormone-induced alterations in macromolecular synthesis (*31,62*).

Explanted salivary glands and imaginal disks reproducibly differentiate in response to physiological concentrations of ecdysteroids *in vitro.* However, the use of these systems for the analysis of the cellular mode of action of steroid hormones is severely limited by the short survival time of these tissues *in vitro* (mass isolated disks survive less than 24 hr in culture). Other drawbacks include the time required to obtain tissue, the limiting amounts of tissue available, the heterogeneity of cells in the tissue, and the presence of endogenous hormones. The development of hormone-responsive cell lines overcomes each of these limitations, allowing better control of growth and endocrine parameters while providing a convenient system for the investigation of hormone action at the cellular level.

III. Ecdysteroid-Responsive Cell Lines

Insect cell lines were first developed because of the interest in propagating arthropod viruses (*37*). The first permanent cell lines were derived from trypsin-dissociated ovarian cells of Saturnid moths (*36*). These lines were established in a medium patterned after insect hemolymph (*66,68*). The development of these insect cell lines resembled that seen in many vertebrate cell culture systems; freshly explanted tissues undergo an initial rapid growth which later slows down or stops. This period is followed by a sudden renewal of growth, occasionally resulting in the establishment of permanent cell lines.

During the late 1960s and early 1970s, cell lines from various insect orders were established using modifications of Grace's medium, usually supplemented with insect hemolymph or mammalian serum (*16*). In 1968, Echalier and Ohanessian reported the establishment of a cell line, the K_C line, from dissociated 6- to 12-hr-old *Drosophila melanogaster* embryos. The K_C line has a number of significant advantages as an experimental system. The line has a stable, predominantly diploid karyotype (*27*). In comparison with other cell lines, the cytological identification of particular chromosomes is facilitated by their small number ($1\ N = 4$). The K_C line can be grown in roller or spinner culture at room temperature in an inexpensive, serum-free medium. In addition, these cells do not metabolize ecdysteroids (*52,63*) and can be cloned to produce homogeneous populations of distinctly different cell types which differ in their responses to ecdysteroids (*50,56*).

There appears to have been no irreversible restriction of developmental potential in the K_C cell genome. It has been shown that nuclei from the K_C line and a number of other *Drosophila* embryonic lines have retained the ability to undergo complete differentiation (*44*). Using nuclear transplantation *in ovo,* Ilmensee (*43*) demonstrated that after more than 5 years in culture, cell line nuclei were able to participate in normal development, forming larval and adult tissues such as gut, fat body, and cuticle. The functional integration of these nuclei was so complete that they were able to undergo normal polytenization.

Although the exact tissue etiology of the K_C line remains unclear, several lines of evidence point to the lateral ectoderm as the source. Moir and Roberts (*54*) have shown that the surface antigens expressed by the K_C line resemble most closely those of imaginal disks and salivary glands, both derivatives of the lateral ectoderm. When surveying the distribution of isozymes in K_C cells, Debec (*25*) found the line similar to imaginal disks and nervous tissues, two ectodermal derivatives. Schneider (*61*) provided further evidence for the disk origin of embryonically derived cell lines. When fragments from primary cultures (up to 10 months old) were implanted into metamorphosing larvae, Schneider found that they were able to form adult imaginal structures. Implanted spheres from any single fragment in the culture always formed the same type of adult structure. Although this ability is lost upon subsequent transfers, it does demonstrate that disklike cells are maintained in the culture for extended periods of time. However, as Moir and Roberts (*54*) point out, lines derived from *Drosophila* embryos may be expressing differentiated functions that are characteristic of cell types found at a later stage of development than the 12-hr-old embryo.

IV. Differentiative Responses of K_C Cells to Ecdysteroids

The administration of physiological concentrations (*14,41*), of 20-OH-ecdysone to the K_C line results in dramatic differentiative changes. Courgeon (*22*) first reported the alterations in K_C cell morphology following hormone exposure. As illustrated in Fig. 1, the normally spherical cells become spindle-shaped, flatten, and emit long, arborized cellular processes within 48 hr of hormone treatment (*9,17,18,22,60*). The role of a microtubular cytoskeleton in producing the morphological change has been established (*1,10*). Cherbas *et al.* (*17*) have shown that the extent of cellular process formation is a quantitatively reliable index of hormone response. For example, they report a hormone dose dependence for process length. The ability of various ecdysteroids to elicit this change in cell shape correlates extremely well with hormonal activity as determined in a wide variety of *in vivo* and *in vitro* assays.

Alterations in enzyme activities in hormone-treated K_C cells have also been observed. Cherbas *et al.* (*18*) reported the induction of acetylcholinesterase (a fiftyfold increase in activity after three days) in response to 10^{-6} *M* 20-OH-ecdysone. Best-Belpomme *et al.* (*12*) found a similar induction of β-galactosidase. Both hormone-inducible enzymes show similar time courses, dose responses, and analogue specificities (*11,13*). The induction of these two enzymes appears to be a specific response to ecdysteroids. Best-Belpomme *et al.* (*13*) assayed 30 different enzymes in 20-OH-ecdysone-treated K_C cells. Of the wide variety of enzymes screened (proteases, aminopeptidases, lipases, kinases, aminoacyl tRNA synthetases) only β-galactosidase and acetylcholinesterase activities increased; the other enzymes showed no change in activity.

Other K_C cell responses to ecdysteroids include increases in cell motility (*18*), aggregation of cells into large clumps (*22,60*), alterations of cell surface components (*53*), and modifications in cellular agglutinability (*60*).

Associated with each of these responses to hormones has been an arrest of K_C cell division (*18,23,60*). Rossett (*60*) showed that K_C cell division stops after 8–10 hr at 22°C and that cell number remains constant in the continued presence of hormone. Rossett concluded [as did Berger *et al.* (*9*)], that this failure of cell number to increase was due to the cessation of cell division rather than a balance between cell growth and death. Viability, as measured by trypan blue exclusion, is not affected by hormone exposure, even at late times (at least up to 1 week; *60,63*). Rossett (*60*) and Berger *et al.* (*9*) have shown that [^{3}H]thymidine incor-

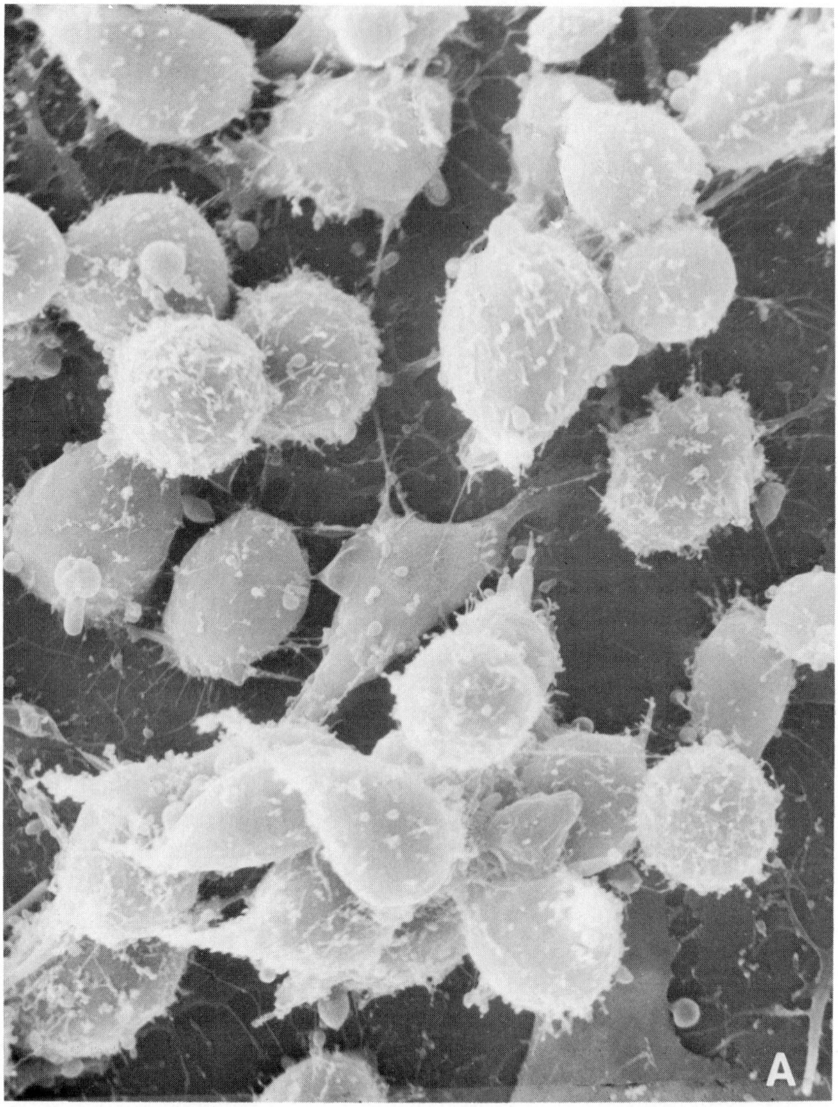

Fig. 1. Morphological response of K_c cells to ecdysteroids. (A) Scanning electronmicrograph of K_c-derived clone 2G5. Magnification, 3150 ×. (B) Scanning electron micrograph of clone 2G5 treated with $10^{-7}M$ 20-OH-ecdysone for 48 hr. Magnification, 2800 ×.

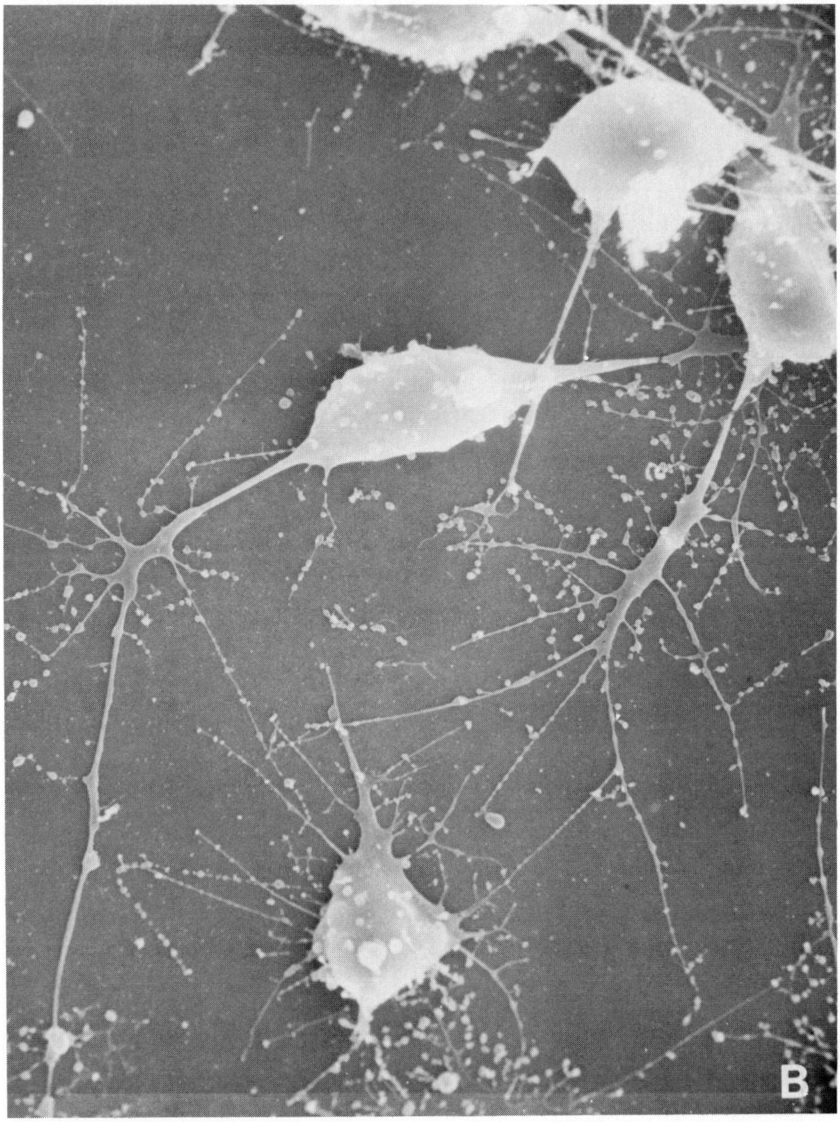
B

poration decreases after 8–10 hr of hormone treatment, indicating that the arrest of division results from a decreased rate of DNA replication in the culture. Berger *et al.* (*9*), in reporting a constancy of cell number and viability after hormone exposure, were the first to suggest a lengthening of the cell cycle in response to ecdysteroids.

The effect of 2×10^{-7} *M* 20-OH-ecdysone on the growth of a K_C-derived clone is shown in Fig. 2. In the control culture, cell number increases logarithmically, not approaching stationary phase until the population has increased eightfold. The population doubling time at 22°C is approximately 30 hr, in agreement with previously reported values (*28*). In the presence of 20-OH-ecdysone, division ceases for 7 days and then resumes in a logarithmic fashion. This apparent cessation of cell division is not an artifact of hemacytometer sampling in aggregating cultures; DNA values measured by the mithramycin assay (*40*) agree with the cell count data (*1,63*).

Effects of ecdysteroids on cell division have been reported in a number of other insect systems. Marks *et al.* (*49,51*) first demonstrated the ability of 20-OH-ecdysone to inhibit cell division in cockroach leg regenerates. Cohen *et al.* (*21*) and Lanir and Cohen (*48*) have reported the inhibition of multiplication in a mosquito embryonic line by physiological concen-

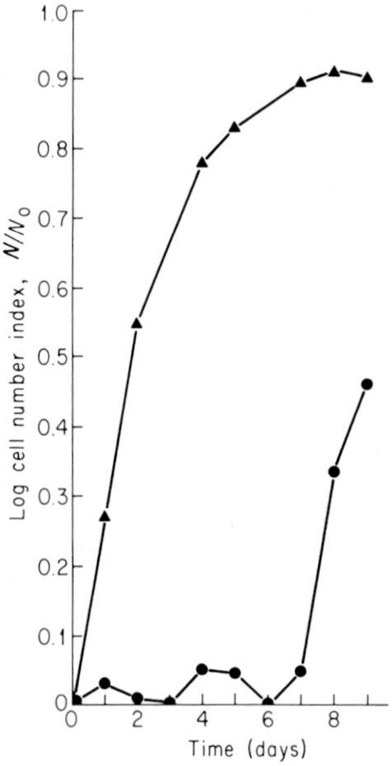

Fig. 2. Effect of 20-OH-ecdysone on K_C cell growth. K_C-derived clone 7C4 was grown in the absence (▲) or presence (●) of 2×10^{-7} *M* 20-OH-ecdysone.

trations of 20-OH-ecdysone. Similar effects have been observed in a moth ovarian line (*58*).

V. Ecdysteroid-Induced Alterations in the K_C Cell Cycle

In vertebrate systems many differentiative events couple an induced biochemical or morphological alteration with a prereplicative cell cycle arrest (G_0 or G_1; *57*). A number of investigators have documented a G_0 or G_1 arrest in stationary-phase populations of mammalian tissue-culture lines (*8*). The same type of pre-S phase arrest has been observed in stationary-phase *Drosophila* cell lines (*59,63*). We felt it was of interest to determine whether the inhibitory effect of ecdysteroids on cell division followed the same pattern. The development of flow cytometry (*2*) has provided a rapid method for the precise monitoring of the cell cycle distribution of batch cultures (*24*).

Figure 3a shows the cell cycle distribution of an exponentially growing population of a K_C-derived clone. The K_C cell cycle is very different from that typically seen in mammalian cells (in which G_1 accounts for

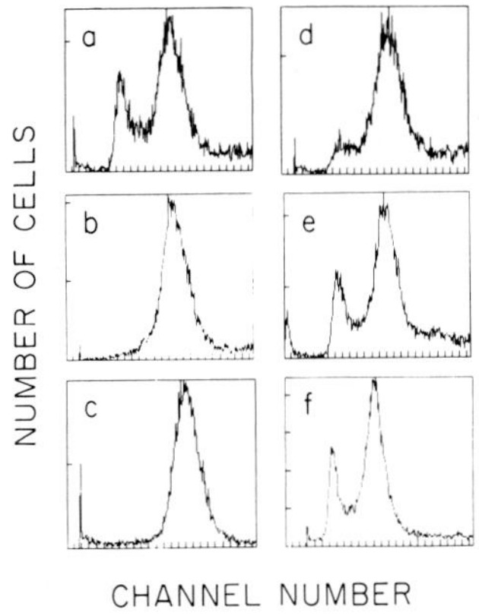

Fig. 3. Ability of ecdysteroids to cause G_2 arrest in K_C cells. Cells were fixed in 70% ethanol, rinsed in *Drosophila* saline, digested with 1 mg/ml RNase and stained with 1 μg/ml propium iodide as described (*64*). Flow cytometry was performed with a Becton-Dickinson Fluorescence-Activated Cell Sorter IV (FACS IV). (a) Histogram of fluorescence of a logarithmically growing population of K_C clone 7C4. (The left peak represents cells in G_1, the right one cells in G_2 + M. The intervening trough represents cells in S.) (b) Same cells as in (a), treated with 10^{-7} *M* 20-OH-ecdysone for 24 hr. (c) Same cells as in (a), treated with 10^{-9} *M* ponasterone A for 24 hr. (d) Same cells as in (a), treated with 10^{-5} *M* ecdysone for 24 hr. (e) Same cells as in (a), treated with 10^{-3} *M* triol for 24 hr. (f) K_C subline Br^-, treated with 10^{-7} *M* 20-OH-ecdysone for 24 hr.

the majority of the cell cycle (*57*). The near equivalence of the G_1 and G_2M peak heights indicates that G_2 is quite long in K_C cells and comprises the greater portion of the K_C cell cycle. Earlier reports by Dolfini *et al.* (*28*) had suggested the G_2 period of the K_C cell cycle to be substantially longer than G_1.

In screening for hormonally induced perturbations in the K_C cell cycle, we were surprised to find that cells are reversibly arrested in G_2. Figures 3b–d show the effect of physiological concentrations of three biologically active ecdysteroids on the cell cycle of a K_C-derived clone. The G_2 arrest, shown by the reduction or disappearance of the smaller peak at the left, appears to be obligatory to any subsequent morphological or enzymatic differentiation. All K_C clones that respond to ecdysteroids with a change in morphology or enzymatic activity are 100% G_2-arrested after 12 hr of hormone exposure at 22°C (B. Stevens, unpublished data). Figure 3f shows the failure of the hormone-insensitive Br^- line to respond to 20-OH-ecdysone with a G_2 arrest. This K_C subline has been shown to be devoid of measurable cytosolic and nuclear ecdysteroid receptors (*56,63*).

It has been determined by fluorescence microscopy with DAPI (*63*), colchicine block experiments, and three-dimensional flow cytometry (B. Stevens and J. D. O'Connor, unpublished data) that hormonally treated cells are arrested in G_2 rather than metaphase. This unusual cell cycle arrest does not appear to be unique to K_C cells. Recently, Evans *et al.* (*29*) identified a plant hormone, trigonelline, that causes a G_2 arrest in several plant tissues.

The hormonally induced G_2 arrest has the same characteristics as a number of other ecdysteroid responses. The relative activities of the three biologically active ecdysteroids (ponasterone A > 20-OH-ecdysone > ecdysone) and the inactivity of the 2β, 3β, 14α-trihydroxy-5-cholest-7-ene-6-one (triol) analogue (Fig. 3e) are consistent with those seen for K_C cellular process extension (*1,17*), enzyme induction (*13*), imaginal disk evagination (*33*), and polytene chromosome puffing (*4*). The dose–response curves for these hormone effects are also similar; in fact, those for the G_2 arrest (*63*) and process extension (*17*) are virtually identical. The doses of these ecdysteroids required to induce a half-maximal response are consistent with their affinities for both the K_C ecdysteroid receptor (*52*) and the imaginal disk ecdysteroid receptor (*70*).

The onset of the ecdysteroid-induced G_2 arrest is shown in Fig. 4. This figure demonstrates the accumulation of cells in G_2 at increasing times of hormone exposure. The time course of the G_2 arrest is similar to that seen for the evagination of imaginal disks *in vitro* (*32*), which show a maximal response 15 hr after the addition of hormone.

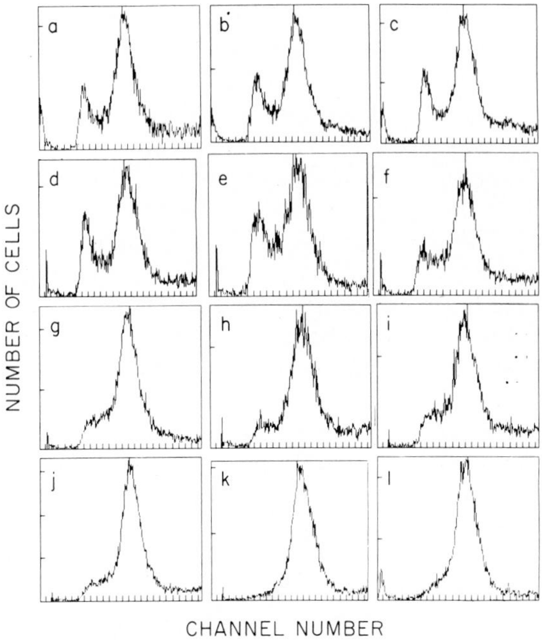

Fig. 4. Onset of ecdysteroid-induced G_2 arrest. The response of the K_C clone 7C4 to the addition of 10^{-7} *M* 20-OH-ecdysone. Cells were fixed at the indicated times and analyzed by flow cytometry on the Becton-Dickinson FACSIV as described (*64*). (a) Control logarithmically growing population of K_C clone 7C4. (b) Same cells as in (a), treated with 10^{-7} *M* 20-OH-ecdysone for 1 hr; (c) treated for 2 hr; (d) treated for 3 hr; (e) treated for 4 hr; (f) treated for 5 hr; (g) treated for 7 hr; (h) treated for 8 hr; (i) treated for 9 hr; (j) treated for 10 hr; (k) treated for 24 hr; (l) treated for 38 hr.

The G_2 arrest is a reversible effect. A series of hormone wash-out experiments have demonstrated that constant exposure to hormone is required to maintain the cells in a G_2-arrested state (*63*). The same requirement for continuous hormone exposure is seen for process extension (*22,63*), and ecdysteroid-induced alterations in macromolecular synthesis (*39*).

VI. Acquisition of Resistance to Ecdysteroids in K_C Cells

In the continued presence of 20-OH-ecdysone, cells eventually escape the G_2 block and reenter the cell cycle. This reentry can be detected by both cell counting (Fig. 2) and flow cytometric monitoring of the reappearance of cells in G_1 and S. After 8 days of hormone exposure, K_C

cells have resumed division. By the ninth day, the cell cycle is indistinguishable from that of control populations.

This resumption of division is not due to the breakdown of hormone by K_C cells. As mentioned previously, K_C cells do not metabolize ecdysteroids. This has been shown by thin layer chromatography (*52*), radioimmunoassay, and the ability of hormone-containing media from 10-day-old cultures to induce a G_2 arrest when added to hormone-naive populations (*63*). Although the effective concentration of hormone is still present in the medium, G_2-arrested cells resume division after 7 days.

Analysis of the hormonal sensitivity of the K_C cells that have escaped the G_2 block and have been transferred to hormone-free medium has shown that these cells are refractory to subsequent hormonal stimulation (*63*). Figure 5 illustrates the inability of two hormonally withdrawn clones to respond to a second ecdysteroid exposure with a G_2 arrest. Insensitivity to a second hormonal challenge is also seen for cellular process formation (*1,63*) and enzymatic induction (R. Hodgetts, personal communication). Populations of previously stimulated cells are unaffected by doses of ecdysteroids several orders of magnitude greater than those required to G_2 arrest a hormonally naive population. Resistance has been maintained in hormonally withdrawn populations for several months (*63*).

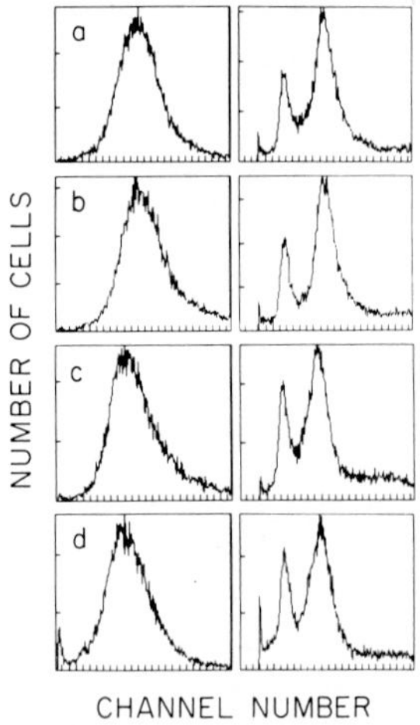

Fig. 5. Inability of K_C-derived clones to undergo and second G_2 arrest. Cytometry was performed as described (*64*). (a) Fluorescent distribution of withdrawn 7C4 cells (7C4w). Cells were treated with 10^{-7} *M* 20-OH-ecdysone for 10 days, then centrifuged and resuspended in fresh D-20 medium. Cells were passaged biweekly to ensure exponential growth. (b) Same cells as in (a), treated with 10^{-7} *M* 20-OH-ecdysone for 48 hr. (c) Withdrawn clone 2G5w, obtained as desribed in (a). (d) Same cells as in (c) treated with 10^{-7} *M* 20-OH-ecdysone for 48 hr.

TABLE I

Nuclear Ecdysteroid Binding in K_C Cells

Clone/cell line	Treatment[a]	[^{3}H]Ponasterone A binding (dpm bound/10^9 cells)(±SEM)	Percentage of control binding
7C4	Control	14,352 ± 287	100
7C4	Withdrawn	4,562 ± 948	32
7C4	Plus excess unlabeled hormone	1,870 ± 410	13
2G5	Control	18,030 ± 1,150	100
2G5	Withdrawn	5,200 ± 284	28
2G5	Plus excess unlabeled hormone	1,780 ± 433	10
7E10	Control	9,667 ± 296	100
7E10	Withdrawn	4,114 ± 276	42
7E10	Plus excess unlabeled hormone	1,820 ± 420	19
Br^-	Control	1,950 ± 380	—

[a] Withdrawn cells were exposed to 10^{-7} *M* 20-OH-ecdysone for 10 days, then washed free of hormone. Binding measurements were made 2–3 months after hormone removal. Nuclear binding was determined as described (63). Results are presented as of the mean ± standard error for three separate experiments. Nonspecific binding was determined in the presence of one-hundredfold excess unlabeled ponasterone A.

TABLE II

Cytosolic Ecdysteroid Binding in K_C Cells[a]

Clone	Treatment	Percentage of maximal binding of [^{3}H]ponasterone A
7C4	Control	100
7C4	Withdrawn	36
7C4	Plus excess unlabeled hormone	22

[a] Withdrawn cells were obtained and nonspecific binding was determined as described in Table I. Cytosolic ecdysteroid binding was determined as described (*63*). Maximal binding represents binding of 18.2 ± 1.7% of ligand (mean ± standard error). Results are presented as the mean of three separate experiments.

Correlated with the long-term refractoriness to hormonal restimulation has been a decrease in K_C ecdysteroid binding. Table I shows the extensive loss (58–72% decrease) of nuclear ecdysteroid-binding activity in hormonally withdrawn cells as compared to naive cells. This decrease has been observed in four different K_C-derived clones. The same loss of binding after 10 days of hormone exposure is seen for the cytosolic ecdysteroid receptor (Table II) (*63*) and whole-cell ecdysteroid binding (B. Stevens, unpublished data).

In attempting to understand the apparent hormone-induced loss of ecdysteroid receptors in K_C cells, it is of paramount importance to determine whether the hormonal exposure is merely selecting for resistant cells within the clonal population. Resistant cells could be responsible for an apparent resumption of division. We have studied this question carefully and have found that there are no detectable resistant (cycling) cells throughout the G_2 arrest period. As mentioned above, cell viability remains high during the period of hormonal exposure. It appears as though the reentry of cells into the cell cycle is a response of the total clonal population rather than a selection for preexisting hormonally resistant cells. We are currently investigating the characteristics of the residual population of ecdysteroid receptors in hormonally withdrawn cells (*63a*).

VII. Conclusions and Future Directions

The results from our laboratory demonstrate that G_2 comprises a much greater portion of the cell cycle in K_C cells than in other continuous lines. We have observed the same cell cycle distribution in a second

Drosophila melanogaster embryonic line, Schneider line 2, which is not affected by ecdysteroids (B. Stevens, unpublished data). However, a number of other *Drosophila* cell lines [Schneider line 3, Dübendorfer line 1, and Wyss' MDR 3 (*69*)] are growth arrested by ecdysteroids. It will be of interest to determine whether these lines are also G_2 arrested.

The physiological significance of the G_2 arrest is not presently understood. These results suggest that G_2 is the stage of the *Drosophila* cell cycle which is associated with hormonally induced differentiative events. Although this is contrary to the more commonly observed differentiation during G_0 or G_1, it is consistent with the *in vivo* behavior of several *Drosophila* larval tissues. A number of larval cell types replicate their DNA but fail to divide (polytenize) prior to an ecdysteroid-induced differentiative event (*5,46*).

In order to determine whether the hormonally induced G_2 arrest has an *in vivo* correlate, we are studying the cell cycle of imaginal disks throughout *Drosophila* development (*29a*). Figure 6 shows the cell cycle distribution of *Drosophila melanogaster* third instar wing and haltere disks in comparison to K_C cells. The resemblance is striking. However, at present it is not possible to differentiate between changes in the duration of particular phases of the imaginal disk cell cycle and shifts between proliferative and nonproliferative stages. Accurate cell number determinations in disks of staged larvae should allow more precise conclusions (M. Fain, personal communication). Cell division in the wing and eye disk is reduced prior to metamorphosis (*55*). Analysis of the

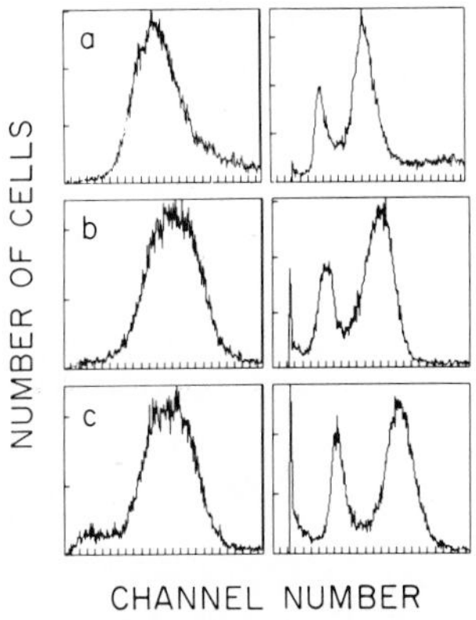

Fig. 6. Cell cycle distribution of *Drosophila melanogaster* imaginal disks. (a) Histogram of fluorescence of exponentially growing K_C clone 7C4. (b) Histogram of fluorescence of wild type (Oregon-R) *D. melanogaster* third instar wing disk nuclei. (c) Histogram of fluorescence of third instar haltere disks.

disk cell cycle at this time may help to determine the *in vivo* relevance of the K_C response.

Although we do not know whether the *in vitro* responses of the K_C line represent normal differentiation, it is certainly an excellent cell line for the investigation of the mode of action of steroid hormones. Many of the hormonal responses of the K_C line can be observed in intact insects (*52*). The apparent identity of the K_C ecdysteroid receptor (*52*) and the imaginal disk receptor (*70*) suggests that K_C ecdysteroid responses may not simply be artifacts of tissue culture. Furthermore, the apparent down regulation of ecdysteroid receptor in hormone-stimulated cells provides an ideal system for the examination of receptor regulation at the molecular level.

Acknowledgments

This work was supported by grants from the N.I.H. and N.S.F. B.S. was supported by U.S. Public Health Service National Research Award 07104.

References

1. Alvarez, C. M. (1980). Characterization of clonal K_C response to ecdysteroids: An approach to understanding hormone action. Ph.D. thesis, Univ. of California, Los Angeles.
2. Arndt-Jovan, D. J., and Jovan, T. M. (1978). Automated cell sorting with flow systems. *Annu. Rev. Biophys. Bioeng.* **7,** 527–558.
3. Ashburner, M. (1972). Patterns of puffing activity in the salivary gland chromosomes of *Drosophila*. VI. Induction by ecdysone in salivary glands of *D. melanogaster* cultured *in vitro*. *Chromosoma* **38,** 255–281.
4. Ashburner, M. (1973). Sequential gene activation by ecdysone in polytene chromosomes of *Drosophila melanogaster*. I. Dependence upon ecdysone concentration. *Dev. Biol.* **35** 47–61.
5. Ashburner, M., Chihara, C., Meltzer, P., and Richards, G. (1973). Temporal control of puffing activity in polytene chromosomes. *Cold Spring Harbor Symp. Quant. Biol.* **38,** 655–662.
6. Ashburner, M., Chihara, C., Meltzer, P., and Richards, G. P. (1974). Temporal control of puffing activity in polytene chromosomes. *Cold Spring Harbor Symp. Quant. Biol.* **38,** 655–662.
7. Ashburner, M., and Richards, G. (1976). The role of ecdysone in the control of gene activity in the polytene chromosomes of *Drosophila*. *In* "Insect Development" (P. A. Lawrence, ed.), pp. 203–225. Wiley, New York.
8. Baserga, R., and Nicolini, C. (1976). Chromatin structure and function in proliferating cells. *Biochim. Biophys. Acta* **458,** 109–134.
9. Berger, E. M., Ringler, R., Alahiotis, S., and Frank, M. (1978). Ecdysone-induced

changes in morphology and protein synthesis in *Drosophila* cell cultures. *Dev. Biol.* **62,** 498–511.
10. Berger, E. M., Sloboda, R. D., and Ireland, R. C. (1980). Tubulin content and synthesis in differentiating cells in culture. *Cell Motility* **1,** 113–129.
11. Best-Belpomme, M., and Courgeon, A. M. (1977). Ecdysterone and acetylcholinesterase activity in cultured *Drosophila* cells. *FEBS Lett.* **82**(2), 345–347.
12. Best-Belpomme, M., Courgeon, A. M., and Rambach, A. (1978). β-Galactosidase is induced by hormone in *Drosophila melanogaster* cell cultures. *Proc. Natl. Acad. Sci. U.S.A.* **75**(12), 6102–6106.
13. Best-Belpomme, M., Courgeon, A. -M., and Echalier, G. (1980). Development of a model system for the study of ecdysteroid action: *Drosophila melanogaster* cells established *in vitro*. *In* "Progress in Ecdysone Research" (J. A. Hoffman, ed.), pp. 379–392. Elsevier/North-Holland, Amsterdam.
14. Borst, D. W., Bollenbacher, W. E., O'Connor, J. D., King, D. S., and Fristrom, J. W. (1974). Ecdysone levels during metamorphosis of *Drosophila melanogaster*. *Dev. Biol.* **39,** 308–316.
15. Bowers, W. S. (1972). Juvenile hormones. *In* "Naturally Occurring Insecticides" (M. Jacobson and D. G. Crosby, eds.), pp. 307–332. Dekker, New York.
16. Brooks, M. A., and Kurtti, T. J. (1971). Insect cell and tissue culture. *Annu. Rev. Entomol.* **16,** 27–52.
17. Cherbas, L., Yonge, C. D., Cherbas, P., and Williams, C. M. (1980). The morphological response of K_c-H cells to ecdysteroids: Hormonal specificity. *Wilhelm Roux' Arch. Dev. Biol.* **189,** 1–15.
18. Cherbas, P., Cherbas, L., and Williams, C. M. (1977). Induction of acetylcholinesterase activity by β-ecdysone in a *Drosophila* cell line. *Science* **197,** 275–277.
19. Chihara, C. J., Petri, W. H., Fristrom, J. W., and King, D. S. (1972). The assay of ecdysones and juvenile hormones on *Drosophila* imaginal disks *in vitro*. *J. Insect Physiol.* **18,** 1115–1123.
20. Clever, V., and Karlson, P. (1960). Induktion von Puff-Veränderungen in den Speicheldrüsenchromosomen von *Chironomus tentans* durch Ecdyson. *Exp. Cell Res.* **56,** 1470–1476.
21. Cohen, E., Lanir, N., and Englander, E. (1976). Morphological effects and metabolism of the molting hormone in *Aedes aegypti* cultured cells. *Insect Biochem.* **6,** 433–439.
22. Courgeon, A. -M. (1972). Action of insect hormones at the cellular level. *Exp. Cell Res.* **74,** 327–336.
23. Courgeon, A. -M. (1972). Effect of α- and β-ecdysone on *in vitro* diploid cell multiplication in *Drosophila melanogaster*. *Nature, (London), New Biol.* **238,** 250–251.
24. Crissman, H. A., Mullaney, P. F., and Steinkamp, J. A. (1975). Methods and applications of flow systems for analysis and sorting of mammalian cells. *Methods in Cell Biol.* **9,** 179–246.
25. Debec, A. (1974). Isozymic patterns and functional states of *in vitro* cultivated cell lines of *Drosophila melanogaster*. *Wilhelm Roux Arch. Entwicklungsmech. Org.* **174,** 1–9.
26. Doane, W. W. (1973). Role of hormones in insect development. *In* "Developmental Systems: Insects" (S. J. Counce and C. H. Waddington, eds.), Vol. 2, pp. 291–497. Academic Press, New York.
27. Dolfini, S. (1971). Karyotype polymorphism in a cell population of *Drosophila melanogaster* cultured *in vitro*. *Chromosoma* **33,** 196–208.
28. Dolfini, S., Courgeon, A. M., and Tiepolo, L. (1970). The cell cycle of an established line of *Drosophila melanogaster* cells *in vitro*. *Experientia* **269,** 1020–1021.

28a. Echalier, G., and Ohanessian, A. (1970). *In vitro* culture of *Drosophila melanogaster* embryonic cells. *In Vitro* **5**; 162–172.

29. Evans, L. S., Almeida, M. S., Lynn, D. G., and Nakanishi, K. (1979). Chemical characterization of a hormone that promotes cell arrest in G_2 in complex tissues. *Science* **203,** 1122–1123.

29a. Fain, M., and Stevens, B. (1982). Alterations in the cell cycle of *Drosophila* imaginal disc cells precede metamorphosis. *Dev. Biol.* (in press).

30. Fristrom, J. (1972). The biochemistry of imaginal disc development. *Results Probl Cell Differ.* **5,** 109–154.

31. Fristrom, J. W., Gregg, T. L., and Siegel, J. (1974). The effect of β-ecdysone on protein synthesis in imaginal discs of *Drosophila melanogaster* cultured *in vitro*. I. The effect on total protein synthesis. *Dev. Biol.* **41,** 301–313.

32. Fristrom, J. W., Logan, W. R., and Murphy, C. (1973). The synthetic and minimal requirements for evagination of imaginal discs of *Drosophila melanogaster in vitro*. *Dev. Biol.* **33,** 441–456.

33. Fristrom, J. W., and Yund, M. A. (1976). Characteristics of the action of ecdysones on *Drosophila* imaginal discs cultured *in vitro*. *In* "Invertebrate Tissue Culture" (K. Maramorosch, ed.), pp. 161–178. Academic Press, New York.

34. Gehring, W. (1972). The stability of the determined state in cultures of imaginal discs in *Drosophila*. *In* "The Biology of Imaginal Discs" (H. Ursprung and R. Nöthiger, eds.), pp. 35–58. Springer-Verlag, Berlin and New York.

35. Gilbert, L. I., and King, D. S. (1974). Physiology of growth and development: Endocrine aspects. *In* "The Physiology of Insecta" (M. Rockstein, ed.), 2nd ed., Vol. 2, pp. 250–368. Academic Press, New York.

36. Grace, T. D. C. (1962). Establishment of four strains of cells from insect tissues grown *in vitro*. *Nature (London)* **195,** 788–789.

37. Grace, T. D. C. (1967). Insect cell culture and virus research. *In Vitro* **3,** 104–107.

38. Gronemeyer, H., and Pongs, O. (1980). Localization of ecdystrone on polytene chromosomes of *Drosophila melanogaster*. *Proc. Natl. Acad. Sci. U.S.A.* **77,**(4), 2108–2112.

39. Gvozdev, V. A., Kakpakov, V. T., Mukhovatova, L. M., Polokarova, L., and Tarantul, V. Z. (1973). Effect of ecdysterone on cell growth and macromolecular synthesis in established embryonic lines of *Drosophila melanogaster*. *Ontogenez* **5,** 33–42.

40. Hill, B. T., and Whatley, S. (1975). A simple, rapid microassay for DNA. *FEBS Lett.* **56,** 20–23.

41. Hodgetts, R. B., Sage, B. A., and O'Connor, J. D. (1977). Ecdysone titers during postembryonic development of *Drosophila melanogaster*. *Dev. Biol.* **60,** 310–317.

42. Horn, D. H. S. (1972). The ecdysones. *In* "Naturally Occurring Insecticides" (M. Jacobson and D. G. Crosby, eds.), pp. 333–459. Dekker, New York.

43. Ilmensee, K. (1976). Nuclear and cytoplasmic transplantation in *Drosophila*. *In* "Insect Development" (P. A. Lawrence, ed.), pp. 76–96. Wiley, New York.

44. Ilmensee, K. (1978). *Drosophila* chimeras and the problem of determination. *In* "Genetic Mosaics and Cell Differentiation" (W. J. Gehring, ed.), pp. 51–69. Springer-Verlag, Berlin and New York.

45. Kopec, S. (1922). Studies on the necessity of the brain for the inception of insect metamorphosis. *Biol. Bull. (Woods Hole, Mass.)* **42,** 323–342.

46. Kraminsky, G. P., Clark, W. C., Estelle, M. A., Gretz, R. D., Sage, B. A., O'Connor, J. D., and Hodgetts, R. B. (1980). Induction of translatable mRNA for dopa decarboxylase in *Drosophila*: An early response to ecdysterone. *Proc. Natl. Acad. Sci. U.S.A.* **71,** 4175–4179.

47. Lane, N. J., and Carter, Y. R. (1972). Puffs and salivary gland function: The fine structure of the larval and prepupal salivary glands of *Drosophila melanogaster. Wilhelm Roux' Arch. Entwicklungsmech. Org.* **169,** 216–238.
48. Lanir, N., and Cohen, E. (1977). Studies on the effect of the molting hormone in a mosquito line. *J. Insect Physiol.* **24,** 613–621.
49. Marks, E. P. (1970). The action of hormones in insect cell and organ cultures. *Gen. Comp. Endocrinol.* **15,** 289–302.
50. Marks, E. P. (1980). Insect tissue culture: An overview. 1971–1978. *Annu. Rev. Entomol.* **25,** 73–101.
51. Marks, E. P., Reinecke, J. P., and Caldwell, J. M. (1967). Cockroach tissue *in vitro*: A system for the study of insect cell biology. *In Vitro* **3,** 85–92.
52. Maroy, P., Dennis, R., Beckers, C., Sage, B. A., and O'Connor, J. D. (1978). Demonstration of an ecdysteroid receptor in a cultured cell line of *Drosophila melanogaster. Proc. Natl. Acad. Sci. U.S.A.* **75,** 6035–6038.
53. Metakovskii, E. V., Kakpakov, V. T., and Gvozdev, V. A. (1975). Effect of ecdysterone on subcultured cells from *Drosophila melanogaster*: Stimulation of high molecular weight polypeptide synthesis and change in cell surface properties. *Dokl. Akad. Nauk. SSSR* **221,** 960–963.
54. Moir, A., and Roberts, D. B. (1976). Distribution of antigens in established cell lines of *Drosophila melanogaster. J. Insect Physiol.* **22,** 299–307.
55. Nöthiger, R. (1972). The larval development of imaginal discs. *In* "The Biology of Imaginal Discs" (H. Ursprung and R. Nöthiger, eds.), pp. 1–34. Springer-Verlag, Berlin and New York.
56. O'Connor, J. D., Maroy, P., Beckers, C., Dennis, R., Alvarez, C. M., and Sage, B. A. (1980). Ecdysteroid receptors in cultured *Drosophila* cells. *In* "Gene Regulation by Steroid Hormones" (A. K. Roy and J. H. Clark, eds.), pp. 263–277. Plenum, New York.
57. Pardee, A. B., Dubrow, P., Hamlin, J. L., and Klietzen, R. F. (1978). Animal cell cycle. *Annu. Rev. Biochem.* **47,** 715–750.
58. Reinecke, J. P., and Robbins, J. D. (1971). Reaction of an insect cell line to ecdysterone. *Exp. Cell Res.* **64,** 335–338.
59. Rizzino, A., and Blumenthal, A. B. (1978). Synthronization of *Drosophila* cells in culture. *In Vitro* **14**(5), 437–442.
60. Rosset, R. (1978). Effects of ecdysterone on a *Drosophila* cell line. *Exp. Cell Res.* **111,** 31–36.
61. Schneider, I. (1972). Cell lines derived from late embryonic stages of *Drosophila melanogaster. J. Embryol. Exp. Morphol.* **27**(2), 353–365.
62. Siegel, J. E., and Fristrom, J. W. (1974). The effect of β-ecdysone on protein synthesis in imaginal discs of *Drosophila melanogaster* cultured *in vitro*. II. Effects on synthesis in specific cell fractions. *Dev. Biol.* **41,** 314–330.
63. Stevens, B., Alvarez, C. M., Bohman, R., and O'Connor, J. D. (1980). An ecdysteroid-induced alteration in the cell cycle of cultured *Drosophila* cells. *Cell* **22,** 675–682.
63a. Stevens, B., and O'Connor, J. D. (1982). The acquisition of resistance to ecdysteroids in cultured *Drosophila* cells. *Dev. Biol.* (submitted for publication).
64. Whitten, J. (1968). Metamorphic changes in insects. *In* "Metamorphosis" (W. Etkin and L. I. Gilbert, eds.), pp. 43–105. Appleton, New York.
65. Wigglesworth, V. B. (1964). The hormonal regulation of growth and reproduction in insects. *Adv. Insect Physiol.* **2,** 243–332.
66. Wyatt, G. R. (1961). The biochemistry of insect hemolymph. *Annu. Rev. Entomol.* **6,** 75–102.

67. Wyatt, G. R. (1968). Biochemistry of insect metamorphosis. *In* ''Metamorphosis'' (W. Etkin and L. I. Gilbert, eds.), pp. 143–184. Appleton, New York.
68. Wyatt, S. S. (1956). Culture *in vitro* of tissue from the silk worm, *Bombyx mori*. *J. Gen. Physiol.* **39,** 841–852.
69. Wyss, C. (1980). Loss of ecdysterone sensitivity of a *Drosophila* cell line after hybridization with embryonic cells. *Exp. Cell Res.* **125,** 121–126.
70. Yund, M. A., King, D. S., and Fristrom, J. W. (1978). Ecdysteroid receptors in imaginal discs of *Drosophila melanogaster*. *Proc. Natl. Acad. Sci. U.S.A.* **75**(120, 6039–6043.

12

Interferon as a Modulator of Human Fibroblast Proliferation and Growth

LAWRENCE M. PFEFFER, EUGENIA WANG, JERROLD FRIED, JAMES S. MURPHY, AND IGOR TAMM

I. Introduction

Interferons are a group of inducible proteins that interact with various animal cells and render them resistant to infection by a wide variety of RNA and DNA viruses. Although interferons were originally characterized as exclusively antiviral substances, interferons appear to have a broader biologic role as potent modulators of cellular structure and func-

GENETIC EXPRESSION IN THE CELL CYCLE

ISBN 0-12-543720-X

tion. In 1962, Paucker and colleagues (*47*) showed that mouse interferon preparations inhibited the proliferation of mouse L cells in a dose-dependent manner, as indicated by the antiviral activity titers of the preparations used. Interferons have been demonstrated to inhibit the proliferation of numerous normal and transformed cells in culture and *in vivo;* however, the mechanism of action is unclear. Extremely low concentrations of interferon ($\sim 10^{-12} M$) have been shown to inhibit viral and cellular replication substantially; however, even concentrations of interferon several orders of magnitude greater than the 50% inhibitory dose do not suppress viral and cellular replication entirely.

Evidence has accumulated that discrete interferon binding sites may exist on the cell surface (*19,62*). However, the precise structure of these binding sites has not yet been established. Several studies have indicated that interferon binding involves a membrane receptor that is composed in part of a ganglioside-like structure (*7,8,60*) and in part of glycoprotein (*25*). If these two components constitute the interferon receptor, then the interferon receptor would be structurally analogous to the membrane receptors for peptide hormones.

Furthermore, several investigators have suggested that the initial steps in interferon action closely resemble steps previously described in the action of peptide hormones (*9,20,50*). The mechanisms of signal generation and transmission activated by peptide hormones involve a series of cellular events which include the following: binding to specific membrane receptors, perturbation in the state of the plasma membrane reflected in altered cell surface determinants, and alterations in intracellular levels of cyclic nucleotides. Treatment of human cells with interferon also results in perturbation of membrane structure, including an increased rigidity of the plasma membrane lipid bilayer (*48,51*). Interferon has also been reported to cause transient increases of cyclic nucleotides (*21, 42,65*); however, activation of membrane-associated cyclases by interferon has not been demonstrated.

After treatment of cells with interferon, cellular metabolic activity is a necessary prerequisite for the development of the antiviral state (*22,33,34,58*). It is generally accepted that interferons are not directly antiviral but rather induce, after a lag of 3–5 hr, the production of several translation-regulatory enzymes which appear to play a role in the establishment of the antiviral state in certain virus–cell systems. Interferons also induce several cellular proteins whose roles have not yet been identified (*16,26,32*).

Many of the biologic effects of interferons were originally attributed to contaminating substances present in partially purified interferon preparations. However, with the advent of pure homogeneous interferon preparations, it has become accepted that antiviral activity is but one expression of interferons' pleiotropic action on cells. Besides inhibiting the replication of viruses and cells, interferons have been shown to modulate many aspects of the humoral and cellular immune systems. Recently, interferon has received much attention as a possible antitumor agent (*24,56*).

Interferons inhibit the proliferation of normal and tumor cells *in vitro* and *in vivo* without causing apparent toxic or degenerative changes in the cells. However, it is unclear whether interferons act directly as antiproliferative drugs or indirectly as cellular effectors that render cells more sensitive to the mechanism that controls cell cycling (*37,50*).

This chapter reviews our studies of the inhibitory action of human fibroblast interferon on the proliferation of human fibroblasts in monolayer culture (*50,52*). We have demonstrated that the reduced overall rate of proliferation of interferon-treated fibroblasts in monolayer culture reflects both a prolonged intermitotic interval in a majority of treated cells and a failure of a fraction of the cells to divide again. At a cell density at which control cells merely slow down, many of the interferon-treated fibroblasts cease to divide. The rates of DNA, RNA, and protein synthesis are somewhat reduced. However, treated fibroblasts grow larger than control cells as indicated by three parameters: cell surface area, cell volume, and cell mass. Cell cycle analysis has indicated that an increased fraction of the interferon-treated population has a DNA content of G_1-phase cells. The increase in G_1-phase cells is accompanied by a decrease in the proportion of S-phase cells. The mobility of interferon-treated fibroblasts is markedly reduced, as are membrane ruffling and intracellular saltatory movements.

In addition, we have characterized the morphological phenotype of interferon-treated cells (*52*). The enlarged and well-spread fibroblasts commonly have enlarged lobed nuclei and contain numerous stress fibers in the cell cytoplasm. The fraction of binucleated cells is increased five-fold. The actin-containing microfilament bundles are markedly increased in size and number. Fibronectin on the cell surface appears to be redistributed into arrays of long fibers covering the cell surface of interferon-treated fibroblasts.

The data we have obtained so far strongly suggest that interferon treatment of fibroblasts elicits a coordinated response that involves the cell membrane with its associated proteins and also the cytoplasmic microfilaments (*52*). In the final section of this paper, we will attempt to relate the findings described above to what is known of the biology of cultured human fibroblasts.

Two diploid cell strains were used in most experiments: FS-4, derived from neonatal foreskin (*61*), and ME (*49*), derived from a subepidermoid biopsy of a young male. Human fibroblasts were grown in monolayer cultures at 37°C in Eagle's minimal essential medium supplemented with 10% fetal calf serum (FCS). Fibroblast cell cultures were used between the tenth and twentieth tissue-culture passages made at a split ratio of 4:1. Thus, the fibroblast cultures have undergone between 20 and 40 population doublings. The interferon preparations, produced in human fibroblasts superinduced with poly(I):poly(C), were from two sources. Partially purified fibroblast interferon was generously provided by Drs. W. Carter and J. Horoszewicz, Roswell Park Memorial Institute, Buffalo, N.Y. and had a specific activity of 2×10^7 U/mg protein. Preparations of fibroblast (β_1) interferon, purified to homogeneity (specific activity $> 2 \times 10^8$ U/mg protein) were gifts of Dr. E. Knight, Jr., E. I. duPont de Nemours and Co., Inc., Wilmington, Del. Interferon activity was assayed by a microtitration procedure with vesicular stomatitis virus, as previously described (*27*), using the W.H.O. international human fibroblast reference standard for comparison, and is expressed in terms of international reference units (U)/ml.

II. Relationship between Interferon Concentration and Antiproliferative Effect of Interferon

Figure 1 illustrates the overall proliferation kinetics of the ME strain of human fibroblasts in the presence of increasing concentrations of interferon as determined by serial cell counts in multiwell dishes (*50,51*). The initial counts were in the range $2–4 \times 10^3$ cells/cm^2. The interferon concentration required for a significant reduction in the overall rate of cell proliferation lies between 40 and 160 U/ml. A near-maximal reduction in the overall rate of proliferation is obtained at 640 U/ml. Interferon treatment results in some deviation from exponentiality in the cell increase curve 2 or more days after the beginning of treatment.

We have computed population doubling times on the basis of the increases in cell number between 24 and 72 hr after the beginning of treatment (*50*). The population doubling times (25–43 hr) obtained for control cultures in medium supplemented with 5% FCS are generally comparable to the values previously reported under similar growth conditions. The reciprocals of doubling times, expressed as generations per hour, are equivalent to exponential growth-rate constants. Such growth-rate constants are used to construct dose–response curves, which are shown in Fig. 2. It should be noted first that human fibroblast interferon has no detectable effect on the proliferation of mouse L-929 cells and only a slight effect on a cloned line of African green monkey CV-1 cells.

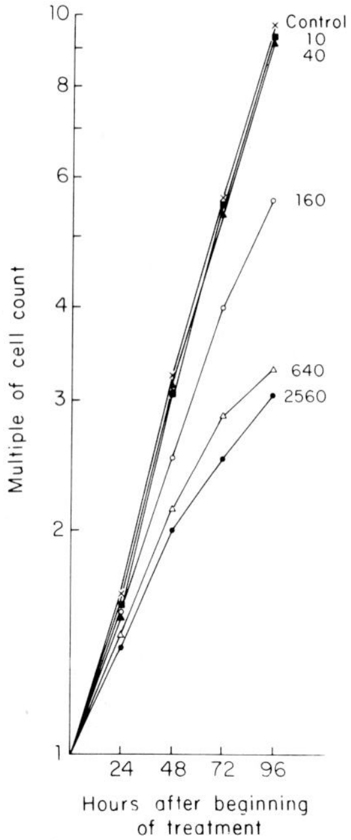

Fig. 1. Proliferation of human fibroblasts in the presence or absence of human fibroblast interferon (based on serial cell counts). The ME cells were seeded onto multiwell tissue-culture plates. At 1 day after plating, cultures were fed again with fresh medium containing no interferon or interferon at varying concentrations (the numbers in the graphs refer to U/ml). The cultures were photographed under phase-contrast optics after refeeding and at 24-hr intervals thereafter over a 5-day period. From Pfeffer *et al.* (*50*).

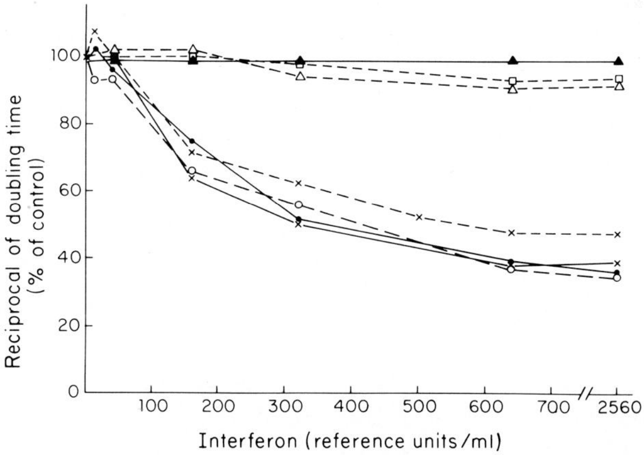

Fig. 2. Relationship between concentration of human fibroblast interferon and reduction in proliferation rate of human, monkey, and mouse cells, determined between 24 and 72 hr after the beginning of treatment. (×– –×) ME (human); (×——×) HeLa-S_3 (human); (●——●) FS-4 (human); (○– –○) GM-258 (human); (□– –□) HeLa monolayer (human); (△– –△) CV-1 (monkey); and (▲——▲) L-929 (mouse). From Pfeffer *et al.* (*50*).

The uncloned human HeLa tumor monolayer line used in our laboratory is also little affected; however, the S-3 subclone of HeLa cells grown in suspension is as sensitive to interferon as the human fibroblast strains that we have used, i.e., FS-4, ME, GM-258. The overall proliferation rates for sensitive human cells, determined between 24 and 72 hr after the beginning of treatment with interferon at 640 U/ml, were decreased by 52–64%. Increasing the interferon concentration to 2560 U/ml was not associated with a significant further increase in the antiproliferative effect. It is important to note that the highest doses of interferon used in these experiments permitted considerable proliferation of human cells. The approximately exponential, rather than linear, dose–response relationship between the concentration and antiproliferative effect of interferon is similar to the relationship obtained for the inhibition of virus multiplication. The antiproliferative effect of interferon was independent of the serum concentration used to supplement the tissue-culture medium, which was varied over a range from 5 to 20%.

III. Relationship between the Duration of Interferon Treatment and the Antiproliferative Effect

The inhibitory action of interferon on the proliferation of human fibroblasts is not rapidly reversible upon withdrawal of interferon from the medium (*50*). Fibroblast cultures of ME cells were treated with interferon for 1, 2, or 24 hr, and the proliferation curves were determined by daily cell counts in multiwell dishes. Appropriate controls were also examined. Dose–response curves were constructed on the basis of proliferation rates between 24 and 72 hr after the beginning of treatment. As shown in Fig. 3, there was no difference in the depression of the rate of proliferation regardless of the length of exposure to interferon. These data suggest that interferon's effect is not reversed within a period of 3 days and that interferon can in a short period cause a long-lasting effect. A short exposure of Daudi lymphoblastoid cells to fibroblast interferon was recently shown to suppress the rate of cell proliferation to a similar extent as continuous exposure of the cells to interferon (*30*).

IV. Time-Lapse Cinemicrographic Analysis of the Kinetics of Proliferation of Control and Interferon-Treated Fibroblasts

Time-lapse cinemicrography permits the study of large numbers of individual cells through several generations. Data from such observations

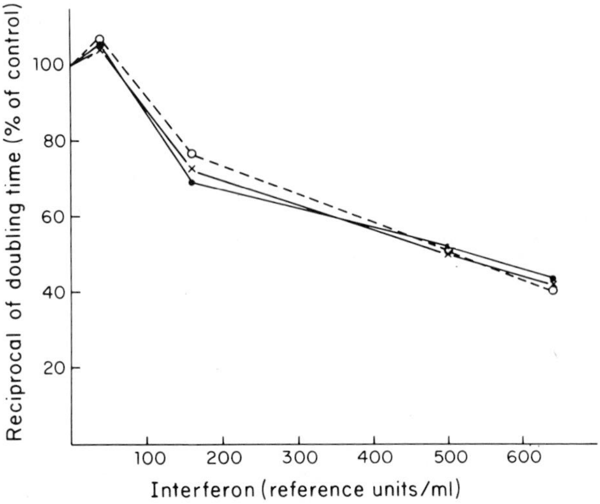

Fig. 3. The effect of the duration of interferon treatment on the relationship between interferon concentration and reduction in proliferation rate of human fibroblasts. ME cells were seeded onto multiwell tissue-culture plates. At 1 day after plating, cultures were treated with interferon at varying concentrations for 1 hr (○– –○) or 24 hr (●——●). At the end of the treatment period, cultures were washed three times with medium, fed again with medium containing 5% FCS, and incubated for 4 days after the beginning of treatment. Other cultures were treated continuously with interferon (×——×). Dose–response curves were constructed as described in the text. Based on Pfeffer *et al.* (*50*).

are essential for the understanding of the mechanisms involved in the antiproliferative effect of interferon observed in the experiments described above.

A. Population Kinetics

Mitotic cells were scored by examination of time-lapse films under a binocular dissecting microscope. Whenever a mitosis was found, one cell was added to the cumulative cell count. Figure 4 illustrates the gradual slowing of cell proliferation over the course of 5 days in the control and in the interferon-treated (640 U/ml) cultures. The cell proliferation curves shown in Fig. 4 were then analyzed by least-squares linear regression. The slopes from the linear regression were used to calculate the doubling times tabulated in Table I. Around 36 hr from refeeding with fresh tissue-culture medium, the doubling time for the control culture was ~18 hr, but by 84 hr from refeeding, the doubling time had increased to 37 hr. The slowing down of the rate of proliferation of control cells illustrates

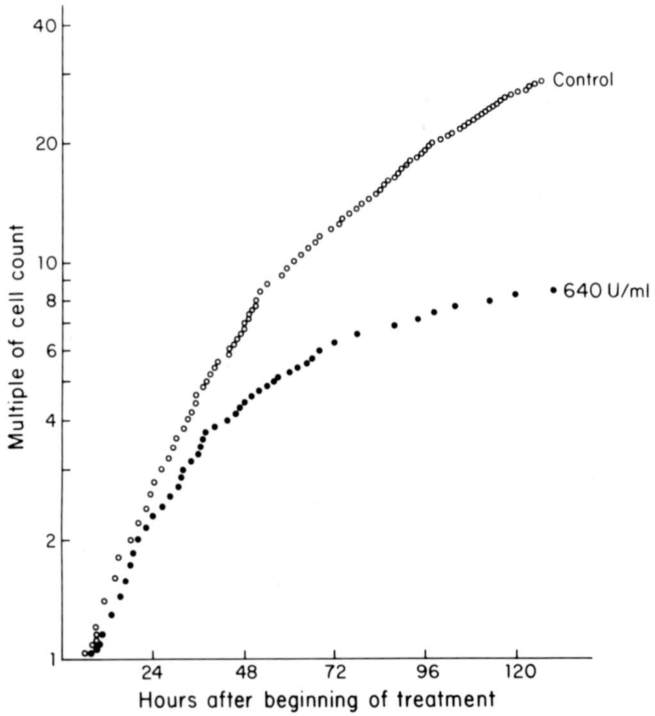

Fig. 4. Proliferation curves of human fibroblasts in the presence or absence of human fibroblast interferon (640 U/ml) (based on scoring mitoses). The ME cells were planted at a density of 2×10^3 cells/cm^2 in medium supplemented with 15% FCS. One day later, cultures were fed again with fresh medium with or without interferon and photographed at 2-min intervals by time-lapse cinemicrography. Using a $2.5 \times$ objective, a relatively large number of cells (20–40) could be followed per photographic field. Every fifth mitosis only was plotted for clarity. From Pfeffer *et al.* (*50*).

the well-known phenomenon of density-dependent inhibition of fibroblast proliferation in cell culture.

In the interferon-treated cultures, the doubling time at around 36 hr after the beginning of treatment was somewhat prolonged (~21 hr), but by 84 hr the doubling time had increased to approximately 126 hr as compared to the doubling time of 37 hr for control cells at this time. Thus, interferon-treated cells appeared to have slowed down sooner and at a lower cell density than control cells as the population density increased. Furthermore, the rate of deceleration of interferon-treated cells was greater than that of control cells; the mean doubling of the interferon-treated culture was 1.16 times the control value at 36 hr and reached 3.43 times the control value at 84 hr after the beginning of treatment.

TABLE I

Effect of Human Fibroblast Interferon (640 U/ml) on the Rate of Proliferation of Human Fibroblasts (ME Strain) (based on scoring mitoses)[a]

	Population doubling time (hr) during consecutive 24-hr intervals			
	24–48	48–72	72–96	96–120
Control	18.0	29.4	36.7	41.8
Interferon	21.0	46.8	125.9	130.3
	Proliferation rate (generations/hr) (% of control)			
Interferon	86	63	29	32

[a] Cell proliferation curves in Fig. 4 were analyzed by least-squares linear regression. The slopes of the linear regression curves for 24-hr intervals were used to calculate the population doubling times. Proliferation rates represent reciprocals of doubling times. Based on Pfeffer *et al.* (*50*).

An important limitation to the population studies which measure mean doubling times is that they do not distinguish between interferon effects on the duration of the cell cycle and effects on the number of cells cycling. As will be discussed below, studies on intermitotic intervals revealed that interferon treatment causes changes in both parameters.

B. Intermitotic Intervals within Cell Pedigrees

In the second series of time-lapse cinemicrographic experiments, we followed the mitotic behavior of individual fibroblasts and their daughter cells (*50*). The time intervals between mitoses were recorded for consecutive descendants. The first intermitotic interval after the beginning of treatment with interferon was prolonged in about two-thirds of the treated ME cells. The increase in the intermitotic interval in interferon-treated fibroblasts was progressive and became more marked with succeeding generations of cells that continued to divide. It should be noted that about one-half of the descendants of the treated fibroblasts that had a lengthened first intermitotic interval, did not divide again throughout the course of the experiment. However, in about one-third of the interferon-treated fibroblasts, the first intermitotic period was similar to that observed for control cells (13–18 hr). With succeeding generations, cells dividing with a normal cell cycle time represented a progressively smaller fraction of the interferon-treated population, which is evidence of a delayed effect of interferon on the proliferation of some fibroblasts.

Time-lapse cinemicrography of mouse EMT6 tumor cells has also shown that mouse interferon causes a progressive increase in the intermitotic interval of these cells (*13*). Furthermore, although 98% of the mouse cells divided during the first two cell generations after the beginning of treatment, by the fourth cell generation 24% of the treated cells had ceased to divide.

It is important to note that although interferon inhibits the proliferation of fibroblasts, it is not directly cytotoxic. Time-lapse cinemicrography provided no evidence of degenerative changes (abnormal cell rounding or contraction) or of cell lysis in interferon-treated fibroblast cultures.

V. Cell Surface Area and Nuclear Characteristics

The inhibition of proliferation of human fibroblasts by interferon is associated with an increase in cell surface area (*52*). After treatment for 3 days at 640 U/ml, approximately 55% of the fibroblasts showed increased cell surface area as compared to control cells. However, a significant fraction of the cells in the treated culture had a surface area similar to that of control cells. The mean attachment-surface area of interferon-treated cells was 4200 μm^2, representing a 65% increase. Taking into account the reduced cell number in interferon-treated cultures (see Fig. 3), it can be computed that 3 days after beginning of treatment, interferon-treated fibroblasts occupy a similar portion of the available growth surface (52%) that control cells do (60%).

An increase in nuclear size accompanied the increased mean cell surface area of interferon-treated cells (*52*). Table II summarizes the results obtained by scoring randomly selected fibroblasts in interferon-treated and control cultures for certain nuclear characteristics. Approximately 20% of the cells in the interferon-treated cell population have normal nuclei. Many of the interferon-treated fibroblasts contain enlarged nuclei that are polymorphic and exhibit varying degrees of lobation. In addition, the interferon-treated culture showed a fivefold increase in the frequency of binucleated cells. We have determined by time-lapse cinemicrography that these binucleated fibroblasts appear to arise as a result of a failure in cytokinesis rather than from fusion of adjacent fibroblasts.

VI. Cell Volume

To characterize further the enlargement of fibroblasts associated with the reduced rate of cell proliferation after interferon treatment, the vol-

TABLE II

Effect of Human Fibroblast Interferon (640 U/ml) on the Mean Surface Area, Mean Volume, and Nuclear Characteristics of Human Fibroblasts (ME Strain)

	Mean surface area[a]		Mean volume[b]		Nuclear characteristics frequency (%)[a]			
	(μm^2)	(% increase)	(μm^3)	(% increase)	Binucleate	Large	Intermediate	Normal
Control	2540	—	3030	—	1.7	4.3	11.2	82.8
Interferon	4200	65	3970	31	8.4	33.9	38.9	18.8

[a] Cells were grown on cover glasses and photographed 3 days after the beginning of treatment. Between 150 and 200 control and interferon-treated cells were photographed through a × 16 phase-contrast objective (total magnification, × 200), and 8 × 11 in. prints were made. The cut-out areas representing individual cells were weighed and the weights converted to surface areas. Based on Pfeffer *et al.* (*52*).

[b] Human fibroblasts were grown in tissue-culture flasks. After 3 days, control and interferon-treated cells were removed with trypsin and fixed in 4% glutaraldehyde in PBS. Cell volumes were determined electronically using a Coulter Channelyzer Model H4. The cell volume analysis was performed by Dr. Robert Zucker, Papanicolaou Cancer Research Institute, Miami, Fla. From Pfeffer *et al.* (*52*).

[a] Cells were grown on cover glasses and examined through a × 63 phase-contrast objective (total magnification: × 900) 3 days after beginning of treatment. A total of 340 control cells and 417 interferon-treated cells were scored for the presence of two nuclei and for nuclear size. From Pfeffer *et al.* (*52*).

ume distribution of interferon-treated and control cells was determined (*52*). Cells were detached by trypsinization after 3 days at 640 U/ml and fixed with glutaraldehyde. Cell volumes were determined electronically using a Coulter Channelyzer model H4. The interferon-treated cell population was found to be much more heterogeneous than the control population, with coefficient of variations of 21.5 and 11.5, respectively. Moreover, the mean volume of interferon-treated fibroblasts, which had been dispersed and suspended, was 3.97×10^3 μm^3, whereas that of control cells was 3.03×10^3 μm^3, which represents a 31% increase (Table II). However, it was also evident that there is considerable overlap in the cell volume distributions between interferon-treated and control fibroblast cultures. The values obtained for mean cell volumes do not represent actual volumes of the fibroblasts in monolayer culture, as they were obtained after removal of the cells from their growth substrate and fixation.

VII. Macromolecular Synthesis and Cellular Content of Macromolecules

Although interferon markedly inhibits the proliferation of fibroblasts, cellular macromolecular synthesis is only slightly depressed (*50*). Human fibroblast cultures were pulse labeled with radioactive precursors at daily intervals after the beginning of treatment. Interferon had a slight inhibitory effect on the uptake of [^{3}H]thymidine. Reduced nucleoside uptake has been previously observed in other cell systems after interferon treatment. After correction for reduced nucleoside uptake, the rate of DNA synthesis in interferon-treated cells shows a gradual decline to 86% of the control value after treatment at 640 U/ml for 3 days (Table III). Interferon also had a slight inhibitory effect on the uptake of [^{3}H]uridine into fibroblasts and a somewhat greater effect on the rate of RNA synthesis, which was 75% of control after 3 days of treatment. Interferon did not inhibit [^{3}H]leucine uptake into cells but did reduce the rate of protein synthesis to 64% of the control value after 3 days of treatment. The observed inhibitions of the rates of cellular DNA, RNA, and protein synthesis by interferon probably have only minor significance. It has been established in several cell systems that a decreased rate of cell proliferation is accompanied by a step-down in macromolecular synthesis (*29*).

However, a more interesting observation came from an examination of the effect of interferon on the cellular content of macromolecules: Although the rate of macromolecular synthesis is somewhat depressed,

TABLE III

Rate of Biosynthesis and Content of Macromolecules in Human Fibroblasts after Treatment with Human Fibroblast Interferon (640 U/ml)[a]

	Interferon treated (% of control)		
	24 hr	48 hr	72 hr
Rate of synthesis			
DNA	95	91	86
RNA	99	80	75
Protein	81	70	64
Cellular content			
DNA	97	122	130
Protein	99	135	150

[a] Based on Pfeffer *et al.* (*50*).

interferon-treated cells increase in mass (*50*). As shown in Table III, after 3 days of interferon treatment, the protein content per fibroblast was close to 150% of control. Clearly, interferon-treated fibroblasts continue to grow while their proliferation is curtailed. We therefore conclude that interferon does not prevent the biosynthesis of macromolecules or the growth of fibroblasts but does interfere with progression of cells through the cell cycle. Fuse and Kuwata (*23*) have also observed an increase in protein content after treatment of human cells with leukocyte interferon.

VIII. Cell Cycle Phase Distribution

In a series of experiments using the technique of flow cytometry in combination with determinations of the [^{3}H]thymidine labeling index, we have investigated the question of whether interferon treatment blocks the progression of fibroblasts through the cell cycle in a particular phase. Cultures were planted at a relatively low cell density, which allowed several cycles of exponential cell proliferation. By labeling index determination, we can, in some experiments, detect a slight reduction in the number of S-phase cells in interferon-treated cultures already on the second day after the beginning of treatment, by which time the flow cytometric profile has not yet changed significantly. After 2 days of treatment (640 U/ml), the overall rate of proliferation of interferon-treated cells is reduced by only ~25%. As shown in Table IV, after interferon treatment of FS-4 cells for 3 days the proportion of G_1-phase cells is

TABLE IV

Cell Cycle Phase Distributions of Control and Interferon-Treated (640 U/ml) Fibroblasts[a]

	Frequency (%)		
	G_1	S	G_2 + M
Control, 72 hr	49	27	24
Interferon, 72 hr	67	13	20
Control, 120 hr	71	13	16
Interferon, 120 hr	75	11	14

[a] Surface-attached cells were stained with a propidium iodide solution containing 0.1% Triton X-100, without the use of trypsin (*18*). Cell fluorescence was measured on a Cytofluorograf, Model 4802 (Ortho Instruments Inc., Westwood, Mass.). Data were analyzed with a PDP 11/70 computer by the method of Fried and Mandel (*17*).

increased compared to the fraction of such cells observed in exponentially dividing control cultures. The increase in the fraction of G_1-phase cells is accompanied by a decrease in the proportion of S-phase cells. The altered cell cycle phase distribution of cells treated with interferon for 3 days or longer is comparable to the shift seen in control fibroblast cultures under conditions of density-dependent inhibition of cell proliferation. Considering that interferon-treated cells are larger than control cells, it may be expected that the treated cells would become subject to density-dependent inhibition of proliferation at lower cell density values (i.e., number of cells per culture) than control cells. Density-dependent inhibition of proliferation of interferon-treated cells may thus become superimposed on interferon-induced inhibition, which may in part explain the fact that inhibition of proliferation in the treated cultures is progressive.

The increased proportion of G_1-phase cells and the decreased proportion of S-phase cells in the interferon-treated cell population appear to contradict the finding, described above, of an increased mean DNA content per cell by the third day after treatment. However, since the staining procedure removes the cell membrane and cytoplasm, the flow-cytometric analysis measures the DNA content per single isolated nucleus, whereas the chemical determination of DNA content refers to the amount of DNA per cell. Thus, the increased DNA content per cell (see Table III) reflects the increased proportion of binucleated cells in the interferon-treated population.

Interferon treatment has been shown to affect the progression of cells through the G_1 phase of the cell cycle in several cell systems. However, interferon treatment apparently alters cell cycle progression in a complex manner, as delays in traversing S, G_2, and M have also been reported. The evidence that has accumulated suggests that interferon treatment affects not only the commitment of cells to DNA synthesis but also perturbs functions during other cell cycle phases.

Interferon treatment of asynchronously growing mouse L1210 (*31*), and Friend leukemia cells (*39*) and human tumor MCF-7 cells (*4*) has been demonstrated to delay the progression of cells through G_1 and G_2 to a similar extent. A slight prolongation of S was also reported for interferon-treated MCF-7 cells (*4*). However, mouse interferon treatment of Ehrlich ascites tumor cells markedly prolonged G_2, S, and mitosis, but shortened G_1 only slightly (*45*). Interferon treatment of actively growing human melanoma cells has been reported to block the traverse of cells through G_1, S, and G_2 + M of the cell cycle (*11*).

In other attempts to probe the effect of interferon on cells, cell cultures have been synchronized in G_0/G_1 by culturing in medium containing low or no serum and then stimulated to growth with serum in the presence or absence of interferon. A prolongation of G_1 as well as S + G_2 was found after interferon treatment of mouse BALB/c 3T3 and Swiss 3T3K cells and human lung HEL27 fibroblasts (*3*). However, in a separate study, interferon treatment of mouse BALB/c 3T3 cells was reported to affect only the rate of entry into S phase (*55*). Interferon treatment of several strains of serum-stimulated human fibroblasts prolonged the G_1 phase and diminished the rate of DNA replication during S phase (*35*). Interferon-treated human melanoma cells exhibited a decreased transition rate from G_0/G_1 into S as well as a prolonged S phase during the first cell cycle following serum stimulation (*11*). Apparently, interferon-treated melanoma cells could slowly traverse one cell cycle, return to G_0/G_1, and fail to enter a new cycle.

IX. Cell Locomotion

Interferon treatment of fibroblasts results in reduced cell locomotion and intracellular movement (*10,52*). The tumor-induced motility of bovine capillary epithelial cells has also been shown to be inhibited by interferon (*10*). Using time-lapse cinemicrography, we have tracked the movement of randomly selected fibroblasts in control and interferon-treated cell cultures during consecutive 12-hr time intervals (*52*). A mean rate of cell locomotion of 0.13 μm/min was determined for control fibroblasts throughout the 96-hr course of experiments, with rates varying between 0.11 to 0.14 μm/min as shown in Fig. 5. During the first 24 hr after the

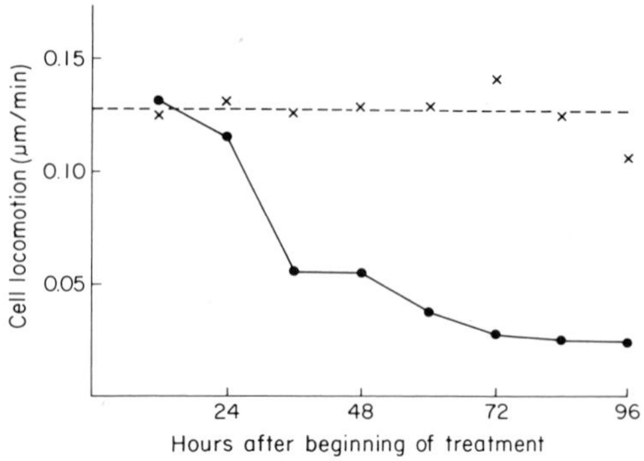

Fig. 5. Time course of inhibition of human fibroblast locomotion by human fibroblast interferon (640 U/ml). The ME cells were processed for time-lapse cinemicrography as described in the legend to Fig. 4, except that the cells were photographed under phase-contrast optics using a × 10 objective. The paths of movement of individual cells in cinemicrographs were traced onto paper with the aid of a stop-motion projector. The paths were measured with a distance-tracking device, and the measurements were divided by the total magnification. The average rates of cell locomotion for 40–62 control fibroblasts (×----- ×) and for 31–39 interferon-treated fibroblasts (●——●) are plotted for consecutive 12-hr time intervals. Based on Pfeffer *et al.* (*52*).

beginning of treatment, the rate of locomotion of interferon-treated cells was similar to that of control cells. However, for the period from 24 to 36 hr after the beginning of treatment, the rate of locomotion decreased to 44% of the mean control value and it leveled off at approximately 20% at 60–72 hr after the beginning of treatment. High-resolution time-lapse cinemicrography of fibroblasts has indicated a marked decrease in the saltatory movement of intracellular granules and in membrane ruffling in interferon-treated fibroblasts.

X. Cytoskeletal Components

As described above, after 3 days of treatment of human fibroblasts with 640 U/ml, more than one-half of the treated cells appear both enlarged and flattened out. Upon further examination by phase-contrast microscopy, interferon-treated cells were observed to contain large stress fibers throughout the cytoplasm (*52*). These structures are believed to be composed of contractile proteins, including actin and myosin. Such large fibers were rarely observed in control cultures of exponentially proliferating fibroblasts. However, a subpopulation of the interferon-treated cultures appeared indistinguishable from control cells in general morphological characteristics. Regulation of the size, spreading, and shape of cells and of the distribution of various cell surface components is believed to involve the organization and distribution of microfilaments, 10-nm filaments, and microtubules.

The cytoskeletal components were characterized by immunofluorescence staining of fibroblasts for actin, 10-nm filaments, and tubulin by methods described in detail elsewhere (*51,52*). Figure 6 illustrates the distribution of intracellular actin in control and interferon-treated fibroblasts. Control cells exhibit a fairly diffuse staining pattern with few prominent fibers present in the cytoplasm (Fig. 6A). Fewer than 5% of control fibroblasts have large actin-containing cables 4000–8000 Å in width. In contrast, over 40% of the interferon-treated fibroblasts exhibit large actin-containing fibers, which commonly span the cell in multiple parallel arrays (Fig. 6B). Most of these fibers appear to lie in the plane adjacent to the attachment surface of the fibroblasts and would thus be considered as part of the submembraneous microfilaments. We have also enumerated actin-containing fibers per unit area of cell surface and have found that, in interferon-treated fibroblasts, there is an 82% increase in the number of actin-containing microfilament bundles as compared to control cells. Thus, interferon treatment alters the organization and distribution of intracellular actin. The abnormally increased organization of microfilaments into bundles in interferon-treated fibroblasts may reflect altered dynamics of actin filament assembly and disassembly. This may

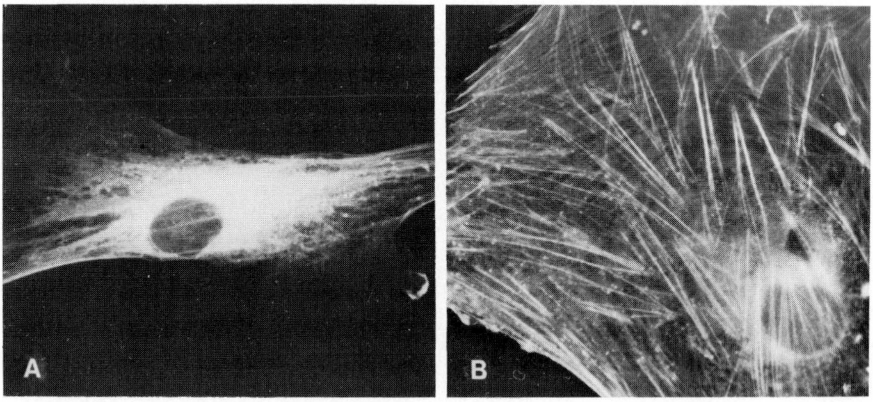

Fig. 6. Effect of interferon treatment (640 U/ml, 3 days) on the distribution of actin-containing microfilaments in human fibroblasts. Cells were grown on cover glasses and stained for intracellular actin by the indirect immunofluorescence technique. (A) Control ME cells; (B) interferon-treated ME cells. Similar results were obtained with FS-4 cells. × 900. From Pfeffer *et al.* (*52*).

adversely affect the construction of microfilaments that function in the cleavage furrow in cytokinesis.

Microtubules and 10-nm filaments were also found to be more abundant on a per-cell basis in interferon-treated cultures. However, in contrast to the findings with actin-containing microfilament bundles, there was apparently little change in the organization of microtubules and 10-nm filaments or in the number of these cytoskeletal structures per unit surface area in interferon-treated fibroblasts as compared to control cells.

XI. Cell Surface Fibronectin

Corresponding to the striking changes observed in the intracellular distribution and organization of actin, the distribution of cell surface fibronectin is also markedly altered after interferon treatment of human fibroblasts (*52*). Figures 7A and B show that in control fibroblasts, fibronectin is distributed as a network of fibers found in pericellular areas, particularly in regions of extensive cell-to-cell contact. After interferon treatment, cellular fibronectin is redistributed into arrays of long filaments covering most portions of the fibroblast cell surface (Figs. 7C and D). The extracellular fibers of fibronectin appear to be aligned with the underlying actin-containing microfilament bundles found in abundance in the cytoplasm of interferon-treated fibroblasts. Preliminary data obtained by lactoperoxidase catalyzed iodination of cell surface components and analysis of iodoproteins on SDS–polyacrylamide gels, indicate an increased amount of cell surface fibronectin in interferon-treated cells on a per-cell basis. However, when fibronectin is expressed as a fraction of total cell protein, the fibronectin protein distribution is similar in control and in interferon-treated cells.

XII. Conclusions and General Comments

Human fibroblast interferon inhibits the proliferation of normal diploid human fibroblasts in a dose-dependent manner. The inhibitory effect is approximately proportional to the logarithm of the dose of interferon between 40 and 640 U/ml. At 640 U/ml, the overall rate of cell proliferation is decreased by ~60%, and increasing the concentration fourfold results in no further increase in the mean doubling time. The antiproliferative effect of interferon is established by a short exposure of fibroblasts to interferon and is not rapidly reversible under the experimental conditions employed.

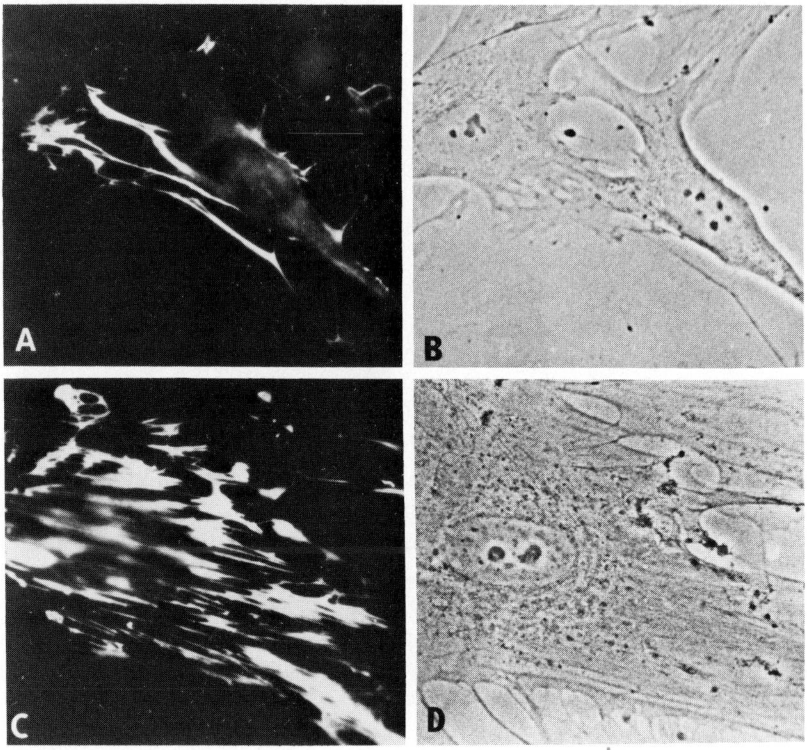

Fig. 7. Effect of interferon treatment (640 U/ml, 3 days) on the extracellular distribution of fibronectin on human fibroblasts. Cells were stained by the indirect immunofluorescence technique and examined by epifluorescence [(A) and (C)] and by phase-contrast optics [(B) and (D)]. (A) and (B) Control ME cells; (C) and (D) interferon-treated ME cells. × 900. From Pfeffer *et al.* (*52*).

Time-lapse cinemicrography has revealed a progressive lengthening of the intermitotic interval and a decreased division potential in most interferon-treated cells. As a result, the proliferation curve for the interferon-treated population deviates markedly from that of the control fibroblasts. However, a subpopulation of close to one-third of the interferon-treated fibroblasts divides at a rate similar to that of control cells.

It has been suggested that the action of interferon may be functionally analogous to that of polypeptide hormones and growth factors, in which interaction with cell surface initiates a cascade of cellular events. However, phenotypically the action of interferon appears to be in opposition to that of growth factors and other cellular effectors (L. M. Pfeffer and I. Tamm, unpublished observations; *32a, 44a, 58a*). In short, interferon acts as an antigrowth factor. Our evidence shows that interferon treatment of fibroblasts results in an enlargement of cells and an accumulation of DNA and protein. Interferon-treated fibroblasts exhibit extraordinarily large actin-containing fibers in the cytoplasm, long filamentous arrays of fibronectin on the cell surface, and decreased cell locomotion and intracellular movements. At present, it is unclear whether some of these alterations are secondary to the decreased proliferative activity of interferon-treated fibroblasts or directly related to interferon action. In either case, the alterations in the constituents of the cell surface and the cytoskeleton can be expected to have consequences for cell proliferation and locomotion.

We have observed a heterogeneous response of fibroblast cultures to interferon treatment, with respect to cell size and volume, nuclear size, distribution and organization of cytoplasmic microfilaments, and, most importantly, cell proliferation. We observe a subpopulation of the interferon-treated fibroblasts with characteristics that are indistinguishable from those of control cells. We do not have evidence at the present time whether this subpopulation is genetically different from the bulk of fibroblasts or whether it represents one end of a broad phenotypic spectrum of responses to the cytostatic action of interferon. It appears that this subpopulation may be equivalent to a proliferative pool of cells. The question arises whether it corresponds to the subfraction of the fibroblast population with the highest division potential (*38*); further work is needed to answer this question.

It is clear from our studies that the inhibition of proliferation in interferon-treated cultures is progressive and becomes maximal only after several cell generations.

Several studies have demonstrated that the original population of diploid fibroblasts is made up of subpopulations of cells that differ in proliferative capacity (*12,28,38,43*). When mass cultures of human fibroblasts are cloned, a bimodal distribution of proliferative capacities is found for the clonal populations. The clones that are capable of few cell divisions consist of cells which are larger and more epitheloid and contain an abundance of stress fibers in contrast to those clones with a greater proliferative capacity (*38*). As mentioned above, the heterogeneity of interferon-treated cultures may reflect the underlying heterogeneity of human fibroblast populations *in vitro*.

Several studies have suggested that for human fibroblasts in culture an inverse relationship may exist between the rate of proliferation and cell size (*1,12,44*). An increase in cell size has been observed in fibroblasts undergoing senescence *in vitro,* treated with interferon, maintained in low serum-containing medium, maintained at a decreased incubation temperature, or treated with hydroxyurea (*44*). Thus, increased cell size is a common cellular response to inhibition of proliferation in the absence of degenerative changes.

We have also shown that, associated with the antiproliferative action of interferon, there is an abnormal increase in the organization of the cytoplasmic actin-containing fibers and an altered distribution of cell surface fibronectin. These interferon-induced changes are opposite to those associated with the unregulated proliferation of transformed cells. An extensive literature exists on the loss or reduction of actin-containing fibers on cell transformation by SV40 or Rous sarcoma virus (*15,41, 53,63*). The amount of fibronectin on the cell surface is commonly reduced upon oncogenic transformation and appears to relate directly to the maintenance of the structure of actin-containing fibers (*2,67*). The microfilaments and fibronectin are cellular elements that are subject to profound changes as a cell expresses its proliferation-related phenotype. Interferon-induced inhibition of fibroblast proliferation is associated with phenotypic changes at one end of the spectrum, whereas changes associated with the unregulated proliferation of tumor virus-transformed cells reflect the other end. These phenotypic changes probably are part of a coordinated cellular response to agents that affect both the shape and the proliferative potential of cells.

In addition, interferon-treated human fibroblasts behave phenotypically like fibroblasts undergoing senescence after numerous passaged *in vitro*. Extensive studies of senescing fibroblasts have revealed many characteristics in common with interferon-treated fibroblasts as recorded in Table V. Thus, it appears that a common cellular response is evoked when cell proliferation becomes restricted either through interferon treatment or through the life-span mechanism operating *in vitro* (*52*).

The phenotype of late-passage diploid mouse fibroblasts also differs strikingly from that of early-passage cells (*59*). The cell surface area and the nuclear size and lobation of late-passage fibroblasts are increased, and there is an abundance of cytoplasmic microfilament bundles. These observations provide additional evidence pointing to a similarity between the major phenotypic features of interferon-treated cells and cells whose proliferative capacity has declined through the life-span mechanism. Van Gansen *et al.* (*59*) have emphasized that the microfilaments within the bundles are poorly organized in the "terminally differentiated" post-mitotic mouse fibroblasts. It will be important to investigate the organization of microfilaments within the thick and long bundles in interferon-treated human fibroblasts, because this would have a direct bearing on the question of the mechanism underlying the impaired motile functions of interferon-treated cells.

TABLE V

Phenotypic Features of Senescent Human Fibroblasts, Which Are Shared by Interferon-Treated Fibroblasts

Feature	Reference
1. Progressive decline in proliferative capacity	(*28*)
2. Marked heterogeneity in the proliferative capacity of cells	(*38*)
3. Reduction in the saturation density achieved by the monolayer	(*36*)
4. Increased proportion of cells in the G_1 phase of the cell cycle	(*54*); (*68*)
5. Decreased cell locomotion	(*1*)
6. Increased mean cell size and mass	(*64*)
7. Increased heterogeneity of cell sizes	(*54*)
8. Increased nuclear size	(*44*)
9. Increased frequency of lobed nuclei	(*40*)
10. Increased frequency of binucleated cells	(*40*)
11. Numerous stress fibers in the cytoplasm of cells	(*38*)

The proliferation of fibroblasts appears to be controlled mainly in the G_0–G_1 phase of the cell cycle (*5,46*). Cells transformed by oncogenic viruses are able to traverse G_1 into S at high cell densities and at low serum concentrations, at which untransformed cells fail to do so (*6, 14,57,66*). Senescent fibroblasts accumulate in G_1 (*54,68*). Serum-arrested fibroblasts treated with interferon, when stimulated to divide by serum replenishment, show inhibition of an early G_1 process (*35*). As was discussed above, we have observed an increase in the fraction of G_1 cells after interferon treatment of human fibroblasts.

In summary, we propose that a common response pathway operates in fibroblasts whose proliferation is impaired by either interferon treatment or by the life-span mechanism. We also propose that changes of an opposite nature affect the same pathway in cells induced by tumor viruses to undergo uncontrolled proliferation. These coordinated cellular responses involve major changes in the plasma membrane, cell surface fibronectin, and cytoplasmic microfilaments.

Acknowledgments

We thank Drs. W. A. Carter, J. S. Horoszewicz, and E. Knight, Jr., for providing human fibroblast interferon, and Dr. R. M. Zucker for performing cell volume analysis. We thank Ms. J. Peters, Ms. E. Clausnitzer, Mr. C. Hellmann, Ms. W. Poppe, Ms. D. Gunderson, and Mr. R. Berkowitz for technical assistance and Ms. A. Cruz for typing this manuscript.

This work was supported by research grants CA-18608 and CA-16757 and program project grant CA-18213 from the National Institute of Health. L. M. Pfeffer was a Postdoctoral Fellow under the Institutional National Research Service Award CA-09256.

References

1. Absher, P. M., Absher, R. G., and Barnes, W. D. (1974). Genealogues of clones of diploid fibroblasts. Cinemicrophotographic observations of all division patterns in relation to population size. *Exp. Cell Res.* **88,** 95–104.
2. Ali, I. U., Mautner, V., Lanza, R., and Hynes, R. O. (1977). Restoration of normal morphology, adhesion, and cytoskeleton in transformed cells by addition of a transformation sensitive surface protein. *Cell* **11,** 115–126.
3. Balkwill, F. R., and Taylor-Papadimitriou, J. (1978). Interferon affects both G_1 and S + G_2 in cells stimulated from quiescence to growth. *Nature* (*London*) **274,** 798–800.
4. Balkwill, F. R., Watling, D., and Taylor-Papadimitriou, J. (1979). The effect of interferon on cell growth and the cell cycle. *In* "Antiviral Mechanisms in the Control of Neoplasia" (P. Chandra, ed), pp. 712–728. Plenum, New York.
5. Baserga, R. (1976). "Multiplication and Division in Mammalian Cells." Dekker, New York.
6. Bell, T. G., Wyke, J. A., and Macpherson, I. A. (1975). Transformation by a temperature sensitive mutant of Rous sarcoma virus in the absence of serum. *J. Gen. Virol.* **27,** 127–134.
7. Besançon, F., and Ankel, H. (1974). Inhibition of interferon action by plant lectin. *Nature* (*London*) **250,** 784–786.
8. Besançon, F., and Ankel, M. (1974). Binding of interferon to gangliosides. *Nature* (*London*) **252,** 478–480.
9. Blalock, J., and Stanton, J. D. (1980). Common pathways of interferon and hormonal action. *Nature* (*London*) **283,** 406–408.
10. Brouty-Boye, D., and Zetter, B. R. (1980). Inhibition of cell motility by interferon. *Science* **208,** 516–518.
11. Creasey, A. A., Bartholomew, J. C., and Merigan, T. C. (1980). Role of G_0–G_1 arrest in the inhibition of tumor cell growth by interferon, *Proc. Natl. Acad. Sci. U.S.A.* **77,** 1471–1475.
12. Cristofalo, V. J., and Sharf, B. B. (1973). Cellular senescence and DNA synthesis. Thymidine incorporation as a measure of population age in human diploid cells. *Exp. Cell Res.* **16,** 419–427.
13. d'Hooghe, C. M., Brouty-Boyé, D., Malaise, E. F., and Gresser, I. (1977). Interferon and cell division. XII. Prolongation by interferon of the intermitotic time of mouse mammary tumor cells *in vitro*. Microcinematographic analysis. *Exp. Cell Res.* **105,** 73–76.
14. Dulbecco, R., Hartwell, L. M., and Vogt, M. (1965). Induction of cellular DNA synthesis by polyoma virus. *Proc. Natl. Acad. Sci. U.S.A.* **53,** 403–410.
15. Edelman, G. M., and Yahara, I. (1976). Temperature-sensitive changes in surface modulating assemblies of fibroblasts transformed by mutants of Rous sarcoma virus. *Proc. Natl. Acad. Sci. U.S.A.* **73,** 2047–2051.
16. Farrell, P. J., Broeze, B. J., and Lengyel, P. (1979). Accumulation of mRNA and protein in interferon-treated Ehrlich ascites tumor cells. *Nature* (*London*) **279,** 523–525.
17. Fried, J., and Mandel, M. (1979). Multi-user system for analysis of data from flow cytometry. *Comput. Programs Biomed.* **10,** 218–230.

18. Fried, J., Perez, A., and Clarkson, B. D. (1978). Rapid hypotonic method for flow cytofluorometry of monolayer cell cultures. Some pitfalls in staining and data analysis. *J. Histochem. Cytochem.* **26,** 921–933.
19. Friedman, R. M. (1967). Interferon binding: The first step in the establishment of antiviral activity. *Science* **156,** 1760–1761.
20. Friedman, R. M. (1977). Antiviral activity of interferon. *Bacteriol. Rev.* **41,** 543–567.
21. Friedman, R. M., and Pastan, I. (1969). Interferon and cyclic 3′5′adenosine monophosphate potentiation of antiviral activity. *Biochem. Biophys. Res. Commun.* **36,** 735–739.
22. Friedman, R. M., and Sonnabend, J. A. (1965). Inhibition of interferon action by puromycin. *J. Immunol.* **95,** 696–703.
23. Fuse, A., and Kuwata, T., (1976). Effects of interferon on the human clonal cell line, RSa: Inhibition of macromolecular synthesis. *J. Gen. Virol.* **33,** 17–24.
24. Gresser, I., and Tovey, M. G. (1978). Antitumor effects of interferon. *Biochim. Biophys. Acta* **458,** 73–107.
25. Grollman, E. F., Lee, G., Ramos, S., Lazo, P. S., Kaback, H. R., Friedman, R., and Kohn, L. D. (1978). Relationships of the structure and function of the interferon receptor to hormone receptors and the establishment of the antiviral state. *Cancer Res.* **38,** 4172–4185.
26. Gupta, S. L., Rubin, B. Y., and Holmes, S. L. (1979). Interferon action: Induction of specific proteins in mouse and human cells by homologous interferons. *Proc. Natl. Acad. Sci. U.S.A.* **76,** 4817–4821.
27. Havell, E. A., and Vilček, J. (1972). Production of high-titered interferon in cultures of human diploid cells. *Antimicrob. Agents Chemother.* **2,** 476–484.
28. Hayflick, L., and Moorhead, P. S. (1961). The serial cultivation of human diploid cell strains. *Exp. Cell Res.* **25,** 585–621.
29. Holley, R. (1974). Serum factors and growth control. *In* "Control of Proliferation in Animal Cells" (B. Clarkson and R. Baserga, eds.), pp. 13–18. Cold Spring Harbor Lab., Cold Spring Harbor, New York.
30. Horoszewicz, J. S., Leong, S. S., and Carter, W. A. (1979). Noncycling tumor cells are sensitive targets for the antiproliferative activity of human interferon. *Science* **206,** 1091–1093.
31. Killander, D., Lindahl, P., Lundin, L., Leary, P., and Gresser, I. (1976). Relationship between the enhanced expression of histocompatibility antigens of interferon-treated L1210 cells and their position in the cell cycle. *Eur. J. Immunol.* **6,** 56–59.
32. Knight, E., Jr., and Korant, B. D. (1979). Fibroblast interferon induces synthesis of four proteins in human fibroblast cells. *Proc. Natl. Acad. Sci. U.S.A.* **76,** 1824–1827.
32a. Lin, S. L., Ts'o, P. O. P., and Hollenberg, M. D. (1980). Effects of interferon on epidermal growth factor action. *Biochem. Biophys. Res. Commun.* **96,** 168–174.
33. Lindenmann, J., Burke, D., and Isaacs, A. (1957). Studies on the production, mode of action and properties of interferon. *Br. J. Exp. Pathol.* **38,** 551–562.
34. Lockart, R. Z., Jr. (1964). The necessity for cellular RNA and protein synthesis for viral inhibition resulting from interferon. *Biochem. Biophys. Res. Commun.* **15,** 513–518.
35. Lundgren, E., Larsson, I., Miörner, H., and Stannegard, Ö. (1979). Effects of leukocyte and fibroblast interferon on events in the fibroblast cell cycle. *J. Gen. Virol.* **42,** 589–595.
36. Macieira-Coehlo, A. (1973). Aging and cell division. *Front. Matrix Biol.* **1,** 46–65.
37. Macieira-Coehlo, A., Brouty-Boye, D., Thomas, M. T., and Gresser, I. (1971). Interferon on the division cycle of L1210 cells *in vitro*. *J. Cell Biol.* **48,** 415–419.
38. Martin, G. M., Sprague, C. A., Norwood, T. M., and Pendergrass, W. R. (1974).

Clonal selection, attenuation and differentiation in an *in vitro* model of hyperplasia. *Am. J. Pathol.* **74,** 137–150.
39. Matarese, G. P., and Rossi, G. B. (1977). Effect of interferon on growth and division cycle of Friend erythroleukemic murine cells *in vitro*. *J. Cell. Biol.* **75,** 344–354.
40. Matsumara, T., Zerrudo, Z., and Hayflick, L. (1979). Senescent human diploid cells in culture: Survival, DNA synthesis and morphology. *J. Gerontol.* **34,** 328–334.
41. McNutt, N. S., Culp, L. A., and Black, P. H. (1973). Contact-inhibited revertant cell lines isolated from SV40-transformed cells. IV. Microfilament distribution and cell shape in untransformed, transformed and revertant BALB/c 3T3 cells. *J. Cell Biol.* **56,** 412–428.
42. Meldolesi, M. F., Friedman, R. M., and Kohn, L. D. (1977). An interferon-induced increase in cyclic AMP level precedes the establishment of the antiviral state. *Biochem. Biophys. Res. Commun.* **79,** 239–246.
43. Merz, G. S., Jr., and Ross, J. D. (1969). Viability of human diploid cells as a function of *in vitro* age. *J. Cell. Physiol.* **74,** 219–221.
44. Mitsui, Y., and Schneider, E. L. (1976). Relationship between cell replication and volume in senescent human diploid fibroblasts. *Mech. Ageing Dev.* **5,** 45–56.
44a. Oleszak, E., and Inglot, A. D. (1981). Platelet derived growth factor (PDGF) inhibits antiviral and anticellular action of interferon in synchronized mouse or human cells. *J. Interferon Res.* **1,** 37–48.
45. Panniers, L. R. V., and Clemens, M. J. (1980). Inhibition of cell division by interferon: Changes in cell cycle characteristics and in morphology of Ehrlich ascites tumor cells in culture. *J. Cell Sci.* **48,** 259–279.
46. Pardee, A. B., Dubrow, R., Hamlin, J. L., and Kletzien, R. F. (1978). Animal cell cycle. *Annu. Rev. Biochem.* **47,** 715–750.
47. Paucker, K., Cantell, K., and Henle, W. (1962). Quantitative studies on viral interference in suspended L cells. III. Effect of interfering viruses and interferon on the growth rate of cells. *Virology* **17,** 324–334.
48. Pfeffer, L. M., Landsberger, F. R., and Tamm, I. (1981). β-Interferon-induced time-dependent changes in the plasma membrane lipid bilayer of cultured cells. *J. Interferon Res.*, **1,** 613–620.
49. Pfeffer, L. M., Lipkin, M., Stutman, O., and Kopelovich, L. (1976). Growth abnormalities of cultural human skin fibroblasts derived from individuals with hereditary adenomatosis of the colon and rectum. *J. Cell. Physiol.* **89,** 29–38.
50. Pfeffer, L. M., Murphy, J. S., and Tamm, I. (1979). Interferon effects on the growth and division of human fibroblasts. *Exp. Cell Res.* **121,** 111–120.
51. Pfeffer, L. M., Wang, E., Landsberger, F. R., and Tamm, I. (1981). Assays to measure plasma membrane and cytoskeletal changes in interferon-treated cells. *In* "Methods in Enzymology" **79B,** 461–473.
52. Pfeffer, L. M., Wang, E., and Tamm, I. (1980). Interferon effects on microfilament organization, cellular fibronectin distribution, and cell motility in human fibroblasts. *J. Cell Biol.* **85,** 9–17.
53. Pollack, R., and Rifkin, D. (1975). Actin-containing cables within anchorage-dependent rat embryo cells are dissociated by plasmin and trypsin. *Cell* **6,** 495–506.
54. Schneider, E. L., and Fowlkes, B. J. (1976). Measurement of DNA content and cell volume in senescent human fibroblasts utilizing flow multiparameter single cell analysis. *Exp. Cell Res.* **98,** 298–302.
55. Sokawa, Y., Watanabe, Y., and Kawade, Y. (1977). Suppressive effect of interferon on the transition from quiescent to a growing state in 3T3 cells. *Nature (London)* **268,** 236–238.
56. Strander, H. (1977). Interferons: Anti-neoplastic drugs? *Blut* **35,** 277–288.

57. Stohl, W. A. (1973). Alteration in hamster cell regulatory mechanisms resulting from abortive infection with an oncogenic adenovirus. *Prog. Exp. Tumor Res.* **18,** 199–239.
58. Taylor, J. (1964). Inhibition of interferon action by actinomycin. *Biochem. Biophys. Res. Commun.* **14,** 447–453.
58a. Taylor-Papadimitriou, J., Shearer, M., and Rozengurt, E. (1981). Inhibitory effect of interferon on cellular DNA synthesis: Modulation by pure mitogenic factors. *J. Interferon Res.* **1,** 401–410.
59. Van Gansen, P., Devos, L., Ororan, Y., and Roxburgh, C. (1979). Phenotypes des fibroblastes d'embryons de soures viellissant *in vitro* (SEM et TEM). *Biol. Cell* **34,** 255–270.
60. Vengris, V. E., Reynolds, F. M., Jr., Hollenberg, M. D., and Pitha, P. M. (1976). Interferon action: Role of membrane gangliosides. *Virology* **72,** 486–493.
61. Vilček, J., and Havell, E. A. (1973). Stabilization of interferon messenger RNA activity by treatment of cells with metabolic inhibitors and lowering of the incubation temperature. *Proc. Natl. Acad. Sci. U.S.A.* **70,** 3909–3913.
62. Vilček, J., and Rada, B. (1962). Studies on an interferon from tick-borne encephalitis virus infected cells. III. Antiviral action of IF. *Acta Virol.* (*Engl. Ed.*) **6,** 9–15.
63. Wang, E., and Goldberg, A. R. (1976). Changes in microfilament organization and surface topography upon transformation of chick embryo fibroblasts with Rous sarcoma virus. *Proc. Natl. Acad. Sci. U.S.A.* **73,** 4065–4069.
64. Wang, K. M., Rose, N. R., Bartholomew, E. A., Balzar, M., Berde, K., and Foldvary, M. (1970). Changes of enzymatic activities in human diploid cell line WI-38 at various passages. *Exp. Cell Res.* **61,** 357–364.
65. Weber, J. M., and Stewart, R. B. (1975). Cyclic AMP potentiation of interferon antiviral activity and effect of interferon on cellular cyclic AMP levels. *J. Gen. Virol.* **28,** 363–372.
66. Weil, R., Salomon, C., May, E., and May, P. A. (1974). A simplifying concept in tumor virology: Virus-specific pleiotropic-effectors. *Cold Spring Harbor Symp. Quant. Biol.* **39,** 381–395.
67. Yamada, K. M., Yamada, S. S., and Pastan, I. (1976). Cell surface protein partially restores morphology, adhesiveness and contact inhibition of movement of transformed fibroblasts. *Proc. Natl. Acad. Sci. U.S.A.* **73,** 1217–1221.
68. Yanishevsky, R., Mendelsohn, M. L., Mayall, M., and Cristofalo, V. J. (1974). Proliferative capacity and DNA content of aging diploid cells in culture: A cytophotometric and autoradiographic analysis. *J. Cell. Physiol.* **84,** 165–170.

13

Different Sequences of Events Regulate the Initiation of DNA Replication in Cultured Mouse Cells

ANGELA M. OTTO AND LUIS JIMENEZ DE ASUA

I. Introduction

In spite of many different approaches taken to unravel the enigma of growth regulation, the cell has retained most of its secrets as to the mechanisms for regulating the initiation of chromosomal DNA replication and cell division. One approach has been to add growth-stimulating compounds to cultures of mammalian cell lines that, by their restricted growth criteria, are considered to be models for normal cells *in vivo* (*1–3,10*).

In particular, Swiss mouse 3T3 cells offer a convenient system to study the stimulation of growth under defined environmental conditions. These cells cease to divide upon deprivation of serum or upon attaining a "saturating density" (*10*). Such quiescent cultures can be stimulated to

GENETIC EXPRESSION IN THE CELL CYCLE

ISBN 0-12-543720-X

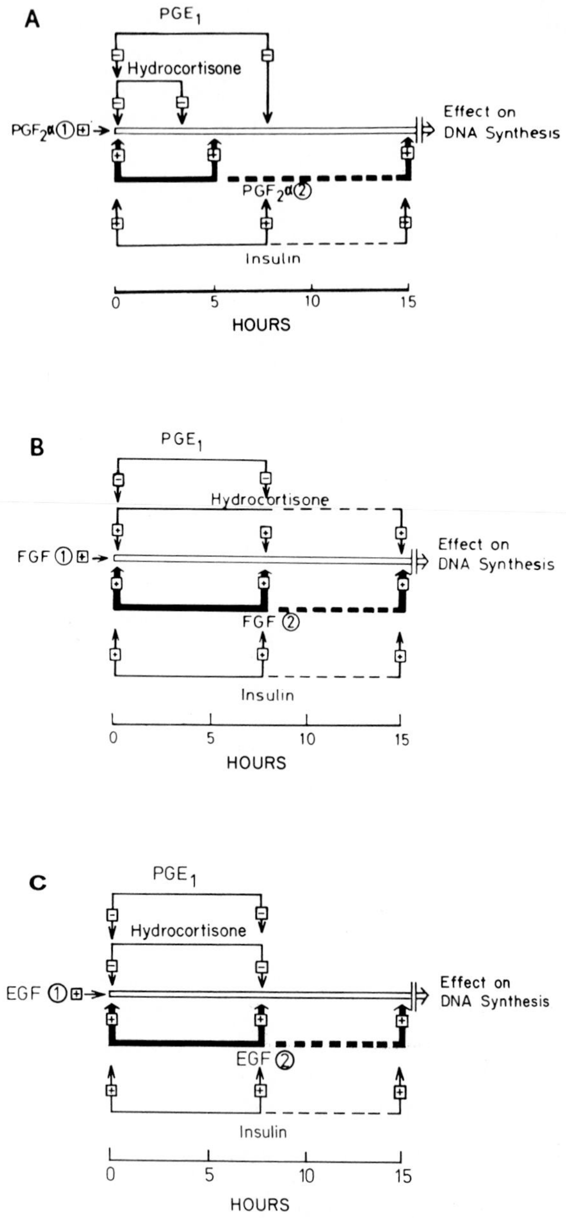

Fig. 1. Diagrammatic representation of the times when growth factors and hormones are required for changes in the kinetics of the initiation of DNA synthesis in Swiss 3T3 cells. (A) Interaction between $PGF_{2\alpha}$, insulin, hydrocortisone, and PGE_1. (B) Interaction between FGF, insulin, hydrocortisone, and PGE_1. (C) Interaction between EGF, insulin,

resume proliferation upon addition of serum (*4*) or defined growth factors such as prostaglandin $F_{2\alpha}$ ($PGF_{2\alpha}$) (*12,13*), fibroblast growth factor (FGF) (*9,25*), or epidermal growth factor (EGF) (*22,23*). This results in two consistent phenomena: (1) a constant prereplicative phase (lag phase) of about 15 hr, which is independent of the concentration of the growth factor above a minimal essential concentration; and (2) a subsequent abrupt increase in the rate at which the cell population initiates DNA synthesis, which is dependent on the growth factor concentration (*12, 14,15*). The latter process appears to follow first-order kinetics and can thus be conveniently quantified by a rate constant k (*5,13,27,30,31*). The rate of initiation of DNA synthesis can be modulated by additions of a growth factor or nonmitogenic compounds later in the lag phase (*13, 14,23–25*). These and other studies provided evidence that the initiation of DNA synthesis is preceded and regulated by a temporal sequence of events (*27,29*). The effects of some nonmitogenic compounds depend on the growth factor used to stimulate the cells and the time of interaction. For example, hydrocortisone inhibits the stimulatory effect of $PGF_{2\alpha}$ and EGF only when added within the first 5–8 hr of the lag phase (*13,22,23*), but with FGF it has a synergistic effect at any time (*25*). Also, microtubule-disrupting drugs such as colchicine and Colcemid have a synergistic effect that is dependent on the growth factor (*7,21,24*): The enhancement with FGF is twice that obtained with either $PGF_{2\alpha}$ or EGF alone, which requires insulin to achieve the same stimulatory effect. Some of these interactions are summarized in Fig. 1.

In this chapter, we show how the kinetics of initiation of DNA synthesis are modulated (1) upon stimulation of quiescent cells by additions of a single growth factor, (2) upon interaction of nonmitogenic compounds with cells stimulated by a single growth factor, and (3) when another growth factor interacts with the stimulated cells. The results provide evidence that each growth factor triggers a different program of events leading to the initiation of DNA synthesis. Parts of the program can be diminished or enhanced by nonmitogenic compounds acting through secondary mechanisms, thereby modulating the rate of initiation of DNA synthesis.

---

hydrocortisone, and PGE_1. In (A), (B), and (C), the solid lines represent the times of addition of the growth factors and hormones which give positive (+) or negative (−) alterations in the value of the rate constant k at the end of the lag phase. Broken lines represent the time of the additions which, at the end of the lag phase, give nonlinear increases in the apparent first-order kinetics for entry into S. These become linear after a few hours. The numbers represent the hypothetical signals delivered by $PGF_{2\alpha}$, FGF, or EGF and/or hormones. (A) Reprinted from Jimenez de Asua *et al.* (13); (B) reprinted from Richmond *et al.* (25); and (C) reprinted from Otto *et al.* (23).

II. Experimental System

A. Materials

We obtained EGF (*6*) from Collaborative Research, and $PGF_{2\alpha}$ was a generous gift of J. Pike, Upjohn Company. Crystalline insulin and thymidine were purchased from Sigma; [methyl-^{3}H]thymidine was obtained from the Radiochemical Center, Amersham, England.

B. Cell Cultures

Subconfluent cultures of Swiss mouse 3T3 cells (*33*) were grown in 90-mm petri dishes containing Dulbecco-Vogt's modified Eagle's medium (DME) supplemented with penicillin (100 units/ml), streptomycin (100 μg/ml), and 10% fetal calf serum (FCS). Cultures were kept at 37°C in a 10% CO_2 atmosphere and routinely monitored for the absence of mycoplasma contamination. New stocks of frozen cells were thawed at 2- to 3-month intervals.

C. Assay for the Initiation of DNA Synthesis and Determination of the Rate Constant for Entry into S Phase

Cells were plated at 1.5×10^5 in 30-mm dishes in 2 ml DME supplemented with 6% FCS and low-molecular-weight nutrients as described before (*12,20*). Cultures were allowed to become confluent and quiescent 3–4 days after an intermediate change with the same medium. They were used when no mitotic cells were observed, giving a very low labeling index (~ 0.8%) after continuous exposure to [methyl-^{3}H]thymidine for 28 hr.

To determine the labeling index of stimulated cultures, duplicates were radioactively labeled by exposure to [methyl-^{3}H]thymidine (3 μCi/ml, 1 μ*M*) from the time of the initial addition of a growth factor to the culture medium (in which the cells became quiescent) until the times indicated in each experiment. Dishes were then processed for autoradiography, and the percentage of radioactively labeled or unlabeled cells was determined after counting about 1800 cells per dish.

For determination of the apparent first-order rate constant (k), the percentage of unlabeled cells (y) in a given time (t) was plotted as the $\log_{10} y$ against t in hr. Straight lines given by $\log y = a - bt$ fit the data well (*12*). The value of k was then calculated from the slope of the curves (b), since $k = \ln 10\ b$. The length of the lag phase was estimated to within 1 hr. To determine k for quiescent cultures, cells were exposed to [methyl-^{3}H]thymidine for 7 days.

III. Action of a Growth Factor Alone

Adding EGF at a subsaturating concentration to the conditioned medium of confluent, quiescent Swiss 3T3 cells increased the basal rate of initiation of DNA synthesis after a lag phase of 15 hr (Fig. 2A). A saturating concentration of EGF increased the rate constant further without changing the length of the lag phase. When the subsaturating amount of EGF was supplemented with a saturating amount at 8 hr, the rate constant increased abruptly at the end of the lag phase to a value similar

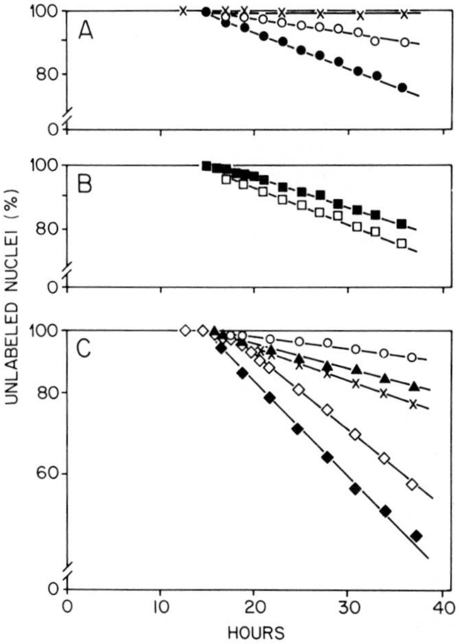

Fig. 2. Initiation of DNA synthesis by different concentrations of EGF alone or with insulin. (A) (×) no additions; (○) EGF (4 ng/ml); and (●) EGF (20 ng/ml). Values of k were 0.04, 0.88, and 1.67 × 10^{-2}/hr, respectively. (B) EGF (4 ng/ml) with EGF (16 ng/ml) added at (□) 8 hr or (■) 15 hr. The values of k were 1.66 and 1.60 × 10^{-2}/hr, respectively. (C) (○) EGF (2 ng/ml); (▲) EGF (20 ng/ml); (×) EGF (2 ng/ml) + insulin; (◇) EGF (2 ng/ml) + insulin, with EGF (18 ng/ml) added at 14 hr; (◆) EGF (20 ng/ml) + insulin. Final values of k were 0.4, 0.9, 1.2, 2.7, and 3.4 × 10^{-2}/hr, respectively. Insulin was added at 50 ng/ml. The value of k for no additions was calculated by exposing quiescent cultures to [methyl-^{3}H]thymidine over a period of 7 days for autoradiography. The length of the lag phase was 14.5 hr. [Reprinted from Otto *et al.* (23).]

to that obtained when the saturating amount of EGF had been present from the beginning. If, however, the saturating concentration of EGF was added at the end of the lag phase, it took 5 hr to increase the initial rate constant given by the low concentration of EGF (Fig. 2B). From these results and similar data from experiments with $PGF_{2\alpha}$ and FGF, it has been concluded that a growth factor delivers at least two different signals: signal 1 to induce the progression through the lag phase independent of the growth factor concentration, and signal 2 to regulate the rate of initiation of DNA synthesis according to the concentrations of the growth factor (*15*). Since about 5 hr are required when the same growth factor is added at the end of the lag phase to increase the value of the rate constant given by the initial low concentration of growth factor, it has been postulated that there is a "rate-limiting step" governing the final value of k about 5 hr before completion of the lag phase, i.e., at around 10 hr (*15*). Such a rate-limiting step has also been proposed from experiments using serum-stimulated cells by Brooks (*4*).

IV. Interaction of a Growth Factor with a Nonmitogenic Compound

One growth factor alone usually does not stimulate the maximal rate of initiation of DNA synthesis in Swiss 3T3 cells, and the differences between subsaturating and saturating concentrations appear relatively small. The results in Fig. 2C, however, show that the effect of EGF can be synergistically enhanced by insulin. Insulin alone at physiological concentrations does not stimulate DNA synthesis in confluent quiescent Swiss 3T3 cells. In the presence of insulin, a subsaturating amount of EGF can be supplemented by a saturating amount of EGF at 15 hr of the lag phase, which results in a marked increase in the rate constant at 20 hr, i.e., 5 hr later. Therefore, insulin does not alter the length of the lag phase or the time required to increase the rate constant, but it does have a synergistic effect on the rate of initiation of DNA synthesis.

Likewise, other nonmitogenic compounds, such as hydrocortisone and microtubule-disrupting drugs, alter the stimulatory effect of a growth factor by modulating the value of k. Colcemid added together with EGF produced the same increase in the rate constant as insulin, and it did not affect supplementary additions of EGF or insulin at the end of the lag phase, since the delay before the second increase in the value of k remained 5 hr (Fig. 3). However, Colcemid does not act identically to insulin, because it also enhanced the synergistic effect of EGF and insulin. Also in contrast to insulin, Colcemid and colchicine exert their

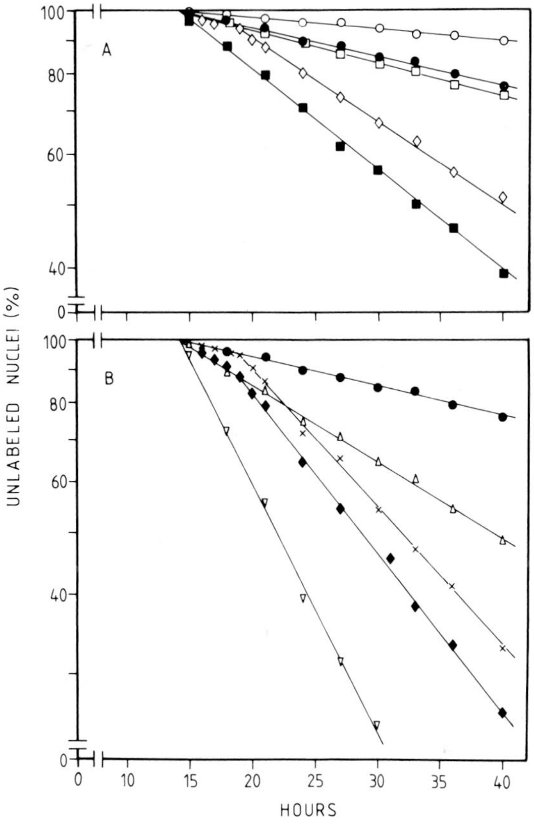

Fig. 3. Nonsynchronous additions of EGF in the presence of Colcemid and insulin. Concentrations used: Colcemid (1 μ*M*); and insulin (50 ng/ml). (A) Effect of Colcemid on second addition of EGF. (○) EGF (2 ng/ml); (□) EGF (20 ng/ml); (●) EGF (2 ng/ml) + Colcemid alone and (◇) with EGF (18 ng/ml) added at 14 hr; (■) EGF (20 ng/ml) + Colcemid. Final values of k were 0.4, 1.2, 1.0, 3.0, and 3.6 × 10^{-2}/hr, respectively. (B) Effect of Colcemid and insulin on second addition of EGF. (●) EGF (2 ng/ml) + Colcemid together and (×) with EGF (18 ng/ml) + insulin added at 14 hr; (△) EGF (2 ng/ml) + Colcemid + insulin together and (◆) with EGF (18 ng/ml) added at 14 hr; (▽) EGF (20 ng/ml) + Colcemid + insulin. Final values of k were 1.0, 5.0, 2.8, 5.7, and 9.0 × 10^{-2}/hr, respectively. The length of the lag phase in (A) and (B) was 14.2 hr. [Methyl-^{3}H]thymidine was present from the initial addition until the times indicated. [Reprinted from Otto *et al.* (24).]

enhancing effect only when present during the first 8 hr of the lag phase (*8,24*).

The interaction of hydrocortisone with one or two growth factors and its comparison with insulin will be presented in Section V,B,2.

V. Interaction between Growth Factors

Normal mammalian cells usually exist in an environment containing several different growth factors to which they could respond (*10,11,28*). This raises the question of how different growth factors interact to regulate the cell's mitogenic response: Does each growth factor stimulate the initiation of DNA synthesis by the same pathway, or does each act through a unique sequence of events? If their mechanisms are different, do two different growth factors cooperate with each other, or does one inhibit the other? In what follows, we will compare the kinetics of initiation of DNA synthesis of EGF, alone or plus insulin, with those of adding EGF and $PGF_{2\alpha}$ together to quiescent Swiss 3T3 cells. Their pattern of interaction with nonmitogenic compounds appears to be the same (Fig. 1); thus, these two growth factors seem to share a common sequence of time-dependent events. On the other hand, they differ profoundly in their chemical structure, so that it may be assumed that the initial event requires interaction with different cellular/molecular components (*14*).

A. Synergistic Effects between EGF, $PGF_{2\alpha}$, and Insulin

The addition of increasing concentration of one growth factor with a saturation concentration of the other is shown in Fig. 4. For EGF or $PGF_{2\alpha}$ alone, the dose–response curve reached a plateau of 15% labeled nuclei 28 hr after stimulation. Addition of insulin with EGF or $PGF_{2\alpha}$ increased the labeling index to 40% and lowered the saturation concentration of the growth factors. When EGF and $PGF_{2\alpha}$ were added together, the labeling index increased further, reaching a plateau at 55%, and even lower concentrations of either growth factor were required for a maximal effect. Insulin enhanced the synergistic effect between the two growth factors to a labeling index of 80% within 28 hr (Figs. 4A and B). Similar results were obtained when both EGF and $PGF_{2\alpha}$ were added at subsaturating concentrations, with or without insulin (Table I). At a concentration that gave very little stimulation for either growth factor, the two growth factors together gave almost the same labeling index as at saturating concentrations.

The synergistic effect between the growth factors and insulin, which by itself does not stimulate DNA synthesis, suggested that insulin is acting through mechanisms or events different from those induced by the growth factors and that these insulin-sensitive events have a cooperative effect on some regulatory events leading to the initiation of DNA

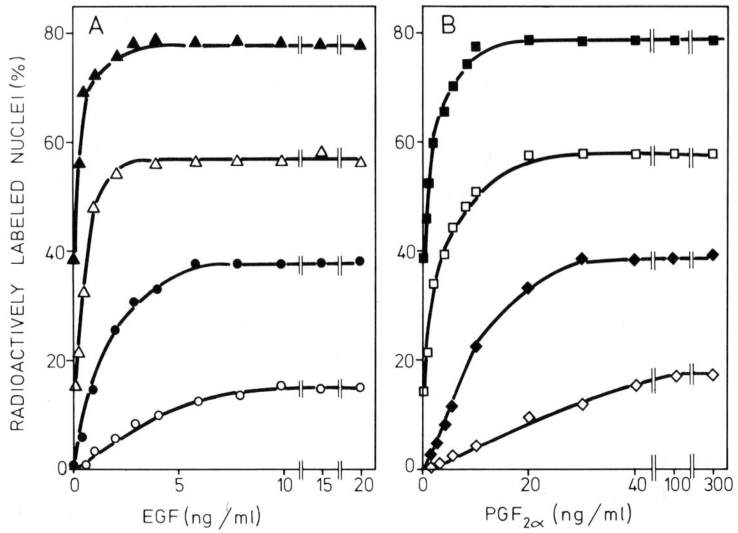

Fig. 4. Synergy between EGF, $PGF_{2\alpha}$, and insulin on the initiation of DNA synthesis. (A) (○) EGF; (●) EGF + insulin (50 ng/ml); (△) EGF + $PGF_{2\alpha}$ (300 ng/ml); (▲) EGF + insulin (50 ng/ml) + $PGF_{2\alpha}$ (300 ng/ml). (B) (◇) $PGF_{2\alpha}$; (◆) $PGF_{2\alpha}$ + insulin (50 ng/ml); (□) $PGF_{2\alpha}$ + EGF (20 ng/ml); (■) $PGF_{2\alpha}$ + insulin (50 ng/ml) + EGF (20 ng/ml). [Methyl-^{3}H]thymidine was added to the culture medium from 0 hr until 28 hr after additions. [Reprinted from Jiminez de Asua *et al.* (16).]

TABLE I

Synergy between Subsaturating Concentrations of EGF and $PGF_{2\alpha}$ with or without Insulin[a]

Additions	Labeling Index (%) Without insulin	With insulin
None	0.5	0.8
EGF (2 ng/ml)	5.0	26.1
$PGF_{2\alpha}$ (30 ng/ml)	9.9	45.1
EGF (2 ng/ml) + $PGF_{2\alpha}$(30 ng/ml)	49.1	73.1
EGF (20 ng/ml)	13.2	45.1
$PGF_{2\alpha}$ (300 ng/ml)	15.2	47.4
EGF (20 ng/ml) + $PGF_{2\alpha}$(300 ng/ml)	56.1	80.1
Serum (10%)	92.0	98.0

[a] The labeling index was determined as in Fig. 4. Cultures were exposed to [methyl-^{3}H]thymidine 0–28 hr after additions and then processed for autoradiography. Insulin was added at 50 ng/ml. [Reprinted from Jimenez de Asua *et al.* (*16*).]

synthesis. If insulin were only stimulating events also induced by a growth factor, no synergistic effect would be expected. The synergistic effect between EGF and $PGF_{2\alpha}$, therefore, indicates that these two growth factors also act through different programs of events. Nevertheless, they must have some events in common, at least those in which the two pathways are integrated and which ultimately regulate the initiation of DNA synthesis.

In a physiological environment, PGE_1 and PGE_2, molecules closely related to $PGF_{2\alpha}$, may be present, since they are products of the metabolism of unsaturated fatty acids. Does EGF have a synergistic effect with related prostaglandins or with fatty acids, which are precursors of prostaglandins? The results in Table II show that at a low concentration (30 ng/ml) only $PGF_{2\alpha}$ was able to act synergistically (*16*). At higher concentrations (300 ng/ml), PGE_1 and PGE_2 have some synergistic effect with EGF, possibly by interacting with the putative $PGF_{2\alpha}$ receptor. The

TABLE II

Effect of Different Prostaglandins and Fatty Acids on the Initiation of DNA Synthesis[a]

Additions	Labeling index (%)
None	0.6
PGE_1 (300 ng/ml)	0.5
PGE_2 (30 ng/ml)	0.4
PGE_2 (300 ng/ml)	1.0
$PGF_{2\alpha}$ (30 ng/ml)	10.1
$PGF_{2\alpha}$ (300 ng/ml)	15.1
Arachidonic acid (300 ng/ml)	0.5
Linoleic acid (300 ng/ml)	0.7
EGF (30 ng/ml)	12.1
+ Arachidonic acid (300 ng/ml)	12.7
+ Linoleic acid (300 ng/ml)	13.1
+ PGE_1 (30 ng/ml)	15.3
+ PGE_1 (300 ng/ml)	22.2
+ PGE_2 (30 ng/ml)	12.1
+ PGE_2 (300 ng/ml)	33.4
+ $PGF_{2\alpha}$ (30 ng/ml)	55.1
+ $PGF_{2\alpha}$ (300 ng/ml)	56.7
Serum	92.1

[a] Labeling index was determined as in Fig. 4. Cultures were exposed to [methyl-^{3}H]thymidine 0–28 hr after additions and then processed for autoradiography. Prostaglandins and fatty acids were dissolved in absolute ethanol and diluted so that the final concentration of ethanol in the culture medium was 0.01%. [Reprinted from Jimenez de Asua *et al.* (*16*).]

synergistic effects of $PGF_{2\alpha}$, thus, seem to be related somehow to its action as a growth factor in these cells.

B. Temporal Interactions during the Lag Phase

1. *EGF and $PGF_{2\alpha}$*

How do EGF and $PGF_{2\alpha}$ interact to change the kinetics of initiation of DNA synthesis? Adding EGF and $PGF_{2\alpha}$ together at saturating concentrations results in a dramatic increase in the rate of initiation of DNA synthesis after a lag phase of 14.5 hr (Fig. 5). Thus, the length of the lag phase was not changed as compared to the kinetics of either growth factor alone; the synergistic effect was observed only in the rate constant.

Can a heterologous growth factor exert its synergistic effect when added at any time in the lag phase, or can it interact only at specific

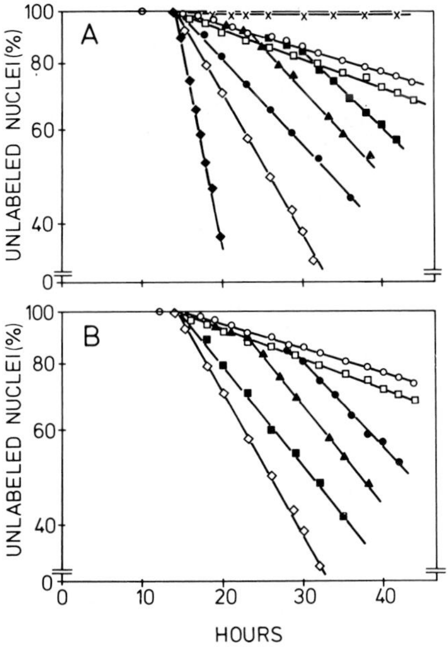

Fig. 5. Fraction of cells that remain unlabeled after synchronous or nonsynchronous additions of EGF (20 ng/ml) and $PGF_{2\alpha}$ (300 ng/ml) and fetal calf serum (10%). (A) (×) no additions; (○) EGF; (□) $PGF_{2\alpha}$; EGF with $PGF_{2\alpha}$ added (◇) at 0 hr, (●) at 6 hr, (▲) at 10 hr, or (■) at 15 hr; (◆) serum. The values of k were 0.06, 1.1, 1.3, 6.1, 3.1, 3.4, 3.3, and 24.3 × 10^{-2}/hr, respectively. The value of k for no addition was determined as in Fig. 2. (B) (□) $PGF_{2\alpha}$; (○) EGF; $PGF_{2\alpha}$ with EGF added (◇) at 0 hr, (■) at 6 hr, (▲) at 10 hr, or (●) at 15 hr. The values of k were 1.3, 1.1, 6.1, 4.3, 4.3, and 3.7 × 10^{-2}/hr, respectively. The length of the lag phase in (A) and (B) was 14.5 hr. [Reprinted from Jimenez de Asua *et al.* (16).]

times? Addition of $PGF_{2\alpha}$ 6 hr after EGF produced a synergistic effect on the rate of entry into the S phase upon completion of the lag phase, although the final value of k was somewhat less than when EGF and $PGF_{2\alpha}$ were added together (Fig. 5A). However, when $PGF_{2\alpha}$ was added 10 or 15 hr after EGF, a synergistic effect on the value of k was not observed until 15 hr after the second addition. Upon completion of the lag phase of 14.5 hr set by EGF, the initial rate of entry into S phase followed apparent first-order kinetics, with a value of k similar to that of EGF alone. Then, at 25 or 30 hr, respectively, the rate constant increased abruptly to a value similar to that obtained when $PGF_{2\alpha}$ is added at 6 hr of the lag phase. The same pattern of interaction was observed when EGF was added 6, 10, or 15 hr after $PGF_{2\alpha}$ (Fig. 5B). In either case, the synergistic effect is expressed at the end of the lag phase only when the second growth factor is added within 6 hr of the first. When the second growth factor is added 10 or 15 hr later, the synergistic effect is delayed, as if the second growth factor had to complete its own sequence of events requiring 15 hr before the final rate of initiation of DNA synthesis was adjusted.

What are the kinetics for the initiation of DNA synthesis when cells stimulated by a subsaturating concentration of a growth factor receive a saturating amount of another growth factor at a later time of the lag phase? Simultaneous additions of a subsaturating amount of EGF and a saturating amount of $PGF_{2\alpha}$ produced the same synergistic effect upon completion of the lag phase of 14.5 hr as if both the growth factors had been added at saturating concentrations (Fig. 6A). When cells stimulated with a subsaturating level of EGF received the saturating amount of $PGF_{2\alpha}$ at 15 hr, i.e., at the end of the lag phase, the synergistic effect was not observed until 15 hr later, i.e., at 30 hr. The kinetics were similar in the reverse case, in which a saturating level of EGF interacted with cells stimulated by a subsaturating concentration of $PGF_{2\alpha}$ (Fig. 6B). Thus, a subsaturating dose of the first growth factor did not result in changes in the times of interaction for the second growth factor.

2. *Effects of Insulin*

Does insulin affect the time of interactions between two growth factors? Insulin added at physiological concentrations with either EGF or $PGF_{2\alpha}$ markedly increased the rate of entry into S phase as shown previously (Fig. 2C; *12,13,22*). Addition of insulin with EGF and $PGF_{2\alpha}$ together also produced a dramatic enhancement of the rate constant (Fig. 7). However, the stimulation was still less than with serum. When $PGF_{2\alpha}$ was added at 6 hr to cells stimulated by EGF and insulin, the synergistic effect was similar to the one obtained when all three compounds were

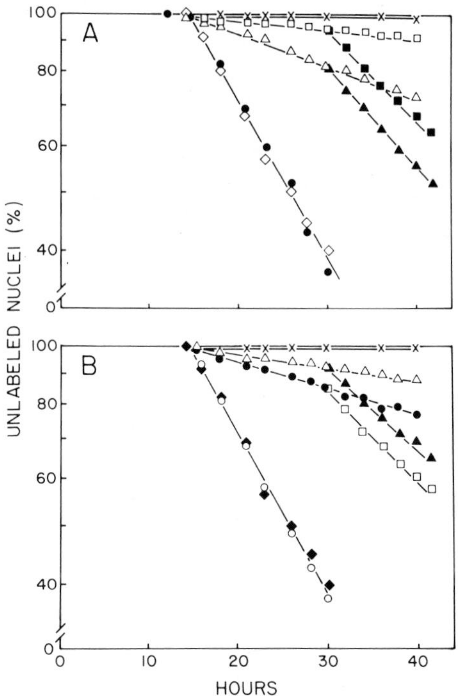

Fig. 6. Fraction of cells that remain unlabeled after synchronous and nonsynchronous additions of subsaturating and saturating concentrations of EGF and $PGF_{2\alpha}$. (A) (□) EGF (3.0 ng/ml); (△) $PGF_{2\alpha}$ (300 ng/ml); (●) EGF (3.0 ng/ml) + $PGF_{2\alpha}$ (300 ng/ml); (◇) EGF (20 ng/ml) + $PGF_{2\alpha}$ (300 ng/ml); (■) EGF (3.0 ng/ml) plus $PGF_{2\alpha}$ (300 ng/ml) at 15 hr; and (▲) EGF (20 ng/ml) plus $PGF_{2\alpha}$ (300 ng/ml) at 15 hr. The values of k were 0.3, 1.3, 6.5, 6.5, 3.5, and 3.7 × 10^{-2}, respectively. (B) (△) $PGF_{2\alpha}$ (60 ng/ml); (●) EGF (20 ng/ml); (○) $PGF_{2\alpha}$ (60 ng/ml) + EGF (20 ng/ml); (◆) $PGF_{2\alpha}$ (300 ng/ml) + EGF (20.0 ng/ml); (▲) $PGF_{2\alpha}$ (60 ng/ml) plus EGF (20 ng/ml) at 15 hr; and (□) $PGF_{2\alpha}$ (300 ng/ml) plus EGF (20 ng/ml) at 15 hr. The values of k were 0.64, 1.0, 6.5, 6.5, 3.1, and 3.5 × 10^{-2}/hr, respectively. The length of the lag phase was 14.5 hr in (A) and (B). The rate constant k was calculated from the final slope of the straight lines as described in Section II.

added together (Fig. 6A). This is in contrast to the lower synergistic effect observed when $PGF_{2\alpha}$ was added later to cells stimulated by EGF without insulin (Fig. 4). Nevertheless, as in the absence of insulin, when $PGF_{2\alpha}$ was added at 15 hr to cells stimulated by EGF and insulin, there was a further delay of 15 hr before the synergistic effect was obtained (Fig. 7A). The same pattern of interaction was obtained when EGF was added at 6 or 15 hr to cells stimulated by $PGF_{2\alpha}$ and insulin (Fig. 7B). Thus, the presence of insulin did not alter the basic pattern of interaction between these two growth factors.

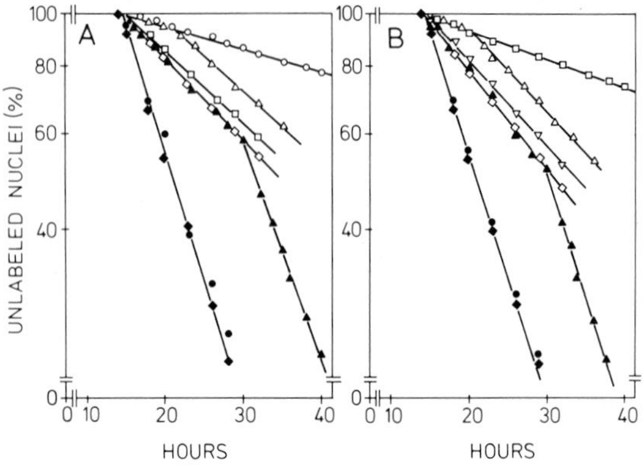

Fig. 7. Fraction of cells that remain unlabeled after synchronous or nonsynchronous additions of EGF (20 ng/ml), $PGF_{2\alpha}$ (300 ng/ml), and insulin (50 ng/ml). (A) EGF (○); EGF with insulin added (◇) at 0 hr, (□) at 6 hr, or (△) at 15 hr; EGF + insulin with $PGF_{2\alpha}$ added (♦) at 0 hr, (●) at 6 hr, or (▲) at 15 hr. The values of k were 1.0, 3.5, 3.1, 3.1, 10.9, 10.9, and 9.3 × 10^{-2}/hr, respectively. (B) $PGF_{2\alpha}$ (□); $PGF_{2\alpha}$ with insulin added (◇) at 0 hr, (▽) at 6 hr, or (△) at 15 hr; $PGF_{2\alpha}$ + insulin with EGF added (♦) at 0 hr, (●) at 6 hr, or (▲) at 15 hr. The values of k were 1.3, 4.2, 3.6, 3.6, 10.9, 10.9, and 10.2 × 10^{-2}, respectively. The length of the lag phase in (A) and (B) was 14.5 hr. [Reprinted from Jimenez de Asua *et al.* (16).]

How does the pattern of interaction between two growth factors compare with that between insulin and one or both growth factors? Insulin added at 6 hr after either EGF or $PGF_{2\alpha}$ alone resulted in a slight loss of the synergistic effect. But, in contrast to the later additions of a heterologous growth factor, addition of insulin at 15 hr after either EGF or $PGF_{2\alpha}$ resulted in a synergistic effect after a delay of only 5 hr (Fig. 4). When insulin was added 6 hr after EGF and $PGF_{2\alpha}$ together, there was no loss of synergism upon completion of the lag phase (Fig. 8). Adding insulin at the end of the lag phase set by EGF and $PGF_{2\alpha}$ together again resulted in a delay of 5 hr before the synergistic effect was observed. The effect of insulin added at different times in the lag phase is basically the same whether the cells were stimulated by one or by two growth factors.

In contrast to insulin, hydrocortisone inhibited the mitogenic effect of EGF or $PGF_{2\alpha}$ by decreasing the rate constant. Furthermore, this glucocorticoid exerted its effect only during the first 8 hr of the lag phase. When added together with both EGF and $PGF_{2\alpha}$, hydrocortisone had a marked inhibitory effect on the rate of entry into S phase without changing the length of the lag phase. As with a single growth factor, hydrocortisone added at the end of the lag phase or at a time when cells had already initiated DNA synthesis had no effect (Table III). The results

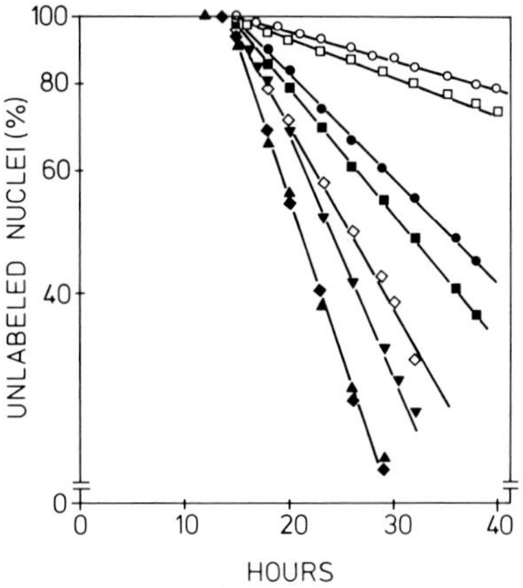

Fig. 8. Fraction of cells remaining unlabeled after synchronous and nonsynchronous additions of insulin (50 ng/ml) with EGF (20 ng/ml) and $PGF_{2\alpha}$ (300 ng/ml) together. (○) EGF; (□) $PGF_{2\alpha}$; (●) EGF + insulin; (■) $PGF_{2\alpha}$ + insulin; (◇) EGF + $PGF_{2\alpha}$; EGF + $PGF_{2\alpha}$ with insulin added at (◆) 0 hr, (▲) 6 hr, or (▼) 15 hr. The values of k were 1.0, 1.3, 3.5, 4.2, 6.1, 10.2, 10.2, and 8.1 × 10^{-2}/hr, respectively. The length of the lag phase was 14 hr. [Reprinted from Jimenez de Asua *et al.* (16).]

TABLE III

Effect of Hydrocortisone on the Rate of Initiation of DNA Synthesis in Swiss 3T3 Cells Stimulated by EGF and $PGF_{2\alpha}$ with or without Insulin[a]

Additions	Rate constant k (× 10^{-2}/hr)
None	0.1
EGF + $PGF_{2\alpha}$	6.5
EGF + $PGF_{2\alpha}$	
+ hydrocortisone at 0 hr	2.7
at 9 hr	6.5
at 15 hr	6.6
EGF + $PGF_{2\alpha}$ + insulin	10.8
EGF + PGF + insulin	
+ hydrocortisone at 0 hr	4.3
at 9 hr	11.0
at 15 hr	10.9
Serum 10%	24.3

[a] The value of the rate constant k was calculated as described in Section II. The length of the lag phase was 14.5 hr in all cases. The value for no additions was calculated as in Fig. 2. Concentrations were as follows: EGF (20 ng/ml); $PGF_{2\alpha}$ (300 ng/ml); and hydrocortisone (30 ng/ml). [Reprinted from Jimenez de Asua *et al.* (*16*).]

of the interactions of insulin and hydrocortisone with EGF and $PGF_{2\alpha}$ together have been interpreted as evidence for the existence of some common events stimulated by both EGF and $PGF_{2\alpha}$.

VI. Possible Interpretations

The interactions of a homologous growth factor (Fig. 2) and between heterologous growth factors (Fig. 5) could be thought of as the result of differently responding cell populations. However, the kinetics of initiation of DNA synthesis, as a result of interactions at different times during the lag phase, cannot be simply explained in that way. When a supplementary amount of EGF was added at the end of the lag phase set by a subsaturating concentration of EGF, it required only about 5 hr to increase the rate of entry into S phase (Fig. 2). If a new fraction of cells were stimulated by the saturating addition of EGF, one would expect a delay of 15 hr before an increase in the rate constant was observed. Furthermore, insulin, which alone does not stimulate DNA synthesis, has a synergistic effect with EGF, even when added at the end of the lag phase. It is more likely that EGF and insulin are interacting with the same cells to modulate the events regulating the rate of initiation of DNA synthesis. When EGF and $PGF_{2\alpha}$ are added together, there is also a synergistic effect upon completion of the lag phase (Fig. 4). If each growth factor were to stimulate a different population in the culture, merely additive effects on the number of cells entering S phase in a given time could be expected. Also, when one growth factor is added 6 hr after the other, the synergistic effect is still observed at the end of the lag phase. Only when the heterologous growth factor is added at later times is a delay in the synergistic effect observed. Although this result may be compatible with the idea of the existence of different populations, the other results, indicating a temporal interaction, require a more comprehensive interpretation.

As mentioned above (Section V,B), the synergistic effects between EGF and $PGF_{2\alpha}$ indicate that these two growth factors act through different programs of events. Nevertheless, these two programs may have in common those events which are enhanced by insulin and inhibited by hydrocortisone. Some common events are a prerequisite for the cooperation between the two programs resulting in the enhancement of the rate of initiation of DNA synthesis; and only when the two programs can interact simultaneously is the full synergistic effect observed. When one growth factor is added 6 hr after the other, which means that the induction of the second program is delayed, these two programs can apparently still cooperate, since the synergistic effect is observed upon completion of the lag phase. However, the synergistic effect is less than the one observed when the two growth factors are added together, which seems to indicate the existence of fewer events in common when there is a time gap of 6 hr between the addition of the two growth factors. When the heterologous growth factor is added at 10 or 15 hr, i.e., at the end of the lag phase, the events of the second growth factor can no longer be integrated with those of the first growth factor, which set in motion the program of the lag phase alone. Instead, it seems that the program of events set by the second growth factor, requiring 15 hr, needs to be completed before a synergistic effect is observed (Fig. 4). Some biochemical effects of the first growth factor persisting for longer times are possibly still interacting with some regulatory events triggered by the second factor so as to allow for a synergistic enhancement of the rate of entry into S phase.

This pattern of interaction between two different growth factors is in contrast to the pattern of interaction observed with different concentrations of a homologous growth factor (Fig. 2) or between a growth factor and insulin (Fig. 6). When insulin or a saturating concentration of a growth factor was added at the end of the lag phase, there was a delay of only 5 hr before the rate of initiation of DNA synthesis was observed. This latter pattern has also been observed with $PGF_{2\alpha}$ and FGF and seems to reflect a more general phenomenon intrinsic to the cell (*13,25*).

If one postulates that upon stimulation by a growth factor the same program of events is expressed in all cells, then one should expect that there are marked changes in specific biochemical events during the lag phase which correlate with the rate of initiation of DNA synthesis. The synergistic effects of EGF, $PGF_{2\alpha}$, and insulin are reflected in 2-deoxyglucose uptake, but only in the protein synthesis-dependent phase measured at 6 hr (*16*). Also, a number of different nuclear non-histone proteins appear upon stimulation by $PGF_{2\alpha}$ and insulin. In particular, one protein appears around 15 hr, which is markedly reduced when the stimulatory effect of $PGF_{2\alpha}$ and insulin is inhibited by hydrocortisone (*14*). This may be an example of the regulation of genetic expression induced upon stimulation by $PGF_{2\alpha}$ and insulin with hydrocortisone. Other reports have shown changes in the metabolism of specific proteins upon serum stimulation, some of which appear to be under transcriptional control (*26,32*). Also, in other mammalian cell lines, temperature-sensitive mutants have been isolated which become arrested near the G_1/S transition. In one case, three polypeptides that accumulate at the nonpermissive temperature have been shown to be under transcriptional control (*18,19*). In another study, two temperature-sensitive functions present in the cytoplasm can complement each other to allow the transition from G_1 to the S phase; they are apparently also under transcriptional control (*17*). It can be postulated that a growth factor could activate the transcription of all the genes required for progression through the lag phase. But it may not induce all the functions regulating the initiation of DNA synthesis with equal efficiency, resulting in relatively low stimulation. Nonmitogenic compounds could act by affecting the expression of gene products at the transcriptional or translational level, or they may induce new proteins, which could interact in a negative or cooperative way with the regulatory proteins expressed by the growth factor. Furthermore, another growth factor may differ in its transcriptional control of specific proteins essential for the initiation of DNA synthesis. The interaction of two different growth factors could then result in a more efficient or cooperative expression of the regulatory functions, which could explain the observed synergistic effects.

As yet, most of the experiments concerning the appearance of specific proteins or functions have been done with serum-stimulated cells. Stimulating quiescent cells with a specific growth factor, however, will help us to understand how the transcriptional and translational control of specific proteins regulates the initiation of DNA synthesis.

Acknowledgments

A. M. O. and L. J. de A. are grateful to Dr. Robert Holley and Marit Nielsen-Hamilton for encouragement, stimulation, and creative discussions while being Visiting Scientists at the Salk Institute, where part of this manuscript was written. We thank Dr. James A. Smith for helping us to clarify some of our concepts, Dr. George Thomas for useful discussions about the possible role of transcription and translational control involved in regulating DNA replication, and Dr. M. Eschenbruch, Dr. D. Monard, and J. Martin-Pérez for critically reading the manuscript. We also thank M.-O. Ulrich for her consistent and skillful assistance.

References

1. Armelin, H. A., Nishikawa, K., and Sato, G. H. (1973). Control of mammalian cell growth in culture: The action of protein and steroid hormones as effector substances. *In* "Control of Proliferation in Animal Cells" (B. Clarkson and R. Baserga, eds.). pp. 97–104. Cold Spring Harbor Lab., Cold Spring Harbor, New York.
2. Baserga, R. (1976). "Multiplication and Division of Animal Cells." Dekker, New York.
3. Baserga, R. (1978). Resting cells and the G_1 phase of the cell cycle. *J. Cell. Physiol.* **95,** 377–386.
4. Brooks, R. F. (1976). Regulation of the fibroblast cell cycle by serum. *Nature* (*London*) **160,** 248–250.

5. Brooks, R. F. (1977). Continuous protein synthesis is required to maintain the probability of entry into S phase. *Cell* **12,** 311–317.
6. Carpenter, G., and Cohen, S. (1979). Epidermal growth factor. *Annu. Rev. Biochem.* **48,** 193–216.
7. Friedkin, M. E., Legg, A., and Rozengurt, E. (1979). Antitubulin agents enhance the stimulation of DNA synthesis by polypeptide growth factors in 3T3 mouse fibroblasts. *Proc. Natl. Acad. Sci. U.S.A.* **76,** 3909–3912.
8. Friedkin, M. E., Legg, A., and Rozengurt, E. (1980). Enhancement of DNA synthesis by colchicine in 3T3 mouse fibroblasts stimulated with growth factors. *Exp. Cell Res.* **129,** 23–30.
9. Holley, R. W., and Kiernan, J. A. (1974). Control of the initiation of DNA synthesis in 3T3 cells: Serum factors. *Proc. Natl. Acad. Sci. U.S.A.* **71,** 2908–2911.
10. Holley, R. W. (1975). Control of growth of mammalian cells in culture. *Nature* (*London*) **258,** 487–490.
11. Holley, R. W. (1980). Control of growth of kidney epithelial cells. *In* "Control Mechanisms in Animal Cells" (L. Jimenez de Asua, R. Levi-Montalcini, R. Shields, and S. Iacobelli, eds.), pp. 15–25. Raven, New York.
12. Jimenez de Asua, L., O'Farrell, M. K., Bennett, D., Clingan, D., and Rudland, P. S. (1977). Interaction of two hormones and their effect on observed rate of initiation of DNA synthesis in 3T3 cells. *Nature* (*London*) **265,** 151–153.
13. Jimenez de Asua, L., O'Farrell, M. K., Clingan, D., and Rudland, P. S. (1977). Temporal sequence of hormonal interactions during the prereplicative phase in quiescent cultured fibroblasts. *Proc. Natl. Acad. Sci. U.S.A.* **74,** 3845–3849.
14. Jimenez de Asua, L., Richmond, K. M. V., O'Farrell, M. K., Otto, A. M., Kubler, A. M., and Rudland, P. S. (1979). Growth factors and hormones interact in a series of temporal steps to regulate the rate of initiation of DNA synthesis in mouse fibroblasts. *In* "Hormones and Cell Culture" (G. H. Sato and Ross, R., eds.), pp. 403–424. Cold Spring Harbor Lab., Cold Spring Harbor, New York.
15. Jimenez de Asua, L. (1980). An ordered sequence of temporal steps regulates the rate of initiation of DNA synthesis in cultured mouse cells. *In* "Control Mechanisms in Animal Cells" (L. Jimenez de Asua, R. Levi-Montalcini, R. Shields, and S. Iacobelli, eds.), pp. 173–197. Raven, New York.
16. Jimenez de Asua, L., Richmond, K. M. V., and Otto, A. M. (1981). Two growth factors and two hormones regulate initiation of DNA synthesis in cultured mouse cells through different pathways of events. *Proc. Natl. Acad. Sci. U.S.A.* **78,** 1004–1008.
17. Jonak, G. J., and Baserga, R. (1979). Cytoplasmic Regulation of two G_1-specific temperature-sensitive functions. *Cell* **18,** 117–123.
18. Melero, J. A., and Fincham, V. (1978). Enhancement of the synthesis of specific cellular polypeptides in a temperature-sensitive Chinese hamster cell line (K12) defective for entry into S phase. *J. Cell. Physiol.* **95,** 295–306.
19. Melero, J. A., and Smith, A, (1978). Possible transcriptional control of three polypeptides which accumulate in a temperature-sensitive mammalian cell line. *Nature* (*London*) **272,** 725–727.
20. O'Farrell, M. K., Clingan, D., Rudland, P. S., and Jimenez de Asua, L. (1979). Stimulation of the initiation of DNA synthesis and cell division in several cultured mouse cell types. Effect of growth promoting hormones and nutrients. *Exp. Cell Res.* **118,** 311–321.
21. Otto, A. M., Zumbé, A., Gibson, L., Kubler, A. M., and Jimenez de Asua, L. (1979). Cytoskeleton-disrupting drugs enhance effect of growth factors and hormones on initiation of DNA synthesis. *Proc. Natl. Acad. Sci. U.S.A.* **76,** 6435–6438.

22. Otto, A. M., Natoli, C., Richmond, K. M. V., Iacobelli, S., and Jimenez de Asua, L. (1981). Glucocorticoids inhibit the stimulatory effect of epidermal growth factor on the initiation of DNA synthesis. *J. Cell. Physiol.* **107,** 155–163.
23. Otto, A. M., Ulrich, M. O., and Jimenez de Asua, L. (1981). Epidermal growth factor initiates DNA synthesis after a time-dependent sequence of regulatory events in Swiss 3T3 cells. Interaction with hormones and growth factors. *J. Cell. Physiol.* **108,** 145–153.
24. Otto, A. M., Ulrich, M. O., Zumbé, A., and Jimenez de Asua, L. (1981). Microtubule-disrupting agents affect two different events regulating the initiation of DNA synthesis in Swiss 3T3 cells. *Proc. Natl. Acad. Sci. U.S.A.* **78,** 3063–3067.
25. Richmond, K. M. V., Kubler, A. M., Martin, F., and Jimenez de Asua, L. (1980). The stimulation of the initiation of DNA synthesis by fibroblasts growth factor in Swiss 3T3 cells: Interactions with hormones during the prereplicative phase. *J. Cell. Physiol.* **103,** 77–85.
26. Riddle, V. G. H., Dubrow, R., and Pardee, A. B. (1979). Changes in the synthesis of actin and other cell proteins after stimulation of serum-arrested cells. *Proc. Natl. Acad. Sci. U.S.A.* **76,** 1298–1302.
27. Rudland, P. S., and Jimenez de Asua, L. (1979). Action of growth factors in the cell cycle. *Biochim. Biophys. Acta* **560,** 91–133.
28. Sato, G. H. (1980). Cell culture, hormones and growth factors. *In* "Control Mechanisms in Animal Cells" (L. Jimenez de Asua, R. Levi-Montalcini, R. Shields, and S. Iacobelli, eds.), pp. 1–5. Raven, New York.
29. Scher, C. D., Shephard, R. C., Antoniades, H. N., and Stiles, C. D. (1979). Platelet-derived growth factor and the regulation of the mammalian fibroblast cell cycle. *Biochim. Biophys. Acta* **560,** 217–241.
30. Smith, J. A., and Martin, L. (1973). Do cells cycle? *Proc. Natl. Acad. Sci. U.S.A.* **70,** 1263–1267.
31. Smith, J. A., and Martin, L. (1974). Regulation of cell proliferation. *In* "Cell Cycle Controls" (G. M. Padilla, I. L. Cameron, and A. Zimmerman, eds.), pp. 43–60. Academic Press, New York.
32. Thomas, G., Thomas, G., and Luther, H. (1981). Transcriptional and translational control of cytoplasmic proteins following serum stimulation of quiescent Swiss 3T3 cells. *Proc. Natl. Acad. Sci. U.S.A.* **78,** 5712–5716.
33. Todaro, G. J., and Green, H. (1963). Quantitative studies of the growth of mouse embryo cells in culture and their environment into established lines. *J. Cell Biol.* **17,** 299–313.

III

Ionic and Membrane Modulations in the Cell Cycle

14

Modulation of Structure and Function of the Plasma Membrane in the Cell Cycle of Neuroblastoma Cells

S. W. DE LAAT AND P. T. VAN DER SAAG

I. Introduction

The plasma membrane of a mammalian cell is mainly composed of: (1) phospholipids, varying in their polar head group and in the degree of saturation of their acyl chains, (2) sterols, predominantly cholesterol, and (3) (glyco-)proteins. The membrane lipids are oriented in a bilayer configuration, forming a fluid matrix in which the (glyco-)proteins are embedded. Based on their degree of interaction with the hydrophobic part of the lipid matrix, intrinsic and extrinsic membrane proteins can be distinguished (*28*). The (glyco-)proteins might have different functions, such as transport sites for specific molecules and ions, receptor sites for external signaling molecules (e.g., growth factors and hormones), and enzymes for controlling the intracellular levels of critical cellular constituents. However, very few membrane (glyco-)proteins have been characterized so far, because of the technical difficulties involved in their purification. The functional properties of the plasma membrane are not

GENETIC EXPRESSION IN THE CELL CYCLE

ISBN 0-12-543720-X

only controlled by the chemical composition of the membrane as such, but also by the dynamic interactions between different membrane molecules, as well as with cellular components, such as cytoskeleton elements (*7*).

Extensive evidence is available implicating the cell surface as a primary site for the control of the cell cycle. A variety of plasma membrane properties have been shown to be modulated during cell cycle progression. These observations concern changes in the following properties: cell morphology and ultrastructural aspects of the cell membrane, lectin-mediated agglutinability, membrane microviscosity, expression of receptor sites, transport of ions and nutrients, and the activity of membrane-bound enzymes. For a review, see Bluemink and de Laat (*2*). Much less is known about the interrelationships between the different observed events and characteristics and their relevance to cell cycle progression.

To separate cause and effect within the complex of plasma membrane events that are possibly involved in cell cycle control and growth regulation, we consider it necessary to study various parameters in parallel in a suitable cell system. For several reasons murine neuroblastoma cell lines provide such a system. Neuroblastoma cells are embryonic tumor cells derived from neural crest neuroblasts. Their normal counterparts will differentiate into the anlagen of sympathetic ganglia or the adrenal medulla. Under *in vitro* conditions, irreversible morphological differentiation is usually induced by intracellular cAMP-elevating agents. This process is accompanied by growth arrest and loss of tumorigenicity of the cells. Depending on the clone, various differentiated functions of mature neurons can be expressed: formation of neurites; development of specialized ion channels and electrical exitability; induction of high activities of neuronal enzymes; and, in some cases, the formation of functional synapses. For a review, see Prasad (*24*). Neuroblastoma cells, like many other cells, can switch from a proliferative program to an irreversible differentiation pathway in G_1 (*23*). In this phase of the cell cycle, they apparently have the capacity to change their fate decisively in response to external stimuli. A proper knowledge of the regulatory circuits that control the cell cycle will therefore simultaneously create a basis for understanding the critical events involved in the initiation of cellular differentiation.

Attracted by their retained capacity to differentiate *in vitro* from a certain phase in the cell cycle, we decided a number of years ago to study the role of the plasma membrane in the regulation of the cell cycle in neuroblastoma cells. The considerations given before led to a rather diverse approach by which we hope to gain insight in the interrelationships and relevance of the structural, physicochemical, and functional membrane modulations in the cell cycle. In this chapter, we describe experiments employing biophysical, biochemical, and ultrastructural methods on synchronized and serum-stimulated murine neuroblastoma cells *in vitro* that, to us, are a first approach toward developing a coherent picture of the membrane-mediated mechanisms underlying the cell cycle.

II. Cell Cycle Kinetics

The two most prominent events in the life history of a replicating cell—mitosis and DNA synthesis—have been used widely to describe its progression in the cell cycle. Experimental determination of these events is relatively easy, and thus could be used to subdivide the cell cycle into its four well-known phases: M, G_1, S, and G_2. More detailed studies of the kinetics of proliferating cell populations and of the effects of experimentally induced changes in population growth rates revealed that variations in intermitotic times between individual cells are mainly due to varying G_1 durations. This has been taken as evidence that controlling events occur predominantly in G_1, and led to mathematical descriptions of the kinetic behavior of proliferating populations based on the occurrence of a random transition in G_1.

In the original model of Smith and Martin (*31*), the cell cycle consists of an A state comprising part of G_1, and a B phase containing the other phases of the cycle. The A state is a quasi-resting state in which cells can stay for a variable period of time, governed by a constant probability per unit of time (λ) of undergoing a transition to a B phase of a determinate duration (T_B). Assuming constant values of λ and T_B for all cells in a population, this model predicts an exponential distribution of the intermitotic times (T_i) within the culture. The parameters λ and T_B can be determined from the variation in T_i of individual cells, as observed in time-lapse films of the culture. Usually, these data are presented as a plot of the logarithm of the fraction of cells with intermitotic times larger than or equal to a certain time t as a function of this time t, which should yield a straight line with a slope equal to $-\lambda$ and an intercept at $t = T_B$ (α curve). However, α curves tend to show an initial curvature before linearity is reached (*27,31,35*), which interferes with an accurate determination of λ. This difficulty, supposed to originate from variability in T_B within the culture, could be overcome by plotting the logarithm of the fraction of sister cells having an absolute difference in intermitotic time (T_s) larger than or equal to an indicated time t, as a function of this time t (β curves), assuming a negligible difference in T_B for sister cells.

In general, a straight line is observed in β curves, with a slope equal to $-\lambda$, which provides strong support for this assumption and for the transition probability model of the cell cycle. Furthermore, it was found that the mean value of T_s ($\bar{T}_s$) is equal to its standard deviation, as predicted from a random transition.

The Neuro-2A cells used in most of our studies were adapted to grow in Dulbecco's modified Eagle's medium without bicarbonate, buffered to pH 7.5 with 25 m*M* *N*-2-hydroxyethylpiperazine-*N*′-2-ethanesulfonic acid (HEPES). Bicarbonate-free culture conditions are in particular useful in experiments requiring extensive manipulations, such as electrophysiological microelectrode studies (see below). Optimal growth rates were obtained by addition to the medium of 0.4 m*M* of each of the amino acids alanine, asparagine, glutamic acid, proline, and cysteine and supplementation with 10% fetal calf serum. Depending on the batch of serum and on the type of culture substratum, their culture generation time varies from 7.5 to 11 hr. In addition to this relatively short generation time, the differential attachment of mitotic and interphase cells to the substratum allows for cell synchronization with a reasonable yield through selective detachment of mitotic cells, without interference with cell metabolism (*10*).

Figure 1A shows a typical example of the progression through the cell cycle of a synchronized Neuro-2A cell population, as determined by [^{3}H]thymidine incorporation and cell counting. At a doubling time of about 10 hr, the duration of M, G_1, S, and G_2 are about 0.5, 2.5, 5, and 2 hr respectively. Besides measuring [^{3}H]thymidine incorporation to trace specific effects on the initiation and progression of DNA synthesis, we analyzed routinely printed sequences of time-lapse films of proliferating cultures in terms of the transition probability model. It appeared that, in particular, the analysis of family trees of proliferating Neuro-2A cells showed features (*37*) that could not be explained by the original Smith and Martin model or by the various modifications of this model presented so far (*6,21,33*). As observed in other cell types (*36*), the mean difference in intermitotic time between cell pairs was found to increase with descending family relationship between the cells, as can be seen from Fig. 1B. Besides the usual α curve, this figure gives β curves determined for sister cells, cousin cells, second cousin cells, and unrelated cells within an exponentially growing Neuro-2A cell population. Evidently, the slope of the β curve decreases with descending family relationship. Taking also into account the observed positively skewed normal distribution of T_i, these observations could be understood within the concept of the single random transition model by additionally assuming that λ and T_B show a variability within the culture, which increases with decreasing

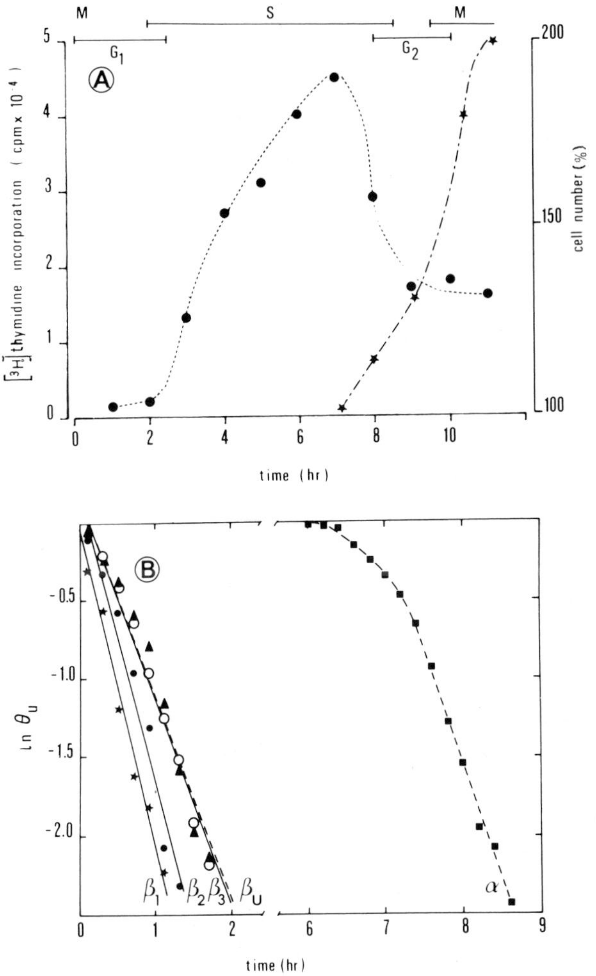

Fig. 1. Cell cycle kinetics of Neuro-2A cells. (A) Synchronization of Neuro-2A cells after selective detachment of mitotic cells: [^{3}H]thymidine incorporation (●) and percentage of cell number (★). From de Laat *et al.* (*10*), with permission of The National Academy of Sciences, Washington, D.C., U.S.A. (B) Analysis of cell proliferation of Neuro-2A cells, as a function of family relationship, according to the transition probability model: α curve (■----■) obtained from 112 intermitotic times, β_1 curve from 56 pairs of sister cells (★——★), β_2 curve of 112 pairs of cousin cells (●——●), β_3 curve of 224 pairs of second cousin cells (○——○), and β_u of 56 pairs of unrelated cells (▲----▲). θ_u represents the fraction of undivided cells. From van Zoelen *et al.* (*37*), with permission of Academic Press, Inc., New York.

relationship between the cells (*37*). Therefore, we formulated an improved mathematical description of the transition probability model taking into account a variability in the two system parameters λ and T_B in terms of a normal distribution, characterized by a mean value and a sample standard deviation (*37*). On the basis of this model, the behavior of the family tree of Neuro-2A cells, the shape of both α and β curves, and the nonexponential distribution of intermitotic times can be understood. In addition, this formulation allows the determination of the fraction of cells in the A state (ϕ_A), a very helpful parameter for comparing cell kinetics of cultures under various experimental conditions. As before, measurements of the intermitotic times of sister cells from time-lapse films provide the data for the analysis. Table I gives an example of a cell cycle analysis based on the modified description by van Zoelen *et al.* (*37*) of an exponentially growing Neuro-2A cell culture under optimal culture conditions. Inspection of these data shows that: (1) the mean intermitotic time is very short when compared to that of most mammalian cells in culture; (2) the value of the transition probability (λ = 2.03 hr^{-1}) is very high compared to that of other cell types (λ = 0.3–0.9 hr^{-1}; see Brooks *et al.* (*6*). The high value of λ indicates that the relatively small variation in T_i will come more from the variability in T_B than from the occurrence of a random transition in the cell cycle. Consequently, the distribution of T_i within the culture is more normal than exponential, whereas the difference in intermitotic times of sister cells clearly obeys an exponential distribution. The short generation time of these cells together with the possibility of using a simple mitotic shake-off procedure

TABLE I

Cell Cycle Analysis of Neuro-2A Cells under Optimal Culture Conditions[a]

Parameter	X	SD (X)	SEM ($\bar{X}$)
Intermitotic time, T_i (hr)	7.57	0.81	—
Difference in T_i between sister cells, T_s (hr)	0.55	0.49	—
Transition probability, λ (hr^{-1})	2.03	—	0.11[c]
Duration of A state, T_A (hr)	0.49	0.49	—
Duration of B state, T_B (hr)	7.08	0.64	—
Fraction of cells in A state, ϕ_A	0.086	—	0.004

[a] Data obtained from analysis of time-lapse film of exponentially growing Neuro-2A cells, frame interval time 0.2 hr, cell density at start of film about 10.000 cells/cm^2. In total, T_i of 56 pairs of sister cells were determined. From van Zoelen *et al.* (*37*), with permission of Academic Press, Inc.

[b] Only for parameters obeying a normal distribution.

[c] Obtained from linear regression analysis of β curve.

for synchronizing them in the cell cycle (*10*) make these cells a suitable system for cell cycle studies.

III. Dynamic Properties of Plasma Membrane Components

Alterations in the dynamic properties of plasma membrane lipids or proteins will affect membrane functions which could be involved in growth control. Alternatively, they might be direct manifestations of alterations in the functional state of particular membrane components by the occurrence of new constraints acting on them.

We have adopted two optical methods to measure dynamic membrane properties directly (*10,11*). The rotational mobility of the fluorescent hydrocarbon 1,6-diphenyl-1,3,5-hexatriene (DPH) was determined by fluorescence polarization measurements on intact synchronized Neuro-2A cells to monitor modulations in the viscosity of the hydrophobic region of the membrane lipid matrix (*10*). Since only the microenvironment of the probe molecule can exert influence on its rotational mobility, the results of such measurements are commonly expressed as microviscosity ($\bar{\eta}$, in poise). We found that $\bar{\eta}$ is maximal during mitosis (3.5 poise) and decreases rapidly in G_1 to reach a minimum (1.9 poise) at the onset of S. It remains constant at this low level during S and increases again during G_2.

An alternative approach to this problem was made by the application of the fluorescence photobleaching recovery method (FPR). By this method, the lateral motion of fluorescently labeled plasma membrane components over distances of a few micrometers can be measured (*1*). Fluorophores within a small area (usually about 3μm^2) are irreversibly photobleached by a short pulse of intense focused laser light. The rate of diffusion (diffusion coefficient, $\mathcal{D}$) of unbleached fluorophores from the surrounding membrane into the bleached region is then determined from the kinetics of the recovery of fluorescence in the bleached region. In this way, we studied the lateral mobility of both membrane lipids and proteins during the cell cycle of Neuro-2A cells (*11*). As fluorescent probes for the lateral mobility of membrane lipids, we used the lipid analogue 3,3′-dioctadecylindocarbocyanine iodide (diI, a gift from A Waggoner) and a fluorescein-labeled ganglioside (F-GM1, a gift from H. Wiegandt). Membrane proteins were labeled with rhodamine-conjugated rabbit antibodies (Fab′ fragments) against mouse El4 lymphoid cells (RaEl4, a gift from G. Edelman). These antibodies showed cross reactivity with an unknown cross section of the surface antigens of Neuro-

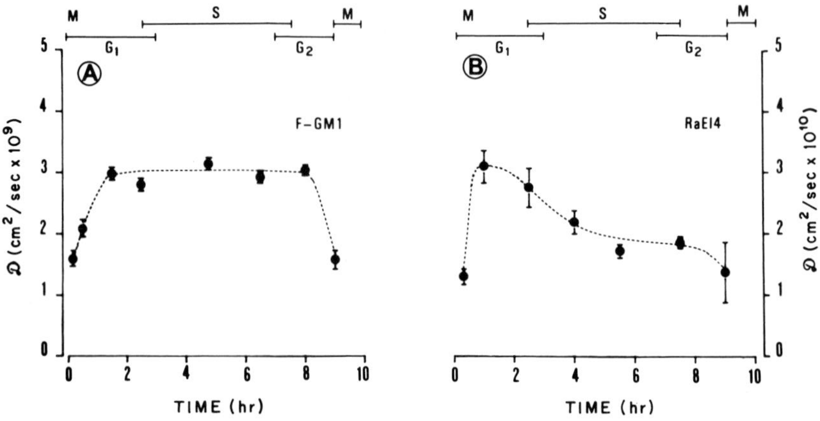

Fig. 2. Lateral mobility of membrane lipids (A) and membrane proteins (B) during the cell cycle of Neuro-2A cells. Diffusion coefficients ($\mathcal{D}$) are given as mean ± SEM. From de Laat *et al.* (*11*), with permission of The National Academy of Sciences, Washington, D.C.

2A cells. The results obtained with F-GM1 and RaEl4 are presented in Fig. 2. The two lipid probes diI and F-GM1 gave essentially identical results: The modulation in their lateral mobility during the cell cycle followed the changes in the apparent membrane lipid fluidity. The lateral mobility of membrane proteins, as revealed by RaEl4 labeling, changed partially in a different way. As found for the lipid probes, $\mathcal{D}$ was minimal in mitosis and increased rapidly during G_1. However, as cells proceeded through S and G_2, the diffusion coefficient for membrane proteins decreased gradually, reaching a minimum again in the next mitosis.

Based on these results we conclude that (1) the lateral mobility of membrane lipids is determined primarily by the fluidity of the membrane lipid matrix; (2) substantial changes in membrane lipid fluidity occur during the cell cycle, in particular around mitosis; (3) the lateral mobility of membrane proteins in M and G_1 is predominantly controlled by the membrane lipid fluidity, whereas other constraints become effective during S, possibly interactions with cytoskeleton elements; (4) the cell cycle-dependent modulations in the dynamic properties of membrane lipids and proteins might reflect changes in membrane functions that are pertinent to growth control.

IV. Structural Features of the Plasma Membrane

A high-resolution visualization of possible intrinsic modulations in the plasma membrane can be obtained by an ultrastructural analysis using freeze-fracture electron microscopy (*5*). In the replica of the fracture plane through the membrane lipid matrix, the observed intramembrane particles (IMP) most likely represent integral membrane (glyco-)proteins (*28*). Analysis of these views into the internal structure of the membrane in quantitative terms transforms this technique into a powerful method to detect subtle changes in the static and dynamic features of the plasma membrane. Alterations in the IMP density, i.e., the number of IMP per unit membrane area and the IMP size distribution, can give information as to, e.g., preferential insertion of (specific) lipids or proteins into the membrane and aggregation or dissociation of macromolecular complexes within the membrane. An analysis of the lateral IMP distribution, i.e., the spatial distribution of IMP within the plane of the membrane, provides a means for the interpretation of static membrane images in terms of the dynamic behavior of membrane components. In the absence of significant directional constraints on the lateral mobility of the molecules represented by the IMP, the IMP will acquire at any given time a random distribution. Directional forces acting on them will induce a nonrandom IMP distribution, whereby aggregated and dispersed (more uniform than random) distributions are distinguishable.

Recently, we have developed an improved method, based on a differential density distribution analysis, to quantify both the numerical and lateral IMP distribution as a function of the IMP diameter (*12*). To this end, the relative coordinates of the widest part of the shadow of each of the IMP in a selected electron micrograph (magnification: 252,000 ×) are digitized with a spatial resolution of 10 Å. They provide the basic data for further computer analysis. The IMP density and the IMP size distribution can be determined by simple counting procedures. The characterization of the lateral distribution of IMP subpopulations is based on a differential density distribution analysis. The electron micrograph is subdivided into square subunits with a given mesh size, and a frequency distribution (F) of the number of IMP of a selectable diameter range per subunit is determined. For a random distribution F will follow a Poisson distribution. Deviations from a random distribution can be characterized by calculating the approximate normal deviate (Z) in a Poisson variance test. Positive and negative Z values indicate IMP aggregation and dispersion, respectively, whereas $Z = 0$ for an ideal random distribution. The 1% significance level corresponds to an absolute Z value of 2.33. An important feature of this method is the selectable mesh size for the subdivision of the visualized membrane area. By varying the mesh size, the nature of the lateral IMP distribution can be characterized at different levels of spatial organization: short-range molecular interactions will express themselves as a nonrandom distribution at a small mesh size, whereas macroscopic heterogeneities in IMP density will become evident at larger mesh sizes.

Modulations in the static and dynamic ultrastructural features of the plasma membrane in Neuro-2A cells were analyzed by freeze-fracture electron microscopy of synchronized cells at 14 time points during the cell cycle (*13*). In general, 10 replicas, each representing 0.5 μm^2 of the P face of the medium-exposed part of the plasma membrane, were selected for each time point, in which the coordinates of in total 72,462 IMP were digitized for the computer analysis, as described above. The IMP diameter ranged from 30 to 170 Å. Although the analysis allows for a detailed analysis of the behavior of IMP subpopulations, with a resolution in the IMP diameter of 10 Å, we will restrict ourselves here to a presentation of the main characteristics. The modulations in the density of small ($\phi < 90$ Å) and large ($\phi \geq 90$ Å) IMP are shown in Figs. 3A and B, and the changes in the lateral distribution of the total IMP population at mesh sizes of 510 Å and 1020 Å, respectively, are shown in Fig. 3C.

For an interpretation of the data, it should be realized that membrane growth during the cell cycle is not linear (*3*). The surface of Neuro-2A cells remains constant from mitosis until the G_1/S transition. From then on, it increases gradually by 30% till mid S phase, after which it remains constant again till G_2. Prior to or during mitosis, the surface area suddenly increases and roughly doubles relative to that during G_1.

Mitosis is characterized by an intermediate total IMP density, greatly varying within a single cell, and a relatively large proportion of small IMP. As we took into account only membrane areas without surface protrusions, and mitotic cells in particular are covered by numerous IMP-free microvilli and surface blebs, the measured IMP density is certainly a gross overestimate. Probably, these protruding membrane areas provide a large storage of segregated, newly incorporated lipids (*2*). The local variations in IMP density result in an aggregated lateral distribution at both mesh sizes, indicating the action of significant directional constraints on the lateral mobility of the molecules represented. This is supported by the measured restricted mobility of membrane lipids and proteins at this stage (see Section III). It is tentative to speculate that the constraints on the mobility are caused by the segregation of lipids. As the cells start to flatten and reattach in early G_1 (0.5 hr), the total surface area remains constant, but the surface protrusions begin to disappear. During this process, a radical reorganization takes place within the plasma membrane. The density of large IMP doubles greatly at the expense of small IMP, as if new macromolecular complexes were being

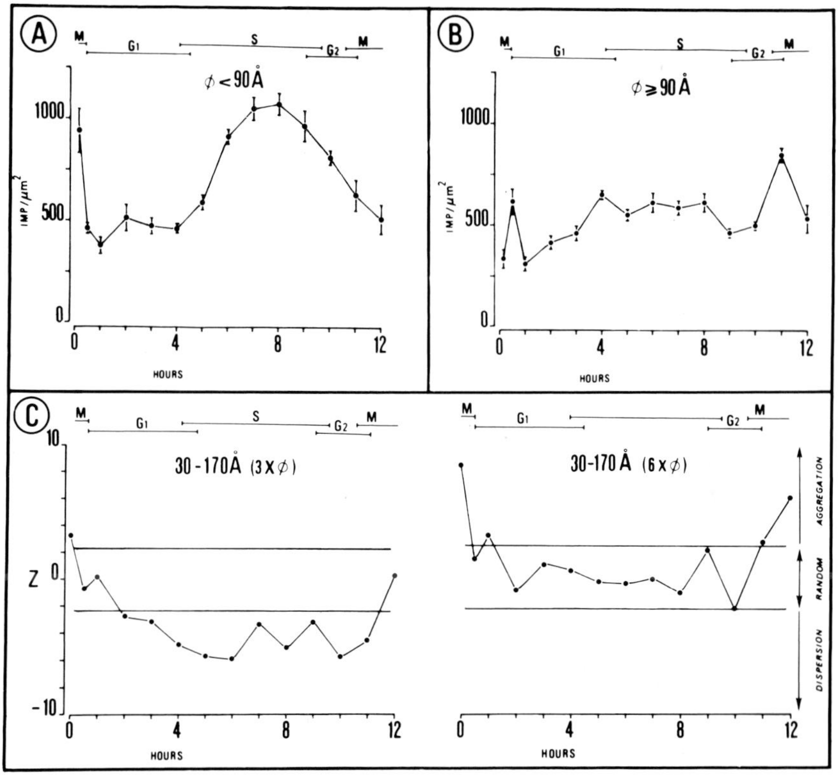

Fig. 3. Numerical and lateral distribution of intramembrane particles (IMP) during the cell cycle of Neuro-2A cells. The density of small IMP (diameter <90Å) and large IMP (diameter ≥90Å) are given as mean ± SEM in (A) and (B), respectively. The lateral distribution of the total IMP population (C) is determined from a differential density distribution at mesh sizes of three and six times the maximum IMP diameter, respectively. The horizontal lines indicate the 1% significance levels for the approximate normal deviate Z, as given by the Poisson variance test.

formed from smaller subunits present earlier. At the same time, the IMP acquire a random lateral distribution, under all the conditions analyzed, while the lateral mobility of both membrane lipids and proteins increases (see Section III). Apparently, the constraints detected in mitosis are released, and the heterogeneities in the plane of the membrane disappear. The increased membrane lipid fluidity could be the origin of this.

During the following 30 min in the cell cycle, the mobility of membrane proteins and lipids reaches a maximum (see Section III), and, correspondingly, the lateral IMP distribution is random under all conditions tested. At present, we have no simple explanation for the specific loss in large IMP observed during this period.

In the remaining part of G_1 (till 4 hr) a preferential insertion of large IMP-forming molecules is observed. Gross spatial heterogeneities in the IMP density distribution remained negligible, as seen by random distribution at the large mesh size. However, new constraints become apparent by the gradual change to a dispersed distribution at the smaller mesh size. Since the lateral mobility of membrane proteins also decreases in this period, as opposed to the lipid mobility (see Section III), anchorage of membrane proteins to cytoskeletal elements is a possible explanation for this observation.

In contrast to G_1, an exclusive increase in the density of the small IMP is seen during S as the surface area grows, indicating a preferential insertion of lipids and small IMP-forming components. The lateral IMP distribution remains similar to that of late G_1 cells.

Despite the inevitably partial desynchronization at the end of the cell cycle, and the resulting relatively poor time resolution, it is clear that a radical reorganization takes place within the plasma membrane just prior to the next mitosis. The surface area increases from 1.4 to about 2 times that of the G_1 cell (*3*), requiring at least a rapid insertion of membrane lipids. Possibly, this leads to the lipid segregation into the surface protrusions of the mitotic cell. Correspondingly, the density of small IMP decreases, but the large IMP show a transient increase in density, for which we do not have a simple interpretation. The dispersive constraints on the IMP are released, since the IMP acquire a random distribution at the smaller mesh size. However, macroscopic heterogeneities in the density distribution are apparent by the IMP aggregation at the larger mesh size. They could be a reflection of the rapid insertion of (specific) lipids, which also could be the origin of the rapid decrease in membrane lipid mobility and the minimal protein mobility.

These results give evidence for preferential insertion of various membrane components in specific phases of the cell cycle. Consequently, during the life history of a cell, the plasma membrane undergoes a continuous reorganization of which the modulations in the structural features of the plasma membrane are direct reflections. The driving forces for the modulations in the spatial organization of the plasma membrane can arise from interactions between molecules within the membrane or from interactions between membrane components and cytoplasmic elements, e.g., the cytoskeleton. The dynamic properties of membrane components (see Section III) and membrane functions (see Section V) could be controlled by similar mechanisms. In this way, the plasma membrane could fulfill its key role in growth control.

V. Cation Transport and Electrical Membrane Properties

The cell surface is the natural barrier between the intracellular environment and the surrounding milieu and the major site for controlling the compositional differences between these aqueous media. Literally, a cell consumes a lot of metabolic energy in doing so. In eukaryotic cells, a major consumer is the plasma membrane-bound Na^+,K^+-ATPase. This membrane-bound enzyme is primarily responsible for setting the intracellular Na^+ and K^+ at their relatively low and high respective levels against their concentration gradients across the plasma membrane, whereby ATP is the source of energy (*29*). From the general occurrence of this metabolically dependent ion translocator, it can be inferred that the resulting intracellular ionic conditions are a prerequisite for the cell to fulfill its functions. The cell makes use of these transmembrane gradients and of the electrical gradient, caused by the charge separation involved, to transport other ions and solutes across the plasma membrane. In addition, basic cellular processes such as protein synthesis depend on the intracellular Na^+ and K^+ concentrations (*14*). Therefore, modulations in the active and passive cationic transport properties can give rise to pleiotropic signals involved in growth control (see Chapter 15).

We studied cation transport during the cell cycle of Neuro-2A cells by electrophysiological and tracer flux methods (*3,4*). A first indication of cell cycle-dependent changes in these properties was found by monitoring the membrane potential (E_m) in synchronized cultures with conventional microelectrodes (Fig. 4A). E_m is maximally hyperpolarized in mitosis (−45 mV), depolarizes rapidly in G_1 to −23 mV, and gradually hyperpolarizes again from the onset of S to −45 mV in mid-S. Throughout the remainder of S and G_2, E_m is constant. E_m is determined predominantly by the K^+ gradient across the membrane and the membrane ion selectivity. Direct measurements of the intracellular K^+ activity (a_K^i), using K^+-selective microelectrodes, revealed that the K^+ gradient is modulated largely in parallel with E_m. Maximal values are observed in mitosis (127 m*M*). During G_1, a_K^i decreases sharply to 80 m*M*, followed by an increase to 118 m*M* in mid-S, after which it remains about constant (Fig. 4B).

Further details were unraveled by studying the efflux and influx of ^{42}K (*3,4,20*). As shown in Fig. 4C, the unidirectional K^+ efflux rate (J_{eff}) is maximal in mitosis at 6.5 pmole $K^+/sec/cm^2$. A sevenfold decrease in J_{eff} is observed upon entering into G_1, followed by a gradual threefold increase during S to 2.9 pmole $K^+/sec/cm^2$. After a transient decrease in G_2, J_{eff} increases largely before or during the next mitosis. As the K^+

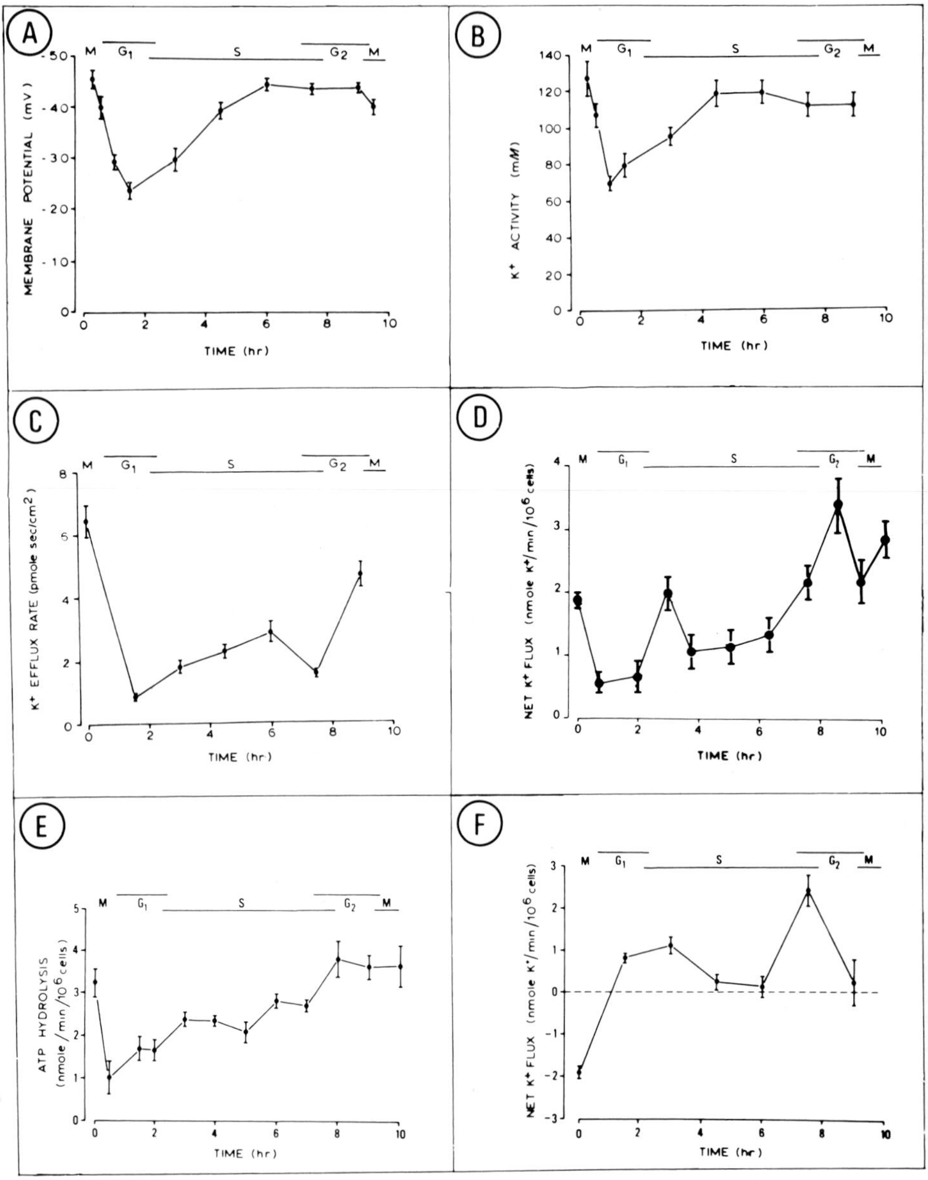

Fig. 4. Electrical membrane properties and K^+ transport properties during the cell cycle of Neuro-2A cells. (A) Membrane potential (E_m, as determined by intracellular microelectrode recording. (B) Intracellular K^+ activity, as determined by intracellular recording with ionselective microelectrodes. (C)K^+ efflux rate, as determined by ^{42}K efflux measurements. (D) Na^+,K^+-ATPase-mediated K^+ influx, as determined by measurements

permeability depends on J_{eff} and E_m, it follows a qualitatively similar pattern during the cell cycle (*3*).

The K^+ influx is made up of at least two components: a ouabain-sensitive part, mediated by the Na^+,K^+-ATPase, and a ouabain-insensitive part, through, e.g., exchange mechanisms. Measurements of ^{42}K influx and of Na^+,K^+-ATPase activity in cell homogenates revealed characteristic modulations of this ion pump mechanism during the cell cycle (*20*). The ouabain-insensitive K^+ influx increases linearly through the cell cycle, as does cellular protein. In contrast, the Na^+K^+-ATPase-mediated K^+ influx is modulated in a characteristic way (Fig. 4D). This component decreases rapidly more than fourfold as cells pass from mitosis to G_1. After a transient sixfold rise on entry into S, a gradual increase occurs with progression through S. Immediately prior to the next mitosis, a small but significant decrease is detectable. Mechanisms of two possible categories can influence the Na^+,K^+-ATPase-mediated K^+ influx: (1) The surface density or the intrinsic properties (substrate affinities) of the Na^+,K^+-ATPase can vary; and (2) the levels of intracellular substrates can change, whereby Na^+ is the most important candidate. A distinction between these modes was obtained by comparing the functional activity of this ion pump, as determined by the ouabain-sensitive K^+ influx (Fig. 4D), with the activity of the Na^+,K^+-ATPase of cell homogenates under optimal substrate conditions (Fig. 4E). Only mechanisms of the first category would modulate activity of the enzyme in cell homogenates. Taking into account that two K^+ ions are transported at the expense of hydrolyzing one ATP molecule, it is evident that in all phases of the cell cycle only part of the full capacity of NA^+,K^+-ATPase is used for translocating K^+ across the plasma membrane. A drastic decrease in activity in early G_1 can be seen under both conditions, indicating a rapid reduction in the number of enzyme copies per cell (cell surface area is not changing) or a conformational change of the enzyme. Based on the rapid changes described earlier in dynamic and structural properties of the plasma membrane in this period, we still favor the latter possibility. However, determination of the number of Na^+,K^+-ATPase copies per cell and of the affinities for various substrates will give the definitive explanation. During further progression in the cell cycle, the functional and optimal activities show a rather different pattern of change, the most prominent

---

of the ouabain-sensitive ^{42} influx. (E) Na^+,K^+-ATPase activity measured by ATP hydrolysis under optimal substrate conditions in cell homogenates. (F) Net K^+ flux determined as the difference between the total K^+ influx and efflux. Positive values represent net influx. In all cases, the data are presented as mean ± S.E.M. (A), (B), and (C) from Boonstra *et al.* (*3*), with permission of Alan R. Liss, Inc., New York. (D), (E), and (F) from Mummery *et al.* (*20*), with permission of Alan R. Liss, Inc., New York.

discrepancy being the transient sixfold increase in ouabain-sensitive influx rate upon entry to S, which is not detectable in cell homogenates under optimal substrate conditions. Preliminary observations indicate that this increased pump rate is caused by the switching on of an amiloride-sensitive Na^+ influx component, which is silent before that time. In its turn, this would lead to an increased level of intracellular Na^+, which acts as substrate for the Na^+,K^+-ATPase on the inner side of the membrane. Consequently, the pump rate will be enhanced. In analogy with our observations on serum stimulation of quiescent cells (see Section VI), we expect that the amiloride-sensitive Na^+ influx component is coupled to a H^+ efflux, so that a change in intracellular pH might occur as the result of these sequential events.

From the observed changes in intracellular K^+ activity (Fig. 4B) and K^+ content [data not shown, see Boonstra *et al.* (*3*)], and the absence of apparent intracellular K^+ sequestering, it can be inferred that there is net K^+ influx in either direction during various periods of the cell cycle. Calculation of the difference between the total unidirectional K^+ influx and efflux rates reveals this clearly (Fig. 4F). A net loss of K^+ occurs with the transition from mitosis to G_1, while K^+ is accumulated with entry into S. In mid-S, net K^+ flux is virtually absent, but a large net influx occurs just prior to the next mitosis.

Having characterized the modulations in these ion transport systems, the question arises whether these phenomena play a regulatory role in the cell cycle. As a first approach, we have analyzed the relevance of the sudden increase in Na^+,K^+ pump activity on entry into S by studying the effects of various concentrations of ouabain on DNA synthesis and cell cycle kinetics (*20*). Complete inhibition of the Na^+,K^+-ATPase activity of exponentially growing cells by addition of 5 m*M* ouabain to the medium leads to a gradual but reversible decrease in DNA synthesis. It takes more than 20 hr, i.e., two normal generation times, before [^{3}H]thymidine incorporation is completely abolished (data not shown). On the other hand, addition of ouabain at the same concentration to synchronized cells in G_1 completely blocks their entry into S. This is in agreement with the results obtained from analyzing the effects of sublimiting concentrations of ouabain on the intermitotic times of sister cells in nonsynchronized cultures (Table II). The concentration-dependent inhibition of the Na^+,K^+ pump by ouabain (K_m = 0.2 m*M*; *38*) results in a more pronounced effect on λ than on T_B, i.e., the cells accumulate in the A state, before the entry into S.

We conclude that the properties of the passive and active transport systems for the principal cell ions K^+ and Na^+ are modulated significantly during the cell cycle and that at least in some critical phases of the cell

TABLE II

Effects of Ouabain on Cell Cycle Kinetics of Neuro-2A Cells[a]

Parameter	Ouabain concentration (mM)		
	0	0.2	0.5
Generation time, T_g (hr)	10.30 ± 0.40	15.16 ± 0.76	20.29 ± 0.86
Transition probability, λ (hr^{-1})	1.39 ± 0.06	0.66 ± 0.06	0.35 ± 0.01
Duration of B state, B_B (hr)	9.60 ± 0.30	13.73 ± 0.69	17.53 ± 0.74
Fraction of cells in A state, ϕ_A	0.09 ± 0.04	0.13 ± 0.01	0.18 ± 0.01
Number of cells	20	34	32

[a] Data obtained from an analysis of time-lapse films, according to van Zoelen *et al.* (*37*), of exponentially growing cultures in the absence or presence of ouabain. Means ± SEM are given. From Mummery *et al.* (*20*) with permission of Alan L. Liss, Inc. New York.

cycle, e.g., G_1/S transition, these modulations provide necessary conditions for a normal further progression of the cells. Two major questions remain as yet largely unresolved: (1) How does the cell control these membrane transport properties, and to what extent are the observed alterations due to the same molecular mechanisms as those involved in the described dynamic and structural membrane modulations; and (2) through which pathways do these transport changes exert their influence on cell cycle-dependent metabolic and synthetic processes? With regard to the first question, the observed synchrony of the described changes in dynamic, structural, and functional plasma membrane properties is especially suggestive of a common molecular basis, but only detailed interdisciplinary studies can shed light on this problem. The second question is even more difficult to answer at present. A direct control of the K^+ and Na^+ transport properties over the transport of "second messenger" ions, such as Ca^{2+} and H^+, and nutrients, such as amino acids and glucose, can be assumed. Interactions with mitochondrial energy metabolism, but also effects of the internal K^+/Na^+ balance on protein synthesis, are to be expected. We hope that a further step-by-step analysis will unravel these pleotropic control mechanisms.

VI. Growth Stimulation and Cation Transport

In the studies described so far, we have been using synchronized Neuro-2A cell cultures to obtain information on cell cycle-related plasma membrane properties. An alternative approach to separate cause and effect at the plasma membrane level, in particular in relation to the G_1/S transition, is the analysis of the sequence of events upon growth stimulation of quiescent cells by serum or growth factors.

It is generally accepted that growth factors initiate their action by binding to specific receptors on the plasma membrane, thereby eliciting a cascade of physiological and biochemical events that ultimately lead to the initiation of DNA synthesis and cell multiplication (see Chapter 13). Within minutes following serum stimulation of quiescent animal cells, a marked stimulation of the Na^+,K^+-ATPase has been observed (*25, 30,34*), which appears to result from a rapid increase in Na^+ influx rate (*15,32*). How these early changes in cation movement are initiated by the ligand–receptor interactions, and through what mechanisms they are related to the initiation of DNA synthesis, is not understood.

Mouse N1E-115 neuroblastoma cells are a favorable system for investigating the mode of action of growth factors in the regulation of growth and differentiation (*18,19*). Under optimal culture conditions, these cells have a doubling time of about 24 hr. Serum deprivation leads to an arrest in G_1 and in the initiation of neuronal differentiation within about 24 hr. As shown in Figs. 5A and B, readdition of 10% fetal calf serum (FCS) or dialyzed FCS (lacking the low-molecular-weight serum components, MW < 1200) results in a synchronous reentry into S after an 8–10 hr lag period, and an increase in cell number 10 hr later. The Na^+-influx inhibitor amiloride, which suppresses DNA synthesis in a variety of cells (*9,26*) including fertilized sea urchin eggs (*8*), is also a potent inhibitor of serum-induced DNA synthesis and cell multiplication in N1E-115 cells (*19*).

The N1E-115 cells are convenient cells for monitoring the very early effects on ion transport, because their large average cell size permits stable penetration by intracellular microelectrodes over prolonged periods (*16,17*). Addition of dialyzed FCS after 24 hr of serum deprivation induces, within seconds, a transient hyperpolarization (H phase) followed by a slow depolarization (D phase) of 5–8 min (Fig. 6), during which the membrane potential gradually repolarizes to a new resting value, which is 6.5 mV less negative than the mean prestimulation value of -38 mV (*18*). A striking change in membrane resistance accompanies the electrical response. Growth-depleted FCS neither induces DNA synthesis nor elicits a similar electrical effect. Furthermore, amiloride and oubain have no effect on this electrical response, which indicates that amiloride acts on electrical silent ion currents and that the Na^+,K^+ pump is not directly involved in mediating the response. A careful analysis of the ionic basis of this response showed that the H phase is mainly due to an increase in K^+ permeability, whereas the D phase reflects a rather unspecific

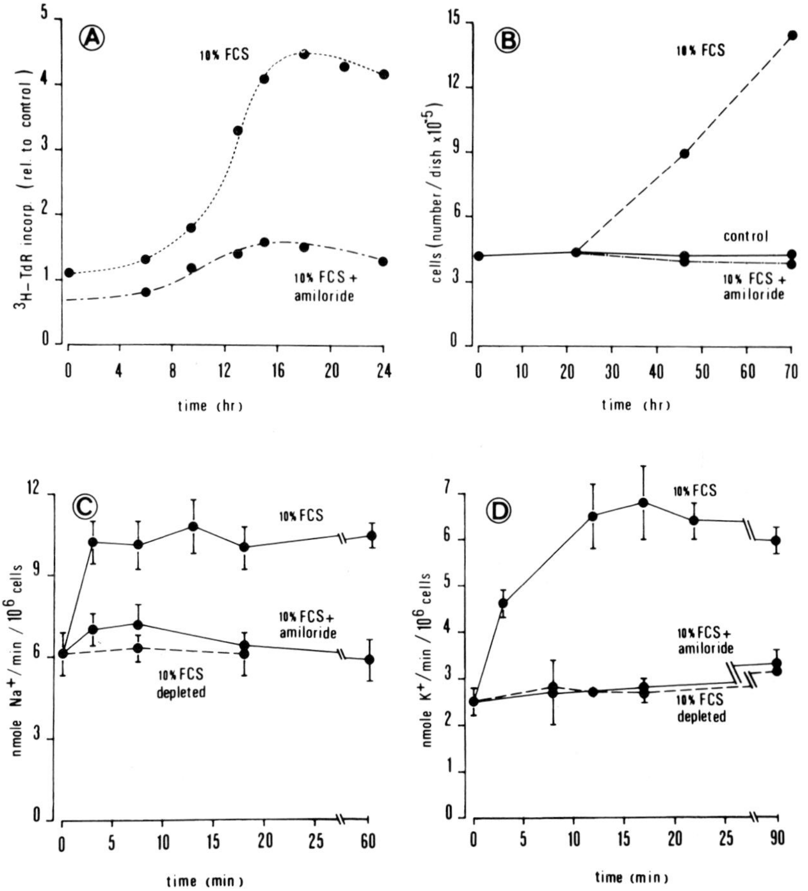

Fig. 5. Serum stimulation of N1E-115 cells, after 24 hr of serum deprivation. Dialyzed serum (FCS), growth-depleted serum (FCS depleted), and amiloride (0.3–0.4 m*M*) were added at time zero and were present continuously. [^{3}H]thymidine incorporation (A) and cell number (B) were assayed to follow growth stimulation by serum in the presence and absence of amiloride. Amiloride inhibits DNA synthesis (A) and cell division (B) but not the rate of thymidine uptake into acid-soluble pools (data not shown). Control values in (B) represent cell numbers in serum-free medium.

(C) Kinetics of stimulation of Na^+ influx as determined by the initial rate of ^{22}Na uptake over 5-min intervals in the presence of 5 m*M* ouabain. (D) Kinetics of stimulation of Na^+, K^+-ATPase-mediated K^+ influx as determined by the ouabain-sensitive ^{86}Rb influx. The error bars in (C) and (D) represent SEM. From Moolenaar *et al.* (*19*), with permission of M.I.T. Press, Cambridge, Mass.

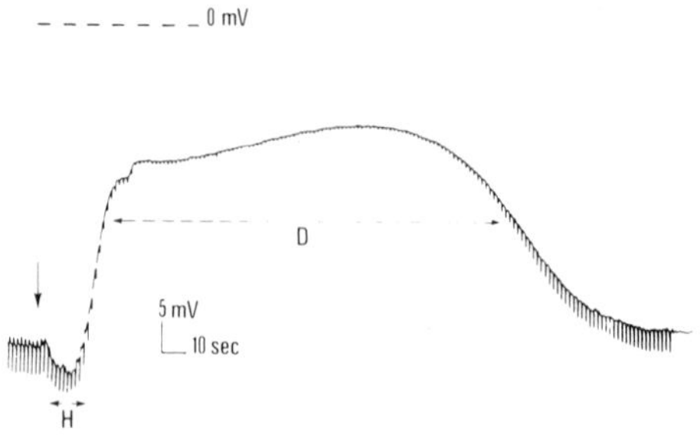

Fig. 6. Membrane potential and resistance changes following addition of 20% dialyzed FCS to an N1E-115 cell (arrow). Membrane resistance was monitored as the voltage response to brief hyperpolarizing current injections (200 msec; 0.2 nA). Dashed line represents zero potential level. H and D indicate the hyperpolarizing and depolarizing phases respectively, as described in the text. From Moolenaar *et al.* (*19*), with permission of M.I.T. Press, Cambridge, Mass.

increase in the permeability for several ionic species, including Na^+ and K^+ and perhaps Ca^{2+} (*18,19*).

To investigate the effects of serum and amiloride on passive Na^+ transport more directly, we determined the unidirectional Na^+ influx from initial rates of uptake of $^{22}Na^+$ (*19*). Whereas growth-depleted FCS is again without effect, addition of dialyzed FCS to N1E-115 cells after 24 hr of serum starvation results in a twofold stimulation within minutes (Fig. 5C). Amiloride blocks the serum-stimulated Na^+ influx almost completely at a concentration that inhibits DNA synthesis (0.4 m*M*), without affecting the basal influx rate. In subsequent experiments (*19*), we provided evidence that, in contrast to the basal Na^+ influx, the serum-stimulated Na^+ influx is not simply of diffusional origin. It is mediated by a separate amiloride-sensitive electroneutral transport system that has an absolute requirement for external Na^+ and is potentiated by anions derived from weak acids. Most likely, serum growth factors switch on an electroneutral Na^+–H^+ exchange mechanism which is not, or hardly, functional in quiescent cells and is blocked by amiloride. Preliminary observations confirm that amiloride inhibits Na^+-dependent H^+ release from N1E-115 cells. Whether and how the electrical response is involved in the stimulation of this Na^+ transport mechanism remains to be clarified. Somewhat later, but still within minutes, addition of dialyzed FCS results in a persisting twofold stimulation of the Na^+,K^+-ATPase-mediated K^+ influx, as measured by the ouabain-sensitive K^+ influx. (Fig. 5D). Again, growth-depleted FCS has no effect. Amiloride completely inhibits the serum-stimulated increase in Na^+,K^+ pump rate, without affecting the basal rate. Mainly from its dependence on external

Na^+, and the effects of Na^+ translocators in the absence of FCS, we conclude that the enhanced pump rate is secondary to and a consequence of the rapid increase in Na^+ influx rate (*19*).

In summary, our results show that serum stimulation of N1E-115 cells rapidly induces a sequence of electrical and ionic events at the plasma membrane which include: (1) transient membrane potential and resistance changes; (2) initiation of an electroneutral, amiloride-sensitive Na^+ transport system, which is probably a Na^+–H^+ exchanger; and (3) stimulation of the rate of the Na^+–K^+ pump as a consequence of the increased Na^+ influx rate. We conclude that these events are due to serum growth factors, since they are not evoked by growth-depleted FCS. We suggest that the initiation of electroneutral Na^+–H^+ exchange is a prerequisite for the initiation of N1E-115 cell proliferation.

VII. Concluding Remarks

In this chapter we have presented the main results of our recent investigations on the role of the plasma membrane in growth regulation of neuroblastoma cells. Two approaches were adopted in these studies: (1) A variety of plasma membrane properties were followed in detail during the course of the cell cycle of synchronized Neuro-2A cells, including the dynamic behavior of membrane lipids and proteins, qualitative and quantitative features of membrane structure, electrical membrane properties, and passive and active cation transport properties; and (2) the early ionic events upon serum stimulation of quiescent N1E-115 cells were analyzed. Although the picture emerging from the cell cycle studies is complex, it is clear that during the life history of a cell the plasma membrane is continuously changing its properties. The process of plasma membrane growth, which is inherent to the cell cycle, is not simply a matter of continuous assembly of membrane components leaving the overall composition and properties unaltered. To the contrary, all our data indicate that specific phases of the cell cycle are characterized by specific compositional, structural, dynamic, and functional plasma membrane properties, whereby the most prominent changes occur around mitosis and at the initiation of DNA synthesis. In particular, the observed modulations in structural and dynamic membrane properties suggest to us that asynchronous assembly of various membrane constituents into the plasma membrane is basic to membrane growth during the cell cycle. Consequently, molecular interactions within the plasma membrane, and between membrane components and intracellular components, such as cytoskeletal elements, are modified continuously, which results in modulations of membrane functions similar to those described for the passive and active cation transport properties.

At present, it is unclear to what extent the described modulations in plasma membrane structure and function apply to other cycling cells. The number of investigations on cell cycle-dependent alterations in membrane structure, dynamic properties, or ion transport properties of other cell types is yet very limited and partly conflicting. Furthermore, in none of these cells has a comparably broad spectrum of membrane properties been studied in parallel. Therefore, we feel that a discussion of other literature data is beyond the scope of this contribution. (See Chapter 15 for further discussion.)

As postulated above, it is very likely that the modulations in membrane functions will provide signals pertinent to growth control, and our data on ion transport changes during the cell cycle and upon serum stimulation provide some evidence for this. The transition from G_1 to S in normally cycling cells, in the presence of growth factors from the serum, is accompanied by a sequence of ionic changes, probably starting with the switching on of an amiloride-sensitive electroneutral Na^+ influx that stimulates transiently the Na^+,K^+-ATPase. Inhibition of these changes blocks the entry of cells into S. A similar sequence of events is detectable within minutes upon serum stimulation of serum-deprived neuroblastoma cells, and again these ionic changes appear to be a prerequisite for the initiation of DNA synthesis. Apparently, these ionic changes are among the first events elicited upon binding of growth factors to their membrane receptor sites. Taking these data into account, it seems plausible that the described alterations in membrane structure and dynamics in G_1 are manifestations of a process of membrane reorganization by which the receptor sites for growth factors become functionally expressed only in late G_1. When this condition at the plasma membrane is fulfilled, the interaction of growth factors with their receptors can evoke the responses, at least partially ionic in nature, necessary for the initiation of DNA synthesis. The frequently reported enhanced expression of various membrane receptor sites at this time in the cell cycle supports this view (see *2*). In the absence of growth factors, cells will lack these responses and consequently become arrested in G_1 (or G_0) or, in case of committed cells like neuroblastoma cells, switch to a differentiation program. In this concept, the molecular basis for the restriction or transition point in G_1 (*22,31*) resides in the postmitotic organization of the plasma membrane.

Acknowledgments

We are deeply indebted to our colleagues at the Hubrecht Laboratory, Drs. J. G. Bluemink, J. Boonstra, W. H. Moolenaar, C. L. Mummery, and E. J. J. van Zoelen, for contributions reported in this chapter, and to Drs. M. Shinitzky, J. Schlessinger (The Weizmann Institute of Science, Rehovot, Israel), and E. L. Elson (Washington University, St. Louis, Mo.) for their cooperation in carrying out part of this research. We also thank Miss A. Feyen, Mrs. W. Miltenburg-Vonk, Mr. P. Meyer, and Mr. L. G. J. Tertoolen for their excellent assistance and Mrs. E. C. Ekelaar for preparing the manuscript. Part of this work was supported by the Koningin Wilhelmina Fonds (Netherlands Cancer Foundation) and Shell International Research Corporation.

References

1. Axelrod, D., Koppel, D. E., Schessinger, J., Elson, E. L., and Webb, W. W. (1976). Mobility measurement by analysis of fluorescence photobleaching recovery kinetics. *Biophys. J.* **16,** 1055–1069.
2. Bluemink, J. G., and de Laat, S. W. (1977). Plasma membrane assembly as related to cell division. *Cell Surf. Rev.* **4,** 403–461.
3. Boonstra, J., Mummery, C. L., Tertoolen, L. G. J., van der Saag, P. T., and de Laat, S. W. (1981). Cation transport and growth regulation in neuroblastoma cells. Modulations of K^+ transport and electrical membrane properties during the cell cycle. *J. Cell Physiol.* **107,** 75–83.
4. Boonstra, J., Mummery, C. L., Tertoolen, L. G. J., van der Saag, P. T., and de Laat, S. W. (1981). Characterization of $^{42}K^+$ and $^{86}Rb^+$ transport and electrical membrane properties in exponentially growing neuroblastoma cells. *Biochim. Biophys. Acta* **643,** 89–100.
5. Branton, D. (1966). Fracture faces of frozen membranes. *Proc. Natl. Acad. Sci. U.S.A.* **55,** 1048–1056.
6. Brooks, R. F., Bennett, D. C., and Smith, J. A. (1980). Mammalian cell cycles need two random transitions. *Cell* **19,** 493–504.
7. Edelman, E. M. (1976). Surface modulation in cell recognition and cell growth. *Science* **192,** 218–226.
8. Johnson, J. D., Epel, D., and Paul, M. (1976). Intracellular pH and the activation of sea urchin eggs after fertilization. *Nature (London)* **262,** 661–664.
9. Koch, K. S., and Leffert, H. L. (1979). Increased sodium ion influx is necessary to initiate rat hepatocyte proliferation. *Cell* **18,** 153–163.
10. Laat, S. W. de, van der Saag, P. T., and Shinitzky, M. (1977). Microviscosity modulation during the cell cycle of neuroblastoma cells. *Proc. Natl. Acad. Sci. U.S.A.* **74,** 4458–4461.
11. Laat, S. W. de, van der Saag, P. T., Elson, E. L., and Schlessinger, J. (1980). Lateral diffusion of membrane lipids and proteins during the cell cycle of neuroblastoma cells. *Proc. Natl. Acad. Sci. U.S.A.* **77,** 1526–1528.
12. Laat, S. W. de, Tertoolen, L. G. J., and Bluemink, J. G. (1981). Quantitative analysis of the numerical and lateral distribution of intramembrane particles in freeze-fractured biological membranes. *Eur. J. Cell Biol.* **23,** 273–279.
13. Laat, S. W. de, Tertoolen, L. G. J., van der Saag, P. T., and Bluemink, J. G. (1981).

Modulations in numerical and lateral distribution of intramembrane particles during the cell cycle of neuroblastoma cells. *J. Cell Biol.* (submitted for publication).
14. Lubin, M. (1967). Intracellular potassium and macromolecular synthesis in mammalian cells. *Nature* (*London*) **213,** 415–453.
15. Mendoza, S. A., Wigglesworth, N. H., Pohjanpelto, P., and Rozengurt, E. (1980). Na entry and Na–K pump activity in murine, hamster and human cells. *J. Cell. Physiol.* **103,** 17–27.
16. Moolenaar, W. H., and Spector, I. (1977). Membrane currents examined under voltage clamp in cultured neuroblastoma cells. *Science* **196,** 331–333.
17. Moolenaar, W. H., and Spector, I. (1978). Ionic currents in cultured neuroblastoma cells under voltage-clamp conditions. *J. Physiol.* (*London*) **278,** 265–268.
18. Moolenaar, W. H., de Laat, S. W., and van der Saag, P. T. (1979). Serum triggers a sequence of rapid ionic conductance changes in quiescent neuroblastoma cells. *Nature* (*London*) **279,** 721–723.
19. Moolenaar, W. H., Mummery, C. L., van der Saag, P. T., and de Laat, S. W. (1981). Rapid ionic events and the initiation of growth in serum-stimulated neuroblastoma cells. *Cell* **23,** 789–798.
20. Mummery, C. L., Boonstra, J., van der Saag, P. T., and de Laat, S. W. (1981). Modulation of functional and optimal (Na^+-K^+)ATPase activity during the cell cycle of neuroblastoma cells. *J. Cell. Physiol.* **107,** 1–9.
21. Murphy, J. A., D'Alisa, R., Gershey, E. L., and Landsberger, F. R. (1978). Kinetics of desynchronization and distribution of generation times in synchronized cell populations. *Proc. Natl. Acad. Sci. U.S.A.* **75,** 4404–4407.
22. Pardee, A. B. (1974). A restriction point for control of normal animal cell proliferation. *Proc. Natl. Acad. Sci. U.S.A.* **71,** 1286–1290.
23. Prasad, K. N., Kumar, S., Gilmer, K., and Vernadakis, A. (1973). Cyclic AMP-induced differentiated neuroblastoma cells: Changes in total nucleic acid and protein contents. *Biochem. Biophys. Res. Commun.* **50,** 973–977.
24. Prasad, K. N. (1975). Differentiation of neuroblastoma cells in culture. *Biol. Rev. Cambridge Philos. Soc.* **50,** 129–165.
25. Rozengurt, E., and Heppel, L. A. (1975). Serum rapidly stimulates ouabain sensitive Rb influx in quiescent 3T3 cells. *Proc. Natl. Acad. Sci. U.S.A.* **72,** 4492–4495.
26. Rozengurt, E., and Mendoza, S. (1980). Monovalent ion fluxes and the control of cell proliferation in cultured fibroblasts. *Ann. N. Y. Acad. Sci.* **339,** 175–190.
27. Shields, R., Brooks, R. F., Riddle, P. N., Capellaro, D. F., and Delia, D. (1978). Cell size, cell cycle and transition probability in mouse fibroblasts. *Cell* **15,** 469–474.
28. Singer, S. J., Nicolson, G. L. (1972). The fluid mosaic model of the structure of cell membranes. *Science* **175,** 720–731.
29. Skou, J. C. (1957). The influence of some cations on the ATPase from peripheral nerves. *Biochim. Biophys. Acta* **23,** 394–401.
30. Smith, G. L. (1977). Increased ouabain-sensitive rubidium uptake after mitogenic stimulation of quiescent chicken embryo fibroblasts with purified multiplication-stimulating activity. *J. Cell Biol.* **73,** 761–767.
31. Smith, J. A., and Martin, L. (1973). Do cells cycle? *Proc. Natl. Acad. Sci. U.S.A.* **70,** 1263–1267.
32. Smith, J. B., and Rozengurt, E. (1978). Serum stimulates the Na^+,K^+ pump in quiescent fibroblasts by increasing Na^+ entry. *Proc. Natl. Acad. Sci. U.S.A.* **75,** 5560–5564.
33. Svetina, S. (1977). An extended transition probability model of the variability of cell generation times. *Cell Tissue Kinet.* **10,** 575–581.

34. Tupper, J. T., Zorgniotti, F., and Mills, B. (1977). Potassium transport and content during G_1 and S phase following serum stimulation of 3T3 cells. *J. Cell. Physiol.* **91,** 429–440.
35. van Wijk, R., van de Poll, K. W., Amesz, W. J. C., and Geilenkirchen, W. L. M. (1977). Studies on the variations in generation times of rat hepatoma cells in culture. *Exp. Cell Res.* **109,** 371–379.
36. van Wijk, R., and van de Poll, K. W. (1979). Variability of cell generation times in a hepatoma cell pedigree. *Cell Tissue Kinet.* **12,** 659–663.
37. van Zoelen, E. J. J., van der Saag, P. T., and de Laat, S. W. (1981). Family tree analysis of a transformed probability model for the cell cycle. *Exp. Cell Res.* **131,** 395–406.
38. van Zoelen, E. J. J., Boonstra, J., van der Saag, P.T., and de Loat, S. W. (1982). Effect of external ATP on the plasma membrane permeability and (Na^+-K^+) ATPase activity of mouse neuroblastoma cells. *Biochim. Biophys. Acta* (in press).

15

The Role of Ions, Ion Fluxes, and Na^+,K^+-ATPase Activity in the Control of Proliferation, Differentiation, and Transformation

R. L. SPARKS, T. B. POOL, N. K. R. SMITH, AND I. L. CAMERON

I. Introduction

Four cations (Na^+, Mg^{2+}, K^+, and Ca^{2+}) have been shown to, or theorized to, play various regulatory roles in cell proliferation (*3,8–10,14–16,18,35,36,38,39,52,58–62,67,68,70,75*). Much is known about the role of Ca^{2+} and Mg^{2+} in many biochemical reactions and processes (*43*) involving secondary messenger functions and cofactor functions,

GENETIC EXPRESSION IN THE CELL CYCLE

ISBN 0-12-543720-X

respectively. The role of these two ions in cell proliferation is still under intense study, but for about the last 10 years, considerable attention has been directed toward the study of the roles of Na^+ and K^+ in cell proliferation. In our laboratory we have directed most of our effort toward the study of changes in transformed cells, and a large portion, but not all, of this review centers around the role of Na^+.

II. The Role of Ions in the Control of Metabolism and of Cell Proliferation

One who has elaborated on the roles of Na^+ and K^+ in cell proliferation, both experimentally and theoretically, is Clarence D. Cone, Jr. His original theory (*15,16*) was based on experimental observations suggesting that a significant correlation may exist between intracellular sodium concentrations $[Na^+]_i$, the level of the electrical transmembrane potential (E_m) difference in somatic cells, and their rate or intensity of mitotic activity (see also *1,6*). Cone's theory linked the activity of the potential generation mechanisms of the cell surface complex (and mitogenic activity) with cellular metabolism and external environmental influences through an explicit system of interacting feedback circuits. An integral part of the theoretical development includes conjecture as to the metabolic and cytogenic etiology and maintenance of the malignant state.

As stated above, Cone's original theory was based on observations of data in the literature and on some of his own preliminary results. He began by correlating E_m levels with the degree of mitotic activity in various somatic cell populations. For example, among various somatic cell types, nerve and muscle cells are characterized by their large E_m (about -79 to -90 mV); and, as noted by Cone (*15*), equally characteristic is the fact that these cells exhibit an exceptionally low degree of mitotic activity. Indeed, mature central nervous system neurons are postmitotic and mature muscle cells are normally postmitotic. To Cone, it seemed significant that a large E_m level (of neurons and muscle cells) is accompanied by a virtual absence of mitotic activity. Cone also noted that the interphase E_m level of other mature somatic cell populations with low mitotic activity (such as liver, lung, and connective tissue) is generally in the range of -50 to -60 mV. Upon dissociation and adaptation to *continuous* proliferation in culture, many somatic cells undergo a decrease (depolarization) in the interphase E_m level. Some cells, such as mature neurons, maintain their large E_m level *in vitro,* and their proliferative activity does not increase beyond the *in vivo* value. Cone was able to induce proliferation of chick neurons cultured *in vitro* by manipulations that were designed to reduce E_m (*19*). Conversely, Cone showed in proliferating cultures of 3T3 and CHO cells (which have a small E_m) that as the cells approached and reached confluence there was a precipitous decline in proliferation with a concomitant increase (hyperpolarization) of the E_m (*17*). Also to this end, Cone used naturally synchronized CHO cells *in vitro* and a medium designed to produce intracellular Na^+, K^+, and Cl^- concentrations to approximate those which would exist at high E_m levels (*20*). Under these conditions, mitosis and DNA synthesis were blocked reversibly by simulated E_m levels of -70 mV and greater. Since the block appeared to be in the latter half of the G_1 period of the cell cycle, Cone and Tongier (*20*) suggested that the prevention of DNA synthesis might be due to metabolic or enzymatic alterations involving blockage of the DNA precursor synthesis or polymerase synthesis or activation. The experimental medium, which was designed to cause a significant hyperpolarization, did inhibit DNA synthesis and mitosis and did cause a decrease in intracellular Na^+ and an increase in K^+. Although the medium was made to stimulate an E_m level of -70 mV, or a net hyperpolarization of 50 to 60 mV, it actually only caused a slight hyperpolarization, 10 mV at most. The fact that the predicted intracellular ion changes occurred with only a small change in E_m led Cone to modify his hypothesis further (*17,18*). In his modified hypothesis, Na^+ assumes a more central role in the regulation of cell proliferation, and E_m becomes functionally less important. (See also Chapter 14.)

As has been previously stated, the metabolic functions of Mg^{2+} and Ca^{2+} have been greatly elaborated on, and even biochemical requirements for K^+ have been shown. The exact role(s) of Na^+ has been more elusive, mechanistically speaking, as far as its function in metabolism and/or cell proliferation regulation is concerned. Some recent data from Piatigorsky's laboratory at the National Institutes of Health and Koch's laboratory in Hamburg imply a direct role for $[Na^+]_i$ as well as for intracellular Na^+/K^+ ratios in controlling the rate, amount, and types of proteins being synthesized. (*34,50*).

Piatigorsky and co-workers (*50*) have shown a correlation between the ratio of synthesis of the larger to the smaller polypeptides of δ-crystallin and the intracellular concentrations of Na^+ and K^+ in cultured embryonic lenses. Opaque or cataractous lenses showed changes in the ratio of synthesis of the δ-crystallin polypeptides and changes in their Na^+ and K^+ concentrations. Manipulation of the external Na^+ and K^+ in the medium of cultured lenses also caused changes in the ratio of synthesis of the same polypeptides. The differential reduction in synthesis of the lower-molecular-weight polypeptides of δ-crystallin was positively correlated with a decrease in the K^+/Na^+ ratio (or an increase in the Na^+K^+

ratio) of the cultured embryonic chick lenses below a value of ~1. Using ouabain to increase $[Na^+]_i$ and decrease $[K^+]_i$ they found that synthesis of the lower-molecular-weight polypeptides is inhibited in lenses treated with ouabain for 24 hr. However, synthesis of the higher-molecular-weight polypeptides of δ-crystallin continued for several days. Studies combining ouabain treatment and manipulation of Na^+ and K^+ in the medium showed that the ratio of synthesis of the δ-crystallin polypeptides varied with the concentration of Na^+ and K^+ in the ouabain-treated lenses. Their data generally showed that intracellular K^+/Na^+ ratios higher than 1 were correlated with the normal ratio of [^{35}S]methionine incorporation into the two bands of δ-crystallin, whereas K^+/Na^+ ratios less than 1 were associated with the differential reduction of isotope incorporation into the smaller polypeptides. Their cell-free experiments suggest that the control of the ratio of synthesis of the δ-crystallin polypeptides varied with the concentration of Na^+ and K^+ in the ouabain-treated lenses. They also found a differential reduction in crystallin synthesis, which was correlated with an increase in the Na^+/K^+ ratio in three types of cataracts. They postulated that changes in ion concentrations will be correlated with changes in protein synthesis in other types of cataract and possibly other diseases.

G. Koch and co-workers have also been working on the role of ions in protein synthesis, specifically translational control (*34*). They have found differential inhibition of different cellular mRNAs by a so-called hypertonic initiation block (HIB) of polypeptide chain initiation. Inducing a rapid increase in intracellular osmolarity of cells *in vitro,* by increasing the osmolarity of the growth medium, caused specific yet differential inhibition of polypeptide chain initiation, or HIB. The inhibition was independent of the solute used to produce hyperosmolarity. They showed that exposing virus-infected cells to hypertonic conditions caused a severe inhibition of host cell mRNA translation, whereas viral mRNA translation was only slightly affected. Their conclusions were that viral mRNAs are more efficient messengers than host mRNAs; they suggested that under conditions that reduce the rate of peptide chain initiation, each mRNA is translated with its own characteristic efficiency. Their working hypothesis is that any mRNA with a high relative translational efficiency (RTE) possesses a higher binding affinity for ribosomes and/or initiation factors and therefore outcompetes mRNAs with lower RTEs, and the competition would be amplified when ribosome and/or initiation factors become limited. Extending their work and theory to the mouse plasmacytoma cell line MPC-11, Koch and co-workers (*34*) found a differential inhibition of peptide synthesis in these cells due to HIB. MPC-11 cells synthesized a heavy and light chain of immunoglobulin gamma

(IgG), which accounted for as much as 20% of newly synthesized peptides. Exposure of these cells to hyperosmotic medium produced a 350–400% increase in the relative incorporation of [^{35}S]methionine into the L chain and a 150% increase into the H chain compared to isotonic conditions. Their results suggest that the mRNAs coding for the specialized IgG polypeptides are more efficient messengers (with higher RTEs) when compared to mRNA species coding for other cellular proteins.

Koch *et al.* (*34*) also showed that the mechanism by which polypeptide chain initiation was inhibited, either by HIB, by amino acid starvation, or upon virus infection, is similar and appears to be membrane mediated. Koch *et al.* (*34*) speculated that all three processes may involve the activation of a normal host cell mechanism used to regulate protein synthesis at the translational level. These investigators showed in three types of cell extracts prepared either: (1) 2 to 3 hr after infection with polio virus; (2) 45 min after transfer of cells to amino acid-deficient medium; or (3) from cells exposed to HIB for 15 min, that none displayed detectable endogenous protein synthesis. Protein synthesis in the extracts could be partially restored after gel filtration, indicating that neither ribosomes nor mRNA were irreversibly inactivated by exposure of cells to any of the three conditions described. Fractions which were eluted late in the separation contained low-molecular-weight substances that inhibited protein synthesis in cell-free extracts from untreated cells. Koch *et al.* (*34*) proposed that the inhibitor is released or activated by a membrane-mediated event and then reversibly bound to ribosomes. The ribosomes with bound inhibitor are less active or are inactive in the formation of initiation complexes.

Koch's group (*34*) also studied changes in protein synthesis during Friend erythroleukemia cell differentiation. Inducers of Friend cell differentiation decrease the initiation of protein synthesis in general. However, globin mRNA translation, a differentiation specific event, still occurred favorably under conditions of reduced initiation. Therefore, the production of a major differentiation-specific protein is potentiated by inhibition of initiation, suggesting a high RTE for globin mRNA. Koch *et al.* (*34*) suggested that the primary importance of translational control, based upon reduced initiation, is that overt changes in the spectrum of proteins being synthesized can occur rapidly and reversibly without the necessity to elaborate a series of specific factors that amplify the translation of specific mRNAs. Also, such membrane-mediated mechanisms could exert a pleiotropic effect on qualitative and quantitative aspects of translation in response to various environmental stimuli. Whereas transcriptional control is the primary regulatory mechanism, it is advantageous to couple transcriptional activation of a gene with translational conditions favorable to the synthesis of that particular gene product (*34*).

Changes in sodium flux across the cell membrane or intracellular sodium levels could affect metabolism in a number of ways, e.g., by the influence of Na^+ on Ca^+ efflux from mitochondria. Sodium has been shown to stimulate the efflux of calcium from the mitochondria of several tissues, including liver, kidney, lung (*29*), adrenal cortex, striated muscle (*21*), and brain and heart (*13,21*). Many cellular processes, including cell division, are known to be regulated by calcium. The mitochondrion is a prime candidate for regulation of intracellular calcium because of its high calcium content and its high-affinity calcium-uptake mechanism as speculated in the work above. It is important to measure the presence of Na^+-induced Ca^{2+} release in the mitochondria of various tissues because of the known role of Ca^{2+} in various cellular processes (*8,43,55*). Some hormones are known to mobilize mitochondrial stored Ca^{2+}, and it is possible that Na^+ is involved in this mobilization process. For example, the glycogenolytic action of vasopressin, angiotensin II, and phenylephrine on liver is accompanied by a depletion of mitochondrial calcium (*7*). It is interesting to note that vasopressin is also a potent mitogen for some cells and causes a rapid influx of Na^+ across the cell membrane (*57*). The work on Na^+-induced Ca^{2+} release in mitochondria has revealed some interesting possibilities for the role of Na^+ in cell metabolism and proliferation. At a recent symposium entitled "Growth Regulation by Ion Fluxes" (*38a*), a unifying idea emerged in which two key ionic events assumed central importance as possible major regulators of cell proliferation. These ionic events are temporally oriented and consist of, first, Na^+–K^+ fluxes and, second, a Ca^{2+}–cyclic adenosine monophosphate (cAMP) coupling (for a review, see *36*). Based on known growth state-dependent changes in Na^+ flux and content in cells as described above and in following sections, Na^+ could possibly play a key role in various Ca^{2+}-dependent cellular processes by regulating mitochondrial Ca^{2+} efflux at critical times.

III. The Role of Na^+ and K^+ in Cell Differentiation

Some of the roles of ions in various cellular processes, including enzyme reactions, DNA synthesis, gene expression, and protein synthesis, were presented in Section II. The role of ions in determining the state of differentiation, proliferation, and transformation will be further illustrated in this and the following two sections, respectively. This section will present some of the evidence obtained to date concerning the role of ions, Na^+ and K^+ in particular, in the induction or suppression of cell differentiation.

Bernstein *et al.* (*5*) have provided compelling evidence that ions and/or ion pump activity play a major role in cell differentiation. One of their initial observations was that ouabain induces differentiation of Friend erythroleukemic cells (*5*). Friend cells can be stimulated by a number of agents *in vitro* to undergo erythroid differentiation. This was an exciting finding by Bernstein's group because, compared to all of the other inducing agents discovered by other workers, ouabain is the only one in which interaction between the inducer and a specific cellular target (namely, Na^+,K^+-ATPase) has been shown to be necessary for induction of Friend cell differentiation. They have carried this work further and have been testing whether alterations in Na^+,K^+-ATPase activity and/or ion concentrations are a common event during Friend cell differentiation. They have now shown that during the induced differentiation of Friend cells, using several different inducing agents, Na^+,K^+-ATPase was inhibited to approximately the same extent as that observed after ouabain incubation (*44*). It was then shown that growth in high-K^+ and low-Na^+ medium (60–90 m*M* each) also induced differentiation (*46*). This showed that Friend cells can be stimulated to differentiate in the absence of a chemical inducing agent simply by manipulating the extracellular levels of Na^+ and K^+. The differentiation of Friend cells involved several events, including the accumulation of globin mRNA, synthesis of heme and hemoglobin, and the appearance of erythrocyte membrane antigen (EMA), all of which were stimulated by changing extracellular Na^+ and K^+. This growth in a high-K^+, low-Na^+ medium, therefore, was shown to stimulate a series of diverse cellular events. Further studies revealed that it probably was not the absolute changes in intracellular Na^+ or K^+ per se but changes in Na^+,K^+-ATPase activity (due to changes in Na^+ and K^+ flux) that induced Friend cell differentiation (*45*). Bernstein *et al.* (*5*) hypothesized that changes in pump activity caused changes in ATP and cAMP levels, which in turn affected certain protein kinases that regulate the phosphorylation of various cell proteins.

Elevated $[K^+]_o$ has also been shown to induce morphological differentiation of embryonic chick dorsal root ganglion neurons *in vitro*. By increasing $[K^+]_o$ from 6 to 40 m*M*, Chalazonitis and Fischbach (*11*) saw a 25% increase in neurons, but not an increase in total cell number. It was determined by cell counts, autoradiography, and time-lapse cinematography that the increase in neurons was due to morphologic differentiation and not cell division. The high K^+ medium also induced a depolarization of the E_m to -22 ± 1 mV (compared to the controls at $-53 \pm$ mV). Chalazonitis and Fischbach (*11*) do not believe that the same mechanisms are involved as those observed by Cone and Cone (*19*), who located some binucleate spinal cord cells following prolonged exposure to ouabain *in vitro*. Chalazonitis and Fischbach (*11*) never observed binucleate sensory neurons in high-K^+ media and believe that their data are more consistent with the hypothesis that high K^+ promotes the differentiation of postmitotic neuroblasts.

Spival *et al.* (*73*) recently reported the effects of ouabain on hematopoietic stem cell proliferation. They showed that at different concentrations ouabain either suppressed or potentiated mouse hematopoietic progenitor cell proliferation. At $10^{-4}M$, ouabain suppressed granulocyte-macrophage colony-forming units (CFU-C), erythroid burst-forming units (BFU-E), and erythroid colony-forming units (CFU-E). They did not assay spleen colony-forming units (CFU-S) at $10^{-4}M$ ouabain. Interestingly, concentrations of ouabain varying from 10^{-6} to $10^{-2}M$ suppressed CFU-S and CFU-C. Spival *et al.* (*73*) theorized that ouabain could have influenced hematopoietic cells, since it is known that ouabain can stimulate cAMP formation (*26*). It might be noted that the cells stimulated to proliferate by ouabain in this case (BFU-E and CFU-E) are in the same series of blood cells (erythroid) as Friend cells, which were stimulated to differentiate by ouabain (*5*); however, only the lower concentrations of ouabain (10^{-6} to $10^{-12}M$) stimulated the hematopoietic erythroid precursors to proliferate (*73*). Higher concentrations (10^{-4} and $10^{-3}M$) inhibited proliferation of erythroid precursors and stimulated differentiation of Friend cells, respectively (*5,73*).

The differential responses (inhibition or stimulation of proliferation or of differentiation) by various cell types to various concentrations of ouabain could be due to any number of things. An attractive possibility is that at varying concentrations ouabain might affect Na^+,K^+-ATPase activity in a dose-dependent fashion, and that the resulting dose-dependent changes in ion fluxes, ion content, ATP hydrolysis, etc. could cause preferential signaling and preferentially determine the growth state of the cell. It is also possible that ouabain may affect cellular processes other than Na^+,K^+-ATPase activity. The ouabain concentration required to inhibit proliferation is one order of magnitude less than that needed to affect cation transport. Segel and Lichtman (*63*) pointed out that, in lymphocytes, the apparent discrepancy between ouabain inhibition of cation transport and of proliferation is due to the time dependence of ouabain binding to lymphocytes. Since it is known that a temporal sequence of events is required in the course of many cellular processes (such as those leading to DNA synthesis), any of these processes that might be influenced by varying ion fluxes, ion content, or Na^+,K^+-ATPase activity could be differentially affected by different concentrations of ouabain. Two additional lines of evidence suggest changes in ion fluxes or ion contents as possible regulators of differential gene expression and differentiation. As mentioned in Section II, changes in $[Na^+]_o$, $[K^+]_o$, or the intracellular Na^+/K^+ ratio can cause differential protein synthesis (*34,51*). The recent discovery of endogenous mammalian Na^+,K^+-ATPase inhibitors (*25,28*), which could affect the intracellular Na^+/K^+ ratio, and the data showing the effects of Na^+ and K^+ on cell differentiation, indicate a direct correlation between changes in Na^+, K^+ (intra- and extracellular), and cell differentiation and a possible regulatory function of their fluxes *in vivo*.

IV. The Role of Ions and Ion Fluxes in the Stimulation of Cell Proliferation

Animal cells respond to various external factors such as hormones or growth factors by the initiation or acceleration of a given set of reactions associated with transport, intermediary metabolism, and the synthesis of macromolecules (*32,35,48,49,66,75*). A number of the early responses after stimulation are independent of one another and of the synthesis of macromolecules (*35,39,77*). Some mitotically quiescent cell populations can be stimulated to undergo DNA synthesis and cell division by various means. For example, *in vivo,* partial hepatectomy stimulates hepatocyte proliferation and injection of 17-β-estradiol stimulates cell proliferation in the basal layer of the vaginal epithelium in ovariectomized animals (*9,39*). *In vitro,* quiescent cell populations (quiescent due to nutrient deprivation, contact inhibition at confluence, etc.) may be stimulated to proliferate by the addition of fresh serum, hormones, growth factors, etc. (*4,33,57,67*). However, the stimulation of cells to undergo DNA synthesis and subsequent cell division is a time-ordered process and it takes hours and even days in some cases for these two events to occur. As stated above, after stimulation occurs, a series of events is required before DNA synthesis and cell division can proceed. A number of laboratories have been trying to determine the earliest events occurring upon stimulation and also which of these events are required for the progression to DNA synthesis and cell division.

The earliest known event occurring after mitogenic stimulation of cells in culture is a rapid influx of Na^+ (*33,35,67*), which begins in a matter of seconds after stimulation. This influx, or "burst," of Na^+ is now being studied as it relates to the events involved in the progression from mitogenic stimulation to DNA synthesis.

Koch and Leffert have investigated the role of Na^+ influx in the stimulation of hepatocytes to proliferate and have shown that an increased

Na^+ influx is necessary to initiate proliferation (*35,37,39*). To understand better how hormones and growth factors stimulate liver regeneration, they have been studying the mechanisms by which these agents promote the entry of resting or quiescent hepatocytes into the DNA synthesis phase of the cell cycle. To this end, they have been trying to identify the rapidly activated cellular processes that lead to DNA synthesis. Various mitogenic hormones and growth factors upon addition to the growth medium have been shown to cause a rapid influx of Na^+ into cells in culture (*35,57*). Using amiloride, a drug known to block the passive influx of Na^+ into cells, Koch and Leffert (*35*) were able to inhibit the mitogenic stimulation of hepatocytes in culture induced by several hormones and growth factors. Control stimulated cultures (no amiloride) showed a rapid influx of Na^+ (as measured by $^{22}Na^+$ uptake) and eventual DNA synthesis and cell division. The mitogen-treated cell cultures showed no influx of $^{22}Na^+$ or entry into DNA synthesis when amiloride was present. The effect of amiloride was shown to be both dose dependent and reversible (*35,39*). Amiloride was only effective when present at the initiation of stimulation and needed to be present during the first 12 hr after stimulation to inhibit DNA synthesis completely. Removal of amiloride at progressively earlier times caused progressively increased growth responses. Amiloride present for only 1 min at the initiation of stimulation or added any time 12 hr or later after stimulation had no effect on DNA synthesis (*35,39*). Similar responses were found with ouabain treatment. The results were interpreted as indicating that drug sensitivity was determined by the initial time of exposure to the drug relative to the application of the stimulus and not by the duration of the exposure. Koch and Leffert (*37*) also suggested that their data indicate that amiloride blocked the event(s) required to initiate DNA synthesis but not DNA replication itself. Koch and Leffert (*35*) also showed *in vivo* that amiloride blocked 70% of the hepatectomy-induced DNA synthesis in the liver. Other prereplicative functions were also blocked by amiloride *in vitro*. Koch and Leffert (*35*) showed that *in vitro* growth-initiating conditions stimulated [^{14}C]aminoisobutyric acid ([^{14}C]AIB) uptake between 6 and 8 hr, a biphasic uptake of [^3H]uridine ([^3H]UdR) into RNA with peaks between 0–4 and 8–12 hr, and [^3H]leucine ([^3H]Leu) uptake into cellular protein between 8 and 12 hr. Amiloride blocked (1) the stimulated [^{14}C]AIB uptake (but not the basal rates), (2) the second peak of [^3H]UdR uptake into RNA (but not the first peak or basal rates), and (3) both the stimulated and basal rates of [^3H]Leu incorporation into cellular protein. Koch and Leffert (*35*) suggested that due to the similar effects of amiloride and ouabain on growth, increased bidirectional Na^+ fluxes appear necessary to initiate hepatocyte DNA synthesis. Parallel animal and culture studies indicate that the amiloride-sensitive and prereplicative intervals directly coincide (at 0–12 hr) (*35*). Interestingly, in the same set of studies these investigators showed that furosemide, which is known to stimulate passive Na^+ influx, also stimulated hepatocyte proliferation. From these sets of studies, Koch and Leffert (*35*) made some general predictions. They suggested that in the liver, agents such as EGF that activate Na^+ flux systems should promote DNA synthesis initiation and that these effects should be potentiated by insulin and glucagon. They also suggested that other ions, such as K^+, Mg^{2+}, and Ca^{2+}, which also play growth-regulating roles (see above), may be needed to maintain proper Na^+ flux system functioning.

One of the more striking changes seen in the plasma membrane of quiescent fibroblasts soon after the addition of growth factors or serum is the activation of Na^+,K^+-ATPase, which in quiescent 3T3 fibroblasts increases about threefold within about 3 min after the addition of fresh serum (for reviews on pump activity, see *32,57,67*). In a classic study, Smith and Rozengurt (*67*) showed that the stimulation of the Na^+,K^+-ATPase by serum in Swiss 3T3 cells was mediated by increasing Na^+ entry into the cells. Simply put, Na^+,K^+-ATPase activation by mitogens is stimulated via increased Na^+ entry. In that report, Smith and Rozengurt (*67*) proposed the following model of Na^+,K^+-ATPase regulation: (1) There is a Na^+ channel in the plasma membrane of fibroblasts; (2) growth factors increase Na^+ entry via the channel; (3) the activity of the Na^+,K^+-ATPase in fibroblasts is limited by the supply of internal Na^+; and (4) growth factors stimulate the Na^+,K^+-ATPase by activating the Na^+ channel, thereby increasing the supply or availability of Na^+ to the Na^+,K^+-ATPase. Smith and Rozengurt (*67*) showed that the addition of fresh serum increased the rate of $^{22}Na^+$ uptake and net Na^+ entry and content in quiescent 3T3 cells. They suggested that the stimulated $^{22}Na^+$ uptake does not appear to enter fibroblasts by cotransport with amino acids because ouabain, which inhibits amino acid uptake, increases $^{22}Na^+$ uptake. They also showed that the Na^+ ionophore monensin enhanced the rate of Na^+ entry, Na^+ content, and Na^+,K^+-ATPase activity. They also showed that the stimulation of DNA synthesis by serum in quiescent 3T3 cells was dependent on the concentration of Na^+ in the medium. Smith and Rozengurt (*67*) hypothesized that an increase in the Na^+ permeability of 3T3 cells (after mitogen stimulation) sets in motion an array of ion redistributions and metabolic events leading to rapid growth (see also Chapter 14).

An interesting parallel between stimulation of normal cell proliferation and proliferation of cancer cells should be mentioned at this point. As stated above, monensin stimulated Na^+ entry and Na^+,K^+-ATPase activity in quiescent 3T3 cells. Smith and Rozengurt (*67*) also mentioned that monensin stimulated glycolysis in these same cells. The parallel is that many tumor cells are known to have extremely high glycolytic activity (*43,54*), high $[Na^+]_i$ (*10,68*), and high Na^+,K^+-ATPase activity (*32,65*). Regulation of this ion or its pump (or lack of regulation) is an area of study that could help to find the lesion or lesions responsible for cancer. Perhaps there is a relationship between $[Na^+]_i$, Na^+ flux rates, Na^+,K^+-ATPase activity, and aerobic glycolysis. The regulation of aerobic glycolysis in tumor cells will be discussed below.

Rozengurt's group has carried out intensive investigations on the stimulation of quiescent cells to proliferate (*48,49,56,57*). They have shown that vasopressin rapidly stimulates Na^+ influx and Na^+,K^+-ATPase activity and is a potent mitogen in Swiss 3T3 cells (*49,57*). They suggest that these results provide further evidence in support of a possible role of monovalent ion fluxes in signalling the initiation of growth stimulation. Recent work suggests that the potent tumor-promoting agent 12-*O*-tetradecanoyl phorbal-13-acetate (TPA) has mechanisms of action and effects similar to vasopressin and their findings of increased Na^+ permeability (influx) and increased Na^+,K^+-ATPase activity in SV40 or polyoma virus transformed Swiss 3T3 cells (*57*), suggest that these two properties are commonalities between rapidly proliferating and transformed cells.

Based on their own work and that of others, Rozengurt and Mendoza (*57*) have speculated about how changes in Na^+ transport could affect cell metabolism in a variety of ways. They liken the activation of Na^+ influx in fibroblasts (one of the earliest membrane changes seen in the action of growth factors) to that of excitable cells such as muscle, in which an inward movement of Na^+ is the primary excitatory process. An induced increased sodium conductance (due to growth factors) would itself be expected to depolarize the membrane. Subsequent stimulation of the Na^+,K^+-ATPase would be expected to cause a hyperpolarization. The net result of these two opposing changes is unpredictable, but any sizable change in E_m could significantly affect or influence cellular function. As indicated in the results described above, $[Na^+]_i$ is a direct regulator of activity of the Na^+,K^+-ATPase, which in turn controls $[K^+]_i$ and affects ATP turnover. Smith and Rozengurt (*67*) suggested that changes in permeability of the cell membrane to Na^+, with concomitant small changes in $[Na^+]_i$, could significantly alter other intracellular ions as well. They suggested that, if a Na^+–H^+ electroneutral exchange, which is known to occur in some cells (*23*), occurs in fibroblasts, then a concomitant rise in intracellular pH, which has been shown by Koch and Leffert (*37*), could occur simultaneously with the increase in Na^+ entry. An increase in intracellular pH (an alkaline "surge") has been speculated to play a role in sea urchin egg activation (see *23*). Changes in pH could be associated with major changes in the activity of a variety of enzymes and with changes in the fixed charge of proteins (for a review, see *38a*). Smith and Rozengurt (*67*) also suggested that another possible way by which $[Na^+]_i$ has a regulatory effect is through the ability of Na^+ to induce mitochondrial Ca^{2+} efflux (as discussed in Section II). Thus, changes in intracellular Na^+ could alter the cytoplasmic Ca^{2+} concentration. We should note that Ca^{2+} has been implicated in the control of cell proliferation (*8,30,77*).

Many other cell types are known to undergo changes in ion fluxes and pump activity after mitogenic stimulation (*33;* for a review, see *32*). For example, Kaplan (*32*) has shown that an increased Na^+,K,-ATPase activity occurs and is required during the stimulation of lymphocytes to proliferate. Similarly, Na^+,K^+-ATPase activity was shown to be essential for cellular processes and cell growth (*47*). These investigators showed that inhibition of Na^+,K^+-ATPase activity with ouabain induced inhibition of cell growth and inhibition of cation transport, but that the inhibition was reversible. Mayhew and Levinson (*47*) suggested that ouabain inhibited cell division by a mechanism related to its inhibition of the membrane transport system(s).

At this point it should be mentioned that natural (endogenous) inhibitors of Na^+,K^+-ATPase have been discovered recently and partially purified (*25,28*). Haupert and Sancho (*28*) isolated a low-molecular-weight (< 2500), basic, nonpeptide factor from bovine hypothalamus that has actions similar to ouabain. Fishman (*25*) isolated a fraction from whole guinea pig brain which contains a substance that blocks the binding of ouabain to Na^+,K^+-ATPase and inhibits the uptake of $^{86}Rb^+$ into human erythrocytes. The substance isolated by Fishman appears to act in a competitive manner with ouabain. Haupert and Sancho's work (*28*) showed the presence of a hypothalamic substance that produces the effects of a natriuretic factor. This substance thus possesses the putative characteristics of a natriuretic factor of hypothalamic origin. As postulated by Fishman (*25*), an endogenous digitalis or "endigen" could regulate ionic fluxes across the membrane, consequently modulating cell volume and transmembrane potential. Endogenous Na^+,K^+-ATPase inhibitors might even act as chalones or regulators of cell proliferation.

The data reviewed in this section describe the movement and role of ions in various aspects of the stimulation of cells to proliferate. As can be interpreted from this brief discussion and from the actual articles reviewed, no actual mechanisms by which changes in Na^+, K^+, and Na^+, K^+-ATPase activity exert a regulatory function have been discovered to date. Direct correlations and requirements for changes in ion fluxes, ion content, and Na^+,K^+-ATPase activity have been documented in the control of cell proliferation, but disappointingly no explicit cause-and-effect relationships have evolved. However, changes in cation flux and Na^+,K^+-ATPase activity are still attractive possibilities as playing key roles in the regulation of cell proliferation, and the work still continues to identify the exact roles of cation flux and Na^+,K^+-ATPase activity in cell proliferation (see also Chapter 14).

V. Comparison of Intracellular Element (Ion) Contents and Na^+,K^+-ATPase Activity of Normal and Cancer Cells

It is known that many cellular processes and reactions require ions such as Ca^{2+}, Mg^{2+}, and K^+. It is known that enzyme reactions and reaction rates are dependent on various conditions such as ionic strength and pH. It therefore seems likely that changes in ion fluxes or intracellular element (ion) concentrations play a key role in the regulatory events concerned with cell proliferation, differentiation, and transformation.

Changes in ion transport and intracellular ion concentrations have been implicated in the control of mitogenesis and oncogenesis. A number of workers have found intracellular concentration differences in Na^+, Cl^-, Ca^{2+}, K^+, and Mg^{2+} between normal and transformed cells. One of the most striking differences found is an extremely high intracellular sodium content in transformed cells. Damadian and co-workers showed, by use of ^{23}Na nuclear magnetic resonance (NMR), that tumor tissues have a higher sodium content than their normal counterparts (*22,27*). These findings were very exciting but had the drawback of being done on whole tissues, which include many cell types and extracellular spaces. Then Smith and co-workers, using the morphoanalytical technique of electron probe X-ray microanalysis, showed a significant increase in Na in the nucleus and the cytoplasm of H6 hepatoma cells compared to normal hepatocytes *in vivo* (*68*). These data supported the hypothesis that intracellular levels of Na and other elements (ions) are directly involved in mitogenesis and/or oncogenesis (*14,15,17,19,20*). In fact, Smith and co-workers showed a 150% increase in Na^+ in the hepatoma cells as well as more than a 100% increase in Cl^- in the hepatoma cells compared to the normal hepatocytes (*68*). They also found that the Mg^{2+} values were 35% lower in the nucleus and 31% lower in the cytoplasm of hepatoma cells as compared to the same compartments of normal hepatocytes. These data are in general agreement with previously reported data that the Mg^{2+} concentration of a whole hepatoma induced by 4-

dimethylaminoazobenzene was lowered to 76% of that for normal liver (*2*). Smith and co-workers (*68*) found no other major differences between normal and tumor cells. However, Anghileri and co-workers (*2*) also reported that the whole hepatoma showed a 229% increase in Na^+, a 541% increase in Ca^{2+}, and a 41% decrease in K^+.

Further studies comparing the intracellular element contents of normal and transformed cells have, in every case tested so far, shown that transformed cells have high Na and Cl concentrations compared to their normal counterpart cells (*9,10,52,68,71,74*). A detailed comparison of the intracellular element contents of four tumor cell types (two hepatomas and two mammary adenocarcinomas) and their normal counterpart cell types was reported by Cameron and co-workers (*10*). The mean of pooled intracellular Na values of the tumor cells was 415 ± 6 mmole/kg dry weight, whereas the mean of pooled Na values of the nontumor counterparts was 138 ± 11 mmole/kg dry weight. These values are statistically different. Chlorine showed the same significant concentration pattern as sodium. Potassium and phosphorus were also shown to be slightly elevated. Trump *et al.* (*74*) showed that, after carcinogen-induced hepatoma formation in BALB/c mice *in vivo,* Na and Cl were greatly increased in the hepatoma cells compared to control hepatocytes. They also found decreases in K and P in the tumor cells. In contrast, Cameron and co-workers have found that the only consistent statistically significant differences between tumor and normal cells were the high Na and Cl content in the tumor cells (*9,10,52,68,71*). But their data were collected on established transplantable tumor lines, whereas Trump's group studied a directly perturbed liver after carcinogen treatment in the host animal.

It has also been shown that intracellular element contents change (during neoplastic transformation) after brief exposure to a carcinogen. Pool *et al.* (*52*) administered the hepatocellular carcinogen hydrazine sulfate to rats at a dosage of 15 mg/kg, twice a day for 5 days. Immediately following this treatment, α-fetoprotein (AFP), which is an indicator of an insult to the liver, could be detected in the serum of the rats by polyacrylamide gel electrophoresis. Pool and co-workers (*52*) showed that after 5 days of treatment the intracellular concentrations of Na and Cl in hepatocytes from animals exposed to the carcinogen were increasing, shifting toward the high levels of those elements found in overtly transformed cells (Morris hepatoma No. 7777). They concluded that elevation of Na and Cl is not only characteristic of transformed cells but that it may somehow be involved in the process of neoplastic transformation.

Ernst and Adams (*24*) found several differences between 3T3 cells (untransformed fibroblasts) and SV40 3T3 cells (virally transformed 3T3 cells). They made an effort to compare values between the two cell types at similar cell densities *in vitro*. They showed slightly higher K^+ and Na^+ values in the SV40 3T3 cells when compared to log-phase 3T3 cells. It might be noted here that Spaggiare *et al.* (*70*), unlike Ernst and Adams (*24*), found slightly less K^+ in SV40 3T3 cells than in 3T3 cells. Ernst and Adams also found a sharp decline in K^+ at confluence in the 3T3 cells but not in the SV40 3T3 cells (which were not contact inhibited and which grew to a very high cell density). Proll *et al.* (*53*) and Pool *et al.* (*52*) have shown similar changes in K levels of 3T3 cells. Ernst and Adams also found a gradual decline in Na^+ during log phase of the 3T3 cells [as did Proll *et al.* (*53*) and Pool *et al.* (*52*)] but not in the SV40 3T3 cells.

Johnson and Weber (*31*) made comparisons between proliferating, untransformed chick embryo fibroblasts and those transformed by Rous sarcoma virus. They found that in transformed fibroblasts K^+ increased by a factor of 1.4, whereas Na^+ rose by a factor of 2.6. When comparing density-inhibited to growing log-phase fibroblasts they found slight decreases in K^+ (44%) and Na^+ (4.4%) in the density-inhibited cells. This slight decrease in K^+ at confluence differs from the large decreases in K^+ seen in untransformed mammalian cells as they reach confluence (as described above).

Recent results with several cell systems show that transformed cells have an increased membrane permeability to Na^+ and a higher Na^+,K^+-ATPase activity (*32,57,65*). It is known that the transport (uptake) of some nutrients is Na^+ dependent (coupled to Na^+ transport) (*12*). The Na^+ gradient across the cell membrane is thought to be maintained by Na^+,K^+-ATPase activity, thus linking cell growth to Na^+,K^+-ATPase activity (*47,64*). Shen and co-workers (*65*), showed that membrane Na^+ permeability increased with tumorigenesis, whereas K^+ and Cl^- perme abilities were unaltered. Their data also suggested an increase in Na^+, K^+-ATPase activity with tumorigenesis. They hypothesized that the uncontrolled growth and metabolism of many transformed cells may require high intracellular Na^+ content, high Na^+ flux levels, and high Na^+,K^+-ATPase activity to meet the need for increased facilitated diffusion of the nutrients coupled to Na^+ transport.

Due to the complexity of interactions, feedbacks, and regulatory processes involved, it is hard to predict the end result of combined changes in membrane permeability to an ion, changes in ion fluxes, changes in E_m, or changes in Na^+,K^+-ATPase activity. For example, changes in external Na^+ and K^+ can alter the coupling ratio of active Na^+ and K^+ transport by the Na^+,K^+-ATPase in Ehrlich ascites tumor cells (*69*). Transport processes and metabolism can be intimately involved with any of the changes mentioned. Changes in E_m, ion fluxes, ion concentrations, membrane ion permeability, and Na^+,K^+-ATPase activity have been found in transformed cells compared to normal cells (as described above). Whether or not any of these changes in transformed cells is causative or directly related to the cancer lesion remains to be determined. The exact cancer lesion(s) may not be completely determined until all ionic and metabolic processes are understood as well as their interrelationship with the host genetic apparatus.

Racker has provided an insight into the relevance of changes in the Na^+,K^+-ATPase of cancer cells in comparison to normal cells (*54*). It is well established that many cancer cells have a higher aerobic glycolysis than do normal cells (*43,54*) and that this increased glycolysis is maintained by generation of ADP and P_i by the plasma membrane Na^+,K^+-ATPase. Racker has shown that the high Na^+,K^+-ATPase activity of Ehrlich ascites tumor cells is caused by a defective pump that operates at a low efficiency. Racker also showed that the rate-limiting step in glycolysis of Ehrlich ascites tumor cells is the regeneration of P_i by ATPase. He showed (*54*) that the addition of ouabain, or the deletion of either Na^+ or K^+ from the medium, inhibited aerobic glycolysis and lactic acid production in Ehrlich cells. This showed a correlation between ion fluxes and pump activity with aerobic glycolysis. Using the flavonoid quercetin, Racker's group has been able to improve the efficiency of the Na^+,K^+-ATPase of Ehrlich cells as measured by the ratio of active Na^+ flux to ATP utilization (*54,72*). Quercetin also did not affect Rb^+ uptake (a measure of pump activity) but did inhibit glycolysis in Ehrlich cells. Racker (*54*) attributed this to quercetin enhancing the efficiency of the pump. The improved efficiency of the Na^+,K^+-ATPase of the Ehrlich cells approached that of normal Na^+,K^+-ATPase isolated from mouse brain or electric eel. The data discussed above show that changes in Na^+,K^+-ATPase activity in tumor cells are related to the high aerobic glycolysis found in these cells. As Racker (*54*) pointed out, in glycolysis the oxidoreduction step is tightly coupled to phosphorylation. Therefore, for each mole of lactic acid that is formed, one mole of P_i and one mole of ADP must be transformed to ATP, and, since in the cell ATP is present in catalytic amounts, one mole of ATP must be cleaved to ADP and P_i. As can be seen from the above data, there is no glycolysis without ATPases. Intracellularly, ATPase activity includes all processes that hydrolyze ATP, such as ion transport, biosynthesis, and futile cycles (*54*).

The data discussed in this section have shown that differences in ion content, ion fluxes, and Na^+,K^+-ATPase activity occur in cancer cells compared to their normal counterparts. The explicit role of any of these changes in tumorigenesis has yet to be determined. However, a relationship between ion content, ion fluxes, Na^+,K^+-ATPase activity, and cellular processes (such as aerobic glycolysis, nutrient uptake, mRNA translation, DNA synthesis, protein synthesis, pH shifts, and mitochondrial Ca^{2+} efflux) has been shown in both normal and transformed cells as described in this and the preceding sections. The differences seen in these ionic, transport, and metabolic activities in transformed cells comprise attractive possibilities as areas of study on the role of the cancer lesion(s), especially since ion content, ion fluxes, and Na^+,K^+-ATPase activity have been shown to be correlated to normal cellular processes and cell proliferation.

VI. The Effects of Amiloride on Normal and Tumor Cell Growth

A. Effects of Amiloride on Transformed Cells *in Vivo*

Amiloride is known to block the passive influx of Na^+ and the proliferation of normal cells (*35,39,57;* Chapter 14). A series of experiments was designed to determine if amiloride would inhibit the proliferation of transformed cells and if a relationship existed between Na content and the rate of proliferation. In the first experiment, three groups of A/J mice bearing H6 transplantable hepatomas were given injections of amiloride every 8 hr. Tumor dimensions were determined daily by making caliper measurements of length and width. A linear regression program was used to determine the slope of tumor growth for each animal. The slope value from each group of mice were subjected to a one-way analysis of variance (ANOVA) to determine when and/or if amiloride affected tumor growth. After 104 hr (14 injections), amiloride at 1.0 μg/g body weight (BW) (slope = 0.109) had suppressed tumor growth compared to the control group (slope = 0.317; $p < .001$) and compared to the group treated with amiloride at 0.1 μg/g BW (slope = 0.276; $p < .01$). Expressing the data in another manner, the cumulative percentage increases in tumor area after 104 hr of treatment were as follows: (1) control (Ringer's injections), 298%; (2) amiloride at 0.1 μg/g BW, 293%; and (3) amiloride at 1.0 μg/g BW, 102%.

To determine if amiloride's influence on tumor growth was due to its diuretic actions which might cause a cell volume change, the diameters of metaphase and interphase hepatoma cells from tissue taken at the termination of the experiment were measured. A one-way ANOVA showed that there were no differences within or between the groups when comparing metaphase and interphase hepatoma cell diameters.

Amiloride was then tested on A/J mice bearing the transplantable DMA/J mammary adenocarcinoma to show that its inhibition of H6 hepatoma growth was not specific for the H6 hepatoma. As in the previous H6 experiment, amiloride at 1.0 μg/g BW inhibited DMA/J tumor growth; however, it required three more injections (a total of 17 injections spaced 8 hr apart). The tumor growth rate of the group treated with amiloride at 1.0 μg/g BW (slope = 0.016) was significantly suppressed when compared to the control group (slope = 0.092; $p < .05$) but not when compared to the group receiving amiloride at 0.1 μg/g BW (slope = 0.042). Expressed as cumulative percentage increases in tumor area from initiation of treatment until sacrifice (128 hr of treatment), the changes were (1) control, 1,033%; (2) amiloride at 0.1 μg/g BW, 226%; and (3) amiloride at 1.0 μg/g BW, 79%. At the end of the experiment, tumor samples were taken and processed for routine histology. No significant differences in metaphase and interphase mean profile cell diameters were found within or between groups. These data and those of Koch and Leffert (*35*) suggest that amiloride's inhibitory actions on cell proliferation *in vivo* are not due to cell water and volume changes.

Further tests were done on the H6 hepatoma *in vivo* to determine whether amiloride's inhibition of growth was correlated to intracellular element (ion) changes. We had shown previously that H6 hepatoma cells have higher elemental Na levels than hepatocytes (*68*). It was postulated that the rate of cell growth may be correlated to the level of intracellular Na (*10,68*). Previous studies on A/J mice indicated that multiple injections of amiloride at 10 μg/g BW or greater were lethal. In this experiment, amiloride was used at concentrations of 0.5, 1.0, and 5.0 μg/g BW, respectively, using the same protocol as described in the previous two experiments. It will be recalled that amiloride at 1.0 μg/g BW was shown to inhibit H6 hepatoma growth after 104 hr of treatment and DMA/J mammary adenocarcinoma growth after 128 hr of treatment. In this experiment, at 5.0 μg/g BW, amiloride inhibited H6 hepatoma growth after only 64 hr of treatment (Fig. 1). The cumulative percentage increases in tumor area were: (1) control, 57%; (2) amiloride at 0.5 μg/g BW, 51%; (3) amiloride at 1.0 μg/g BW, 58%; and (4) amiloride at 5.0 μg/g BW, 5%. One-way ANOVA of tumor growth rates verified that amiloride at 5.0 μg/g BW (slope = 0.092) suppressed tumor growth in comparison to controls (slope = 0.316; $p < .01$) or animals given amiloride at doses of 0.5 μg/g BW (slope = 0.331; $p < .01$) and 1.0 μg/g BW (slope = 0.356; $p < .01$). At the termination of the experiment, pieces of tumor

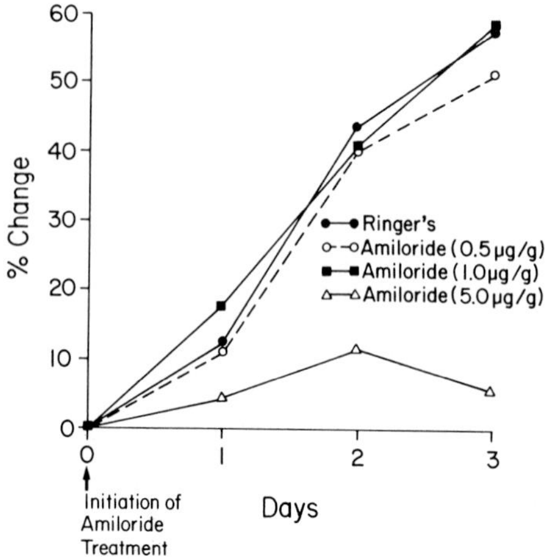

Fig. 1. Effect of amiloride treatment on H6 hepatoma growth *in vivo*. Each day, the mean tumor area (in cm^2) for each group (at least 10 animals per group) was determined and expressed as the cumulative percentage change in the area relative to size at initiation of treatment. Each animal received an injection of amiloride at the indicated dosage every 8 hr for 64 hr.

were removed, rapidly frozen, and processed for electron probe X-ray microanalysis as described (*10,51,68*). Electron probe X-ray microanalysis of the nucleus and cytoplasm of hepatoma cells of the various groups revealed that several changes had occurred due to the amiloride treatment. There were no dose–response relationships in the lowering of intracellular Na in the amiloride-treated groups. In fact, the group receiving the lowest concentration of amiloride (0.5 μg/g BW) showed the greatest reduction in Na content (87 mmole/kg dry weight) compared to the control group (119 mmole/kg dry weight). This demonstrated that a reduction in Na content will not necessarily inhibit tumor growth. Amiloride given at a dose of 5.0 μg/g BW, which inhibited tumor growth, may have done so by inhibiting the Na^+ influx that is associated with cell proliferation (*35,36,39,48,49,57,67*).

There has been speculation regarding the effects of different Na^+/K^+ ratios on gene expression and protein synthesis (*34,45,46,50*). It is possible that the intracellular Na^+/K^+ ratio may play a role in the regulation of cell proliferation. Determinations of Na/K ratios from Table I reveal interesting relationships between the groups. The group receiving the tumor-suppressive dose of amiloride at 5.0 μg/g BW had the lowest intracellular Na/K ratio in the hepatoma cells. There was also a direct correlation between the intracellular Na/K ratios of the hepatoma cells and the cumulative percentage increase in tumor area from the initiation

TABLE I

Effect of Amiloride on Intracellular Element Content of H6 Hepatoma Cells *in Vivo*[a]

Group	Na	Mg	P	S	Cl	K	Ca
Ringer's[b]	119 ± 2	56 ± 1	610 ± 25	214 ± 9	231 ± 6	624 ± 15	10.7 ± 2.1
Amiloride							
0.5 μg/g[c]	87 ± 2	39 ± 3	475 ± 18	200 ± 4	199 ± 11	497 ± 21	6.1 ± 1.4
	(−27)	(−31)	(−22)	(−6.2)	(−14)	(−20)	(−24)
1.0 μg/g[d]	113 ± 8	50 ± 2	554 ± 16	208 ± 6	220 ± 7	566 ± 11	9.4 ± 0.8
	(−5.5)	(−12)	(−9.2)	(−2.5)	(−4.7)	(−9.3)	(−11)
5.0 μg/g[e]	100 ± 1	55 ± 1	564 ± 11	173 ± 2	200 ± 3	611 ± 0.3	8.8 ± 0.02
	(−16)	(−1.3)	(−7.6)	(−19)	(−13)	(−2.1)	(−17)

[a] Content in mmole/kg dry weight (mean ± SEM). Numbers in parentheses represent percentage change from control.

[b] Values represent the mean combined nuclear and cytoplasmic values from two animals and a total of 50 X-ray spectra.

[c] Values represent the mean combined nuclear and cytoplasmic values from four animals and a total of 52 X-ray spectra.

[d] Values represent the mean combined nuclear and cytoplasmic values from three animals and a total of 52 X-ray spectra.

[e] Values represent the mean combined nuclear and cytoplasmic values from one animal and a total of 30 X-ray spectra.

of treatment until sacrifice. The percentage increases in tumor area and the intracellular Na/K ratios listed in descending order are as follows: (1) amiloride at 1.0 μg/g BW, 58%; Na/K = 0.1996; (2) control, 57%; Na/K = 0.1907; (3) amiloride at 0.5 μg/g BW, 51%; Na/K = 0.1751; and (4) amiloride at 5.0 μg/g BW, 5%; Na/K = 0.1637. This project was not specifically designed to study or manipulate intracellular Na/K ratios, so that the Na/K ratios described above are presented solely as an observed correlation. Besides blocking Na^+ influx, it is possible that amiloride may alter intracellular Na^+/K^+ ratios as well in other cell systems. Koch and Leffert (*35*) showed by indirect methods that amiloride caused a slight decrease in $[Na^+]_i$ of hepatocytes in culture. These data indicate that changes in intracellular Na/K ratios may play a role in cell proliferation.

In summary, amiloride suppressed H6 hepatoma growth and caused a partial depletion of the intracellular Na content of H6 hepatoma cells; however, a direct correlation between intracellular Na content and tumor growth after amiloride treatment was not found. This would lead one to assume that the high Na content of tumor cells is a result of the lesion(s) causing cell transformation. It would be that in transformed cells, which, compared to normal cells, have increased cell membrane Na^+ permeability and increased Na^+,K^+-ATPase activity (among other things), the high intracellular Na content is a compensatory result of the lesion and is part of the new steady state.

B. Effects of Amiloride on Transformed Cells *in Vitro*

The effects of amiloride on H6 cell proliferation were tested *in vitro* because the growth conditions could be better controlled than in the *in vivo* H6 hepatoma system previously described. Recent reports dealing with the effects of amiloride on the proliferation of stimulated normal cells *in vitro* agree with our results concerning the range of effective inhibitory doses of amiloride (*35,39,57*).

Amiloride inhibited H6 cell proliferation *in vitro* in a dose-dependent manner. A concentration of 1×10^{-3} *M* amiloride was completely inhibitory to cell proliferation, 5×10^{-4} *M* caused a partial suppression of cell proliferation, and 1×10^{-4} *M* had little or no effect on cell proliferation (Figs. 2 and 3). Results, not shown here, indicate that even at 1×10^{-3} *M*, which completely inhibited cell proliferation, the inhibitory effects were reversible. Others have shown that the inhibitory effects of amiloride on DNA synthesis are reversible (*35,57*). Thus, our studies and those of others indicate that amiloride can inhibit cell proliferation. Tritiated thymidine incorporation was measured by autoradiography to test the effects of amiloride on the DNA labeling index yielding additional information on the inhibitory effects of amiloride (Fig. 3).

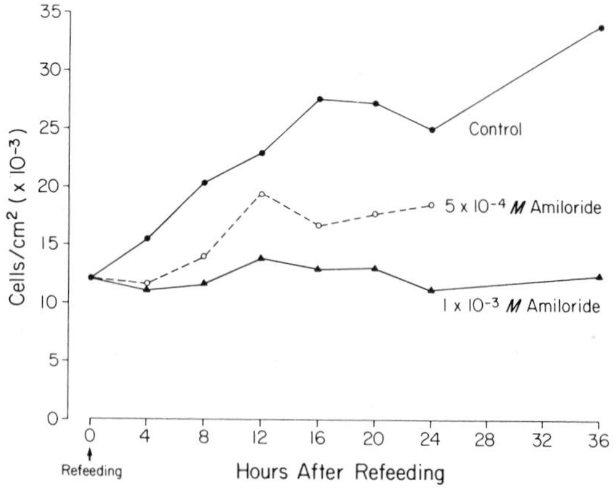

Fig. 2. Effect of amiloride on growth of H6 hepatoma cells *in vitro*. Cells were plated and fed again 24 hr later. Each point represents the mean cell number per cm^2 ($\times 10^{-3}$) of two culture dishes.

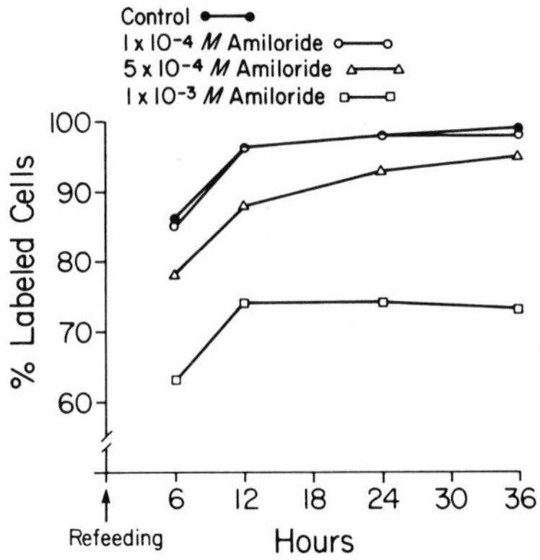

Fig. 3. Effects of amiloride on the cumulative labeling index of H6 hepatoma cells *in vitro* as determined by [^{3}H]TdR autoradiography. All groups of cells were plated and fed again with medium containing [^{3}H]TdR at 1.0 μCi/ml and amiloride at the indicated dosage. At least 500 cells were counted per time point for each dosage group.

Our preliminary work also showed that amiloride inhibits the proliferation of other transformed cells *in vitro* (DMA/J mammary adenocarcinoma cells and the NQT-1 chemically transformed 3T3 cells) at concentrations similar to those that inhibited the H6 cells. Initial studies also suggest that amiloride does not affect the cell volume of H6 cells *in vitro*. Koch and Leffert (*35*) showed that furosemide, a drug which should enhance Na^+ influx, stimulated DNA synthesis of hepatocytes *in vitro*. Our preliminary work with NQT-1 cells suggests that these cells are also stimulated to proliferate *in vitro* by furosemide treatment.

C. Effects of Amiloride on the Proliferation of Primary Liver Cell Cultures

These studies were performed to complement the series of experiments done on normal liver cells *in vivo* (*68*) and on the H6 hepatoma cells both *in vivo* and *in vitro*. Primary fetal liver cultures were chosen as a model to study the effects of amiloride on the proliferation of normal liver cells *in vitro*.

Rat liver (fetal and adult) has been used extensively for primary cultures; therefore, the techniques used to isolate and culture fetal rat liver cells were adapted to isolate and culture fetal mouse liver cells. The procedure of Leffert and Paul (*42*) and Leffert *et al.* (*40*) was adapted to the isolation of A/J mouse fetal liver cells. The plating and growth procedures worked very well in the mouse system. Slight differences were seen in the types of colonies formed in mouse primary liver cell cultures compared to those types of colonies found in rat liver primary cultures.

Basically, in the rat there is one type of colony found, and within it are two populations of hepatocytes (*41*). At the peripheral borders of the rat hepatocyte colony are the less differentiated, more rapidly proliferating cells, and in the central portion of the colony are the more differentiated, slowly proliferating cells (*41*). In the mouse, the two types of cells seem to form separate colonies. The more differentiated and slowly proliferating hepatocytes form colonies of tightly aggregated cells, whereas the less differentiated, rapidly proliferating hepatocytes form colonies of loosely aggregated cells (manuscript in preparation).

Amiloride treatment of the mouse primary liver cultures produced the following results: The colonies of rapidly proliferating cells were far more sensitive to amiloride's inhibitory effects than were the colonies of slowly proliferating hepatocytes (Figs. 4 and 5). Forty-eight hr of amiloride treatment had virtually *no effect* on the rate of increase of labeled cells in any of the slowly proliferating colonies of hepatocytes in any of the amiloride treated groups; however, by 24 hr of treatment with 5×10^{-4} *M* or 1×10^{-3} *M* amiloride, the rapidly proliferating cells that formed the loosely aggregated colonies were dying and lysing. Amiloride was selectively inhibiting and killing the colonies of rapidly proliferating cells. These same concentrations of amiloride were not even toxic to the H6 hepatoma cells over the same time period. In fact, H6 cells continued to proliferate, although somewhat more slowly, when subjected to amiloride at 5×10^{-4} *M* (Figs. 2 and 3). Koch and Leffert (*35*) showed a complete inhibition of [^{3}H]TdR uptake into nuclear DNA, as measured

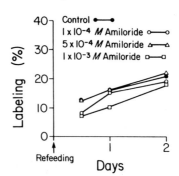

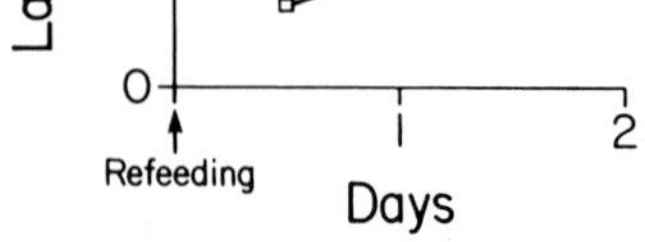

Fig. 4. Effects of amiloride on the cumulative labeling index of mouse fetal liver primary cultures as determined by [^{3}H]TdR autoradiography. The percentage of labeled cells in the colonies of *tightly aggregated* cells were determined from autoradiographs obtained at 12, 24, and 48 hr after refeeding with fresh culture medium containing amiloride (the control group received no amiloride) and [^{3}H]TdR. At least 300 total cells were counted per group at each time point (due to the loss of a sample, only 130 cells were counted for the 1×10^{-3} *M* amiloride group at the 24-hr time point).

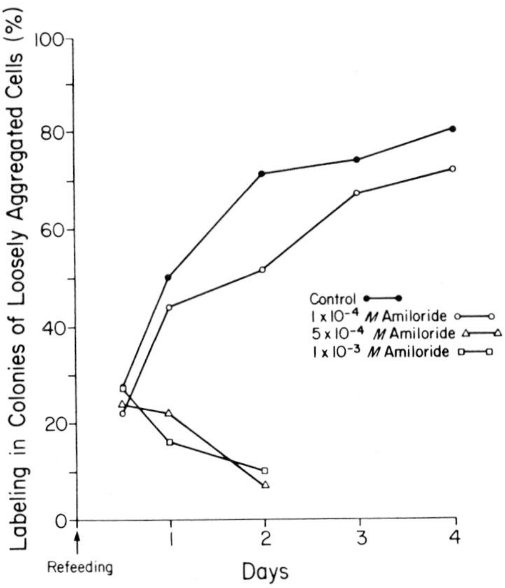

Fig. 5. Effects of amiloride on the cumulative labeling index of mouse fetal liver cultures as determined by [^{3}H]TdR autoradiography. For the control group which received no amiloride and the group that received amiloride at $1 \times 10^{-4}M$, the percent of labeled cells in loosely aggregated colonies was determined at 12, 24, 48, 72, and 96 hr after refeeding with fresh medium and [^{3}H]TdR. At least 500 total cells were counted for each group per time point. The percentage of labeled cells of the two groups that received amiloride at $5 \times 10^{-4}M$ or $1 \times 10^{-3}M$, respectively, were determined only at 12, 24, and 48 hr after drug exposure because at these higher amiloride concentrations the colonies of loosely aggregated cells were soon killed and lysed.

by scintillation counting, by amiloride at 3×10^{-4} M in rat hepatocyte cultures stimulated to resume cell proliferation. Using this technique they did not distinguish which type of hepatocytes was inhibited, but 3×10^{-4} M amiloride appeared to inhibit DNA synthesis of all cells. It appears that rat and mouse liver cultures are inhibited from proliferating by similar, but not exactly the same, concentrations of amiloride. Koch and Leffert (*35*) did not report differential inhibition of proliferation of the two types of hepatocytes after amiloride treatment. Their results were not expressed in a manner to yield information concerning differences between the two types of liver cells.

Transformed cells tend to have high membrane Na^+ permeability and Na^+,K^+-ATPase activity compared to normal cells. We have shown that transformed cells have high Na content compared to normal cells. One question that was asked in this project was whether or not transformed cells might be more sensitive than normal cells to a perturbation that affects Na^+ flux or Na^+ content. In light of the results on amiloride-treated H6 hepatoma cell cultures, transformed cells may not be more sensitive to amiloride's effects. It does appear that rapidly proliferating cells (normal or transformed) are more sensitive than slowly proliferating cells to amiloride's effects, because the concentrations of amiloride that inhibited H6 cell proliferation and inhibited rapidly proliferating liver cells did not inhibit the slowly proliferating liver cells (see also Chapter 14).

VII. Conclusions

Amiloride, a drug known to block passive Na^+ influx, was shown to inhibit H6 hepatoma and DMA/J mammary adenocarcinoma growth *in vivo* and to inhibit H6 cell and DMA/J cell proliferation *in vitro*. Treatment of animals bearing an H6 hepatoma with amiloride caused a decrease in intracellular Na, but not in a dose-dependent manner. In fact, the group receiving the lowest dosage of amiloride had the lowest levels of intracellular Na, but tumor growth had not been significantly inhibited at that dosage. Thus, the intracellular level of Na was not directly related to tumor growth. This would suggest that high intracellular levels of Na are related to oncogenesis, but may be a result of the cancer lesion(s) rather than a cause.

It appears from *in vitro* studies that rapidly proliferating cells (normal or transformed) are more sensitive to amiloride than are slowly proliferating normal cells. This conclusion rests on the findings that H6 hepatoma cells and rapidly proliferating primary liver cells in culture were inhibited from proliferating by concentrations of amiloride that had no effect on the proliferation of slowly proliferating primary liver cells in culture. In accordance with current findings published in the literature, Na^+ influx may play a greater role in the regulation of cell proliferation than does $[Na^+]_i$. Therefore, it may be that amiloride, a known inhibitor of passive Na^+ influx, inhibited proliferation of the different cell types used in this project due to its inhibition of Na^+ influx. This amiloride-sensitive Na^+ influx pathway is probably only active in proliferating cells (*35,57,76*), and the degree of inhibition by amiloride is probably correlated to the degree of proliferative activity of any given cell population.

Acknowledgment

This material is based on work supported by the National Science Foundation under Grant No. PCM 8104084.

References

1. Altman, P. I., and Katz, D. D., eds. (1976). "Cell Biology," Vol. 1, pp. 117–121. Fed. Am. Soc. Exp. Biol.
2. Anghileri, L. J., Heidbreder, M., Weiler, G., and Dermietzel, R. (1977). Hepatocarcinogenesis by thioacetamide: Correlations of histological and biochemical changes and possible role of cell injury. *Exp. Cell Biol.* **45,** 34–47.
3. Balk, S. C., Polimeni, P. I., Hoon, B. S., LeStourgeon, D. N., and Mitchell, R. S. (1979). Proliferation of Rous sarcoma virus-infected, but not or normal, chicken fibroblasts in a medium of reduced calcium and magnesium concentration. *Proc. Natl. Acad. Sci. U.S.A.* **76,** 3913–3916.
4. Barsh, G. S., and Cunningham, D. D. (1977). Nutrient uptake and control of animal cell proliferation. *J. Supramol. Struct.* **7,** 61–77.
5. Bernstein, A., Hunt, M. D., Crichley, V., and Mak, T. W. (1976). Induction by ouabain of hemoglobin synthesis in cultured Friend erythroleukemic cells. *Cell* **9,** 375–381.
6. Binggeli, R., and Cameron, I. L. (1980). Cellular potentials of normal and cancerous fibroblasts and hepatocytes. *Cancer Res.* **40,** 1830–1835.
7. Blackmore, R. F., Dehaye, J. P., and Exton, J. H. (1979). Studies on α-adrenergic activation of hepatic glucose output. *J. Biol. Chem.* **254,** 6945–6950.
8. Boynton, A. L., Whitfield, J. F., Isaacs, R. J., and Tremblay, R. G. (1977). Different extracellular calcium requirements for proliferation of nonneoplastic, preneoplastic and neoplastic mouse cells. *Cancer Res.* **37,** 2657–2661.
9. Cameron, I. L., Smith, N. K. R., Pool, T. B., Grubbs, B. G., Sparks, R. L., and Jeter, J. R., Jr. (1980). Regulation of cell reproduction in normal and cancer cells: The role of Na, Mg, Cl, K, and Ca. *In* "Nuclear–Cytoplasmic Interactions in the Cell Cycle" (G. Whitson, ed.), pp. 250–270. Academic Press, New York.
10. Cameron, I. L., Smith, N. K. R., Pool, T. B., and Sparks, R. L. (1980). Intracellular concentrations of sodium and other elements as related to mitogenesis and to oncogenesis *in vivo*. *Cancer Res.* **40,** 1493–1500.
11. Chalazonitis, A., and Fischbach, G. D. (1980). Elevated potassium induces morphological differentiation of dorsal root ganglionic neurons in dissociated cell culture. *Dev. Biol.* **78,** 173–183.
12. Christensen, H. N. (1972). Electrolyte effects on the transport of cationic amino acids. *In* "Na^+-linked Transport of Organic Solutes" (E. Heinz, ed.), pp. 39–50. Springer-Verlag, Berlin and New York.
13. Clark, A. F., and Roman, I. J. (1980). Mg^{2+} inhibition of Na^+-stimulated Ca^{2+} release from brain mitochondria. *J. Biol. Chem.* **255,** 6556–6558.
14. Cone, C. D., Jr. (1969). Electrosomotic interactions accompanying mitosis initiation in sarcoma cells *in vitro*. *Trans N. Y. Acad. Sci.* [2] **31,** 404–427.
15. Cone, C. D., Jr. (1971). Maintenance of mitotic homeostasis in somatic cell populations. *J. Theor. Biol.* **30,** 183–194.
16. Cone, C. D., Jr. (1971). Unified theory on the basic mechanism of normal mitotic control and oncogenesis. *J. Theor. Biol.* **30,** 151–181.
17. Cone, C. D., Jr. (1974). The role of the surface electrical transmembrane potential in normal and malignant mitogenesis. *Ann. N. Y. Acad. Sci.* **238,** 420–435.
18. Cone, C. D., Jr. (1980). Ionically mediated induction of mitogenesis in CNS neurons. *Ann. N. Y. Acad. Sci.* **339,** 115–131.
19. Cone, C. D., Jr., and Cone, C. M. (1976). Induction of mitosis in mature neurons in central nervous system by sustained depolarization. *Science* **192,** 155–158.
20. Cone, C. D., Jr., and Tongier, M., Jr. (1971). Control of somatic celll mitosis by stimulated changes in the transmembrane potential level. *Oncology* **25,** 168–182.

21. Crompton, M., Moser, R., Ludi, H., and Carafoli, E. (1978). The interrelations between the transport of sodium and calcium in mitochondria of various mammalian tissues. *Eur. J. Biochem.* **82,** 25–31.
22. Damadian, R., and Cope, F. (1974). NMR in cancer. V. Electronic diagnosis of cancer by potassium (^{39}K) nuclear magnetic resonance: Spin signatures and T_1 beat patterns. *Physiol. Chem. Phys.* **6,** 309–322.
23. Epel, D. (1980). Ionic triggers in the fertilization of sea urchin eggs. *Ann. N.Y. Acad. Sci.* **339,** 74–85.
24. Ernst, M., and Adams, G. (1979). Dependence of intracellular alkali-ion concentrations of 3T3 and SV40-3T3 cells on growth density. *Cytobiologie* **18,** 450–459.
25. Fishman, M. C. (1979). Endogenous digitalis-like activity in mammalian brain. *Proc. Natl. Acad. Sci. U.S.A.* **76,** 4661–4663.
26. Gagerman, E., Hellman, B., and Taljedal, I. B. (1979). Effects of ouabain on insulin release, adenosine, 3′5′-monophosphate guanine, 3′5′-monophosphate in pancreatic islets. *Endocrinology* **104,** 1000–1002.
27. Goldsmith, M., and Damadian, R. (1975). NMR in cancer. VII. Sodium-23 magnetic resonance of normal and cancerous tissues. *Physiol. Chem. Phys.* **7,** 263–269.
28. Haupert, G. T., Jr., and Sancho, J. M. (1979). Sodium transport inhibitor from bovine hypothalamus. *Proc. Natl. Acad. Sci. U.S.A.* **76,** 4658–4660.
29. Haworth, R. A., Hunter, D. R., and Berkoff, H. A. (1980). Na^+ releases Ca^{2+} from liver, kidney and lung mitochondria. *FEBS Lett.* **110,** 216–218.
30. Jaffe, L. (1980). Calcium explosions as triggers of development. *Ann. N.Y. Acad. Sci.* **339,** 86–101.
31. Johnson, M. A., and Weber, M. J. (1979). Potassium fluxes and ouabain binding in growing, density-inhibited and Rous-sarcoma virus-transformed chicken embryo cells. *J. Cell Physiol.* **101,** 89–100.
32. Kaplan, J. G. (1978). Membrane cation transport and the control of proliferation of mammalian cells. *Ann. Rev. Physiol.* **40,** 19–41.
33. Kaplan, J. G., and Owens, T. (1980). Activation of lymphocytes of man and mouse: Monovalent cation fluxes. *Ann. N.Y. Acad. Sci.* **339,** 191–200.
34. Koch, G., Bilello, J., Kruppa, J., Koch, F., and Opperman, H. (1980). Amplification of translational control by membrane-mediated events: A pleiotropic effect on cellular and viral gene expression. *Ann. N.Y. Acad. Sci.* **339,** 280–306.
35. Koch, K. S., and Leffert, H. L. (1979). Increased sodium ion influx is necessary to initiate rat hepatocyte proliferation. *Cell* **18,** 153–163.
36. Koch, K. S., and Leffert, H. L. (1979). Ionic landmarks along the mitogenic route. *Nature (London)* **279,** 104–105.
37. Koch, K. S., and Leffert, H. L. (1980). Growth control of differentiated adult rat hepatocytes in primary cultures. *Ann. N.Y. Acad. Sci.* **349,** 111–127.
38. Ledbetter, M. L., and Lubin, M. (1977). Control of protein synthesis in human fibroblasts by intracellular potassium. *Exp. Cell Res.* **105,** 223–236.
38a Leffert, H. L., ed. (1980). "Growth Regulation by Ion Fluxes." *Ann. N.Y. Acad. Sci.* **339.**
39. Leffert, H. L., and Koch, K. S., (1980). Ionic events at the membrane initiate rat liver regeneration. *Ann. N.Y. Acad. Sci.* **339,** 201–205.
40. Leffert, H. L., Koch, K. S., Moran, T., and Williams, M. (1979). Liver cells. *In* "Methods in Enzymology" (W. B. Jacoby and I. H. Pastan, eds.), Vol. 58, pp. 536–544. Academic Press, New York.
41. Leffert, H. L., Koch, K. S., Rubalcava, B., Sell, S., Moran, T., and Boorstein, R. (1978). Hepatocyte growth control: *In vitro* approach to problems of liver regeneration and function. *Natl. Cancer Inst. Monogr.* **48,** 87–101.

42. Leffert, H. L., and Paul, D. (1972). Studies on primary cultures of differentiated fetal liver cells. *J. Cell Biol.* **52,** 559–568.
43. Lehninger, A. (1975). "Biochemistry," 2nd ed. Worth Publ., New York.
44. Mager, D., and Bernstein, A. (1978). Early transport changes during erythroid differentiation of Friend leukemic cells. *J. Cell. Physiol.* **94,** 275–285.
45. Mager, D., and Bernstein, A. (1978). The program of Friend cell erythroid differentiation: Early changes in Na^+/K^+ ATPase function. *J. Supramol. Struct.* **8,** 431–438.
46. Mager, D. L., MacDonald, M. E., and Bernstein, A. (1979). Growth in high-K^+ medium induced Friend cell differentiation. *Dev. Biol.* **70,** 268–273.
47. Mahew, E., and Levinson, C. (1968). Reversibility of ouabain induced inhibition of cell division and cation transport in Ehrlich ascites cells. *J. Cell. Physiol.* **72,** 73–76.
48. Mendoza, S. A., Wigglesworth, N. M., Pohjanpelto, P., and Rozengurt, E. (1980). Na entry and Na–K pump activity in murine, hamster, and human cells—Effect of monesin, serum, platelet extract, and viral transformation. *J. Cell. Physiol.* **103,** 17–27.
49. Mendoza, S. A., Wigglesworth, N. M., and Rozengurt, E. (1980). Vasopressin rapidly stimulates Na entry and Na–K pump activity in quiescent cultures of mouse 3T3 cells. *J. Cell. Physiol.* **105,** 153–162.
50. Piatigorsky, J., Shinohara, T., Bhat, S., Reszelbach, R., Jones, R. E., and Sullivan, M. (1980). Correlated changes in δ-crystallin synthesis and ion concentrations in the embryonic chick lens: Summary current experiments and speculations. *Ann. N.Y. Acad. Sci.* **339,** 265–279.
51. Pool, T. B., Smith, N. K. R., Doyle, K. H., and Cameron, I. L. (1980). Evaluation of a preparative method for X-ray microanalysis of soft tissues. *Cytobios* **28,** 17–33.
52. Pool, T. B., Cameron, I. L., Smith, N. K. R., and Sparks, R. L. (1981). Intracellular sodium and growth control: A comparison of normal and transformed cells. *In* "The Transformed Cell" (I. L. Cameron and T. B. Pools, eds.), pp. 398–420. Academic Press, New York.
53. Proll, M. A., Pool, T. B., and Smith, N. K. R. (1979). Quantitative decreases in the intracellular concentrations of sodium and other elements as BALB/c 3T3 fibroblasts reach confluence. *J. Cell Biol.* **83,** 13a.
54. Racker, E. (1976). Why do tumor cells have a high aerobic glucolysis? *J. Cell. Physiol.* **89,** 697–700.
55. Rasmussen, J., Goodman, D. B. P., and Tenenhouse, A. (1972). The role of cyclic AMP and calcium in cell activation. *CRC Crit. Rev. Biochem.* **1,** 95.
56. Rozengurt, E. (1976). Coordination of early membrane changes in growth stimulation. *J. Cell. Physiol.* **89,** 627–632.
57. Rozengurt, E., and Mendoza, S. (1980). Monovalent ion fluxes and the control of cell proliferation in cultured fibroblasts. *Ann. N.Y. Acad. Sci.* **339,** 175–190.
58. Rubin, H. (1977). Specificity of the requirements for magnesium and calcium in the growth and metabolism of chick embryo fibroblasts. *J. Cell. Physiol.* **91,** 449–458.
59. Rubin, H. (1976). Magnesium deprivation reproduces the coordinate effects of serum or cortisol addition on transport and metabolism in chick embryo fibroblasts. *J. Cell. Physiol.* **89,** 613–626.
60. Rubin, H. (1975). Central role for magnesium in coordinate control of metabolism and growth in animal cells. *Proc. Natl. Acad. Sci. U.S.A.* **72,** 3551–3555.
61. Rubin, A. H., Terasaki, M., and Sanui, H. (1979). Major intracellular cations and growth control. Correspondence among magnesium content, protein synthesis, and the onset of DNA synthesis in BALB/c 3T3 cells. *Proc. Natl. Acad. Sci. U.S.A.* **76,** 3917–3921.
62. Sanui, H., and Rubin, H. (1977). Correlated effects of external magnesium on cation

content and DNA synthesis in cultured chicken embryo fibroblasts. *J. Cell Physiol.* **92,** 23–32.
63. Segel, G. B., and Lichtman, M. A. (1980). The apparent discrepancy of ouabain inhibition of cation transport and of lymphocyte proliferation is explained by time-dependency of ouabain binding. *J. Cell. Physiol.* **104,** 21–26.
64. Shank, B. B., and Smith, N. E. (1976). Regulation of cellular growth by sodium pump activity. *J. Cell. Phys1ol.* **87,** 377–388.
65. Shen, S. S., Hamamoto, S. T., Bern, H. A., and Steinhardt, R. A. (1978). Alteration of sodium transport in mouse mammary epithelium associated with neoplastic transformation. *Cancer Res.* **38,** 1356–1361.
66. Sivak, A. (1977). Induction of cell division in BALB/c-3T3 cells by phorbol myristate acetate or bovine serum: Effects of inhibitors of cyclic AMP phosphodiesterase and Na^+-K^+-ATPase. *In Vitro* **13,** 337–343.
67. Smith, J. B., and Rozengurt, E. (1978). Serum stimulates the Na^+, K^+ pump in quiescent fibroblasts by increasing Na^+ entry. *Proc. Natl. Acad. Sci. U.S.A.* **75,** 5560–5564.
68. Smith, N. R., Sparks, R. L., Pool, T. B., and Cameron, I. L. (1978). Differences in the intracellular concentration of elements in normal and cankerous liver cells as determined by X-ray microanalysis. *Cancer Res.* **38,** 1952–2959.
69. Smith, T. C., and Robinson, S. C. (1981). Variable coupling of active (Na^+ and K^+)-transport in Ehrlich ascites tumor cells: Regulation by external Na^+ and K^+. *J. Cell. Physiol.* **106,** 407–418.
70. Spaggiare, S., Wallach, M. J., and Tupper, J. T. (1976). Potassium transport in normal and transformed mouse 3T3 cells. *J. Cell. Physiol.* **89,** 403–416.
71. Sparks, R. L., and Cameron, I. L. (1979). High intracellular levels of sodium as related to the transformation of mammary epithelium. *J. Cell Biol.* **83,** 2a.
72. Spector, M., O'Neal, S., and Racker, E. (1980). Reconstitution of the Na^+K^+ pump of Ehrlich ascites tumor and enhancement of efficiency by quercetin. *J. Biol. Chem.* **255,** 5504–5507.
73. Spivak, J. L., Misiti, J., Stuart, R., Sharkis, S. J., and Sensenbrenner, L. L. (1980). Suppression and potentiation of mouse hematopoietic progenitor cell proliferation by ouabain. *Blood* **56,** 315–317.
74. Trump, B. F., Berezsky, I. K., Chang, S. H., Pendergrass, R. E., and Mergner, W. J. (1979). The role of ion shifts in cell injury. *Scanning Electron Microsc.* **3,** 1–14.
75. Tupper, J. T., Zorngniotti, F., and Mills, B. (1977). Potassium transport and content during G_1 and S phase following serum stimulation of 3T3 cells. *J. Cell. Physiol.* **91,** 429–440.
76. Villareal, M. (1981). Sodium fluxes in human fibroblasts: Effects of serum, Ca^{2+} and amiloride. *J. Cell. Physiol.* **107,** 359–369.
77. Whitfield, J. F., Boynton, A. L., MacManus, J. P., Rixon, R. H., Sikorska, M., Tsang, B., Walker, P. R., and Swierenga, S. H. H. (1980). The roles of calcium and cyclic AMP in cell proliferation. *Ann. N.Y. Acad. Sci.* **339,** 216–240.

16

The Central Role of Calcium in the Modulation of Cell Division

PAUL A. CHARP AND GARY L. WHITSON

I. Introduction

Interest in the role of calcium as a regulator or modulator of cellular activities has expanded rapidly since Heilbrunn (*17*) suggested that this cation was important to cellular functions. As is well known, calcium regulates many cellular and organismal functions such as muscle contraction (*1*), microtubule regulation (*28*), cell motility (*9*), cell lysis (*11*), macromolecular synthesis (*4,37*), platelet activation (*46*), protein phosphorylation (*53*), and cell secretion (*56*). These processes, although dependent on calcium for their activity, also require protein modifiers. These modifiers are known as calcium-dependent regulatory proteins, with one protein—calmodulin—being of extreme interest. It is beyond the scope of this chapter to discuss all the possible interactions of calcium/calmodulin complexes in biological systems; however, excellent reviews are available (*2,7,8,40–42*).

GENETIC EXPRESSION IN THE CELL CYCLE

ISBN 0-12-543720-X

In this chapter, we discuss the relevant roles of calcium as a regulator of the cell cycle and some of the events prior to and including the course of cell division. The use of a model system, *Tetrahymena pyriformis,* which explores these roles, is described. Some insights are presented as future possibilities and directions one could take to obtain a better understanding of the important role of calcium in the modulation of cell division.

II. General Concepts of Calcium as a Modulator of Diverse Cell Functions

All living cells examined thus far are capable of maintaining a large free calcium gradient across their membranes. Whereas, on the one hand, the extracellular concentration of free calcium in a variety of cell types can range from less than 1 m*M* to greater than 10 m*M* (*3*), on the other hand the intracellular concentration of free calcium is always very low, 0.1 μ*M* or less (*24*).

Intracellular calcium is either present in the bound or free state. The known intracellular binding agents for calcium are phosphates present in the mitochondria (*39*) or cytoplasm (*33*) and cytoplasmic proteins (*24*). This is not to rule out that calcium may also bind certain lipids. It is the cytosolic calcium-binding proteins, however, which ultimately lead to regulation of calcium movements. Calmodulin (CaM) appears to be the most promising of the calcium-binding cytosolic proteins and has been shown to activate Ca^{2+}-ATPase (*59*), phosphodiesterase (*8,10*), adenylate cyclase (*61*), and guanylate cyclase (*22*) as well as many other enzymes (for review, see *8*).

The flux of calcium across membranes of either mitochondrial or plasma membrane origin is regulated by means of a Ca^{2+}, Mg^{2+}-ATPase system. In the erythrocyte system, calcium activates the enzyme carrier complex, which in turn regulates the amount of intracellular calcium (*8,27*). This pump, which is now known to be stimulated by the addition of CaM (*59*), has been shown to decrease in activity with age and is most active in the presence of low calcium concentrations (*27*). It also appears to have asymmetric properties, in that the magnesium and ATP are confined to the intracellular side of the membrane whereas the calcium-binding site is present only on the extracellular side (*59*). Thus, the ability of the pump to maintain low intracellular calcium concentration is due to this spatial arrangement. But how such action is accomplished is still unclear because of this asymmetry. It is possible that calcium, by binding to CaM, activates the pump and thus is the major mechanism whereby control of the activity of the pump occurs from the intracellular side of the membrane.

Whereas calcium movements through the plasma membrane are fairly well understood, the mechanism of how calcium fluxes occur through membranes of the mitochondria is still unresolved. Racker (*39*) states that "the influx of Ca^{2+} into mitochondria is electrophoretic," which is due to ATP hydrolysis associated with mitochodrial metabolism. It is also possible that calcium movements across mitochrondrial membranes may be mediated by small peptides (*39*).

Although the influx of calcium into mitochondria may be electrophoretic, the efflux appears to function in a different manner (*6*). The dye, ruthenium red, blocks the electrophoretic movement of calcium into mitochondria (*29,44*) while it permits or enhances an efflux of calcium (*50*). Therefore, the existence of a second transport mechanism is possible. This second transport system appears to function as an exchange system in which the calcium is pumped from the mitochondria to replace sodium and hydrogen (*6*). Racker (*39*) suggested that there is a two-transport system, a Ca^{2+}/Na^{+} and a Na^{+}/H^{+} transporter thus yielding the Ca^{2+}/Na^{+} transport system.

Movement of calcium across membranes may therefore involve some modulation of transport through lipoprotein interactions, as have been described by Hui and Harmony (*18*) who showed that low-density lipoproteins decrease the ability of mitogen-stimulated lymphocytes to sequester calcium by binding to the lymphocyte cell membrane.

In *Tetrahymena,* the major storage site for calcium appears to be calcium-magnesium pyrophosphate granules and not the mitochondria (*33*). These granules, called volutin granules, appear to be membrane bound (*30*) and may function in the division process (*47*). These volutin granules are normally situated around the posterior portion of the organism. During division, the granules migrate to the region of the division furrow where they may discharge their contents for use in the division process. This, however, is only one speculation. Rosenberg and Munk (*49*) showed that *Tetrahymena,* placed in a phosphate-free buffer, lost calcium from these granules. After the calcium was depleted, magnesium was removed, leaving only the pyrophosphate stores from the granules available for energy production. This interesting possibility requires further investigation. It must be noted, however, that the flux of calcium and magnesium ions was not studied in dividing cells. Therefore, our present interest focuses upon the role of calcium as a modulator for cell division in synchronized *Tetrahymena.*

III. Synchronized *Tetrahymena* as a Model System to Study Calcium Fluxes in Relation to Cell Division

The ciliated protozoan *Tetrahymena pyriformis* is an excellent choice of a laboratory organism to study the biochemical and biophysical events associated with cell division. First, the organism is easily synchronized by means of temperature shifts from 28.5° to 34.5°C for 30 min followed by a downshift to 28.5°C for 30 min. This schedule is repeated throughout a period of 6.5 hr. The resulting cell cultures are synchronized with respect to division, which occurs 75 min after the last heat shock (EHS). The synchronized division, which can approach levels as high as 80%, takes approximately 25 min. Second, large numbers of cells can be grown aseptically in small volumes for chemical determinations.

Using this model system, we have studied the relationship between calcium fluxes and the cell division process. Specific agents known to affect the movement of calcium across membranes have been used in these studies. These included the calcium-chelating agent EGTA, verapamil, a blocker of the inward slow calcium channel (*49*), and the divalent cation ionophore A23187 (*43*). The flux of calcium was measured using ^{45}Ca as the radiotracer by measuring either its uptake or release with standard techniques.

In synchronized *Tetrahymena,* there is an influx of calcium 30 min prior to the initiation of cell division. Beginning at EHS + 60 min, there is an efflux of calcium preceding an 80% increase in cell numbers, which occurs within a 20-min time period (Fig. 1). The decrease in intracellular ^{45}Ca can be explained by the fact that ^{45}Ca appears in the calcium-free buffer (*15*). In other words, the drop in cell calcium (efflux) occurs before cell division is complete. Since a calcium-free buffer was used, both calcium exchange and partitioning can be ruled out, as suggested by Walker and Zeuthen (*60*). They argued that the drop in intracellular calcium is due only to an increase in cell number.

Addition of EGTA to a final concentration of 25 m*M* to synchronized cultures of *Tetrahymena* delays the onset of cell division in a time-dependent manner as well as decreasing the degree of synchrony obtained in control cultures (Fig. 2). The cell cultures are also delayed in reaching the peak of synchronized cell division (Fig. 3). There also does not appear to be a classical transition point to the movement of calcium, i.e., a point at which cell division would have been delayed or affected by the addition of EGTA.

In similar experiments, the ionophore A23187 was added at EHS to synchronized *Tetrahymena* at various concentrations. Low concentrations of the ionophore increased the uptake of ^{45}Ca at the time prior to cell

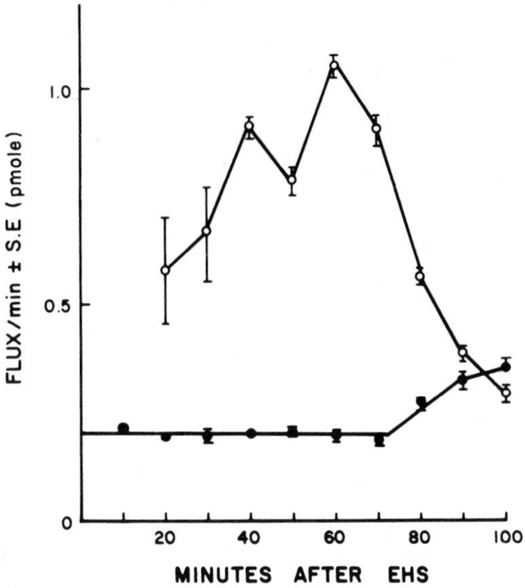

Fig. 1. Flux of ^{45}Ca in synchronized *Tetrahymena*. The flux was determined by converting ^{45}Ca counts to pmole ^{45}Ca and cell numbers to cell volumes. The flux of calcium (○); cell numbers (●).

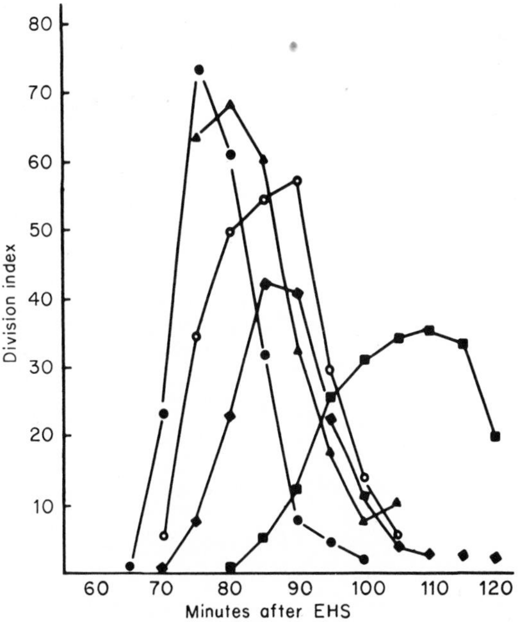

Fig. 2. Division patterns in synchronized *Tetrahymena* after addition of 25 m*M* EGTA. Controls (●); EHS + 70 min; (▲); EHS + 60 min (○); EHS + 50 min (♦); EHS (■).

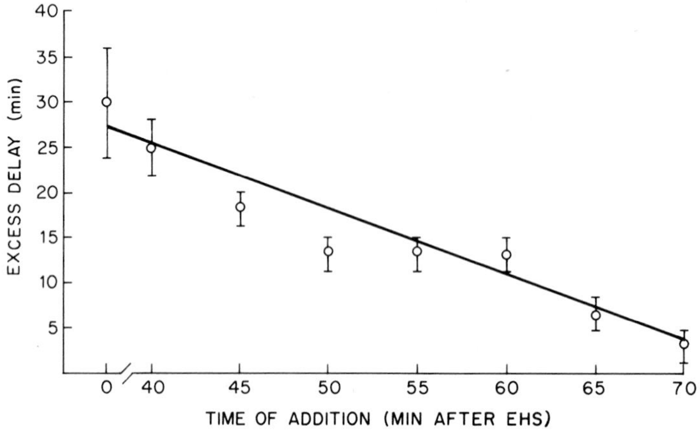

Fig. 3. Minutes excess delay in synchronized *Tetrahymena* treated with 25 m*M* EGTA. Minutes delay is described as the time needed to reach the peak of cell division as compared to the controls.

division (P. A. Charp, unpublished observation). The excess delay in cell division was measured, and these results are shown in Fig. 4. At an ionophore concentration of 60 μ*M*, no cell division was observed for at least 1 hr after control cells divided. It was also observed that the degree of synchronized cell division was depressed. As in the case of EGTA, there does not appear to be a transition point to the ionophore.

The inward flow of calcium into cells can be blocked using the slow calcium channel-blocker verapamil (*49*). Addition of verapamil to syn-

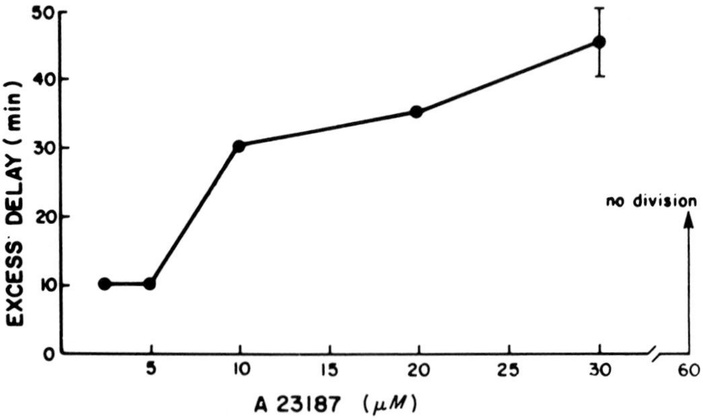

Fig. 4. Minutes excess delay in synchronized *Tetrahymena* treated with various concentrations of A23187. Minutes excess delay is described as the time needed for treated cultures to reach the peak of division as compared to control cultures.

chronized cultures of *Tetrahymena* at EHS led to a decrease in the degree of synchrony in a dose-dependent manner. At a verapamil concentration of 500 μ*M* all cell division was inhibited, indicating that this slow inward flux of calcium is a necessary prerequisite to cell division (Fig. 5).

In summary, our results showed that a calcium influx always preceded and was followed by a calcium efflux before cell division would occur in synchronized cultures of *Tetrahymena*. Blockage of this calcium influx with EGTA resulted in the delay of cell division. Addition of EGTA to synchronized *Tetrahymena* during the process of cell division always increased the time required for completion of cell division. The presumption here is that EGTA prevents the calcium from leaving the cell. Ionophore studies showed that increasing the movement of calcium into the cell resulted in an increased delay in cell division. In the presence of verapamil, the inward flux of calcium is inhibited prior to cell division and thus cell division is halted.

There are very few studies dealing with the possible role for calcium

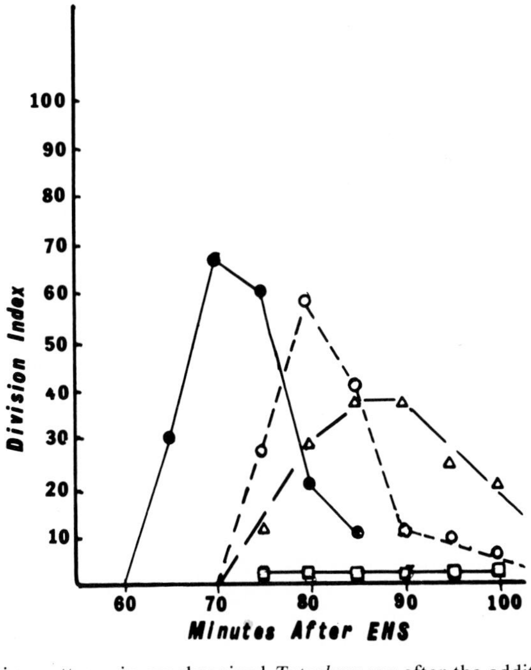

Fig. 5. Division patterns in synchronized *Tetrahymena* after the addition of verapamil at EHS. Controls (■); 100 μ*M* verapamil (○); 200 μ*M* verapamil (△); 500 μ*M* verapamil (□).

as a modulator of cell division. Hazelton and Tupper (*16*) showed that during exponential cell growth, the rate of calcium uptake was less than that of quiescent cell cultures. Paul and Ristow (*36*) reported that 3T3 cell growth was completely blocked in the presence of 2 m*M* EGTA. They also stated that cells grown in low calcium medium were stalled in G_0.

Osborn and co-workers (*35*) treated *Xenopus* embryos with ionophore A23187 and showed that, after a 10-min exposure to the drug, there was a redistribution of intracellular calcium and cortical contractions. They suggested that this redistribution of calcium leading to increased intracellular calcium interfered with normal cell division patterns and normal occurrence of division. Measurement of free calcium changes during the cell cycle of *Xenopus* embryos using microelectrodes was performed by Rink *et al.* (*45*). They detected no change in free calcium during two complete cleavages. They concluded that calcium has no regulatory role in *Xenopus* embryonic cell division.

Using WI 38 cells, Tupper and co-workers (*57*) studied the progression of these cells through G_1 to S phase. In the absence of calcium and/or serum, cell proliferation was inhibited in G_1. Addition of serum in the presence of low calcium did not stimulate cells to enter the S phase of the cell cycle. Further addition of calcium to these cells resulted in normal cell cycle progression.

Petzelt (*38*) reported that sea urchin eggs had a high calcium-activated ATPase after fertilization and this activity decreased prior to nuclear membrane disintegration. After breakdown, the activity increased again, reaching a peak at metaphase of the second division. He also stated that enzyme activity appeared to be concentrated in the mitotic apparatus.

The mechanism of how calcium affects cell division and growth is presently unknown; however, the possible role for calcium-binding proteins can not be overlooked. In the next section, we will discuss the role of CaM as a pivotal modulator of cellular function and its effect on enzyme systems and structural components known to be related to the division process.

IV. Possible Role for Calmodulin as a Modulator of Events Associated with Cell Division

Calmodulin was described by Cheung (*8*) as a heat-stable activator of cAMP phosphodiesterase and confirmed through the work of Kakiuchi and Yamazaki (*21*). Since this time, CaM has been shown to be a prominent calcium binder possessing the ability to modulate many cellular functions (*8*) such as calcium transport mechanisms, cyclases, phosphodiesterases, and microtubule–microfilament interactions.

A. Calcium Transport

From the studies reported in this chapter, it is evident that changes in intracellular calcium concentrations are required before cell division occurs. We have previously discussed the role of the mitochondrion as a regulator of intracellular calcium. We will now discuss the interaction of calmodulin with plasma membrane Ca^{2+},Mg^{2+}-ATPase, which is distinctly different from Na^{+},K^{+}-ATPase (for review, see *23*) (see also Chapter 15).

Larsen and Vincenzi (*26*) showed that the transport of calcium into inside-out red blood cell membranes was stimulated by CaM. The stimulation of the calcium pump would then regulate red blood cell calcium concentrations and lead to regulation of other intracellular functions dependent on calcium concentrations. Niggli and co-workers (*32*) reconstituted purified red blood cell membranes into phospholipid vesicles and showed the ability of the calcium-dependent ATPase to respond to CaM was greater than that of the enzyme without CaM. They also showed that acidic phospholipids were capable of stimulating this enzyme. Although these studies were performed using nonphysiological systems, the authors believe that, in the *in vivo* systems, both CaM and phospholipids could function as regulators of the control of calcium transport through the plasma membrane.

B. Cyclases

Associated with cell division are changes in cyclic nucleotide levels. These changes are mediated through a variety of enzymatic steps leading to the formation and degradation of the cyclic nucleotides cAMP and cGMP. It has now been shown that calcium and CaM can modulate the functions of some of these enzymes. Schultz *et al.* (*54*) reported that accumulation of cGMP in rat ductus deferens could be regulated by the presence of calcium ions and acetylcholine. Removal of the calcium, with the continual presence of acetylcholine, resulted in a large decrease in intracellular cGMP. Rodan and Feinstein (*46*) showed that calcium inhibited adenylate cyclase obtained from blood platelet membranes, whereas A23187 administered to intact platelets resulted in a calcium influx and decreased adenylate cyclase activity. They also showed that calcium stimulated guanylate cyclase as well as the ability of platelets to accumulate calcium from the medium.

Brostrom and co-workers (*5*) described an adenylate cyclase preparation from glial tumor cells that showed a biphasic calcium response. Low levels of free calcium (less than 1 μM) stimulated the enzyme, whereas higher concentrations inhibited the enzyme action. They also added a calcium-dependent regulatory protein to enzyme preparations, which resulted in a greater stimulation of adenylate cyclase than occurred in the presence of calcium alone.

Evain and co-workers (*12*), using CHO cells, showed that adenylate cyclase activity was inhibited by low concentrations of calcium-dependent regulatory protein (600 ng/ml). The ability of the protein to inhibit enzyme activity was lost at concentrations of the regulatory protein above 600 ng/ml. This ability to inhibit adenylate cyclase activity was also dependent on the density of the cell cultures. Adenylate cyclase from low-density cultures was not greatly affected by calcium-dependent regulatory protein, whereas the enzyme from high-density cultures was inhibited to a greater degree by the protein. They also reported that the intracellular concentration of calcium-dependent regulatory protein was inversely proportional to the density of the cell cultures. From their studies, Evain *et al.* (*12*) believe there is no regulatory role for calcium-dependent regulatory protein in the adenylate cyclase system of CHO cells. It is possible, however, that the regulatory protein used in their study was not CaM, for they could find no activation of phosphodiesterase.

Valverde *et al.* (*58*) showed that CaM stimulated pancreatic islet-cell adenylate cyclase in the presence of free calcium. They hypothesize that insulin release from these cells may be modulated by calcium and CaM stimulation of adenylate cyclase, since a rise in cAMP is necessary before insulin is released.

In *Tetrahymena,* a calcium-binding protein was first described by Suzuki *et al.* (*55*) and was shown to possess the ability to stimulate guanylate cyclase (*31*) without stimulation of adenylate cyclase. The stimulation of the binding protein could be inhibited in the presence of EGTA. Jamieson *et al.* (*20*), however, were the first to identify CaM specifically in *Tetrahymena*. The physiochemical properties of CaM in their report are very similar to the properties described by Suzuki and co-workers (*55*). Jamieson and co-workers (*20*) found the highest concentration of CaM associated with *Tetrahymena* cilia and concluded that CaM was an important regulatory agent in the locomotion of *Tetrahymena*.

Charp and Whitson (*7*) postulated that CaM would be a stimulator for guanylate cyclase, and this was shown by Kakiuchi *et al.* (*22*). Calmodulin obtained from *Tetrahymena* was shown to stimulate guanylate cyclase isolated from *Tetrahymena*.

C. Phosphodiesterase Activation

As previously mentioned, CaM was first described as an activator of cyclic nucleotide phosphodiesterase (*8*). There are virtually hundreds of papers dealing with the ability of CaM and calcium-binding proteins to stimulate phosphodiesterase reactions. It is believed that CaM activates phosphodiesterase through a series of bindings whereby calcium binds to CaM forming an active state. The activated CaM then forms a calcium–CaM phosphodiesterase complex. The complex is then capable of forming other calcium–CaM activated complexes (*10*).

At the present time, we are unaware of any work reported dealing with the interactions of CaM and phosphodiesterase during the division of any organism or cell line. However, Gomes *et al.* (*13*) did report that a calcium-binding protein with similar properties to CaM was present in the fungus *Blastocladiella*. After measuring both phosphodiesterase levels and calcium-binding protein levels during the life cycle of the fungus, they reported no relationship between the enzyme and the protein regulator. The activity of phosphodiesterase was lowest during the midlife cycle of the organism and highest during sporulation and during the zoospore stage. The levels of calcium-regulatory protein remained relatively constant throughout the life cycle. Since the calcium-regulatory protein from the fungus did not activate phosphodiesterase, they suggested that the protein has some other regulatory role in this organism.

Obviously, more research must be done in this area to further the understanding of the role of CaM and phosphodiesterase as pertaining to the dividing cell.

D. Microtubule–Microfilament Interactions

Cell division is a specialized form of motility. In order for a cell to divide, the membrane must constrict, break down in the region of the division furrow, and reseal without the loss of cytoplasm. Since the constriction is a type of contraction, Schroeder (*52*) proposed the presence of a ring of contractile proteins concentrated around this division furrow. Sanger (*51*) using fluorescent-labeled meromyosin, located actin in dividing chick fibroblast cells. During telophase, he observed strong fluorescent staining in the region associated with the division furrow. After cell division was completed, the fluorescent pattern in the division furrow decreased and, as cells began to attach to the culture dish, the fluorescent intensity was localized around the periphery of the cell.

Griffith and Pollard (*14*) studied the interaction of actin filaments and microtubular proteins. Their results showed that microtubule-associated proteins were a necessary component in order for any interaction to occur. They stated that their findings are significant in that actin binding to microtubules would be useful as an anchor for microtubule-dependent motility such as mitosis.

The assembly and disassembly of microtubular elements is under the control of calcium and calcium-dependent regulatory protein (CDR). Marcum *et al.* (*28*) showed that in the presence of free-calcium concentrations greater than 10^{-5} M and CDR, the assembly of microtubules was inhibited. In the absence of CDR, disassembly was not as evident. They also showed, using indirect immunofluorescent techniques, that CDR was associated with the mitotic spindle in the region between the spindle poles and chromosomes. Since the CDR showed the ability to disassemble microtubules, they believe that this finding may aid in the understanding of chromosome movements during mitosis and the interaction of calcium with CDR to modulate these events.

The mechanism by which CDR binds to microtubular elements was investigated by Nishida *et al.* (*34*) and Kumagai and Nishida (*25*). They proposed that a calcium–CDR complex binds to the tubulin monomer of microtubules, which then becomes nonpolymerizable. The reaction is dependent on the concentration of calcium and is reversible. The calcium–CDR complex did not bind to microtubule-associated proteins.

V. Conclusions and Perspectives

It is tempting to say that, of all things, calcium is the single most important regulator of cell division in eukaryotic cells. There is little proof to support this contention. There is good evidence now, however, that this divalent cation evokes many diverse cellular responses that have a common basic change in, or a stimulus for, motility. Cell division is certainly a response that involves a motility change and contractional elements in the eukaryotic cell. One of the most current widely accepted biochemical mechanisms for calcium-induced changes is in the function of the calcium-binding protein CaM, and its evolution is perhaps one of the most conserved in eukaryotic cells. There is good evidence that "stimulus-induced fluxes in free intracellular Ca^{2+} concentrations are coupled to cellular regulation" (*20*). Calmodulin has earned for itself high marks as one of the significant structurally related calcium-dependent regulatory proteins that are involved in physiological responses involving cell motility.

Some of the possible ways in which CaM may function in stimulus-induced fluxes and coupling have been presented in this chapter (for instance, the initial influx of calcium associated with the onset of cell division in synchronized *Tetrahymena*). We do not know whether the initial influx of calcium involves CaM directly. We must assert, however, that this protein could be involved, as there are many metabolic changes associated with cyclic nucleotide levels that follow in relation to the calcium influx.

We have shown that after the calcium influx there is a surge or raising of the internal levels of free calcium. According to Jaffe (*19*) and others, the initial "calcium signal" could act as a calcium "detonator" or "bomb" which could involve internal membrane changes associated with depolarizing events and calcium increases. Calmodulin could play a role in regulating or restoring calcium levels to normal internal levels. One possible way would involve the activation of calcium-dependent ATPases in the cell. Such identified activity changes, however, have yet to be established in dividing cells. Any prevention in the efflux of calcium after the initial influx, however, does inhibit or delay cell division.

One attractive hypothesis, other than the fluctuation of cyclic nucleotide levels in dividing cells associated with changing calcium levels at this time, would involve a model for activation of microtubule disassembly and cortical changes associated with actin-like proteins. How would calcium and/or CaM affect these changes? As we stated, there is evidence that high calcium or calcium in the presence of CaM could depolymerize microtubules. Such a depolymerization of microtubules in the region of the division furrow in *Tetrahymena* would be a necessary prerequisite for cell division. Actin-like proteins in the cortex have yet to be discovered in *Tetrahymena*. This does not mean that they are not there. The high quantities of both ciliary and cortical microtubules may mask such elements in *Tetrahymena*. Some researchers claim that calcium and even CaM may involve the phosphorylation of specific proteins (activation) which involve contractional mechanisms. Such is certainly true in striated and smooth muscle.

There are numerous possibilities for the role of calcium fluxes and CaM in the regulation of cell division. More experimentation and especially greater imagination is required to devise novel approaches to this fundamental question of how division is regulated. One perspective would involve the purification of *Tetrahymena* CaM and preparation of monoclonal antibodies to this protein. Then, a necessary set of precise experiments must follow to locate the interactions of calcium and CaM in dividing cells. Perhaps these questions and perspectives are too simple minded to be resolved at this time. Nevertheless, *Tetrahymena* represents a good test system for future studies along these kinds of investigations.

References

1. Ashley, C. C., and Ridgway, E. B. (1970). On the relationships between membrane potential calcium transient and tension in single barnacle muscle fibers. *J. Physiol. (London)* **209,** 105–130.
2. Berridge, M. J. (1975). The interaction of cyclic nucleotides and calcium in the control of cellular activity. *Adv. Cyclic Nucleotide Res.* **6,** 1–98.
3. Borle, A. B., and Anderson, J. H. (1976). A cybernetic view of cell calcium metabolism. *Symp. Soc. Exp. Biol.* **30,** 141–160.
4. Boynton, A. L., Whitfield, J. F., and MacManus, J. P. (1980). Calmodulin stimulates DNA synthesis by rat liver cells. *Biochem. Biophys. Res. Commun.* **95,** 745–749.
5. Brostrom, M. A., Brostrom, C. O., Breckenridge, B. M., and Wolff, D. F. (1976). Regulation of adenylate cyclase from glial tumor cells by calcium and a calcium binding protein. *J. Biol. Chem.* **251,** 4744–4750.
6. Carafoli, E. (1979). The calcium cycle of mitochondria. *FEBS Lett.* **104,** 1–5.
7. Charp, P. A., and Whitson, G. L. (1980). Calcium and cyclic nucleotide interactions during the cell cycle. *In* "Nuclear–Cytoplasmic Interactions in the Cell Cycle" (G. L. Whitson, ed.), pp. 309–333. Academic Press, New York.
8. Cheung, W. Y. (1980). Calmodulin plays a pivotal role in cellular regulation. *Science* **207,** 19–27.
9. Dedman, J. R., Brinkley, B. R., and Means, A. R. (1979). Regulation of microfilaments and microtubules by calcium and cyclic AMP. *Adv. Cyclic Nucleotide Res.* **11,** 131–174.
10. Dedman, J. R., Potter, J. D., Jackson, R. L., Johnson, J. D., and Means, A. R. (1977). Physicochemical properties of rat testes Ca^{2+}-dependent regulator protein of cyclic nucleotide phosphodiesterase. Relationship of Ca^{2+}-binding conformation changes and phosphodiesterase activity. *J. Biol. Chem.* **252,** 8415–8422.
11. Durant, S., Homo, F., and Duval, D. (1980). Calcium and A23187-induced cytolysis of mouse thymocytes. *Biochem. Biophys. Res. Commun.* **93,** 385–391.
12. Evain, D., Klee, C., and Anderson, W. B. (1979). Chinese hamster ovary cell population density affects intracellular concentrations of calcium-dependent regulator and ability of regulator to inhibit adenylate cyclase activity. *Proc. Natl. Acad. Sci. U.S.A.* **76,** 3962–3966.
13. Gomes, S. L., Mennucci, L., and Maia, J. C. C. (1979). A calcium dependent protein activator of mammalian cyclic nucleotide phosphodiesterase from *Blastocladiella emersonii. FEBS Lett.* **99,** 39–42.
14. Griffith, L. M., and Pollard, T. D. (1978). Evidence for actin filament–microtubule interaction mediated by microtubule-associated proteins. *J. Cell Biol.* **78,** 958–965.
15. Hamburger, K., and Zeuthen, E. (1957). Synchronous divisions in *T. pyriformis* as studied in an inorganic medium. *Exp. Cell Res.* **13,** 443–453.
16. Hazelton, B. J., and Tupper, J. T. (1979). Calcium transport and exchange in mouse 3T3 and SV 40-3T3 cells. *J. Cell Biol.* **81,** 538–542.
17. Heilbrunn, L. V. (1956). "The Dynamics of Living Protoplasm." Academic Press, New York.

18. Hui, D. Y., and Harmony, J. A. K. (1980). Inhibition of calcium accumulation in mitogen-activated lymphocytes: Role of membrane-bound liposomes. *Proc. Natl. Acad. Sci. U.S.A.* **77,** 4764–4768.
19. Jaffe, L. F. (1980). Calcium explosions as triggers of development. *Ann. N.Y. Acad. Sci.* **339,** 86–101.
20. Jamieson, G. A., Jr., Vanaman, T. C., and Blum, J. J. (1979). Presence of calmodulin in *Tetrahymena. Proc. Natl. Acad. Sci. U.S.A.* **76,** 6471–6475.
21. Kakiuchi, S., and Yamazaki, R. (1970). Calcium dependent phosphodiesterase activity and its activating factor from brain. *Biochem. Biophys. Res. Commun.* **41,** 1104–1110.
22. Kakiuchi, S., Sobue, K., Yamazaki, R., Nagao, S., Umeki, S., Nazawa, Y., Yazawa, M., and Yagi, K. (1981). Ca^{2+}-dependent modulator proteins from *Tetrahymena pyriformis,* sea anemone, and scallop and guanylate cyclase activation. *J. Biol. Chem.* **256,** 19–22.
23. Kretsinger, R. H. (1976). Evolution and function of calcium binding proteins. *Int. Rev. Cytol.* **46,** 323–393.
24. Kretsinger, R. H. (1979). The informational role of calcium in the cytosol. *Adv. Cyclic Nucleotide Res.* **11,** 2–26.
25. Kumagai, H., and Nishida, E. (1979). The interactions between calcium-dependent regulator protein of cyclic nucleotide phosphodiesterase and microtubule protein. II. Association of calcium-dependent with tubulin dimers. *J. Biochem.* (*Tokyo*) **85,** 1267–1274.
26. Larsen, F. L., and Vincenzi, F. F. (1979). Calcium transport across the plasma membrane: Stimulation by calmodulin. *Science* **204,** 306–309.
27. Luthra, M. G., and Kim, H. D. (1980). (Ca^{2+} + Mg^{2+})-ATPase of density-separated human red cells. Effects of calcium and a soluble cytoplasmic activator (calmodulin). *Biochim. Biophys. Acta* **600,** 480–488.
28. Marcum, J. M., Dedman, J. R., Brinkley, B. R., and Means, A. R. (1978). Control of microtubule assembly–disassembly by calcium-dependent regulator protein. *Proc. Natl. Acad. Sci. U.S.A.* **75,** 3771–3775.
29. Moore, C. L. (1971). Specific inhibition of mitochondrial Ca^{++} transport by ruthenium red. *Biochem. Biophys. Res. Commun.* **42,** 298–305.
30. Munk, N., and Rosenberg, H. (1969). On the deposition and utilization of inorganic pyrophosphate in *Tetrahymena pyriformis. Biochim. Biophys. Acta* **117,** 629–640.
31. Nagao, S., Suzuki, Y., Watanabe, Y., and Nozawa, Y. (1979). Activation by a calcium-binding protein of guanylate cyclase in *Tetrahymena pyriformis. Biochem. Biophys. Res. Commun.* **90,** 261–268.
32. Niggli, V., Adunyah, E. S., Penniston, J. T., and Carafoli, E. (1981). Purified (Ca^{2+}-Mg^{2+})-ATPase of the erythrocyte membrane. *J. Biol. Chem.* **256,** 395–401.
33. Nillson, J. R., and Coleman, J. R. (1977). Calcium-rich refractile granules in *Tetrahymena pyriformis* and their possible role in intracellular ion regulation. *J. Cell Sci.* **24,** 311–325.
34. Nishida, E., Kumagai, H., Ohtsuki, I., and Sakai, H. (1979). The interaction between calcium dependent regulator protein of cyclic nucleotide phosphodiesterase and microtuble proteins. I. Effect of calcium-dependent regulator protein on the calcium sensitivity of microtubule assembly. *J. Biochem.* (*Tokyo*) **85,** 1257–1266.
35. Osborn, J. C., Duncan, C. J., and Smith, J. L. (1979). Role of calcium ions in the control of embryogenesis of *Xenopus*: Changes in the subcellular distribution of calcium in early cleavage embryos after treatment with the ionophore A23187. *J. Cell Biol.* **80,** 589–604.

36. Paul, D., and Ristow, H. J. (1979). Cell cycle control by Ca^{++} ions in mouse 3T3 cells and in transformed 3T3 cells. *J. Cell. Physiol.* **98,** 31–40.
37. Perchellet, J. P., and Sharma, R. K. (1979). Mediatory role of calcium and guanosine 3′, 5′ monophosphate in adrenocortiotropin-induced steroidogenesis by adrenal cells. *Science* **203,** 1259–1261.
38. Petzelt, C. (1972). Ca^{2+}-activated ATPase during the cell cycle of the sea urchin *Strongylocentrotus purpuratus. Exp. Cell Res.* **70,** 333–339.
39. Racker, E. (1980). Fluxes of calcium and concepts. *Fed. Proc., Fed. Am. Soc. Exp. Biol.* **39,** 2422–2426.
40. Rasmussen, H. (1970). Cell communication, calcium ions and cyclic adenosine monophosphate. *Science* **170,** 404–412.
41. Rasmussen, H., Goodman, D. B. P., and Tenenhouse, A. (1972). The role of cyclic AMP and calcium in cell activation. *CRC Crit. Rev. Biochem.* **1,** 95–148.
42. Rebhun, L. I. (1977). Cyclic nucleotides, calcium and cell division. *Int. Rev. Cytol.* **49,** 1–54.
43. Reed, P. W., and Lardy, H. A. (1972). A23187: A divalent cation ionophore. *J. Biol. Chem.* **247,** 6970–6977.
44. Rigoni, F., Mathien-Shire, Y., and Deana, R. (1980). Effect of ruthenium red on calcium efflux from rat liver mitochondria. *FEBS Lett.* **120,** 255–258.
45. Rink, T. J., Tsien, R. Y., and Warner, A. E. (1980). Free calcium in *Xenopus* embryos measured with ion-selective microelectrodes. *Nature (London)* **283,** 658–660.
46. Rodan, G. A., and Feinstein, M. B. (1976). Interrelationships between Ca^{2+} and adenylate and guanylate cyclases in the control of platelet secretion and aggregation. *Proc. Natl. Acad. Sci. U.S.A.* **73,** 1829–1833.
47. Rosenberg, H. (1966). The isolation and identification of "Volutin" granules from *Tetrahymena. Exp. Cell Res.* **41,** 397–409.
48. Rosenberg, H., and Munk, N. (1969). Phenomena associated with the deposition and disappearance of pyrophosphate granules in *Tetrahymena pyriformis. Biochim. Biophys. Acta* **184,** 191–197.
49. Rosenberger, L., and Triggle, D. J. (1978). Calcium, calcium translocation and specific calcium antagonists. *In* "Calcium and Drug Action" (G. B. Weiss, ed.), pp. 3–31. Plenum, New York.
50. Rossi, C. S., Vasington, F. D., and Carafoli, E. (1973). The effect of ruthenium red on the uptake and release of Ca^{2+} by mitochondria. *Biochem. Biophys. Res. Commun.* **50,** 846–852.
51. Sanger, J. W. (1975). Changing patterns of actin localization during cell division. *Proc. Natl. Acad. Sci. U.S.A.* **72,** 1913–1916.
52. Schroeder, T. E. (1973). Cell constriction: Contractile role of microfilaments in division and development. *Am. Zool.* **13,** 949–960.
53. Schulman, H., and Greengard, P. (1978). Stimulation of brain membrane protein phosphorylation by calcium and an endogenous heat stable protein. *Nature (London)* **271,** 478–479.
54. Shultz, G., Hardman, J. G., Schultz, K., Baird, C. E., and Sutherland, E. W. (1973). The importance of calcium ions for the regulation of guanosine 3′:5′-cyclic monophosphate levels. *Proc. Natl. Acad. Sci. U.S.A.* **70,** 3889–3893.
55. Suzuki, Y., Hirabayashi, T., and Watanabe, Y. (1979). Isolation and electrophoretic properties of a calcium-binding protein from the ciliate *Tetrahymena pyriformis. Biochem. Biophys. Res. Commun.* **90,** 253–260.
56. Theoharides, T. C., and Douglas, W. W. (1978). Secretion in mast cells by calcium entrapped within phospholipid vesicles. *Science* **201,** 1143–1145.

57. Tupper, J. T., Kaufman, L., and Bodine, P. V. (1980). Related effects of calcium and serum on the G_1 phase of the human WI 38 fibroblast. *J. Cell. Physiol.* **104,** 97–103.
58. Valverde, I., Vandermeers, S., Anjaneyulum, R., and Malaisse, W. J. (1979). Calmodulin activation of adenylate cyclase in pancreatic islets. *Science* **206,** 225–227.
59. Vincenzi, F. F., and Larsen, F. L. (1980). The plasma membrane calcium pump: Regulation by a soluble Ca^{2+} binding protein. *Fed. Proc., Fed. Am. Soc. Exp. Biol.* **39,** 2427–2431.
60. Walker, G. M., and Zeuthen, E. (1980). Changes in calcium and magnesium levels during heat-shock synchronized cell division in *Tetrahymena*. *Exp. Cell Res.* **127,** 487–490.
61. Wolff, D. J., and Brostrom, C. O. (1979). Properties and functions of the calcium-dependent regulator protein. *Adv. Cyclic Nucleotide Res.* **11,** 27–88.

17

Univalent Cation Concentration and Regulation of the BALB/c-3T3 Growth Cycle

CHRISTOPHER N. FRANTZ

GENETIC EXPRESSION IN THE CELL CYCLE

ISBN 0-12-543720-X

I. Introduction

When cells are growth arrested in the G_0/G_1 phase of the cell growth cycle due to lack of hormones or nutrients, reinitiation of growth is associated with increases in the flux of monovalent cations across the cell membrane. In addition, growing cells (necessarily in the presence of adequate nutrients and hormones) have greater monovalent fluxes than growth-arrested cells. Growing cells also have different monovalent cation content and concentration than growth-arrested cells. The significance of the monovalent cation fluxes and concentrations has not been determined, but two similar, related hypotheses have been proposed. First, a number of investigators have proposed that the ion fluxes reflect the mechanism of hormone action on the cell membrane, based on the model of acetylcholine (ACh). Second, it has been proposed that the ion fluxes per se may be a secondary result of hormone action, but that the increased monovalent cation flux is a critical, required step in the sequence of events between binding of hormone to receptor and cell growth. Thus, there are two basic questions: (1) Do increased monovalent cation fluxes cause events necessary for cell growth? (2) Do increased monovalent cation fluxes reflect membrane events necessary for cell growth? There is also a corollary question: Do increased monovalent cation fluxes result directly from the hormone receptor modulating an ion channel?

A. Measurement of Cell Growth

It has proved difficult to investigate the mechanisms of action of growth-regulating hormones in G_0/G_1, because the endpoint, usually measured as the onset of DNA synthesis, may not occur until many hours after exposure to the hormonal stimulus and requires participation of all the complex macromolecular synthetic apparatus of the cell. The mammalian cell growth cycle is defined by two identifiable events, DNA synthesis (S phase) and mitosis (M phase). The period between S and M is termed G_2, and the period between M and S is termed G_0/G_1. In a given cell type, the duration of S, G_2, and M is relatively invariant, but the length of G_0/G_1 is subject to regulation by hormones and nutrient availability (*5*). Thus, hormones regulate the rate of cell growth in G_0/G_1 (*5,42*). Therefore, studies of the mechanisms of hormonal regulation of growth, including the role of monovalent cation flux, have focused on events in the G_0/G_1 phase of the cell cycle, and entry into the S phase has been used as the measure of successful completion of the important growth regulatory events.

B. Acetylcholine Model of Ion Flux

Current understanding of membrane permeability and the flux of cations across cell membranes has emerged from the intensive study of nerve excitation and conduction pioneered by Huxley and Katz (*27*). The study of nerve membranes provides a basis for understanding ionic events in nonexcitable cells. The ACh–receptor–ionophore system is an excellent model of hormone–receptor–effector action, because the response to the hormonal signal (ACh) is immediate and subject to detailed electrophysiologic analysis. The ACh is released from a nerve and binds to specific ACh receptors on an adjacent nerve or other excitable tissue. The ACh–receptor complex effects an alteration in an ion channel, changing the cell membrane potential. The type of ion channel that is altered varies markedly between tissues and species, but a change in membrane potential is always noted in the cell after ACh binds (*72*). In nerve fibers, depolarization of the cell membrane potential results in the rapid opening of ion channels, which allow movement first of Na^+ and then of K^+ across the cell membrane. The resulting Na^+ influx results in depolarization of adjacent areas of membrane, so a wave of depolarization is transferred down the length of the nerve fiber. This wave of membrane depolarization constitutes a signal that may then be transferred via neurotransmitters to additional nerves or other tissue. Membrane potential is determined by both the distribution of ions across the membrane and the permeabilities to those ions. Thus, the effect of ACh binding to receptor is to increase monovalent cation permeability, which depolarizes the cell membrane. The local depolarization directly alters membrane properties; depolarization results in opening other channels, which, in nerve signal transmission, causes further depolarization.

In excitable cells other than nerve cells, ACh induces additional events. Binding of the neurotransmitter to membrane receptors on muscle cells results in depolarization of the membrane potential associated with muscle contraction; similarly, ACh binds to adrenal medulla cells and results in membrane depolarization and secretion of catecholamines. In these tissues, ACh stimulation leads to both Na^+-dependent depolarization and influx of Ca^{2+} into the cell. The resultant increased cytoplasmic free-calcium concentration must occur in order to result in contraction or secretion. The entry of Ca^{2+} depends at least partly on depolarization (*14,58*).

Investigation of hormone-induced secretion in other endocrine and exocrine tissues has not consistently revealed a simple mechanism of excitation→depolarization→Ca^{2+}influx→secretion. Rather, the discovery that cAMP is a "second messenger," which transmits the action of many hormones to the inside of the cell, has provided an explanation for many but by no means all mechanisms of hormone action. In addition, convincing evidence has been accumulated that Ca^{2+} participates in the mechanism of action of some hormones, but the increase in cytoplasmic Ca^{2+} comes from intracellular Ca^{2+}-storage sites rather than from extracellular fluid (*30*). The closely interrelated roles of Ca^{2+} and cAMP in the mechanism of action of different hormones vary widely between hormone, tissue, hormone effect measured, and species (*51,72*).

II. Hormonal Regulation of Cell Growth and Ion Flux

Cyclic nucleotides and Ca^{2+} are implicated in the mechanism of action of a variety of hormones. What roles might they play in the regulation of cell growth by hormones? A "second messenger" role may be played by cAMP in some phases of growth of some cells, but no consistent pattern has been defined (*43*). Although investigation of the role of Ca^{2+} as a second messenger or coupling agent in cell growth has been hampered by an inability to measure free cytoplasmic Ca^{2+} concentration accurately in small cells (*10*), important roles for Ca^{2+} and cAMP as mediators of cell growth in late G_1 is strongly suggested (*78*). However, the roles of cAMP and Ca^{2+} in stimulation of growth of cells arrested in G_0 are less clear (*78*). In G_0 cells, altered monovalent cation permeability is frequently associated with hormonal stimulation of growth. Stimulation of growth of G_0 cells is clearly associated with stimulation of monovalent cation flux in fibroblasts, lymphocytes, neuroblastoma cells, and hepatocytes. The extensive literature demonstrating growth-associated changes in monovalent cations is reviewed by Sparks *et al.* in Chapter 15 and specifically for neuroblastoma cells by de Laat and van der Saag in Chapter 14. Therefore, the points of general agreement among investigators will be summarized and disputed issues emphasized, and then recent experiments with one cell type—mouse 3T3 fibroblasts—will be presented.

Marine Eggs and Late-G_1 Growth Arrest

The best studied examples of cells growth arrested in late G_1 are the eggs of a variety of marine species. The large size of the eggs allows the use of penetrating electrodes, to measure membrane potential, and injection of the calcium-sensitive photoprotein aequorin, to measure changes in cytoplasmic calcium. "Activation" (growth stimulation) of the eggs by either sperm or artificial means results in immediate membrane depolarization associated with markedly increased fluxes of Na^+

and K^+ across the cell membrane. Several seconds later, a brief, massive increase in cytoplasmic Ca^{2+} concentration is seen. It is not known whether the Ca^{2+} enters the cytoplasm from the internal surface of the cell membrane, from organelles, or both, but the presence of extracellular Ca^{2+} is not required to observe the increase in cytoplasmic Ca^{2+}, which appears to be necessary for subsequent egg development (*23*). Ca^{2+} activates NAD kinase by means of the calcium-dependent regulatory protein calmodulin (*16*). Additional biochemical events regulated by the increase in cytoplasmic Ca^{2+} have not yet been demonstrated. After the increase in cytoplasmic Ca^{2+}, extrusion of protons results in increased intracellular pH, which appears to be important in further development of some egg species (*15*). The causal relationships between changes in membrane potential, Na^+ and K^+ fluxes, H^+ efflux, and cytoplasmic Ca^{2+} are not clear.

In mammalian somatic cells, both Ca^{2+} and cAMP may play a significant role in growth regulation in late G_1 (*78*). Total cellular calmodulin is increased in late G_1 (*11*). Some small peripheral blood lymphocytes are growth arrested in late G_1; when stimulated to enter S phase, an increase in intracellular cAMP concentration and increased influx and efflux of Ca^{2+} are closely associated. Inhibition of the increase in cAMP is associated with inhibition of entry into S phase. If the lymphocytes are stimulated in medium deficient in Ca^{2+}, increase in cAMP and entry into S phase does not occur; readdition of Ca^{2+} permits entry into S phase and stimulates increase in both Ca^{2+} flux and in cellular cAMP. Similar phenomena occur in hepatocytes *in vivo* and *in vitro* and in cultured fibroblasts (for review, see *78*). Monovalent cation fluxes have not been extensively studied in cells arrested near G_1/S. Relationships between monovalent cation flux, Ca^{2+} and cAMP are speculated upon in Chapter 15.

In rapidly cycling cells that do not enter G_0, increased Na^+,K^+ pumping may occur in late G_1, as described for neuroblasts (*7,41*) (see Chapter 14). Membrane potential and intracellular K^+ activity, measured by direct electrode, and K^+ efflux and influx measured by tracer technique, revealed a steady increase in intracellular K^+ activity throughout G_1 and a marked decrease in K^+ influx in early G_1 but an increase at G_1/S. The decrease is proportional to a decrease in Na^+,K^+ ATPase capacity measured in cell homogenates, but the increase in K^+ influx in late G_1 may result from increased intracellular Na^+ stimulation of the Na^+,K^+ pump (*7,41;* Chapter 14). Similar stimulation of Na^+,K^+ pump activity occurs when G_0 cells are stimulated to grow, as described below. The relationship between increased Na^+,K^+ pump activity during G_1/S in cycling cells and the changes during G_0/G_1 in G_0-arrested cells is unclear.

When growth-stimulating substances are added to G_0-arrested cultured cells, immediate increases are seen in influx of Na^+ and K^+ (*26,29, 32,36,38,39,53–55,64,77*). Investigation in small cells, in which microelectrode puncture and electrophysiologic studies are not easily performed, have used radioactive ion analogues as tracers of ion movement. These tracer experiments must be interpreted carefully, because rate of movement of tracer does not necessarily reflect a net change in ion concentration, as exchange of ions, e.g., $^{22}Na^+$–Na^+, may occur across the cell membrane. Tracer experiments have detected stimulation of monovalent cation flux by many hormones. These fluxes are much slower than those stimulated by ACh, but more rapid changes in membrane potential have not been carefully investigated, as they would not be detectable with the flux techniques. Application of these techniques to study of hormonal regulation of cell growth has been difficult, because many intervening events must occur before cell growth, the experimental endpoint, is expressed. Attempts to relate ion flux to cell growth have therefore relied on inhibition of fluxes and inhibition of cell growth by some agent and have often assumed that the only effect of inhibitor was the observed effect on monovalent cation fluxes.

1. Amiloride-Sensitive Na^+ Influx

Increased Na^+ influx occurs by several routes. Some may result in response to increased Na^+ efflux as Na^+ is pumped out by Na^+,K^+-ATPase; some may result from a very brief generalized increased permeability to many ions, which is associated with a brief drop in membrane potential (*38,39*). However, much of the Na^+ influx occurs by a route that is inhibited by amiloride (*29,32,36,39,55,77*). The amiloride-sensitive Na^+ channel is either absent or very small in G_0 neuroblasts or fibroblasts in serum-free medium (*39,77*). Upon addition of serum, this Na^+ channel is immediately activated and remains active for at least 1 hr, possibly much longer. Serum-depleted of growth-stimulating ability does not stimulate amiloride-sensitive Na^+ influx (*39*). The Na^+ influx is measured as rate of accumulation of cellular $^{22}Na^+$ in the presence of ouabain, which prevents removal of intracellular $^{22}Na^+$ by the Na^+,K^+ pump. It is widely assumed that the rate of $^{22}Na^+$ accumulation in the presence of ouabain is a measure of net Na^+ influx rather than a result of net K^+ efflux, although Na^+ replaces K^+. Amiloride has no effect on the voltage-dependent or tetrodotoxin-sensitive Na^+ channels (*39*), and opening of this Na^+ influx route in fibroblasts appears unrelated to serum or growth stimulation (*46*).

Amiloride is a potent diuretic and inhibitor of Na^+ flux across a wide variety of epithelia (*6*). In some epithelia, amiloride inhibits Na^+–H^+

exchange (*28*). The amiloride-sensitive Na^+ channel operates in an electroneutral fashion; there is no difference in membrane potential whether the channel is open or closed (*39*). Preliminary evidence suggests that H^+ may be the ion exchanged for Na^+ (*39*). Also, "growth stimulation" of G_0 lymphocyte plasma membrane vesicles results in rapid extrusion of protons and elevation of intravesicular pH (*37*), but the effect of amiloride on this process is unknown. Thus, activating the amiloride-sensitive Na^+ channel (during growth stimulation) might allow excess H^+ to be excreted from the cell or might directly increase intracellular pH (pH_i). Cell volume changes may also be mediated by the same or other amiloride-sensitive pathways, because growth stimulation of G_0 cells results in increased cell volume (*7,26,75*), and amiloride decreases the increase in cell volume, measured as final $^{22}Na^+$ content, seen in hepatocytes stimulated to grow in the presence of ouabain (*32*). The effects of amiloride on K^+ fluxes during growth stimulation have not been examined in all cell types, but amiloride completely prevents any serum-stimulated increase in Na^+,K^+ pump activity in G_0 neuroblasts (Chapter 14; *39*).

In growth-stimulated G_0 cells, amiloride inhibits entry into S phase, so it has been proposed that the amiloride-sensitive Na^+ channel must be open to allow cell growth (*29,32,39,54,77*). The amiloride concentration for growth inhibition is similar to the concentration needed to inhibit Na^+ influx. However, examination of the effects of amiloride on 3T3 cells suggests another conclusion. Addition of amiloride to 3T3 cells at the same concentration used in the experiments above resulted in severe inhibition of protein synthesis to levels less than 30% of that seen in G_0 (C. N. Frantz, unpublished observation). Thus, amiloride may prevent DNA synthesis, not by preventing Na^+ influx per se, but rather by inhibiting protein synthesis. Continuous protein synthesis is required for G_0-arrested cells to enter S phase (*8,52*). Moreover, inhibition of protein synthesis in 3T3 cells with either ouabain or cycloheximide results in reversible inhibition of entry into S phase very similar to that seen with hepatocytes in the presence of amiloride (*18*). Also, when mutant MDCK cells were selected for resistance to amiloride toxicity, they were found to be resistant to toxic effects of amiloride on mitochondria, rather than amiloride inhibition of Na^+ influx (*71*). Thus, amiloride inhibition of plasma membrane Na^+ influx may be unrelated to amiloride inhibition of cell growth.

2. *Na^+,K^+ Pump Activity*

Growth stimulation of G_0 cells is associated with an immediate twofold increase in ouabain-inhibitable Na^+,K^+ pumping (*26,39,54*). The change is in V_{max}, not K_m (*26,41,49,53,61*), suggesting no change in affinity for substrate ions. Attention has focused on the enhanced active transport of Na^+ out of and K^+ into the cell by Na^+,K^+-ATPase, because ouabain inhibits both mitogenesis and active fluxes (for further review, see Chapter 15). Inhibition of mitogenesis by ouabain may be prevented by removing ouabain or by elevating extracellular K^+ concentration $[K^+]_o$ as would be expected if ouabain acted by inhibiting Na^+,K^+-ATPase (*7,26,54*). The cell membrane Na^+,K^+ pump, or Na^+,K^+-ATPase, is an enzyme which utilizes ATP to provide energy to pump Na^+ out of the cell and K^+ into the cell, both against concentration gradients. Ouabain is a highly specific inhibitor. Activity of the pump is generally measured (only in intact cells) as the rate of entry of K^+ into the cell minus the rate of K^+ entry in the presence of ouabain. The activity of Na^+,K^+-ATPase, due to the same enzyme, is measured (in cell homogenates) as the rate of ATP breakdown minus the rate of ATP breakdown in the presence of ouabain. Just as the Na^+,K^+-ATPase activity is stimulated by Na^+, lymphocyte and 3T3 cell Na^+,K^+ pumping is stimulated by intracellular Na^+ with a K_m of about 15 mmole/liter (*45,54,59,60*). The possible role of the Na^+,K^+ pump in growth regulation is discussed in Sections III,E and IV below.

The mechanism by which mitogens stimulate the Na^+,K^+ pump has not yet been determined. Three hypothetical mechanisms by which mitogens might increase Na^+,K^+ pumping have been proposed: (1) an increase in ATPase activity; (2) an increase in available cell surface Na^+,K^+ pumps; and (3) stimulation of unchanged Na^+,K^+ pumps by increased $[Na^+]_i$ (*26*). There is now general agreement that PHA treatment of G_0 T lymphocytes does not directly increase ATPase activity (measured as ATP breakdown in cell homogenate or plasma membrane vesicles) and has no effect on affinity of the ATPase for ATP, Mg^{2+}, K^+, or Na^+ (*26,60,61*), in spite of earlier reports to the contrary. Similarly, total cellular Na^+,K^+-ATPase activity does not suddenly increase during growth stimulation of G_0 3T3 cells (*54*) or neuroblasts (*7*). Attempts to determine the number of available cell surface Na^+,K^+ pumps by ouabain-binding studies (*49*) have been complicated by endocytosis of labeled pumps and synthesis and insertion of new pumps into the plasma membrane in response to ouabain (*45*). However, this modulation of pump number appears to require many hours (24 hr) (*45,62*). Modulation of T lymphocyte Na^+,K^+ pump activity during a brief (0–4 hr) period may be due to a 33% increase in $[Na^+]_i$ of 5 mmole/liter (from 15 mmole/liter in resting cells to 20 mmole/liter in PHA-stimulated cells) (*61*). In fibroblasts, increased $[Na^+]_i$ occurs following growth stimulation in some (*19*) but not all fibroblast lines (*36*), and no increase in $[Na^+]_i$ is seen during stimulation of increased Na^+,K^+ pumping in G_0 neuroblasts (*39*). In neuroblasts, amiloride prevents the increase in Na^+,K^+ pumping (*39*), so it may be due to increased Na^+ influx but not increased $[Na^+]_i$ (see Section III,E). The increased Na^+,K^+ pump activity is associated with an increase in $[K^+]_i$ in neuroblasts (*39*) and 3T3 cells (*19,75*), but probably not lymphocytes (*22,26,61*).

III. Rapid Changes in Fibroblast Monovalent Cation Flux

A. Serum and Other Growth-Promoting Agents

Most studies of the relationship of fibroblast growth to ion flux have been performed with mouse 3T3 cells, because these cells provide a good model system for growth regulation. When 3T3 cells reach confluence, they stop growing; this is termed density inhibition of growth. Transfer to fresh serum induced a well-coordinated round of cell replication; cells begin to synthesize DNA after a 10–12 hr lag period, and with maximal stimulation most cells divide by 30 hr. A variety of polypeptides, termed growth factors, can stimulate 3T3 cell growth in the absence of serum. Either serum or the growth factors stimulate increased Na^+ influx and Na^+,K^+ pump activity within a few minutes of addition to quiescent cells (*53,54,64*). Serum maximally stimulates the Na^+,K^+ pump; increasing Na^+ influx still further with high concentrations of monensin does not further increase the rate of $^{86}Rb^+$ uptake. Combinations of the growth factors that may be in serum are required to stimulate cell growth; these include (1) the platelet-derived growth factor (PDGF) or fibroblast growth factor (FGF); (2) somatomedin or insulin; (3) epidermal growth factor (EGF); and (4) as yet unknown additional components (*70*). Combinations of serum polypeptides are also required to stimulate Na^+ influx and the Na^+,K^+ pump maximally (*54*). EGF alone minimally stimulates DNA synthesis and slightly stimulates Na^+ influx and Na^+,K^+ pump activity (*53,54*). Platelet extract containing PDGF stimulates DNA synthesis (*44*), Na^+ influx, and Na^+,K^+ pump activity (Table I; *36*) submaximally. Insulin or somatomedin alone does not stimulate DNA synthesis in 3T3 cells (*70*) and does not stimulate Na^+ influx or the Na^+ pump (*53,64*). The combination of EGF and insulin submaximally stimulates 3T3 cell DNA synthesis and submaximally stimulates the Na^+,K^+ pump (*53*). A combination of PDGF, EGF, and insulin can maximally stimulate Na^+,K^+ pump activity (C. N. Frantz, unpublished). Thus, there is generally a good correlation between growth stimulation and immediate stimulation of the Na^+,K^+ pump, and it is clear that all three serum

TABLE I

Early, Rapid Increase in K^+ Uptake Following Serum Stimulation of 3T3 Cells[a]

	$^{86}Rb^+$ uptake (cpm/μg protein/10 min ± 1 SD)	
	Total	Ouabain-inhibitable
Depleted medium + 10% saline	0.64 ± 0.16	0.24 ± 0.16
Depleted medium + 10% DH serum	0.72 ± 0.12	0.33 ± 0.14
Fresh medium + 10% saline	0.79 ± 0.25	0.57 ± 0.28
Fresh medium + 10% DH serum	2.22 ± 0.34	1.34 ± 0.38
Fresh medium + PE	1.18 ± 0.04	0.75 ± 0.06

[a] Density-inhibited 8-day Swiss 3T3 cell cultures were transferred to Dulbecco's modified Eagle's medium containing no bicarbonate but with the usual Na^+ concentration (161 m*M*) and 10 m*M* HEPES (pH 7.4 at 37°C). After 1 hr incubation at 37°C, cultures were transferred either to the depleted medium, in which the cells had grown to confluence, or to fresh medium, containing $^{86}RbCl$ and 10% (v/v) platelet extract (PE) (*44*), human serum dialyzed against saline (DH serum), or saline and either 1.6 m*M* ouabain in DMSO or DMSO at an equivalent (0.07%) concentration. After 10 min at 37°C, cultures were rinsed rapidly three times at 4°C in 0.35 *M* sucrose 10 m*M* HEPES, pH 7.4; $^{86}Rb^+$ was determined by gamma counter (Beckman Biogamma II) and protein by the Lowry technique.

components are required for maximal stimulation of DNA synthesis or cell growth (*70*) or for maximal stimulation of the Na^+,K^+ pump. We have confirmed that PDGF, EGF, and insulin each do not maximally stimulate the rapid increase in Na^+,K^+ pump activity (C. N. Frantz, unpublished). In contrast, three agents that stimulate cell growth in combination with other factors markedly stimulate Na^+ and K^+ influxes in the absence of additional growth factors: tetradecanoyl phorbol acetate (a tumor promoter), vasopressin, and melittin (a bee venom polypeptide) (*54,55*). The tumor promoter, mellitin, and vasopressin all appear to stimulate growth in a closely related way, because they stimulate DNA synthesis in quiescent 3T3 cells synergistically with all other growth factors but not synergistically with each other (*13,55*).

B. Inhibitor in Depleted Medium

The increase in quiescent fibroblast Na^+,K^+ pumping, which results from transfer to fresh serum-containing medium, is not due to serum alone. In order to determine whether change of medium contributed to "serum-stimulated" Na^+,K^+ pump activity, we compared ouabain-inhibitable $^{86}Rb^+$ uptake in fresh medium with that in the depleted medium in which the cells grew to confluence. Some depleted medium was removed from the culture. With and without ouabain, $^{86}Rb^+$ was added to both depleted and fresh serum-free medium. The K^+ concentration was identical in fresh and depleted media. The medium was added to density-inhibited cultures, and $^{86}Rb^+$ uptake was determined over the subsequent 10 min. Fresh medium increased total and ouabain-inhibitable K^+ uptake consistently, but the difference was not significant (Table I). In order to determine whether any inhibitor of serum-stimulated Na^+,K^+ pumping might be present in depleted medium, it was prepared as above, and $^{86}Rb^+$ uptake measured over the next 10 min. Addition of serum to depleted medium failed to stimulate a significant increase in K^+ influx (Table 1). In contrast, addition of serum to depleted medium resulted in a small but significant stimulation of DNA synthesis measured 24 hr later by autoradiography: 22% of the cells synthesized DNA after addition of serum, but only 3% after addition of saline to control cultures. Thus, serum may stimulate 3T3 cell growth without stimulating any rapid increase in K^+ influx. However, because the degree of growth stimulation is small, the increase in K^+ influx may have also been small or delayed, and not detected. It appears that the depleted medium either lacks a nutritional factor supplied with fresh medium or contains an inhibitor of the serum-stimulated early, rapid K^+ influx. The latter appears to be correct, because addition of dialyzed serum to a solution containing only Na, K, Cl, and HEPES (pH 7.4) is able to stimulate the rapid K^+ influx (data not shown). Thus, nutrients (and even divalent cations) are not necessary for serum stimulation of rapid K^+ influx. A similar finding has been reported in quiescent hepatocytes, in which addition of growth-stimulating peptides to depleted medium fails to stimulate Na^+ influx (*29*). Also, studies in neuroblastoma cells have been performed in fresh medium (*38,39*).

C. Effect of Quiescence and Confluence

Postconfluence may lower the level of Na^+,K^+ pump activity in quiescent cells. It has clearly been shown that Na^+,K^+ pump activity is decreased in confluent, quiescent density-inhibited 3T3 cells compared to subconfluent, growing cells (*17,53,54,65,73*). During attempts to reproduce the findings of Rozengurt, we noted considerable variability in the rapid increase in ouabain-inhibitable $^{86}Rb^+$ uptake within 2 min after addition of serum to quiescent Swiss 3T3 cells in fresh medium. The $^{86}Rb^+$ is handled by cellular ion transport systems in a nearly identical fashion to $^{42}K^+$ and serves as a measure of K^+ influx (*65*). An increase in non-ouabain-inhibitable $^{86}Rb^+$ uptake was also seen. Rozengurt used postconfluent cultures, which were incubated 5–7 days without medium changes (*54*). We found that unstimulated Na^+,K^+ pump activity slowly decreased in cultures that were postconfluent and not growing at an appreciable rate (data not shown). In younger, but still quiescent cultures, the basal Na^+,K^+ pump-mediated and non-pump-mediated K^+ uptake was higher, and transfer to fresh medium containing serum resulted in less increase. In addition, the degree of increase in each of the two components of K^+ influx (ouabain- and non-ouabain-inhibitable) varied relative to each other in younger density-inhibited Swiss 3T3 cultures (data not shown). The relationship of postconfluence to the inhibitor in depleted medium is unknown.

D. Na^+,K^+ Pump Stimulation Is Transient

The rapid increase in K^+ influx is transient. Rozengurt noted the transient nature of this effect (*53,55*) but studied ion fluxes which occur within a few minutes of transfer of cells to medium containing serum or purified growth factors. We measured ouabain-inhibitable $^{86}Rb^+$ influx at intervals after transfer of cultures to fresh medium containing 10% serum. Density-inhibited 3T3 cultures were transferred to fresh medium containing 10% serum. At timed intervals, $^{86}Rb^+$ with or without ouabain was added, and cultures were rinsed 10 min later. The rate of Rb uptake, both total and ouabain-inhibitable, increased markedly compared to controls transferred to depleted medium or serum-free medium. However, within 30 min, both ouabain-inhibitable and total K^+ influx returned to levels just slightly higher than those found in depleted medium (Fig. 1). In another experiment, cultures were transferred to either fresh medium alone or fresh medium containing 10% serum, and K^+ influx was determined 1 hr later. There was no difference in either total K^+ influx or ouabain-inhibitable K^+ influx (data not shown). Thus, serum stimulates a transient increase in K^+ influx. It usually lasts about 15 min and then returns to levels identical to the K^+ influx found in fresh medium without serum. The temporal relationship to serum-stimulated Na^+ influx has not been determined, but serum-stimulated Na^+ influx may last considerably longer (*54,64,77*).

E. Relation of Na^+ Influx to Na^+,K^+ Pump Activity

Rozengurt has proposed that growth factors enhance Na^+,K^+ pump activity by allowing entry of Na^+ into the cells, which in turn stimulates the Na^+,K^+ pump. He showed that the Na^+ ionophore monensin stimulates Na^+ influx and results in stimulation of the Na^+,K^+ pump, and that low concentrations of amphotericin B (which makes holes in the

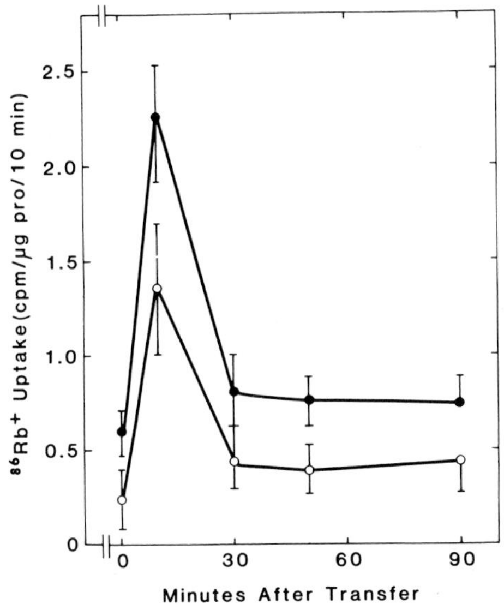

Fig. 1. Serum stimulates a rapid but transient increase in total (●) and ouabain-inhibitable (○) K^+ influx. Seven-day density-inhibited BALB/c-3T3 cultures were transferred to fresh medium with or without 10% heat-inactivated (56°C) human serum. At timed intervals, $^{86}RbCl$ was added with either ouabain in DMSO or DMSO alone (final concentrations 1.6 m*M* ouabain, 0.07% DMSO). Cultures were rinsed 10 min later three times in 0.35 *M* sucrose 10 m*M* HEPES (pH 7.4) at 4°, solubilized in 1 *M* NaOH; $^{86}Rb^+$ was quantitated by gamma counter and protein by the Lowry reaction. Uptake of $^{86}Rb^+$ is plotted at the midpoint of the 10 min interval; the "zero" time point is 10-min uptake of $^{86}Rb^+$ at identical specific activity in depleted medium before transfer to fresh medium with serum.

plasma membrane) or gramicidin (a less specific ionophore) have a similar effect (*54,64*). Also, stimulation of the pump by growth factors is dependent on the extracellular Na^+ concentration. Finally, the degree of stimulation of Na^+,K^+ pump activity by serum correlates with the degree of stimulation of Na^+,K^+ pump activity by monensin across a wide variety of fibroblastic cell lines (*36*). The degree of stimulation by serum or monensin of the Na^+,K^+ pump and Na^+ influx varies considerably between cell lines studied. In some fibroblast lines, serum induces a significant increase in $[Na^+]_i$, whereas in others no change is seen; but an increase in Na^+,K^+ pumping is always seen (*36*). Although the close association of increased Na^+ influx with increased Na^+,K^+ pump activity is amply demonstrated, it is not clear that the increased Na^+ entry directly enhances Na^+,K^+ pump activity. Rozengurt bases his hypothesis on the fact that Na^+,K^+-ATPase (ATP breakdown) activity is regulated in red cell membrane preparations by Na^+ concentration. Yet, he proposes that the increase in Na^+,K^+ pumping is stimulated by increased Na^+ influx, which does not result in a measurable increase in $[Na^+]_i$.

There are many possible explanations for the discrepancy between the measured $[Na^+]_i$ and the postulated effect of Na^+ influx on the Na^+,K^+ pump. First, an increase in $[Na^+]_i$ may actually occur, but has not been successfully measured. Second, the increase in $[Na^+]_i$ may be quite transient; if the growth factor stimulates a transient influx of Na^+ that results in a transient increase in Na^+,K^+ pumping, it might be expected that the $[Na^+]_i$ might rise transiently, stimulate an increase in pump activity, and then pump activity would decrease as the $[Na^+]_i$ was progressively lowered by the Na^+,K^+ pump. $[Na^+]_i$ was measured 20 min after addition of serum (*36,64*); in some fibroblasts the serum-stimulated Na^+,K^+ pump activity may already be returning toward baseline at this point (Fig. 1). Therefore, a transient increase in $[Na^+]_i$ may have been missed. Alternatively, total intracellular Na^+ may not reflect the Na^+ concentration, which regulates Na^+,K^+ pump activity. The important Na^+ concentration for Na^+,K^+ pump stimulation must be the concentration at the internal surface of the cell membrane, where Na^+ interacts with the Na^+,K^+ pump. Very close to the cell membrane, $[Na^+]_i$ may be much higher as a result of the negative charge of the membrane itself produced by negatively charged membrane phospholipids (*34*). Moreover, membrane static charge may change with serum stimulation, because addition of serum to quiescent fibroblasts results in rapid decrease in the amount of Ca^{2+} that may be loosely attached to the external surface of the cell membrane (*74*). The decrease in loosely bound Ca^{2+} suggests a decrease in negative charge at the cell membrane external surface. Serum alters the composition of membrane phospholipids, which could alter membrane charge (*63*). Thus, an increase in $[Na^+]$ at the inner cell membrane surface might stimulate the Na^+,K^+ pump without alteration of total $[Na^+]_i$, or an increase in $[Na^+]_i$ may not have been detected for technical reasons.

It is also possible that some factor other than Na^+ is involved in the mechanism of Na^+,K^+ pump stimulation by serum. Na^+,K^+-ATPase activity is sensitive to pH (*59*); although changes in intracellular pH have not been measured in fibroblasts, mitogen-stimulated changes in H^+ flux may occur in related systems. Addition of mitogen to plasma membrane vesicles of T cells resulted in a rapid increase in intravesicle pH (*37*), and serum-stimulated Na^+ entry in neuroblasts occurs as electroneutral exchange, possibly with H^+ (*39*). Finally, the ionophore monensin catalyzes electroneutral Na^+–H^+ exchange, so the stimulation of the Na^+,K^+

pump by monensin may be due in part to elevation of pH_i, and in part to elevation of $[Na^+]_i$. The effects of pH_i on Na^+,K^+ pump activity have not been determined. Another possible non-Na^+ regulator is the lipid membrane environment of the Na^+,K^+ pump. Addition of serum to quiescent fibroblasts induces rapid changes in membrane phospholipids (*63*), and the phospholipid environment affects Na^+,K^+ pump activity as shown in reconstitution experiments (*50*). In summary, the mechanism by which growth factors stimulate Na^+,K^+ pump activity in quiescent cells has not been clearly defined, but it may result from a closely associated increase in Na^+ influx.

The mechanism by which mitogens induce the increase in amiloride-sensitive Na^+ influx is also unknown. Villereal (*77*) has suggested that Na^+ influx might result from release of intracellular membrane-bound Ca^{2+} into the cytosol, because addition of high concentrations of the Ca^{2+} ionophore A23187, which elevates cytosol $[Ca^{2+}]$, induced amiloride-sensitive Na^+ influx in serum-deprived human fibroblasts. However, A23187 does not stimulate Na^+ influx in 3T3 (*64*) or neuroblastoma cells (*39*).

F. Summary

Serum stimulates a rapid increase in Na^+ influx and Na^+,K^+ pumping in quiescent fibroblasts. In human fibroblasts, serum stimulates Na^+ influx via an amiloride-sensitive pathway not active in cells in serum-free medium. There is a correlation between stimulation of cell growth and stimulation of Na^+,K^+ pump activity by serum-related growth factors, and a good correlation between stimulation of Na^+ influx and Na^+,K^+ pump activity by the growth factors. Maximal stimulation of Na^+,K^+ pumping and cell growth requires a combination of several such growth factors. In contrast, melittin alone maximally stimulates Na^+ influx and Na^+,K^+ pumping but requires other growth factors to stimulate cell growth. Allowing cells to become postconfluent further decreases Na^+,K^+ pump activity, and an inhibitor of serum stimulation of the Na^+,K^+ pump has been detected in depleted medium. Serum initially stimulates only a transient increase in Na^+,K^+ pump activity, and serum stimulation of Na^+ influx may be of longer duration. Although there is a close correlation between increased Na^+ influx and increased Na^+,K^+ pumping, and Na^+,K^+-ATPase and Na^+,K^+ pump activity are markedly increased by appropriate small increases in Na^+ concentration, it is not clear that the serum-stimulated increase in Na^+,K^+ pump activity results directly from increased Na^+ influx.

IV. Later Changes in Monovalent Cations during G_0/G_1

A. K^+ Influx

It has been demonstrated that serum stimulation of density-inhibited 3T3 cell growth is associated with an increase in $[K^+]_i$ that is clearly evident by 2 hr after serum addition. Therefore, an additional component of net K^+ influx must occur, and it must occur during the time of rapid increase in $[K^+]_i$ rather than within a few minutes of serum addition. To detect an increase in K^+ influx when the increase in $[K^+]_i$ was occurring, we examined K^+ influx 1.5 and 4 hr after transfer of density-inhibited Swiss 3T3 cultures to fresh medium with and without various combinations of growth factors. Cultures transferred to fresh medium alone had a stable rate of K^+ influx. However, an increase in total K^+ influx was seen in cultures incubated in medium containing serum (not shown), plasma, or insulin (Table II). The increase in total K^+ influx was sustained for at least 12 hr in cultures transferred to serum. In contrast to plasma or insulin, PDGF did not induce an increase in total K^+ influx. However, prior incubation with PDGF for 3 hr followed by rinsing and incubation with plasma enhanced the increase in total K^+ influx. The increase in total K^+ influx was not inhibited by ouabain; only a small, insignificant

TABLE II

The Sustained Increase in $^{86}Rb^+$ Uptake Later in G_0/G_1[a]

			$^{86}Rb^+$ uptake ± 1 SD (cpm/μg protein/10 min)	
Time (hr)	Pretreatment	Additions	Total	Ouabain-inhibitable
1.5	None	None	1.32 ± 0.12	0.90 ± 0.15
1.5	None	PE	1.16 ± 0.10	0.84 ± 0.13
1.5	None	Plasma	1.78 ± 0.21	1.04 ± 0.28
1.5	None	Insulin	1.68 ± 0.11	0.84 ± 0.18
4	None	PE	1.30 ± 0.10	0.84 ± 0.13
4	None	Plasma	2.18 ± 0.11	1.10 ± 0.15
4	None	Insulin	2.12 ± 0.08	1.08 ± 0.11
4	PE	Plasma	2.62 ± 0.14	1.24 ± 0.18
4	PE	Insulin	1.60 ± 0.16	1.02 ± 0.22

[a] Six-day quiescent, confluent Swiss 3T3 cultures were transferred to medium containing the additions noted. At the times after transfer noted, $^{86}Rb^+$ uptake was determined as described in the foot note to Table I. Some cultures were transferred to medium containing 10% platelet extract (PE) (*44*). Then, 3 hr later, these cultures were rinsed and transferred to medium containing the additions noted. Uptake of $^{86}Rb^+$ was determined 4 hr later.

increase in Na^+,K^+ pump activity was detected, except when quiescent cells were preincubated in PDGF and transferred to plasma. This is also the only condition in Table II in which significant cell growth occurs (*44*).

Thus, in addition to an early, transient rise in K^+ influx, a later prolonged rise in K^+ influx is stimulated by serum growth factors. The late increase in K^+ influx can be induced by plasma or insulin but not by PDGF. In contrast, the rapid, transient increase in K^+ influx is not stimulated by insulin in the absence of other growth factors (*53*). Both the rapid, transient increase and the later, sustained increase consist of two components, a ouabain-inhibitable increase in Na^+,K^+ pumping and an increase in K^+ influx unrelated to Na^+,K^+ pumping. The non-ouabain-inhibitable increase in K^+ influx is substantial. This increase demonstrates that serum induces complex changes in fibroblast monovalent cation flux; the increase in Na^+ influx and Na^+,K^+ pumping are only part of the picture. However, a small late increase in Na^+,K^+ pumping occurs only under conditions under which cell growth occurs.

In addition, a net increase in $[K^+]_i$ occurs only when both PDGF and plasma are added to the culture medium. An increase in K^+ influx does not necessarily result in an increase in $[K^+]_i$. Incubation of density-inhibited 3T3 cells with plasma alone resulted in an increase in K^+ influx but no increase in $[K^+]_i$. Presumably, an increase in K^+ efflux also occurred. However, when cells were first incubated 3 hr with PDGF and then rinsed and transferred to medium containing plasma, both an increase in K^+ influx and $[K^+]_i$ were seen (data not shown). Thus, when measured several hours after growth stimulation, plasma stimulates K^+ influx, but both PDGF and plasma are required for an increase in Na^+,K^+ pumping and 3T3 cell growth (*44*). A sustained increase in $[K^+]_i$ and Na^+,K^+ pump activity are then closely associated with cell growth (*75*).

B. $[K^+]_i$ and Growth Arrest

Subconfluent 3T3 cells replicating at different rates in different serum concentrations have a similar $[K^+]_i$ of about 160–180 mmole/liter (*76*). Ernst and Adam (*17*) suggested that $[K^+]_i$ decreased as 3T3 cells became density inhibited. To learn whether density-dependent inhibition of growth is associated with a decrease in $[K^+]_i$, we plated BALB/c-3T3 (clone A31) cells sparsely in medium containing 10% (v/v) bovine serum, and the total cell number and $[K^+]_i$ were determined at intervals. The $[K^+]_i$ of the subconfluent exponentially replication cultures was about 180 mmole/liter. When the cells reached confluence, the $[K^+]_i$ decreased to 90 mmole/liter (Fig. 2). Thus, in BALB/c-3T3 cells, density-dependent inhibition of growth is associated with a 50% decrease of $[K^+]_i$.

Most of the decrease in $[K^+]_i$ resulted from a decrease in K^+ content per cell. Between days 2 and 5, mean cell volume decreased by 37% from 4300 to 2870 μm^3; however, the K^+ content per cell decreased 65% from 0.83 to 0.26 pmole/cell. During the same interval, K^+ content per gram of protein decreased by < 10% because the protein per cell decreased about as much as K^+ content per cell. Measurement of K^+ content per gram of protein or K^+ content per cell does not accurately reflect $[K^+]_i$, because the relationship between cell protein and cell volume varies at different cell densities (*19*).

We also measured $[Na^+]_i$ as cells grew to confluence. $[Na^+]_i$ decreased from 40 mmole/liter in subconfluent logarithmically growing cells to 15 mmole/liter in confluent cells. Thus, the total univalent cation concentration, $[Na^+ + K^+]_i$, fell from 220 mmole/liter to 112 mmole/liter as cells became confluent. Because cell volume determination is critical to measurement of $[K^+]_i$ and $[Na^+]_i$, cell water was also measured in adherent cells on each day. The 3-O-MG water space varied little in relation to cell volume, averaging 75 ± 13% of total cell volume throughout the experiment. Thus, two independent measurements, cell volume and cell

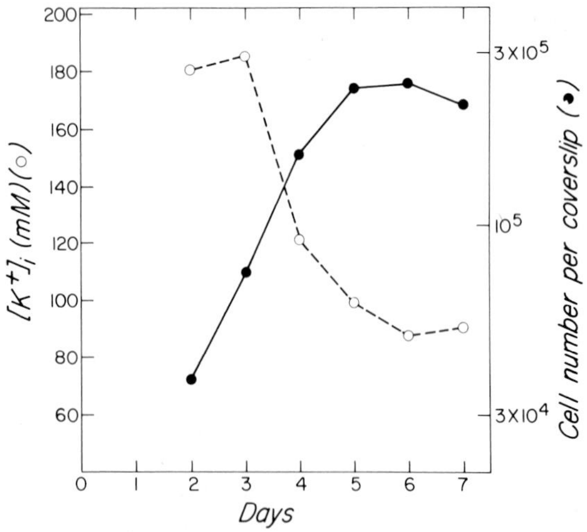

Fig. 2. Density-dependent inhibition of BALB/c-3T3 cell growth is associated with a decrease in $[K^+]_i$. The BALB/c-3T3 cells were planted on acid-treated glass coverslips (area = 1.1 cm^2) in Dulbecco's modified Eagle's medium containing 10% bovine serum. The cell number (●) and $[K^+]_i$ (○) were determined at intervals beginning 48 hr after planting, as described (*19*). Cells became confluent between days 4 and 5. [Reprinted from Frantz *et al.* (*19*) with the permission of the authors and Rockefeller University Press.]

water, demonstrated a marked decrease in $[K^+]_i$ and $[Na^+]_i$ as BALB/c-3T3 cells grew to confluence. The decrease in both $[Na^+]_i$ and $[K^+]_i$ at confluence is consistent with the hypothesis that cells stop growing when these cations decrease and start growing when both increase (*19*). These major changes in $[Na^+]_i$ and $[K^+]_i$ with growth state of 3T3 cells have recently been confirmed by a different technique, energy-dispersive-x-ray microanalysis (*48*).

V. Increased Na^+,K^+ Pumping Is Not Required for Growth of G_0-Arrested 3T3 Cells

It has been proposed that increased Na^+,K^+ pumping is required for cell growth. The mechanism by which mitogens increase Na^+,K^+ pump activity has not been completely defined (see above), but the increase in pump activity per se is implicated in growth control, because mitogens increase Na^+,K^+ pumping, and both Na^+,K^+ pumping and cell growth are specifically inhibited by ouabain. Both increased $[K^+]_i$ and increased ATP utilization have been proposed as mechanisms by which the increased pumping might alter other cellular events. Unlike the rapid changes in Na^+,K^+ pump activity investigated by Rozengurt, the changes in $[K^+]_i$ occur a few hours after growth stimulation, as described above. Utilizing the specificity of the Na^+,K^+ pump inhibitor ouabain, we have shown that, when the increase in pump activity is carefully abolished without inhibiting the pump activity to well below basal levels, cell growth can proceed. Thus, the increase in pump activity is not required for cell growth.

A. Inhibition of Increased K^+ Uptake but Not Cell Growth by Ouabain

To learn whether increased K^+ uptake and increased $[K^+]_i$ are closely associated with growth stimulation of confluent cells, the inhibitory effects of ouabain on K^+ uptake were explored. Density-inhibited BALB/c-3T3 cells were maximally stimulated to synthesize DNA by the addition of both calf serum and partially purified human PDGF (*44*); varying concentrations of ouabain were also present to inhibit K^+ uptake. To quantify DNA synthesis, we added tritiated thymidine ($[^3H]TdR$) to some cultures; the cells were fixed 24 hr later and processed for autoradiography to determine the percentage of cells that entered S phase. Ouabain at 500 μM completely prevented entry into S phase. However, ouabain at 100 μM did not inhibit DNA synthesis (Fig. 3). In the absence of

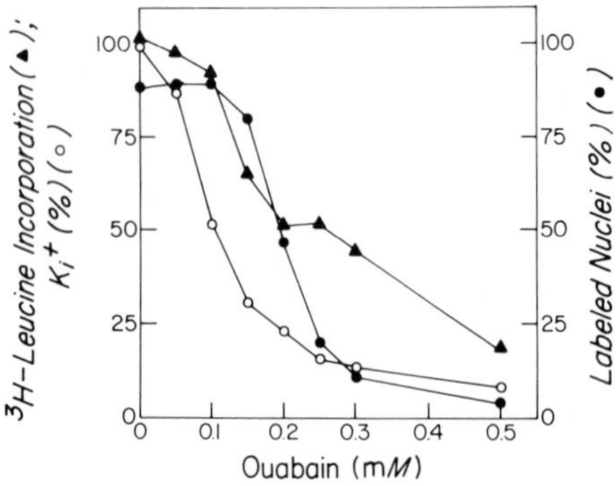

Fig. 3. Increasing concentrations of ouabain progressively decrease $[K^+]_i$ but inhibit protein synthesis and entry into the S phase only after a marked decrease in $[K^+]_i$. Rapid growth of density-inhibited BALB/c-3T3 cultures was stimulated by transfer to fresh medium containing 10% calf serum and 200 μg/ml partially purified PDGF. Ouabain (0–0.5 m*M*) was added simultaneously. [^{3}H]thymidine was added to some cultures, and these were fixed 24 hr later and processed for autoradiography (●). $[K^+]_i$ (○) and [^{3}H]leucine incorporation (▲) were measured 12 hr after growth stimulation as described (*18*). [Reprinted from Frantz *et al.* (*18*) with the permission of the authors and Alan R. Liss, Inc.]

ouabain, $^{86}Rb^+$ uptake increased by about 50% above the level in quiescent cells. Ouabain blocked this increase in $^{86}Rb^+$ transport as a function of concentration. In the presence of 100 μ*M* ouabain, the serum factor-induced increase in ouabain-inhibitable $^{86}Rb^+$ uptake was prevented, but the cells synthesized DNA (*19*).

B. Effect of 100 μ*M* Ouabain on $[K^+]_i$, $[Na^+]_i$, and Entry into S Phase

The increase in $[K^+]_i$ during G_1 correlates with increased Na^+,K^+ pump activity and might constitute the mechanism by which serum-stimulated cell growth occurs. To determine whether serum-stimulated cells could enter S phase without undergoing an increase in $[K^+]_i$, we measured $[K^+]_i$,$[Na^+]_i$, and entry into S phase at timed intervals after growth stimulation. Density-inhibited BALB/c-3T3 cultures were transferred to medium containing calf serum supplemented with partially purified PDGF with or without 100 μ*M* ouabain; the $[K^+]_i$ and the $[Na^+]_i$ were measured at intervals throughout the cell cycle. Cultures were also fixed periodically and processed for autoradiography to determine the rate at which cells entered the S phase. The stimulated cells entered the S phase after a lag of 12 hr at the same rate in the presence and absence of ouabain (Fig. 4a). In the absence of ouabain, $[K^+]_i$ increased from 92 to 104

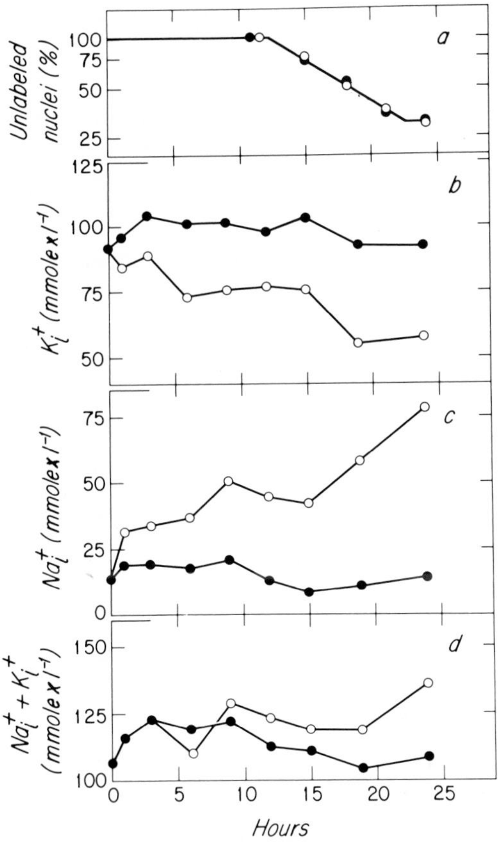

Fig. 4. Serum-stimulated cells with decreased $[K^+]_i$ enter the S phase at a normal rate. Density-inhibited BALB/c-3T3 cells were stimulated to synthesize DNA by the addition of serum and partially purified PDGF in the presence (○) or absence (●) of 100 μM ouabain. (a) Cultures were fixed at intervals, and DNA synthesis was determined by autoradiography. (b) The $[K^+]_i$ was determined at intervals. (c) The $[Na^+]_i$ was determined by flame photometry at intervals as described (*19*). (d) The $[Na^+ + K^+]_i$ was calculated from (b) and (c). [Reprinted from Frantz *et al.* (*19*) with the permission of the authors and Rockefeller University Press.]

mmole/liter within 3 hr, stayed at this level until 15 hr, and then decreased to its initial value. In the presence of ouabain, the $[K^+]_i$ decreased throughout the period of the experiment; at no time did it rise above the level present in quiescent cells (Fig. 4) (*19*). Thus, although the $[K^+]_i$ rises rapidly after the addition of serum growth factors to density-inhibited cells, this increase is not required for DNA synthesis.

The total intracellular univalent cation concentration ($[Na^+ + K^+]_i$) increased in cells stimulated with PDGF and serum. In resting cells, $[Na^+ + K^+]_i$ was 106 mmole/liter and increased by 16% to 123 mmole/liter 3 hr after addition of serum and PDGF, in the presence or absence of ouabain (Fig. 4d). Thus, the ouabain-induced decrease in $[K^+]_i$ was matched by an increase in $[Na^+]_i$ (*19*).

C. Increased $[Na^+ + K^+]_i$ and Cell Volume

Although an increase in $[K^+]_i$ per se is not required for cell growth, cells growing during partial inhibition of the Na^+,K^+ pump by 100 μM ouabain underwent an increase in cell volume and total univalent cation concentration. Since the cells growing in 100 μM ouabain did not have a higher rate of ouabain-inhibitable Na^+,K^+ pumping than quiescent cells, the increase in $[Na^+ + K^+]_i$ and cell volume must have occurred by a different mechanism, although some Na^+,K^+ pump activity is clearly required to allow the increase in $[Na^+ + K^+]_i$ and cell volume to occur, because complete inhibition of the Na^+,K^+ pump with 1 mM ouabain resulted in a decrease in cell volume (data not shown). Cotransport of Na^+ and K^+ described in red cells may be related to the observed increases. Apparently, a single channel (or a group of ion channels coordinately regulated) in erythrocytes catalyzes bidirectional electroneutral exchange across the plasma membrane of Na^+,K^+, Cl^-, and H_2O. It has been best described in turkey erythrocytes and is regulated by cAMP-dependent protein phosphorylation (*2,57*). In avian erythrocytes, this coordinate ion flux is bidirectional, can be driven by a variety of forces, and is inhibited by furosemide. Although furosemide has no effect on the serum-stimulated increases in $[Na^+ + K^+]_i$ and cell volume during G_0/G_1 and S (data not shown), a related ion flux mechanism may be responsible for growth-related changes in mammalian cell $[Na^+ + K^+]_i$ and cell volume. In support of this suggestion, mutant mammalian cells that can grow in low K^+ medium are defective in K^+ and cell volume regulation but not in the Na^+,K^+ pump (*20*).

VI. Mechanism of Ouabain Inhibition of Cell Growth

A. $[K^+]_i$ Protein Synthesis and Entry into S Phase

Preventing growth by ouabain or low $[K^+]_0$ correlates with inhibition of the cell membrane Na^+,K^+ pump, resulting in decreased $[K^+]_i$. Lubin and his colleagues (*9,31*) have presented evidence that intracellular K^+ is required for protein synthesis in both cell-free systems and intact cells.

Because continuous protein synthesis during G_0/G_1 is required for entry into the S phase (*8,52*), we investigated the relationship between ouabain inhibition of DNA synthesis and the inhibition of protein synthesis.

Protein synthesis was measured by the incorporation of [^{3}H]leucine into acid-insoluble material. To demonstrate that [^{3}H]leucine incorporation into acid-insoluble counts is a valid measure of protein synthesis in cells with altered $[K^+]_i$, the uptake of [^{3}H]leucine was measured in the presence and absence of ouabain. No effect of ouabain was found. Because ouabain does not interfere with leucine entry into the cell, [^{3}H]leucine incorporation in ouabain-treated cells reflects protein synthesis (*18,31*).

In order to compare the effects of ouabain on protein synthesis, entry into the S phase, and $[K^+]_i$, density-inhibited cultures were stimulated to grow by the addition of bovine serum and partially purified PDGF; varying concentrations of ouabain were present. In some cultures, the $[K^+]_i$ was determined 12 hr later, whereas in others the incorporation of [^{3}H]leucine into acid-insoluble counts between 11.5 and 12.5 hr later was used to quantify protein synthesis. Protein synthesis and $[K^+]_i$ were measured at 12 hr, just before the cells began to enter S phase. In yet another group of cultures, [^{3}H]thymidine was added at the time of serum addition; these cultures were fixed 24 hr later and processed for autoradiography to determine the percentage of cells that synthesized DNA. The $[K^+]_i$ in the ouabain-treated cells decreased as a function of ouabain concentration; in 0.1 m*M* ouabain, the $[K^+]_i$ was 50% (62 mmole/liter) of that found in untreated cells (107 mmole/liter), while in 0.5 m*M* ouabain it was 10% (Fig. 3). In 0.5 m*M* ouabain, protein synthesis decreased to approximately 10% that found in the absence of ouabain, and DNA synthesis was almost completely prevented; however, in 0.1 m*M* ouabain, there was little, if any, inhibition of either protein or DNA synthesis (Fig. 3). In a similar experiment, $[K^+]_i$ and [^{3}H]leucine incorporation were measured 6 hr after addition of serum and PDGF. The results were similar, although $[K^+]_i$ and protein synthesis were slightly less decreased at each ouabain concentration (data not shown). Thus, there is a threshold of $[K^+]_i$ (approximately 60 mmole/liter) below which inhibition of both protein synthesis and DNA synthesis occurs (*18*).

In addition, a series of experiments were performed that demonstrated that the effect of ouabain on the BALB/c-3T3 cell cycle correlated closely with the effect of cycloheximide, an inhibitor of protein synthesis (*18*).

B. $[K^+]_i$ and Protein Synthesis: Threshold Effect

Inhibition of the Na^+,K^+-ATPase of BALB/c-3T3 cells causes the $[K^+]_i$ to fall below the critical threshold needed for both protein and DNA synthesis. Ouabain gradually lowers $[K^+]_i$ at a rate dependent on ouabain concentration, but no effect is seen on protein synthesis in G_0/G_1 or on subsequent entry into the S phase until $[K^+]_i$ falls below a critical threshold, at which time protein synthesis is inhibited and subsequent DNA synthesis is prevented. Therefore, inhibition of Na^+,K^+-ATPase results in a fall in $[K^+]_i$. When $[K^+]_i$ is low enough, protein synthesis is significantly inhibited, and, as a result, serum-stimulated cells cannot enter the S phase (*18*). Our findings are consistent with those of Rubin and his colleagues, who found that the growth-inhibitory effects of reduced $[K^+]_i$ fall outside the physiological range of $[K^+]_i$ (*40*). Our findings also agree with those of Lubin and his colleagues, who have demonstrated that cellular protein synthesis requires intracellular K^+. In both intact human fibroblasts treated with ouabain (*31*) and reticulocytes made freely permeable to external ions by nystatin (*9*), protein synthesis decreased when $[K^+]_i$ decreased below 60 mmole/liter; a smaller decrease in $[K^+]_i$ did not affect protein synthesis.

VII. Monovalent Cations and Transformation of Fibroblasts

Transformation is characterized by loss of growth regulation; transformed fibroblasts continue to grow in the absence of the growth factors required by normal cells. Progressively, 3T3 cells lose growth control with time in culture. Over many passages, the normal 3T3 cells require progressively less serum for growth and become capable of growing to a much higher cell concentration in a given amount of serum. This loss of growth control is associated with a marked increase in Na^+,K^+ pump activity (*73*). When viral-transformed 3T3 cells are compared to early-passage normal 3T3 cells, the Na^+,K^+ pump activity is higher in the transformed cells, but late-passage 3T3 cells have more pump activity than transformed cells even though they can stop growing, whereas the viral transformed cells cannot (*73*). Unlike transformed cells, normal 3T3 cells stop growing at confluence. Normal 3T3 cells in this density-inhibited state have less Na^+,K^+ pumping activity (*65,73*) and lower $[K^+]_i$ (*17*) than transformed cells at similar density. In contrast, passive K^+ fluxes and membrane potential decrease with increasing cell density in both normal and SV40 transformed 3T3 cells (*1,65*). Decrease in $[K^+]_i$ and Na^+,K^+ pumping also occurs in chick embryo fibroblasts at confluence compared to their Rous sarcoma virus-transformed counterparts at similar densities (*24*). The amount of Na^+,K^+-ATPase activity in cell homogenates does not show a consistent difference between normal and transformed cells. When subconfluent, growing normal and Rous-transformed cells were compared, Na^+,K^+ pumping was similar (*24*), but Na^+ influx was higher in transformed cells (*4*). Similarly, addition of ouabain results in much more rapid accumulation of Na^+ in growing, subconfluent viral-transformed 3T3 cells than in normal 3T3 cells growing at the same rate (*36,54*). In chick cells, increased Na^+ influx appears to be a direct result of Rous sarcoma virus transformation (*4*). Thus, transformed cells may not have more Na^+,K^+ pumps but appear to have a higher rate of Na^+,K^+ pumping. The increase in Na^+ influx in transformed cells may occur via the amiloride-sensitive pathway, because amiloride may preferentially inhibit growth of transformed and tumor cells (Chapter 15; *66*). High rates of Na^+ influx and Na^+,K^+ pumping in Ehrlich ascites tumor cells may be due to phosphorylation of the Na^+,K^+ pump by a "cascade" of protein kinases, which includes a protein kinase antigenically related to the Rous sarcoma virus-transforming gene product *src*, a protein kinase (*69*). Thus, in some viral-transformed cells, increased Na^+ influx and increased Na^+,K^+ pumping may be a direct effect of viral transformation. In addition, growth of transformed cells is less inhibited by reduction of $[K^+]_i$ to a given level than is growth of normal cells (*33*). Thus, three differences in monovalent cations between normal and transformed fibroblasts have been noted: (1) Some growing transformed cells have higher Na^+ influx than their growing normal counterparts, and, probably as a result, a higher $[Na^+]_i$ and higher rate of Na^+,K^+ pumping; (2) unlike transformed cells, normal 3T3 cells are able to become density-inhibited, and this is associated with a decrease in Na^+,K^+ pumping and $[K^+]_i$; and (3) transformed cells are less sensitive to inhibition of cell growth by reduced $[K^+]_i$. For further discussion of transformation and univalent cation fluxes, see Chapter 15.

VIII. Monovalent Cation Flux in Fibroblasts: Current Status

In summary, subconfluent growing 3T3 cells have high intracellular $[Na^+]$ and $[K^+]$ and a large cell volume compared to density-inhibited quiescent 3T3 cells. When the density-inhibited cells are stimulated to grow by the addition of polypeptides, the cells increase intracellular $[Na^+]$ and $[K^+]$ and cell volume. The mechanisms by which the increases occur are not clear, but increased Na^+,K^+ pumping is associated, and the increase in Na^+,K^+ pumping may be due to increased net Na^+ influx.

However, neither increased Na^+,K^+ pumping nor increased $[K^+]_i$ is required for cell growth, because BALB/c-3T3 cells may enter S phase without increasing $[K^+]_i$ or increasing net K^+ influx. Inhibition of cell growth by ouabain, observation of which led to the consideration that Na^+,K^+ pumping might regulate cell growth, results only when ouabain inhibition of the Na^+,K^+ pump results in depletion of K^+ to well below the levels seen in nongrowing density-inhibited 3T3 cells. Furthermore, the observed increase in $[K^+]_i$ may be conceived as an increase in $[Na^+ + K^+]_i$ (and cell volume), which occurs via some other pathway than the Na^+,K^+ pump. The significance of such Na^+,K^+ cotransport in cell growth regulation is unexplored.

Unlike increased Na^+,K^+ pump activity, the serum-stimulated increase in amiloride-inhibitable, non-ouabain-inhibitable Na^+ influx may possibly be required for cell growth. Significance for cell growth regulation has been attributed to this Na^+ influx, because amiloride inhibits cell growth; but amiloride may inhibit cell growth by inhibiting protein synthesis and/or mitochondrial function in some way unrelated to its effect on serum-stimulated Na^+ influx. Thus, there is no evidence that amiloride-sensitive Na^+ influx is required for fibroblast growth. The mechanism by which mitogens stimulate the Na^+ influx is unknown.

IX. The Cell Growth Cycle and Monovalent Cation Flux: Future Directions of Research

A. Synthesis of Events in G_0 Cells

We have summarized the current descriptions of growth-related monovalent cation flux in a variety of cultured cells. In four systems—rat hepatocyte, human T lymphocyte, mouse neuroblastoma cell, and mouse 3T3 fibroblast—cells are arrested in G_0, and the application of mitogens or growth factors to the cells results in both cell growth and increased monovalent cation flux. If we assume that the ion fluxes in all four culture systems are closely related, we may draw many conclusions about the increases in plasma membrane monovalent cation flux associated with growth stimulation:

1. Each cell undergoes a depolarization of the cell membrane and a decrease in membrane resistance, which consists of a bidirectional increase in flux of both Na^+ and K^+. More or less simultaneously, hyperpolarization occurs, presumably from increased K^+ efflux.
2. The cell membrane potential shortly returns to a level slightly less than that of G_0 cells, suggesting that plasma membrane potential per se

plays little role in any subsequent growth events. Subsequently the membrane potential increases over several hours as $[K^+]_i$ increases.

3. An increase in amiloride-sensitive electroneutral Na^+ influx also occurs immediately upon growth stimulation and persists for at least 1 hr. There is no good evidence to suggest that this increase is important for cell growth and not just another epiphenomenon associated with growth stimulation of G_0-arrested cells.

4. In exchange for Na^+, H^+ may be excreted by the amiloride-sensitive pathway, but this has not yet been demonstrated. Net H^+ efflux would result in increased pH_i and may or may not be important for cell growth regulation.

5. Increased Na^+ influx may or may not result in increased $[Na^+]_i$. Increased Na^+,K^+ pumping is transiently stimulated in close association with Na^+ influx. However, Na^+ influx per se may not be the cause of the transient increase in Na^+,K^+ pumping.

6. The significance for growth regulation of these early increases in Na^+ influx, K^+ efflux, and Na^+,K^+ pumping are unknown, but attention is directed to the function of the receptors for growth-promoting agents. Since so many agents cause the early events, they must be caused by some membrane effects that result from the action of many different receptors. It should be noted that many different receptors may utilize a common effector, such as adenylate cyclase.

7. In addition, several hours after growth stimulation, a second increase in Na^+,K^+ pumping occurs. This sustained increase is associated with increasing $[K^+]_i$. The cause is unclear, but it may result from an increase in $[Na^+]_i$.

8. This later increase in Na^+,K^+ pumping is not required for entry into S phase, so it must not be part of the mechanism by which departure from G_0 is accomplished.

9. The increase in $[K^+]_i$ is associated with an increase in total univalent cation concentration and cell volume and may not result directly from an increase in Na^+,K^+ pumping; the relationship of the increase in $[Na^+ + K^+]_i$ and cell volume to hormonal regulation of cell growth has not been investigated. Also, the relationship of putative serum-stimulated Na^+,K^+ cotransport to serum-stimulated Na^+ influx has not been determined; they may be identical.

B. Future Research

1. Mechanisms of Hormone Action

Future research on the rapid, transient monovalent cation fluxes will be directed at (1) the measurement of pH_i and significance of any increased H^+ efflux, (2) the relation to mechanism of hormone action, especially in relation to changes in cAMP, Ca^{2+}, and other yet-to-be-discovered mediators of hormone action, and (3) the relation to changes in membrane phospholipid composition. The significance of the rapid fluxes lies in their relationship to the mechanisms of hormone–receptor–effector function, in analogy with acetylcholine. Plasma membrane vesicles will be used to isolate these rapid events from secondary effects; plasma membrane vesicles are essentially plasma membrane balloons filled with the investigator's choice of solutes. Isolated plasma membrane, in a form in which ionic events may be measured, allows the events directly related to receptor function to be studied. Thus artifacts, due to sequestration of ions and indicators of membrane potential or pH difference within intracellular compartments, may be overcome.

Future investigations of the late, sustained changes in monovalent cation flux, concentration, and cell volume will be directed at a better characterization of the observed changes and at correlations with known mechanisms of hormone action, such as cAMP and Ca^{2+}-mediated membrane protein phosphorylation. Isolation of plasma membrane in vesicle form may also allow conclusive determination of the nature of the membrane change. Recently, hormonal modulation of a novel ion channel has been thoroughly described in avian erythrocytes. Catecholamines stimulate uptake of Na^+,K^+, Cl^-, and H_2O in avian erythrocytes (*57*). The movement of the ions is bidirectional, so that changes by this pathway in cellular ion content and cell volume depend on internal and external ion concentrations. The movement of the ions is also electroneutral, with Cl^- balancing each Na^+ or K^+. A role for H^+ has not been ruled out. The ratio of K^+ to Na^+ moved is about 2:1. Furosemide, but not ouabain or amiloride, inhibits the ion movement in either direction across the cell membrane. Catecholamines induce this specific increased ion permeability by stimulating adenyl cyclase to produce cAMP, resulting in an elevation in cAMP within the erythrocyte, and the same effect on ion flux may be produced by addition of dibutyryl cAMP. The increase in erythrocyte cAMP results in phosphorylation of a specific site on goblin, a 230,000 MW intrinsic membrane protein (*2*). Addition of cholera toxin to turkey erythrocytes results, after a delay of 30–60 min, in an increase in basal cAMP and an increase in cAMP response to catecholamines, as well as increase in basal and catecholamine-stimulated Na^+,K^+, and H_2O flux. Amphibian erythrocytes undergo related changes in cation flux and protein phosphorylation (*56*). Changes in avian erythrocyte monovalent cation flux, which are mediated by cAMP, are remarkably similar to the later, sustained changes seen when G_0-arrested cells are stimulated by hormones to grow. Mammalian cells probably have related pathways (*20*). This and other alternative pathways of ion flux and mechanisms of their regulation need to be investigated in mammalian cell plasma membrane vesicles. Although cyclic nucleotides may not play a significant role in growth stimulation of G_0 cells (*43*), regulation by phosphorylation of specific proteins in the plasma membrane may occur by mechanisms other than cAMP (*21,30,72*).

2. The Na^+,K^+ Pump

Increased Na^+,K^+ pump activity is not required for the transition of G_0 cells to S phase. However, mechanisms of Na^+,K^+ pump regulation during cell growth have not been defined. It is likely that some of the increase in Na^+,K^+ pumping is due to increased $[Na^+]$ at the internal surface of the plasma membrane. This mechanism is doubtful where increased Na^+,K^+ pumping occurs without increased $[Na^+]_i$ and needs to be investigated. Investigation of increased Na^+,K^+ pumping in transformed cells may shed light on the mechanism by which growth-stimulating polypeptides stimulate the Na^+,K^+ pump. Spector *et al.* (*69*) have shown in Ehrlich ascites tumor cells that a series of protein kinases which phosphorylate each other in a cyclic manner on a phosphotyrosine residue, and one of which appears antigenically related to the oncogenic *src* gene product of Rous avian sarcoma virus, phosphorylate a 54,000 MW protein, which is a component of the plasma membrane Na^+,K^+ pump. Phosphorylation decreases the "efficiency" of the pump, resulting in excessive ATPase activity relative to Na^+ pumping (*67,68*). It is not quite clear whether such "inefficiency" might result directly from increased Na^+ influx. It will be of interest to see whether similar phosphorylation of the Na^+,K^+ pump results from polypeptide growth-stimulating hormones. The effect of amiloride on the Ehrlich ascites cell Na^+,K^+ pump has not been investigated. However, addition of quercetin restores the efficiency of the phosphorylated, inefficient Na^+,K^+ pump from tumor cells by an unknown mechanism (*67*). The effect of quercetin on growth-related changes in Na^+,K^+ pumping is unknown. It is also quite possible that the increased Na^+ influx occurs at unrelated sites in different transformed and normal growth-stimulated cells.

The nature of the inhibitor of serum stimulation of the Na^+,K^+ pump found in depleted medium has not yet been explored. One hopes that it will be found to inhibit early membrane events by different means than the mechanism of action of amiloride. The existence of two specific inhibitors that inhibit discrete sequential events may allow the mechanism of mitogen stimulation of membrane ion flux events to be better defined.

3. *The Amiloride-Sensitive* Na^+ *Channel*

Unlike increased Na^+,K^+ pumping, the growth-associated increase in Na^+ influx described by Rozengurt (*36,64*) may possibly be required for cell growth. The serum-stimulated increase in Na^+ influx is partly inhibited by amiloride; however, amiloride may inhibit cell growth by inhibiting protein synthesis rather than by preventing the increase in Na^+ influx (C. N. Frantz, unpublished data). Availability of amiloride analogues (*12*) may allow the effects on protein synthesis and Na^+ influx to be separated and the significance of the amiloride-sensitive Na^+ channel to be determined. Many investigators have suggested that the amiloride-sensitive Na^+ channel may function as an Na^+–H^+ exchange channel and that mitogens increase pH_i by this route (*26,32,39,54,77*). Recently, Zetterberg and Engstrom (*79*) reported that growth of quiescent 3T3 cells may be stimulated by chemical elevation of pH_i; these findings have not been duplicated, although agents that increase pH_i also enhance protein synthesis of quiescent 3T3 cells (C. N. Frantz, unpublished data). Measurement of pH_i is technically difficult, so speculation will undoubtedly continue for several years.

Some transformed cells have a constitutively high rate of Na^+ influx and Na^+,K^+ pump activity compared to their normal counterparts (*36*). The increased Na^+ influx could possibly be a direct effect of activity of the transforming gene product of some viruses (*4,69*). If increased Na^+ influx by some specific pathway is required for cell growth, then in some cells the loss of growth regulation due to transformation may consist of constitutive arrangement of the amiloride-sensitive Na^+ channel (or another Na^+ channel with similar properties) in an open position, so that it cannot be modulated by growth factors.

Several mechanisms by which Na^+ influx might induce cell growth have been suggested (Table III). Stimulation of Na^+,K^+ pumping per se is not required for cell growth (*19;* Section V). Changes in cell membrane potential of cycling cells are minor and related directly to $[K^+]_i$ and K^+ permeability (*7,41*). The rapid brief hyperpolarization and de-

TABLE III

Possible Mechanisms by Which Increased Net Na^+ Influx Might Alter Cell Growth

1. Increase Na^+,K^+ pumping
2. Depolarize cell membrane potential
3. Release Ca^{2+} from mitochondria into cytoplasm
4. Increase H^+ extrusion

polarization seen upon addition of serum to quiescent neuroblastoma cells (*38*) seems unlikely to be required for growth factor action because it is so brief. However, it is conceivable that depolarization activates an enzyme in the plasma membrane that performs its entire function, such as alteration of membrane phospholipids, within a minute, because some membrane protein functions related to ion channels are membrane potential dependent. It has been suggested that increase in $[Na^+]_i$ may result in release of Ca^{2+} from mitochondria, elevating cytosol Ca^{2+} concentration, which in turn might alter specific protein function (*3,30,35,72*). However, the effect of $[Na^+]$ on mitochondrial Ca^{2+} does not occur in all mammalian cells (*35*). In addition, an increase in cytosol $[Ca^{2+}]$ has not been detected during growth stimulation of G_0-arrested mammalian cells, and the technology of Ca^{2+} detection must improve before such an increase can be determined (*10*). Future research will undoubtedly be directed at a possible role of cytosol $[Ca^{2+}]$ in growth regulation when technology permits, whether or not Na^+ influx plays a role in the regulation of cytosol Ca^{2+} or vice versa. It has also been suggested that elevation of cytosol $[Ca^{2+}]$ results in increased amiloride-sensitive electroneutral Na^+ influx (*39,77*). Both increased Na^+,K^+ pumping and release of Ca^{2+} are postulated to result from increased $[Na^+]_i$. However, increased Na^+ influx per se does not stimulate or enhance 3T3 cell growth (*19,54*). As previously noted, several investigators have suggested that the sustained increase in electroneutral amiloride-sensitive Na^+ influx induces or permits exchange of extracellular Na^+ for intracellular H^+, resulting in or permitting an increase in pH_i. However, monensin induces electroneutral Na^+ influx and Na^+–H^+ exchange (*25,47*) but does not stimulate or enhance rate of cell growth (*54*). Moreover, monensin does not replace any serum growth factors required for maximal stimulation of DNA synthesis in 3T3 cells (C. N. Frantz, unpublished data), and increased Na^+ influx through a tetrodotoxin-sensitive Na^+ channel had no effect on cell growth (*46*). Thus, it is clear that Na^+ influx per se is not involved in growth regulation. However, it is possible that cell growth is also inhibited by other toxic effects of monensin, so a requirement of Na^+–H^+ exchange for cell growth cannot be ruled out. From a teleological point of view, the association of increased Na^+–H^+ exchange capacity with growth stimulation might permit cells an efficient pathway for disposal of excessive H^+ created by an unfavorable environment. Similarly, increased rates of transport of glucose, inorganic phosphate, and amino acids associated with growth stimulation are not required for cell growth but may be required in environments more unfavorable than tissue-culture medium. The development of new technology for the measurement of pH_i will allow these issues to be better addressed.

Acknowledgments

This investigation was supported by U.S. Public Health Service Grant Number Ca 26889, awarded by the National Cancer Institute, DHHS. Christopher Frantz is the recipient of an American Cancer Society, Massachusetts Division, Inc., Research Fellowship.

References

1. Adam, G., Ernst, M., and Seher, J.-P. (1979). Regulation of passive membrane permeability for potassium ions by cell density of 3T3 and SV40-3T3 cells. *Exp. Cell Res.* **120,** 127–139.
2. Alper, S. L., Beam, K. G., and Greengard, P. (1980). Hormonal control of Na^+–K^+ cotransport in turkey erythrocytes. *J. Biol. Chem.* **255,** 4864–4871.
3. Al-Shaikhaly, M. H., Nedergard, J., and Cannon, B. (1979). Sodium-induced calcium release from mitochondria in brown adipose tissue. *Proc. Natl. Acad. Sci. U.S.A.* **76,** 2350–2353.
4. Bader, J. P., Ikazaki, T., and Brown, N. R. (1981). Sodium and rubidium uptake in cells transformed by Rous sarcoma virus. *J. Cell. Physiol.* **106,** 235–243.
5. Baserga, R. (1976). "Multiplication and Division in Mammalian Cells." Dekker, New York.
6. Bentley, P. J. (1979). The comparative pharmacology of amiloride. *In* "Amiloride and Epithelial Sodium Transport" (A. W. Cuthbert, G. M. Fanelli, Jr., and A. Scriabine, eds.), pp. 35–40. Urban & Schwarzenberg, Baltimore, Maryland.
7. Boonstra, J., Mummery, C. L., Tertoolen, L. G. J., van der Saag, P. T., and de Laat, S. W. (1981). Cation transport and growth regulation in neuroblastoma cells. Modulations of K^+ transport and electrical membrane properties during the cell cycle. *J. Cell. Physiol.* **107,** 75–83.
8. Brooks, R. F. (1977). Continuous protein synthesis is required to maintain the probability of entry into S phase. *Cell* **12,** 311–317.
9. Cahn, F., and Lubin, M. (1978). Inhibition of elongation steps of protein synthesis at reduced potassium concentrations in reticulocytes and reticulocyte lysate. *J. Biol. Chem.* **253,** 7798–7803.
10. Caswell, A. H. (1979). Methods of measuring intracellular calcium. *Int. Rev. Cytol.* **56,** 145–181.
11. Chafouleas, J. G., Bolton, W. E., Boyd, A. E., III, and Means, A. R. (1981). Calmodulin plays an important role in G_1/S transition. *J. Supramol. Struct., Suppl.* **5,** 233 (abst.).
12. Cragoe, E. J., Jr. (1979). Structure–activity relationships in the amiloride series. *In* "Amiloride and Epithelial Sodium Transport" (A. W. Cuthbert, G. M. Fanelli, Jr., and A. Scriabine, eds.), pp. 1–20. Urban & Schwarzenberg, Baltimore, Maryland.
13. Dicker, P., and Rozengurt, E. (1980). Phorbol esters and vasopressin stimulate DNA synthesis by a common mechanism. *Nature* (*London*) **287,** 607–612.
14. Douglas, W. W. (1975). Secromotor control of adrenal medullary secretion: Synaptic, membrane, and ionic events in stimulus–secretion coupling. *In* "Handbook of Physiology" (H. Blaschko and A. D. Smith, eds.), Sect. 7, Vol. 6, pp. 367–388. Waverly Press, Baltimore, Maryland.
15. Epel, D. (1980). Ionic triggers in the fertilization of sea urchin eggs. *Ann. N.Y. Acad. Sci.* **339,** 74–85.

16. Epel, D., Patton, C., Wallace, R. W., and Cheung, W. Y. (1981). Calmodulin activates NAD kinase of sea urchin eggs: An early event of fertilization. *Cell* **23,** 543–549.
17. Ernst, M., and Adam, G. (1979). Dependence of intracellular alkali-ion concentrations of 3T3 and SV40-3T3 cells on growth density. *Cytobiologie* **18,** 450–459.
18. Frantz, C. N., Stiles, C. D., Pledger, W. J., and Scher, C. D. (1980). Effect of ouabain on growth regulation by serum components in BALB/c-3T3 cells: Inhibition of entry into S phase by decreased protein synthesis. *J. Cell. Physiol.* **105,** 439–448.
19. Frantz, C. N., Nathan, D. G., and Scher, C. D. (1981). Intracellular univalent cations and the regulation of the BALB/c-3T3 cell cycle. *J. Cell Biol.* **88,** 51–56.
20. Gargus, J. J., Miller, I. L., Slayman, C. W., and Adelberg, E. A. (1978). Genetic alteration in potassium transport in L cells. *Proc. Natl. Acad. Sci. U.S.A.* **75,** 5589–5593.
21. Greengard, P. (1978). Phosphorylated proteins as physiological effectors. *Science* **199,** 146–152.
22. Holian, A., Deutsch, C. J., Holian, S. K., Daniele, R. P., and Wilson, D. F. (1979). Lymphocyte response to phytohemagglutinin: Intracellular volume and intracellular [K^+]. *J. Cell. Physiol.* **98,** 137–144.
23. Jaffe, L. F. (1980). Calcium explosions as triggers of development. *Ann. N.Y. Acad. Sci.* **339,** 86–101.
24. Johnson, M. A., and Weber, M. J. (1979). Potassium fluxes and ouabain binding in growing, density-inhibited and Rous sarcoma virus-transformed chicken embryo cells. *J. Cell. Physiol.* **101,** 89–100.
25. Kaback, H. R. (1980). Electrochemical ion gradients and active transport. *Ann. N.Y. Acad. Sci.* **339,** 53–60.
26. Kaplan, J. G., and Owens, T. (1980). Activation of lymphocytes of man and mouse: Monovalent cation fluxes. *Ann. N.Y. Acad. Sci.* **339,** 191–200.
27. Katz, B. (1966). "Nerve, Muscle, and Synapse." McGraw-Hill, New York.
28. Kirschner, L. B. (1979). Extrarenal action of amiloride in aquatic animals. *In* "Amiloride and Epithelial Sodium Transport" (A. W. Cuthbert, G. M. Fanelli, Jr., and A. Scriabine, eds.), pp. 41–49. Urban & Schwarzenberg, Baltimore, Maryland.
29. Koch, K. S., and Leffert, H. L. (1979). Increased sodium ion influx is necessary to initiate rat hepatocyte proliferation. *Cell* **18,** 153–163.
30. Kretsinger, R. H. (1979). The informational role of calcium in the cytosol. *Adv. Cyclic Nucleotide Res.* **11,** 1–26.
31. Ledbetter, M., and Lubin, M. (1977). Control of protein synthesis in human fibroblasts by intracellular potassium. *Exp. Cell Res.* **105,** 223–236.
32. Leffert, H. L., and Koch, K. S. (1980). Ionic events at the membrane initiate rat liver regeneration. *Ann. N.Y. Acad. Sci.* **339,** 201–215.
33. Lubin, M. (1980). Control of growth by intracellular potassium and sodium concentrations is relaxed in transformed 3T3 cells. *Biochem. Biophys. Res. Commun.* **97,** 1060–1067.
34. McLaughlin, S. (1977). Electrostatic potentials at membrane–solution interfaces. *Curr. Top. Membr. Transp.* **9,** 71–144.
35. Mela, L. (1977). Mechanism and physiological significance of calcium transport across mammalian mitochondrial membranes. *Curr. Top. Membr. Transp.* **9,** 321–336.
36. Mendoza, S. A., Wigglesworth, N. M., Pohjanpelto, P., and Rozengurt, E. (1980). Na entry and Na–K pump activity in murine, hamster, and human cells—Effect of monensin, serum, platelet extract, and viral transformation. *J. Cell. Physiol.* **103,** 17–27.
37. Mikkelson, R. B., Schmidt-Ullrich, R., and Wallach, D. F. H. (1980). Concanavalin

A induces an intraluminal alkalinization of thymocyte membrane vesicles. *J. Cell. Physiol.* **102,** 113–117.

38. Moolenaar, W. H., de Laat, S. W., and van der Saag, P. T. (1979). Serum triggers a sequence of ionic conductance changes in quiescent neuroblastoma cells. *Nature (London)* **279,** 721–723.
39. Moolenaar, W. H., Mummery, C. L., van der Saag, P. T., and de Laat, S. W. (1981). Rapid ionic events and the initiation of growth in serum-stimulated neuroblastoma cells. *Cell* **23,** 789–798.
40. Moscatelli, D., Sanui, H., and Rubin, A. H. (1979). Effects of depletion of K^+, Na^+, or Ca^{++} on DNA synthesis and cell cation content in chick embryo fibroblasts. *J. Cell. Physiol.* **101,** 117–128.
41. Mummery, C. L., Boonstra, J., van der Saag, P. T., and de Laat, S. W. (1981). Modulation of functional and optimal (Na^+-K^+) ATPase activity during the cell cycle of neuroblastoma cells. *J. Cell. Physiol.* **107,** 1–9.
42. Pardee, A. B., Dubrow, R., Hamlin, J. L., and Kletzien, R. L. (1978). Animal cell cycle. *Annu. Rev. Biochem.* **47,** 715–750.
43. Pastan, I. H., Johnson, G. S., and Anderson, W. B. (1975). Role of cyclic nucleotides in growth control. *Annu. Rev. Biochem.* **44,** 491–522.
44. Pledger, W. J., Stiles, C. D., Antoniades, H. N., and Scher, C. D. (1977). Induction of DNA synthesis in BALB/c-3T3 cells by serum components: Reevaluation of the commitment process. *Proc. Natl. Acad. Sci. U.S.A.* **74,** 4481–4485.
45. Pollack, L. R., Tate, E. H., and Cook, J. S. (1981). Na^+,K^+-ATPase in HeLa cells after prolonged growth in low-K^+ or ouabain. *J. Cell. Physiol.* **106,** 85–97.
46. Pouyssegur, J., Jacques, Y., and Lazdunski, M. (1980). Identification of a tetrodotoxin-sensitive Na^+ channel in a variety of fibroblast lines. *Nature (London)* **286,** 162–164.
47. Pressman, B. C. (1976). Biological applications of ionophores. *Annu. Rev. Biochem.* **45,** 501–530.
48. Proll, M. A., Pool, T. B., and Smith, N. K. R. (1979). Quantitative decreases in the intracellular concentrations of sodium and other elements as BALB/c-3T3 fibroblasts reach confluence. *J. Cell Biol.* **83,** 13a (abstract).
49. Quastel, M. R., and Kaplan, J. G. (1975). Ouabain binding to intact lymphocytes. Enhancement by phytohemagglutinin and leucoagglutinin. *Exp. Cell Res.* **94,** 351–362.
50. Racker, E., Miyamoto, H., Mogerman, J., Simons, J., and O'Neal, S. (1980). Cation transport in reconstituted systems. *Ann. N.Y. Acad. Sci.* **358,** 64–72.
51. Rasmussen, H., and Goodman, D. B. P. (1977). Relationships between calcium and cyclic nucleotides in cell activation. *Physiol. Rev.* **57,** 421–509.
52. Rossow, P. W., Riddle, V. G. H., and Pardee, A. B. (1979). Synthesis of labile, serum-dependent protein in early G_1 controls animal cell growth. *Proc. Natl. Acad. Sci. U.S.A.* **76,** 4446–4450.
53. Rozengurt, E., and Heppel, L. A. (1975). Serum rapidly stimulates ouabain-sensitive $^{86}Rb^+$ influx in quiescent 3T3 cells. *Proc. Natl. Acad. Sci. U.S.A.* **72,** 4492–4495.
54. Rozengurt, E., and Mendoza, S. (1980). Monovalent cation fluxes and the control of cell proliferation in cultured fibroblasts. *Ann. N.Y. Acad. Sci.* **339,** 175–190.
55. Rozengurt, E., Gelehrter, T. D., Legg, A., and Pettican, P. (1981). Melittin stimulates Na^+ entry, Na^+–K^+ pump activity and DNA synthesis in quiescent cultures of mouse cells. *Cell* **23,** 781–788.
56. Rudolph, S. A., and Greengard, P. (1980). Effects of catecholamines and prostaglandin E_1 on cyclic AMP, cation fluxes, and protein phosphorylation in the frog erythrocyte. *J. Biol. Chem.* **255,** 8534–8540.
57. Rudolph, S. A., Schafer, D. D., and Greengard, P. (1977). Effects of cholera enterotoxin

on catecholamine-stimulated changes in cation fluxes, cell volume, and cyclic AMP levels in the turkey erythrocyte. *J. Biol. Chem.* **252,** 7132–7139.
58. Schneider, A. S., Herz, R., and Rosenheck, K. (1977). Stimulus–secretion coupling in chromaffin cells isolated from bovine adrenal medulla. *Proc. Natl. Acad. Sci. U.S.A.* **77,** 5036–5040.
59. Schwartz, A., Lindenmayer, G. E., and Allen, J. C. (1972). The Na^+–K^+ ATPase membrane transport system: Importance in cellular function. *Curr. Top. Membr. Transp.* **3,** 1–82.
60. Segel, G. B., Simon, W., and Lichtman, M. A. (1979). Regulation of sodium and potassium transport in phytohemagglutinin-stimulated human blood lymphocytes. *J. Clin. Invest.* **64,** 834–840.
61. Segel, G. B., Kovach, G., and Lichtman, M. A. (1979). Sodium–potassium adenosine triphosphatase activity of human lymphocyte membrane vesicles: Kinetic parameters, substrate specificity, and effects of phytohemagglutinin. *J. Cell. Physiol.* **100,** 109–118.
62. Segel, G. B., and Lichtman, M. A. (1980). The apparent discrepancy of ouabain inhibition of cation transport and of lymphocyte proliferation is explained by time dependency of ouabain binding. *J. Cell. Physiol.* **104,** 21–26.
63. Shier, W. T. (1979). Serum stimulation of phospholipase A_2 and prostaglandin release in 3T3 cells is associated with platelet-derived growth-promoting activity. *Proc. Natl. Acad. Sci. U.S.A.* **77,** 137–141.
64. Smith, J. B., and Rozengurt, E. (1978). Serum stimulates the Na^+–K^+ pump in quiescent fibroblasts by increasing Na^+ entry. *Proc. Natl. Acad. Sci. U.S.A.* **75,** 5560–5564.
65. Spaggiare, S., Wallach, M. J., and Tupper, J. T. (1976). Potassium transport in normal and transformed mouse 3T3 cells. *J. Cell. Physiol.* **89,** 403–416.
66. Sparks, R. L., and Cameron, I. L. (1980). The effect of amiloride on proliferation and intracellular ion concentrations of transformed cells. *J. Cell Biol.* **87,** 7a (abstr.).
67. Spector, M., O'Neal, S., and Racker, E. (1980). Reconstitution of the Na^+–K^+ pump of Ehrlich ascites tumor and enhancement of efficiency by quercetin. *J. Biol. Chem.* **255,** 5504–5507.
68. Spector, M., O'Neal, S., and Racker, E. (1980). Phosphorylation of β-subunit of Na^+-K^+ ATPase in Ehrlich ascites tumor by a membrane-bound protein kinase. *J. Biol. Chem.* **255,** 8370–8373.
69. Spector, M., O'Neal, S., and Racker, E. (1981). Regulation of phosphorylation of the β-subunit of the Ehrlich ascites tumor Na^+-K^+ ATPase by a protein kinase cascade. *J. Biol. Chem.* **256,** 4219–4227.
70. Stiles, C. D., Capone, G. T., Scher, C. D., Antoniades, H. N., Van Wyk, J. J., and Pledger, W. J. (1979). Dual control of cell growth by somatomedins and platelet-derived growth factor. *Proc. Natl. Acad. Sci. U.S.A.* **76,** 1279–1283.
71. Taub, M., and Saier, M. H., Jr. (1981). Amiloride-resistant Madin-Darby canine kidney (MDCK) cells exhibit decreased cation transport. *J. Cell. Physiol.* **106,** 191–199.
72. Triggle, D. J. (1980). Receptor–hormone interrelationships. *In* "Membrane Structure and Function" (E. E. Bittar, ed.), Vol. 3, pp. 1–58. Wiley, New York.
73. Tupper, J. T. (1977). Variation in potassium transport properties of mouse 3T3 cells as a result of subcultivation. *J. Cell. Physiol.* **93,** 303–308.
74. Tupper, J. T., and Zorgniotti, F. (1977). Calcium content and distribution as a function of growth and transformation in the mouse 3T3 cell. *J. Cell Biol.* **75,** 12–22.
75. Tupper, J. T., Zorgniotti, F., and Mills, B. (1977). Potassium transport and content during G_1 and S phase following serum stimulation of 3T3 cells. *J. Cell. Physiol.* **91,** 429–440.
76. Tupper, J. T., and Zografos, L. (1979). Effect of imposed serum deprivation on growth

of the mouse 3T3 cell. Dissociation from changes in potassium ion transport as measured from [^{86}Rb]rubidium ion uptake. *Biochem. J.* **174,** 1063–1065.
77. Villereal, M. L. (1981). Sodium fluxes in human fibroblasts: Effect of serum, Ca^{2+}, and amiloride. *J. Cell. Physiol.* **107,** 359–369.
78. Whitfield, J. F., Boynton, A. L., MacManus, J. P., Rixon, R. H., Sikorska, M., Tsang, B., and Walker, P. R. (1980). The roles of calcium and cyclic AMP in cell proliferation. *Ann. N.Y. Acad. Sci.* **339,** 216–240.
79. Zetterberg, A., and Engstrom, W. (1980). Mitogenic effect of alkaline treatment on serum starved 3T3 cells. *Eur. J. Cell Biol.* **22,** 488 (abstr.).

Index

CELL BIOLOGY: A Series of Monographs

EDITORS

G. M. Padilla, G. L. Whitson, and I. L. Cameron (editors). THE CELL CYCLE: Gene-Enzyme Interactions, 1969

A. M. Zimmerman (editor). HIGH PRESSURE EFFECTS ON CELLULAR PROCESSES, 1970

I. L. Cameron and J. D. Thrasher (editors). CELLULAR AND MOLECULAR RENEWAL IN THE MAMMALIAN BODY, 1971

I. L. Cameron, G. M. Padilla, and A. M. Zimmerman (editors). DEVELOPMENTAL ASPECTS OF THE CELL CYCLE, 1971

P. F. Smith. The BIOLOGY OF MYCOPLASMAS, 1971

Gary L. Whitson (editor). CONCEPTS IN RADIATION CELL BIOLOGY, 1972

Donald L. Hill. THE BIOCHEMISTRY AND PHYSIOLOGY OF *TETRAHYMENA*, 1972

Kwang W. Jeon (editor). THE BIOLOGY OF AMOEBA, 1973

Dean F. Martin and George M. Padilla (editors). MARINE PHARMACOGNOSY: Action of Marine Biotoxins at the Cellular Level, 1973

Joseph A. Erwin (editor). LIPIDS AND BIOMEMBRANES OF EUKARYOTIC MICROORGANISMS, 1973

A. M. Zimmerman, G. M. Padilla, and I. L. Cameron (editors). DRUGS AND THE CELL CYCLE, 1973

Stuart Coward (editor). DEVELOPMENTAL REGULATION: Aspects of Cell Differentiation, 1973

I. L. Cameron and J. R. Jeter, Jr. (editors). ACIDIC PROTEINS OF THE NUCLEUS, 1974

Govindjee (editor). BIOENERGETICS OF PHOTOSYNTHESIS, 1975

James R. Jeter, Jr., Ivan L. Cameron, George M. Padilla, and Arthur M. Zimmerman (editors). CELL CYCLE REGULATION, 1978

Gary L. Whitson (editor). NUCLEAR–CYTOPLASMIC INTERACTIONS IN THE CELL CYCLE, 1980

Danton H. O'Day and Paul A. Horgen (editors). SEXUAL INTERACTIONS IN EUKARYOTIC MICROBES, 1981

Ivan L. Cameron and Thomas B. Pool (editors). THE TRANSFORMED CELL, 1981

Arthur M. Zimmerman and Arthur Forer (editors). MITOSIS/CYTOKINESIS, 1981

Ian R. Brown (editor). MOLECULAR APPROACHES TO NEUROBIOLOGY, 1982

Henry C. Aldrich and John W. Daniel (editors). CELL BIOLOGY OF *PHYSARUM* AND *DIDYMIUM*, Volume I: Organisms, Nucleus, and Cell Cycle, 1982

John A. Heddle (editor). MUTAGENICITY: New Horizons in Genetic Toxicology, 1982

Potu N. Rao, Robert T. Johnson, and Karl Sperling (editors). PREMATURE CHROMOSOME CONDENSATION: Application in Basic, Clinical, and Mutation Research, 1982

George M. Padilla and Kenneth S. McCarty, Sr. (editors). GENETIC EXPRESSION IN THE CELL CYCLE, 1982

In preparation

Henry C. Aldrich and John W. Daniel (editors). CELL BIOLOGY OF *PHYSARUM* AND *DIDYMIUM*, Volume II: Differentiation, Metabolism, and Methodology, 1982.

David S. McDevitt (editor). CELL BIOLOGY OF THE EYE, 1982

Govindjee (editor). PHOTOSYNTHESIS, Volume I: Energy Conversion by Plants and Bacteria, 1982; Volume II: Development, Carbon Metabolism, and Plant Productivity, 1982

P. Michael Conn (editor) CELLULAR REGULATION OF SECRETION AND RELEASE, 1982